Changxing Miao, Bo Zhang, Jiqiang Zheng
Harmonic Analysis Methods in Partial Differential Equations

De Gruyter Studies in Mathematics

Volume 102

Changxing Miao, Bo Zhang, Jiqiang Zheng

Harmonic Analysis Methods in Partial Differential Equations

DE GRUYTER

Mathematics Subject Classification 2020
58J50, 42B20, 35J10, 33C05

Authors
Prof. Changxing Miao
Institute of Applied Physics
and Computational Mathematics
National Key Laboratory of Computational Physics
6 Huayuan Road
100088 Beijing
P.R. China
miao_changxing@iapcm.ac.cn

Prof. Jiqiang Zheng
Institute of Applied Physics
and Computational Mathematics
National Key Laboratory of Computational Physics
6 Huayuan Road
100088 Beijing
P.R. China
zheng_jiqiang@iapcm.ac.cn

Prof. Bo Zhang
State Key Laboratory of Mathematical Sciences
Academy of Mathematics and Systems Science
Chinese Academy of Sciences
100190 Beijing
P.R. China
b.zhang@amt.ac.cn

ISBN 978-3-11-138451-1
e-ISBN (PDF) 978-3-11-138472-6
e-ISBN (EPUB) 978-3-11-138486-3
ISSN 0179-0986

Library of Congress Control Number: 2025932116

Bibliographic information published by the Deutsche Nationalbibliothek
The Deutsche Nationalbibliothek lists this publication in the Deutsche Nationalbibliografie; detailed bibliographic data are available on the Internet at http://dnb.dnb.de.

Typesetting: VTeX UAB, Lithuania

www.degruyter.com
Questions about General Product Safety Regulation:
productsafety@degruyterbrill.com

Preface

The history of modern harmonic analysis dates back to Fourier's solution of the heat equation in the early 19th century. After more than 200 years of development, it has now become one of the core research areas in modern mathematics. In particular, it has emerged as a powerful tool in the studies of partial differential equations, analytic number theory, mathematical physics and engineering sciences. Throughout the history and development of partial differential equations, many classical results in harmonic analysis have proven to be the most powerful tools and techniques for solving essential problems of partial differential equations. From the following results, we may appreciate the significant role played by harmonic analysis in the study of partial differential equations.

(i) *$\mathcal{H}_1$ and* BMO *spaces*. The introduction and application of the BMO space (the dual space of $\mathcal{H}_1$) play an essential part in establishing the L^p- and C^α-theories for solutions of elliptic and parabolic equations. With the L^p- and C^α-theories, the regularity of solutions to elliptic and parabolic boundary value problems can be established. As alternatives of the L^1 and L^∞ spaces, the Hardy space $\mathcal{H}_1$ and the BMO space play a very important role in interpolation theory and the study of boundedness of operators.

(ii) *The classical (first generation) Calderón–Zygmund theory of singular integrals.* Typical examples include the well-known Hilbert and Riesz transforms, which are important applications in the study of various problems of mathematical physics. In the context of partial differential equations (PDEs), it plays an important part in the study of positive symmetric systems of hyperbolic equations, estimates of integral potentials (single-layer and double-layer potentials) and elliptic boundary value problems. On the other hand, utilizing Bessel and Riesz potentials, we can extend Sobolev spaces of the integral order to the fractional-order case.

(iii) *The second generation of Calderón–Zygmund singular integral operators.* This plays an essential role in the theory of pseudodifferential operators; in particular, the L^p theory of pseudodifferential operators can be regarded as a generalization of classical potential theory and is one of the basic methods for the study of general elliptic boundary value problems. Certainly, the Fourier integral operators, as a further development of pseudodifferential operators, can be regarded as essential second-type oscillatory integrals.

(iv) *The Hardy–Littlewood maximal function theory.* This theory plays an important role in the study of boundedness of operators and pointwise convergence of functions, particularly in the characterization of boundaries in elliptic boundary value problems.

(v) *Spaces of differentiable functions such as Besov spaces, Triebel–Lizorkin spaces, and the usual Sobolev spaces.* These spaces, in particular, their Littlewood–Paley characterization, atomic and molecular characterization, Gaussian kernel characterization and Poisson kernel characterization and other harmonic analysis tech-

https://doi.org/10.1515/9783111384726-202

niques provide not only working spaces for the study of partial differential equations, but also effective tools for deriving linear and nonlinear estimates.

(vi) *Theory of spherical harmonic functions.* It is a classical approach for various well-posedness problems of partial differential equations, in particular, the foundation of the compact methods based on approximation, and provides a tool for the numerical solution approaches. Certainly, it also plays an important role in the study of such core direction of harmonic analysis as the theory of singular integrals (e. g., the theory of singular integral operators).

(vii) *Interpolation theory and method.* Both the real and complex methods of interpolation, and the Stein method of interpolation are the main tools of studying the theory of function spaces, operator theory and nonlinear estimates. For example, the Marcinkiewicz interpolation theorem changes end-point weak-type operator estimates into interior point strong-type operator estimates, which is of essential significance in the study of partial differential equations.

(viii) *The Hörmander translation invariance operator theory, the Littlewood–Paley g-function method and the Calderón–Stein g_λ^*-function method.* These are the foundation of the multiplier theory. The multiplier theory can be used to determine the well-posedness of linear evolution equations (i. e., whether or not the solution operator generates a C_0-semigroup in the Banach space X considered), which essentially provides appropriate solution spaces for studying the solvability problem of the corresponding nonlinear partial differential equations.

(ix) *The Littlewood–Paley decomposition theory.* This theory has demonstrated significant potential in various areas, including the characterization of function spaces, the estimation of nonlinear functions in fractional-order Sobolev spaces, and many other aspects. A typical example, based on the Littlewood–Paley decomposition theory, is Bony's twice microlocal decompositions and estimates of fractional derivatives.

(x) *Oscillatory integral estimates and restriction estimates on geometric surfaces of Fourier transforms.* These estimates serve as the foundation for establishing various types of estimates, including L^p–L^q estimates, the Strichartz estimates, regular Strichartz estimates and reverse space-time estimates (maximum norm estimates) for solutions of linear evolution equations. All these results provide a powerful tool for studying both the well-posedness theory of nonlinear evolution equations and the scattering theory for wave and dispersive wave equations. Over the past three decades, numerous breakthroughs in the study of partial differential equations, particularly evolution equations, have been attributed to these estimates. Notable contributions in this area include the work of Stein, Ginibre-Velo, Brenner, Bourgain, Kenig, Klainerman, Ponce and Tao, etc.

Invited by Professors Gang Tian and Ling Hsiao, one of the authors, Changxing Miao, delivered special mathematics lectures at Peking University and the Morningside Mathematics Centre of the Chinese Academy of Sciences in 2000. Afterwards, he gave 12 lectures on harmonic analysis methods for partial differential equations in the Department of Mathematics at Zhejiang University invited by Professors Fanghua Lin and Daoyuan Fang. Then, supported by the Royal Society of UK through a Royal Fellowship, Changxing Miao visited the School of Mathematical and Information Sciences at Coventry University, UK, for one year from January 2001, where the first two authors collaborated to work on this book and completed the first draft based on the series of lectures given previously by Changxing Miao at the Chinese Academy of Sciences and Zhejing University. Invited by Professors Zhouping Xin, Changxing Miao gave a series of lectures on parts of the results of this book in the Institute of Mathematical Sciences at the Chinese University of Hong Kong during January and March 2003. In addition, Changxing Miao also lectures on parts of the results of this book in the Institute of Mathematics at Fudan University, Shanghai and the Department of Mathematics at Nanjing University, Nanjing. This book is extended, based on the above lectures. The purpose of this book is to provide pure and applied mathematicians and mathematics postgraduates with a monograph on harmonic analysis methods for partial differential equations, emphasizing in particular the core role played by harmonic analysis methods. An emphasis will be on nonlinear evolution equations, with a particular focus on well-known problems in mathematical physics such as parabolic equations, the Navier–Stokes equations, the nonlinear Schrödinger equations, and the nonlinear wave equations.

The book consists of six chapters. Chapter 1 gives a general review of the classical spaces of differentiable functions as well as several classical results in harmonic analysis. It then briefly discusses the research methods and progress in elliptic boundary value problems. The reader can thus appreciate the role of the second and third generations of the Calderón–Zygmund singular integral operators in the study of elliptic boundary value problems, and in particular, the key role played by the harmonic analysis methods in dealing with the case with Lipschitz boundaries. Meanwhile, this chapter also provides a general overview of the research background of harmonic analysis methods for evolution equations, multipliers estimates and the multiplier characterization of operator semigroups, the relationship between the restriction estimates on compact smooth surfaces or surfaces with nonzero Gaussian curvature of the Fourier transforms and the space-time estimates of solutions to linear evolution equations, conjecturing and constructing well-suited space-time Banach spaces for nonlinear evolution equations by use of the scaling argument.

Chapter 2 mainly discusses the well-posedness of the Cauchy problem for semilinear parabolic equations. To achieve this, we introduce the concepts of admissible and generalized admissible triplets. Similar to the wave and dispersive wave equations, we first establish space-time estimates for the corresponding linear equations and then propose an effective and unified method to deal with semilinear parabolic equations with quite general nonlinear terms. The method not only simplifies proofs of many previously known important results, but also leads with the help of a scaling argument and nonlinear estimates in Besov spaces to local and small global well-posedness theories in fractional-order Sobolev spaces as well as in Besov spaces.

Chapter 3 focused on the Navier–Stokes equations. In Section 3.1, a comprehensive and historical review of the Leray–Hopf weak solution and its study are presented, including von Wahl's abstract method, the Solonnikov estimates and the regularity theory of Serrin and von Wahl. In Section 3.2, using the space-time estimate method, we prove the local existence result and the small global well-posedness result of Kato. These results can be extended to the negative-order Besov spaces. Section 3.3 introduces some recent results of Meyer and Cannone. By using the Littlewood–Paley theory, we give a characterization of the so-called well-suited Banach space X, and thus establish the local well-posedness of the Navier–Stokes equations in this space. It should be pointed out that the well-suited Banach space X includes all the noncritical solution spaces used for the study of the Navier–Stokes equations. Section 3.4 mainly presents the method developed by Koch and Tataru. Based on the heat kernel characterization of the BMO space, we define a larger Banach space BMO^{-1} of degree −1 and establish the global well-posedness in $X = \mathrm{BMO}^{-1}$ of the small solution of the Navier–Stokes equations in the case when the initial value function $\varphi \in \mathrm{BMO}^{-1}$. This result generalizes previous well-posedness results in the known critical spaces such as L^n and $\dot{B}^{-n/p+1}_{p,q}$.

Chapter 4 deals with the well-posedness and scattering theory of the Cauchy problem for the nonlinear Schrödinger equations, in particular, the outstanding work of Bourgain and I-team (Colliander–Keel–Staffilani–Takaoka–Tao). In Section 4.1, we present basic estimates for the linear Schrödinger equations, including the Strichartz estimates, the maximal norm estimates, Kato's local smoothness effects, and particularly the endpoint Strichartz estimates of Keel and Tao. Section 4.2 gives a historical overview of the study and discusses the main issues regarding the well-posedness and scattering theory of the nonlinear Schrödinger equations. It also shows briefly the role played by the Strichartz estimates in such studies. Section 4.3 introduces the Fourier truncation method of Bourgain. By combining the Bourgain spaces with the Littlewood–Paley dyadic decomposition, we establish the bilinear Strichartz estimates and then used them to prove the well-posedness as well as the scattering theory (in the radially symmetrical case) for the cubic nonlinear Schrödinger equations in fractional order Sobolev spaces. In Section 4.4, we introduce I-team's I-energy method. It is another effective method in dealing with low regularity problems, and can be applied directly to study the scattering theory. Section 4.5 is devoted to the Cauchy problem and scattering theory of the nonlinear Schrödinger equations with H^1-critical nonlinear growth in the case when the initial function $\varphi(x) = \varphi(|x|) \in H^1$. It is important to understand how Bourgain makes use of the Morawetz estimates to get rid of the infinitely many times "concentration" effects. Here, we adopt a new method of proof by Tao. It is worth noting that for general H^1 initial functions, Tao and his collaborators resolved the Cauchy problem and scattering theory of H^1-critical nonlinear Schrödinger equations in $\mathbb{R}^3$.

In Chapter 5, we aim to study the well-posedness and scattering theories of the Cauchy problem for the wave equation. In Section 5.1, we introduce the restriction estimates of Fourier transforms to manifolds and the classical Strichartz space-time estimates. Section 5.2 presents the bilinear method and the endpoint Strichartz space-time estimates of Keel and Tao. In Section 5.3, we discuss the global well-posedness in the energy space of semilinear wave equations. The methods we will use are the Strichartz estimates and the contraction mapping principle. Section 5.4 discusses the well-posedness of global smooth solutions to the wave equation in $\mathbb{R}^3$ with both subcritical and critical nonlinear growth. By employing the Lagrange density function, we can establish the so-called Morawetz estimates, and then prove that the energy cannot be focused at one point so that the well-posedness of the global smooth solutions can thus be established. Section 5.5 addresses the global well-posedness below the energy norm for the Cauchy problem of the Klein–Gordon equation. Here, we mainly make use of Bourgain's technique of initial value decomposition, along with the nonlinear estimates in Besov spaces.

In Chapter 6, we discuss the well-posedness and scattering theories of the Cauchy problem for the defocusing energy-critical Schrödinger equation with inverse square potential. The proof employs the induction on energy argument, which was first introduced by Bourgain in [22] and later refined by the I-team in [48], and subsequently employed by Kenig and Merle in [132]. Assuming that the solution does not scatter, one first demonstrates the existence of a minimal counterexample, that is, a solution with infinite space-time norm and minimal energy among all such solutions. The (concentration) compactness argument used to prove this existence shows more, namely that such a minimal blowup solution must be almost periodic (cf. Theorem 6.3.1 in Section 6.3). The second half of the argument uses monotonicity formulae and/or conservation laws to rule out the existence of such solutions in Section 6.4. Chapter 6 developed from the work jointed with Rowan Killip, Monica Visan and Junyong Zhang. We are grateful for their help on that project and for allowing us to use that material.

Finally, the authors would like to thank their young colleagues: Qionglei Chen, Xing Cheng, Yanfang Gao, Baishun Lai, Junfeng Li, Haifeng Shang, Zhiyong Wang, Gang Wu, Haigen Wu, Guixiang Xu, Xiaojing Xu, Liutang Xue, Qingying Xue, Jianwei Yang, Yaojun Ye, Baoquan Yuan, Jia Yuan, Lifeng Zhao, Junyong Zhang, Xiaoyi Zhang, Zhifei Zhang, Xiaoxin Zheng and their students. They have provided many helpful and constructive comments and suggestions.

This book is supported by the National Key Research and Development Program of China (No. 2022YFA1005700).

Beijing, China, January 2025

Changxing Miao
Bo Zhang
Jiqiang Zheng

Contents

1 Elliptic boundary value problems and harmonic analysis methods for abstract evolution equations

1.1 Basic function spaces and classical results from harmonic analysis

In this section, we introduce some basic function spaces, discuss their properties, and give some remarks as well as references for further study. In addition, we present certain useful classical results from harmonic analysis for later use, which are as elementary as possible.

We use standard notations: $\mathbb{R}^n$ denotes the n-dimensional Euclidean space, $\mathbb{N}$ denotes the set of all natural numbers and $\mathbb{N}_0 = \mathbb{N} \cup \{0\}$. Denote by $C^\infty(\mathbb{R}^n)$ the set of all infinite-times continuously differentiable functions on $\mathbb{R}^n$, and by $C_c^\infty(\mathbb{R}^n)$ the set of all functions from $C^\infty(\mathbb{R}^n)$, which have compact supports. Define the Schwartz space

$$\mathcal{S}(\mathbb{R}^n) = \Big\{\varphi \in C^\infty(\mathbb{R}^n) \mid \|\varphi\|_{(\alpha,\beta)} = \sup_{x\in\mathbb{R}^n} |x^\alpha \partial^\beta \varphi| < \infty, \ \forall \alpha, \beta \in \mathbb{N}_0^n\Big\},$$

which is a Fréchet space with the seminorms $\|\cdot\|_{(\alpha,\beta)}$. It is clear that

$$C_c^\infty(\mathbb{R}^n) \subset \mathcal{S}(\mathbb{R}^n) \subset C^\infty(\mathbb{R}^n).$$

$\mathcal{S}'(\mathbb{R}^n)$ stands for the topological dual space to $\mathcal{S}(\mathbb{R}^n)$, which is the usual space of Schwartz distributions or space of tempered distributions. We refer to [119, 255, 293] for distributions. Similarly, we can define the corresponding spaces of functions defined on a general domain $\Omega \subset \mathbb{R}^n$ (see [273, 274] for details). For simplicity, we focus on function spaces defined on $\mathbb{R}^n$. For $k \in \mathbb{N}$, let

$$C^k(\mathbb{R}^n) = \Big\{f \in C(\mathbb{R}^n) \ \Big|\ \|f; C^k\| = \sum_{|\alpha|\le k} \sup_{x\in\mathbb{R}^n} |\partial^\alpha f(x)| < \infty\Big\},$$

where

$$\partial^\alpha f(x) = \frac{\partial^{|\alpha|} f(x)}{\partial x_1^{\alpha_1} \cdots \partial x_n^{\alpha_n}}, \qquad \alpha = (\alpha_1, \cdots, \alpha_n) \in \mathbb{N}_0^n, \ |\alpha| = \alpha_1 + \cdots \alpha_n.$$

Write $C^0(\mathbb{R}^n) = C(\mathbb{R}^n)$. Denote by m the Lebesgue measure. If f is a measurable function on $\mathbb{R}^n$, then its distributional function is defined as

$$f_*(\alpha) = m\{x \in \mathbb{R}^n \ : \ |f(x)| > \alpha\}.$$

(I) The Lebesgue spaces $L^p(\mathbb{R}^n)$

For $1 \le p < \infty$, the Lebesgue space $L^p(\mathbb{R}^n)$ is defined as

https://doi.org/10.1515/9783111384726-001

$$L^p(\mathbb{R}^n) = \left\{ f \in \mathcal{S}'(\mathbb{R}^n) \;\middle|\; \|f\|_p = \Big(\int_{\mathbb{R}^n} |f(x)|^p dx \Big)^{\frac{1}{p}} < \infty \right\},$$
$$L^\infty(\mathbb{R}^n) = \left\{ f \in \mathcal{S}'(\mathbb{R}^n) \mid \|f\|_\infty = \operatorname*{ess\,sup}_{x \in \mathbb{R}^n} |f(x)| < \infty \right\}.$$

Remark 1.1.1. The Lebesgue space $L^p(\mathbb{R}^n)$ has the following properties:

$$\|f\|_p = \Big(p \int_0^\infty a^{p-1} f_*(a) da \Big)^{\frac{1}{p}}, \quad 1 \le p < \infty,$$
$$f_*(a) \le a^{-p} \|f\|_p^p, \quad 1 \le p < \infty, \quad \text{Chebyschev's inequality},$$
$$\|f g\|_p \le \|f\|_q \|g\|_r, \quad \frac{1}{p} = \frac{1}{q} + \frac{1}{r}, \quad \text{Hölder's inequality},$$
$$\Big\| \int_{\mathbb{R}^n} f(x,y) dx \Big\|_{L_y^p} \le \int_{\mathbb{R}^n} \|f(x,y)\|_{L_y^p} dx, \quad \text{Minkowski's inequality}.$$

Additionally, let $L_w^p(\mathbb{R}^n)$ denote the weak L^p space induced by the Chebyschev inequality, that is,

$$L_w^p(\mathbb{R}^n) = \left\{ f \mid f(x) \text{ is measurable, } \|f; L_w^p\| = \sup_{a>0} a (f_*(a))^{\frac{1}{p}} < \infty \right\}.$$

This space, also known as the Marcinkiewicz space, plays a pivotal role in the Marcinkiewicz interpolation theorem.

(II) The Hölder–Zygmund space

For any $s \in \mathbb{R}^+$, we have $s = [s] + \{s\} = [s]^- + \{s\}^+$, where $[s]$ and $[s]^-$ are integers, and

$$0 \le \{s\} < 1, \quad 0 < \{s\}^+ \le 1.$$

With this, we define the Hölder space $C^s(\mathbb{R}^n)$, $0 < s \notin \mathbb{Z}$ as follows:

$$C^s(\mathbb{R}^n) = \left\{ f \in C^{[s]}(\mathbb{R}^n) \;\middle|\; \|f; C^s\| = \|f; C^{[s]}\| + \sum_{|\alpha|=[s]} \|\partial^\alpha f; C^{\{s\}}\| < \infty \right\},$$

where

$$\|f; C^\sigma\| = \sup_{\substack{x,y \in \mathbb{R}^n \\ x \ne y}} \frac{|f(x) - f(y)|}{|x-y|^\sigma}, \quad 0 < \sigma < 1.$$

Let

$$\Delta_h^k f(x) = \sum_{j=0}^{k} (-1)^{k-j} \binom{k}{j} f(x + hj), \quad k \in \mathbb{N},\ h \in \mathbb{R}^n,$$

where $\binom{k}{j}$ is the binomial coefficient. Similarly, we introduce the Zygmund space:

$$\mathcal{C}^s(\mathbb{R}^n) = \left\{ f \in C^{[s]^-}(\mathbb{R}^n) \;\middle|\; \|f; \mathcal{C}^s\| = \|f; C^{[s]^-}\| + \sum_{|\alpha|=[s]^-} \sup_{0\neq h\in\mathbb{R}^n} |h|^{-\{s\}^+} \|\Delta_h^2 \partial^\alpha f; C\| < \infty \right\}.$$

Remark 1.1.2.

(i) For $k = 1, 2, \ldots, \mathcal{C}^k(\mathbb{R}^n)$ is called as the Zygmund space and satisfies the embedding relation

$$C^k(\mathbb{R}^n) \subset \mathcal{C}^k(\mathbb{R}^n), \quad C^k(\mathbb{R}^n) \neq \mathcal{C}^k(\mathbb{R}^n).$$

(ii) When s is not an integer, the Hölder space coincides with the Zygmund space, that is,

$$\mathcal{C}^s(\mathbb{R}^n) = C^s(\mathbb{R}^n), \quad s \in \mathbb{R}^+,\ s \notin \mathbb{Z}.$$

(iii) Let $s > 0$, $k \in \mathbb{N}_0$, $k < s$ and let $m \in \mathbb{N}$ satisfy that $m > s - k$. The Zygmund space $\mathcal{C}^s(\mathbb{R}^n)$ has the following equivalent norm:

$$\|f, \mathcal{C}^s(\mathbb{R}^n)\| = \|f; C^k(\mathbb{R}^n)\| + \sum_{|\alpha|=k} \sup_{0\neq h\in\mathbb{R}^n} h^{-(s-k)} \|\Delta_h^m \partial^\alpha f; C(\mathbb{R}^n)\|.$$

To introduce spaces of differentiable functions, we need some notation. Write $D_j = i^{-1}\partial_{x_j}$ and let $\mathcal{F}$ and $\mathcal{F}^{-1}$ denote the Fourier transform and inverse Fourier transform, respectively, that is,

$$\mathcal{F}\varphi = (2\pi)^{-\frac{n}{2}} \int_{\mathbb{R}^n} e^{-ix\cdot\xi} \varphi(x)dx = \hat{\varphi}(\xi),$$

$$\mathcal{F}^{-1}\psi = (2\pi)^{-\frac{n}{2}} \int_{\mathbb{R}^n} e^{ix\cdot\xi} \psi(\xi)d\xi = \check{\psi}(x).$$

Clearly, $\mathcal{F}^{-1}\varphi(\xi) = (\mathcal{F}\varphi)(-\xi) = \hat{\varphi}(-\xi)$.

(III) The Sobolev spaces

Let $1 \le p \le \infty$. The Sobolev spaces are defined by

$$W^{k,p} = \{f \in \mathcal{S}' \mid \partial^\alpha f \in L^p,\ |\alpha| \le k,\ k \in \mathbb{N}_0\}$$

equipped with the norm

$$\|f; W^{k,p}\| = \Big(\sum_{|\alpha|\le k} \|\partial^\alpha f\|_p^p\Big)^{\frac{1}{p}}, \quad 1 \le p < \infty,$$

$$\|f; W^{k,\infty}\| = \max_{0\le|\alpha|\le k} \|\partial^\alpha f\|_\infty.$$

Remark 1.1.3. When $k = 0$, $W^{0,p} = L^p$. The case $1 < p < \infty$ is frequently used in the study of partial differential equations. When $p = 1$ or $p = \infty$, the Sobolev space defined above remains a Banach space without reflectivity.

(IV) The Nikolskij spaces and the Slobodeckij spaces

These two types of function spaces are the earliest generalizations to the fractional-order Sobolev spaces of the usual Sobolev spaces. Subsequently, a variety of gneralizations and characterizations of the fractional-order Sobolev spaces were given using the Bessel potentials, the Riesz potentials and the Littlewood-Paley decomposition.

The Nikolskij spaces

For $0 < s \notin \mathbb{Z}$ and $1 < p < \infty$, the Nikolskij space is defined by

$$B^s_{p,\infty} = \Big\{f \in \mathcal{S}'(\mathbb{R}^n) \;\Big|\; \|f; B^s_{p,\infty}\| = \|f; W^{[s],p}\| + \sum_{|\alpha|=[s]} \sup_{0\ne h\in\mathbb{R}^n} |h|^{-\{s\}} \|\Delta_h \partial^\alpha f\|_p < \infty\Big\},$$

where Δ_h is the first-order difference Δ^1_h.

The Slobedeckij spaces

For $0 < s \notin \mathbb{Z}$ and $1 < p < \infty$, the Slobedeckij space is defined as

$$B^s_{p,p} = \Big\{f \in \mathcal{S}'(\mathbb{R}^n) \;\Big|\; \|f; B^s_{p,p}\| = \|f; W^{[s],p}\| + \sum_{|\alpha|=[s]} \Big(\int_{\mathbb{R}^n} |h|^{-\{s\}p} \|\Delta_h \partial^\alpha f\|_p^p \frac{dh}{|h|^n}\Big)^{\frac{1}{p}} < \infty\Big\}.$$

Remark 1.1.4. The Nikolskij and Slobodeckij spaces are special cases of the Besov spaces defined below.

(V) The Besov spaces

Applying the second- and higher-order differences in the fractional-order Sobolev spaces in a way similar to that used in the definition of the Zygmund spaces, one can get general spaces of differentiable functions, that is, the Besov spaces.

For $s > 0, 1 \le p \le \infty$ and $1 \le q \le \infty$, the Besov space is defined as

$$B^s_{p,q} = \Big\{f \in \mathcal{S}'(\mathbb{R}^n) \;\Big|\; \|f; B^s_{p,q}\| = \|f; W^{[s]^-,p}\| + \sum_{|\alpha|=[s]^-} \Big(\int_{\mathbb{R}^n} |h|^{-\{s\}^+ q} \|\Delta_h^2 \partial^\alpha f\|_p^q \frac{dh}{|h|^n}\Big)^{\frac{1}{q}} < \infty\Big\},$$

$$B^s_{p,\infty} = \Big\{ f \in \mathcal{S}'(\mathbb{R}^n) \;\Big|\; \|f; B^s_{p,\infty}\| = \|f; W^{[s]^-,p}\| + \sum_{|\alpha|=[s]^-} \sup_{0\neq h\in\mathbb{R}^n} |h|^{-\{s\}^+} \|\Delta_h^2 \partial^\alpha f\|_p < \infty \Big\},$$

where $s = [s]^- + \{s\}^+$.

Remark 1.1.5.

(i) As Zygmund spaces, when s is not an integer, the second-order difference can be replaced by the first-order difference Δ_h in the definition of Besov spaces. Clearly, the Nikolskij and Slobodeckij spaces are special cases of Besov spaces.

(ii) The Besov spaces have the following equivalent norms:

(a) Let $k \in \mathbb{N}_0$, $m \in \mathbb{N}$ satisfy $m > s - k$. Then

$$\|f; B^s_{p,q}\| = \|f; W^{k,p}\| + \sum_{|\alpha|=k} \Big(\int_{\mathbb{R}^n} |h|^{-(s-k)q} \|\Delta_h^m \partial^\alpha f\|_p^q \frac{dh}{|h|^n} \Big)^{\frac{1}{q}},$$

$$\|f; B^s_{p,\infty}\| = \|f; W^{k,p}\| + \sum_{|\alpha|=k} \sup_{0\neq h\in\mathbb{R}^n} |h|^{-(s-k)} \|\Delta_h^m \partial^\alpha f\|_p.$$

(b) While establishing nonlinear estimates, the following equivalent norms in the polar coordinates are often used:

$$\|f; B^s_{p,q}\| = \|f\|_p + \sum_{|\alpha|=[s]^-} \Big(\int_0^\infty t^{-q\{s\}^+} \sup_{|y|<t} \|\Delta_y^2 \partial^\alpha f\|_p^q \frac{dt}{t} \Big)^{\frac{1}{q}},$$

$$\|f; B^s_{p,\infty}\| = \|f\|_p + \sum_{|\alpha|=[s]^-} \sup_{|y|<t,\, t>0} t^{-\{s\}^+} \|\Delta_y^2 \partial^\alpha f\|_p,$$

where $s > 0, 1 \le p \le \infty, 1 \le q < \infty$. In particular, when $s \notin \mathbb{Z}$, the above two equalities can be written in an equivalent form as follows:

$$\|f; B^s_{p,q}\| = \|f\|_p + \sum_{|\alpha|=[s]} \Big(\int_0^\infty t^{-q\{s\}} \sup_{|y|<t} \|\Delta_y \partial^\alpha f\|_p^q \frac{dt}{t} \Big)^{\frac{1}{q}},$$

$$\|f; B^s_{p,\infty}\| = \|f\|_p + \sum_{|\alpha|=[s]} \sup_{|y|<t,\, t>0} t^{-\{s\}} \|\Delta_y \partial^\alpha f\|_p.$$

(iii) The Besov spaces can be generalized to the case $0 < p < 1$. In this case, when $p < 1$ or $q < 1$, the Besov spaces are not Banach spaces but quasi-Banach spaces, so they are rarely used in the study of partial differential equations. The interested reader is referred to Triebel [273, 274].

(iv) Denote by $\dot{W}^{k,p}$ and $\dot{B}^s_{p,q}$ the homogeneous spaces associated with the Sobolev space $W^{k,p}$ and the Besov space $B^s_{p,q}$, respectively, which are defined as follows:

$$\dot{W}^{k,p} = \Big\{ f \in \mathcal{S}'(\mathbb{R}^n) \;\Big|\; \|f; \dot{W}^{k,p}\| = \Big(\sum_{|\alpha|=k} \|\partial^\alpha f\|_p^p \Big)^{\frac{1}{p}} < \infty \Big\},$$

$$\dot{B}^s_{p,q} = \{f \in \mathcal{S}'(\mathbb{R}^n) \mid \|f; \dot{B}^s_{p,q}\| < \infty\},$$

where we have used the equivalent characterization of the spaces and

$$\|f; \dot{B}^s_{p,q}\| = \sum_{|\alpha|=[s]^-} \left(\int_0^\infty t^{-\{s\}^+ q} \sup_{|y|<t} \|\Delta^2_y \partial^\alpha f\|^q_p \frac{dt}{t} \right)^{\frac{1}{q}}.$$

In the case $q = \infty$, we only need to make modification accordingly.

(VI) The spaces of Bessel and Riesz potentials

The classical space of Bessel potentials is defined by

$$H^{s,p}(\mathbb{R}^n) = \{f \in \mathcal{S}'(\mathbb{R}^n) \mid \|f; H^{s,p}\| = \|(I - \Delta)^{\frac{s}{2}} f\|_p = \|\mathcal{F}^{-1}((1 + |\xi|^2)^{\frac{s}{2}} \hat{f}(\xi))\|_p < \infty\},$$

where $s \in \mathbb{R}$, $1 \le p \le \infty$. The corresponding homogeneous space is exactly the space of Riesz potentials:

$$\dot{H}^{s,p} = \{f \in \mathcal{S}'(\mathbb{R}^n) \mid \|f; \dot{H}^{s,p}\| = \|(-\Delta)^{\frac{s}{2}} f\|_p = \|\mathcal{F}^{-1}(|\xi|^s \hat{f})\|_p < \infty\}.$$

The multiplier operators defined by the multipliers $(1 + |\xi|^2)^{-\frac{s}{2}}$ and $|\xi|^{-s}$, respectively, are usually denoted by

$$J_s = (I - \Delta)^{-\frac{s}{2}} = \mathcal{F}^{-1}(1 + |\xi|^2)^{-\frac{s}{2}} *,$$
$$I_s = (-\Delta)^{-\frac{s}{2}} = \mathcal{F}^{-1}|\xi|^{-s} *,$$

which are called as the Bessel-potential operator and the Riesz-potential operator, respectively, since $\mathcal{F}^{-1}(1 + |\xi|^2)^{-\frac{s}{2}}$ corresponds to the Bessel function. It is easy to see that

$$H^{s,p}(\mathbb{R}^n) = J_s L^p(\mathbb{R}^n), \quad \dot{H}^{s,p}(\mathbb{R}^n) = I_s L^p(\mathbb{R}^n).$$

Remark 1.1.6.

(i) A direct calculation shows that for any $k \in \mathbb{N}_0$,

$$H^{k,p} = W^{k,p}, \quad 1 < p < \infty.$$

(ii) Let $s, \sigma \in \mathbb{R}$, $1 < p < \infty$. Then J_s is a homeomorphic mapping from $H^{\sigma-s,p}$ onto $H^{\sigma,p}$, and I_s is a homeomorphic mapping from $\dot{H}^{\sigma-s,p}$ onto $\dot{H}^{\sigma,p}$, that is,

$$J_s H^{\sigma-s,p} = H^{\sigma,p}, \quad I_s \dot{H}^{\sigma-s,p} = \dot{H}^{\sigma,p}.$$

(iii) Utilizing the theory of Fourier multipliers, one can generalize the Sobolev spaces to the case of negative indices. Similarly, this method can also be used to generalize the Besov space $B^s_{p,q}$ and the Zygmund space C^s to the case of negative indices. For $s, \sigma \in \mathbb{R}, 1 < p < \infty, 1 \leq q \leq \infty$, we have

$$J_s B^{\sigma-s}_{p,q} = B^{\sigma}_{p,q}, \quad J_s C^{\sigma-s} = C^{\sigma},$$
$$I_s \dot{B}^{\sigma-s}_{p,q} = \dot{B}^{\sigma}_{p,q}, \quad I_s \dot{C}^{\sigma-s} = \dot{C}^{\sigma}.$$

(iv) When $p = 2$, $H^{s,2} \equiv H^s$ is the classical Sobolev space of Hilbert type and

$$B^s_{2,2} = H^s, \quad \dot{B}^s_{2,2} = \dot{H}^s.$$

To introduce more spaces of differentiable functions, we need the Littlewood–Paley theory. Let $\varphi_0(\xi) \in C_c^\infty(\mathbb{R}^n)$ be a radial bump function satisfying that

$$\begin{cases} \varphi_0(\xi) = 1, & |\xi| \leq 1, \\ \varphi_0(\xi) = 0, & |\xi| \geq 2. \end{cases}$$

Let

$$\begin{cases} \varphi_j(\xi) = \varphi_0(2^{-j}\xi) - \varphi_0(2^{-j+1}\xi), & j \in \mathbb{N}, \\ \psi_j(\xi) = \varphi_0(2^{-j}\xi) - \varphi_0(2^{-j+1}\xi), & j \in \mathbb{Z}. \end{cases}$$

Then $\varphi_j(\xi)$ and $\psi_j(\xi)$ are also radial bump functions and

$$\sup_{\xi \in \mathbb{R}^n} 2^{j|\alpha|} |\partial^\alpha \psi_j(\xi)| < \infty, \quad j \in \mathbb{Z}.$$

Thus, we have the following two dyadic decompositions of unity:

$$\sum_{j=0}^{\infty} \varphi_j(\xi) = 1, \quad \xi \in \mathbb{R}^n,$$
$$\sum_{j \in \mathbb{Z}} \psi_j(\xi) = 1, \quad \xi \in \mathbb{R}^n \setminus \{0\}.$$

For any $f \in L^2(\mathbb{R}^n)$, we have the following decomposition:

$$f = \sum_{j=0}^{\infty} \mathcal{F}^{-1}(\varphi_j \mathcal{F} f) = \sum_{j=0}^{\infty} \check{\varphi}_j * f,$$
$$f = \sum_{j \in \mathbb{Z}} \mathcal{F}^{-1}(\psi_j \mathcal{F} f) = \sum_{j \in \mathbb{Z}} \check{\psi}_j * f,$$

where the series converges in the L^2-norm. For convenience, we set

$$\psi_j(D) f = \mathcal{F}^{-1}(\psi_j \mathcal{F} f) = \check{\psi}_j * f, \quad \forall j \in \mathbb{Z}, \tag{1.1}$$
$$\varphi_j(D) f = \mathcal{F}^{-1}(\varphi_j \mathcal{F} f) = \check{\varphi}_j * f, \quad \forall j \in \mathbb{N}_0. \tag{1.2}$$

Since $\psi_j \in C_c^\infty(\mathbb{R}^n)$, $\varphi_j \in C_c^\infty(\mathbb{R}^n)$ and

$$\operatorname{supp}(\psi_j(\xi)) \subset \{\xi \mid 2^{j-1} \le |\xi| \le 2^{j+1}\}, \quad j \in \mathbb{Z},$$
$$\operatorname{supp}(\varphi_j(\xi)) \subset \{\xi \mid 2^{j-1} \le |\xi| \le 2^{j+1}\}, \quad j \in \mathbb{N},$$

then for any $f \in \mathcal{S}'(\mathbb{R}^n)$, (1.1) and (1.2) are still well-defined. Let

$$Sf = \Big(\sum_{j\in\mathbb{Z}} |\psi_j(D)f|^2\Big)^{\frac{1}{2}}. \tag{1.3}$$

Then by the almost orthogonal property of $\{\psi_j(D)f\}_{j\in\mathbb{Z}}$ and the orthogonality on $L^2([0,1])$ of the Rademacher function $\{r_\lambda(t)\}$, it is easy to deduce that

$$C_p^{-1}\|f\|_p \le \|Sf\|_p \le C_p\|f\|_p, \quad 1 < p < \infty. \tag{1.4}$$

Further, by Minkowski's inequality, we obtain that

$$\|f\|_p \le C_p\Big(\sum_{j\in\mathbb{Z}} \|\psi_j(D)f\|_p^2\Big)^{\frac{1}{2}}, \quad 2 \le p < \infty, \tag{1.5}$$

$$\Big(\sum_{j\in\mathbb{Z}} \|\psi_j(D)f\|_p^2\Big)^{\frac{1}{2}} \le C_p\|f\|_p, \quad 1 < p \le 2. \tag{1.6}$$

We will see below that (1.4) means $H^{s,p}(\mathbb{R}^n) \sim F^s_{p,2}(\mathbb{R}^n)$ and that both (1.5) and (1.6) mean that

$$B^s_{p,p}(\mathbb{R}^n) \hookrightarrow F^s_{p,2}(\mathbb{R}^n), \quad 1 < p \le 2,$$
$$F^s_{p,2}(\mathbb{R}^n) \hookrightarrow B^s_{p,p}(\mathbb{R}^n), \quad 2 \le p < \infty.$$

The inequalities (1.4), (1.5) and (1.6) establish a bridge between the Besov spaces and the Triebel–Lizorkin spaces.

We now use the Littlewood–Paley decomposition to characterize the functional spaces defined above. In fact, the Littlewood–Paley characterization of Besov and Zygmund spaces essentially extends these spaces to the case of negative indices. For the Sobolev spaces, we have

$$\begin{aligned} H^{s,p}(\mathbb{R}^n) = \Bigg\{ f \in \mathcal{S}'(\mathbb{R}^n) \;\Bigg|\; \|f\|_{H^{s,p}} &= \Bigg\| \Big(\sum_{j=0}^{\infty} 2^{2js} |\check{\varphi}_j * f|^2\Big)^{\frac{1}{2}} \Bigg\|_p \\ &= \Bigg\| \Big(\sum_{j=0}^{\infty} 2^{2js} |\varphi_j(D)f|^2\Big)^{\frac{1}{2}} \Bigg\|_p < \infty \Bigg\}, \end{aligned}$$

$$\dot{H}^{s,p}(\mathbb{R}^n) = \Big\{ f \in \mathcal{S}'(\mathbb{R}^n) \;\Big|\; \|f\|_{\dot{H}^{s,p}} = \Big\| \Big(\sum_{j\in\mathbb{Z}} 2^{2js} |\check{\psi}_j * f|^2 \Big)^{\frac{1}{2}} \Big\|_p = \Big\| \Big(\sum_{j\in\mathbb{Z}} 2^{2js} |\psi_j(D)f|^2 \Big)^{\frac{1}{2}} \Big\|_p < \infty \Big\},$$

where $1 \le p < \infty$ and $s \in \mathbb{R}$. When $p = \infty$, we only need to modify the definition of the norms above. The Besov spaces are defined as

$$B^s_{p,q} = \Big\{ f \in \mathcal{S}'(\mathbb{R}^n) \;\Big|\; \|f\|_{B^s_{p,q}} = \Big(\sum_{j=0}^{\infty} 2^{jsq} \|\check{\varphi}_j * f\|_p^q \Big)^{\frac{1}{q}} = \Big(\sum_{j=0}^{\infty} 2^{jsq} \|\varphi_j(D)f\|_p^q \Big)^{\frac{1}{q}} < \infty \Big\}, \tag{1.7}$$

$$\dot{B}^s_{p,q} = \Big\{ f \in \mathcal{S}'(\mathbb{R}^n) \;\Big|\; \|f\|_{\dot{B}^s_{p,q}} = \Big(\sum_{j\in\mathbb{Z}} 2^{jsq} \|\check{\psi}_j * f\|_p^q \Big)^{\frac{1}{q}} = \Big(\sum_{j\in\mathbb{Z}} 2^{jsq} \|\psi_j(D)f\|_p^q \Big)^{\frac{1}{q}} < \infty \Big\}, \tag{1.8}$$

where $1 \le p \le \infty$, $1 \le q < \infty$ and $s \in \mathbb{R}$. When $q = \infty$, we only need to modify the definition of the norms above. In particular, $B^s_{\infty,\infty} = C^s$, that is,

$$C^s = \big\{ f \in \mathcal{S}'(\mathbb{R}^n) \mid \|f\|_{C^s} = \sup_{x\in\mathbb{R}^n, j\in\mathbb{N}_0} 2^{js} |\varphi_j(D)f| < \infty \big\},$$

$$\dot{C}^s = \big\{ f \in \mathcal{S}'(\mathbb{R}^n) \mid \|f\|_{\dot{C}^s} = \sup_{x\in\mathbb{R}^n, j\in\mathbb{Z}} 2^{js} |\psi_j(D)f| < \infty \big\}.$$

Remark 1.1.7.
(i) Actually, the Littlewood–Paley characterization extends the Besov spaces to the case with $0 < p \le \infty$, $0 < q \le \infty$ and $s \in \mathbb{R}$.
(ii) In the Littlewood–Paley characterization of Besov spaces, if the $l^q(L^p)$-norm is replaced by the $L^p(l^q)$-norm, then we obtain the well-known Triebel–Lizorkin spaces (see below).

(VII) The Triebel–Lizorkin spaces

For $s \in \mathbb{R}$, $1 \le p < \infty$, $1 \le q < \infty$, we define

$$F^s_{p,q} = \Big\{ f \in \mathcal{S}'(\mathbb{R}^n) \;\Big|\; \|f\|_{F^s_{pq}} = \Big\| \Big(\sum_{j=0}^{\infty} 2^{jsq} |\varphi_j(D)f|^q \Big)^{\frac{1}{q}} \Big\|_p < \infty \Big\},$$

$$\dot{F}^s_{p,q} = \left\{ f \in \mathcal{S}'(\mathbb{R}^n) \,\middle|\, \|f\|_{\dot{F}^s_{p,q}} = \left\| \left(\sum_{j\in\mathbb{Z}} 2^{jsq} |\psi_j(D)f|^q \right)^{\frac{1}{q}} \right\|_p < \infty \right\}.$$

(VIII) The Hardy space $\mathcal{H}_p$

It is well known that the Hardy space $\mathcal{H}_p$, $0 < p < \infty$ has many characterizations such as the method of harmonic functions of multivariables of Stein–Weiss [244] and the real-variable method of Fefferman–Stein [90]. We use the Littlewood–Paley decomposition to characterize the Hardy spaces here.

Definition of the Hardy space

Let $0 < p < \infty$ and let $\{\varphi_j\}_{j=0}^{\infty}$, $\{\psi_j(\xi)\}_{j\in\mathbb{Z}}$ be the Littlewood–Paley decomposition. Then the Hardy space is defined by

$$\mathcal{H}_p = \left\{ f \in \mathcal{S}'(\mathbb{R}^n) \,\middle|\, \|f; \mathcal{H}_p\| = \left\| \left(\sum_{j\in\mathbb{Z}} |\psi_j(D)f|^2 \right)^{\frac{1}{2}} \right\|_p < \infty \right\}.$$

From the definition of the Hardy space, it is known that $\mathcal{H}_p$ is homogeneous. The corresponding inhomogeneous space h_p is called as the local Hardy space:

$$h_p = \left\{ f \in \mathcal{S}'(\mathbb{R}^n) \,\middle|\, \|f; h_p\| = \left\| \left(\sum_{j=0}^{\infty} |\varphi_j(D)f|^2 \right)^{\frac{1}{2}} \right\|_p < \infty \right\}.$$

Remark 1.1.8.

(i) Using the Littlewood–Paley characterization of the function spaces, it can be shown that

$$h_p = \mathcal{H}_p = L^p = F^0_{p,2}, \quad 1 < p < \infty.$$

When $p = 1$, $\mathcal{H}_1$ is the classical Hardy space.

(ii) Note that the space $F^s_{\infty,q}$ is not defined. We now use the Littlewood–Paley method to characterize this space, which is essentially the dual space to $F^{-s}_{1,q'}$. In particular, $F^0_{\infty,2}$ is the dual space to the Hardy space $\mathcal{H}_1$.

Definition of $F^s_{\infty,q}$

Let $\{\varphi_j\}_{j=0}^{\infty}$ be the inhomogeneous Littlewood–Paley decomposition of the unity. For $1 \le q < \infty$ and $s \in \mathbb{R}$, define

$$F^s_{\infty,q} = \left\{ f \in \mathcal{S}'(\mathbb{R}^n) \,\middle|\, \exists \{f_j\}_{j=0}^{\infty} \subset L^\infty \text{ such that } f \overset{\mathcal{S}'(\mathbb{R}^n)}{=\!=} \sum_{j=0}^{\infty} \varphi_j(D) f_j \right.$$

$$\left. \text{and } \|f; F^s_{\infty,q}\| = \sup_{x\in\mathbb{R}^n} \left(\sum_{j=0}^{\infty} 2^{jsq} |\varphi_j(D) f_j|^q \right)^{\frac{1}{q}} < \infty \right\}.$$

When $q = \infty$, suitable modification is needed in the definition of the norm above.

(IX) The BMO space

The BMO space is a class of important function spaces in the study of PDEs and is the dual space to the Hardy space $\mathcal{H}_1$ (see Fefferman–Stein [90]). In the interpolation theory, the Hardy space $\mathcal{H}_1$ is the substitute of L^1, while the BMO space is the substitute of L^∞.

Definition of the BMO space

Let Q be an arbitrary cube in $\mathbb{R}^n$ and let f be locally integrable with

$$f_Q = \frac{1}{|Q|} \int_Q f(x)dx$$

being the average of f on Q. Define

$$\mathrm{BMO} = \left\{ f \in L^1_{\mathrm{loc}}(\mathbb{R}^n) \,\middle|\, \|f\|_{\mathrm{BMO}} = \sup_{Q\subset\mathbb{R}^n} \frac{1}{|Q|} \int_Q |f - f_Q| dx < \infty \right\}.$$

The associated inhomogeneous space is

$$\mathrm{bmo} = \left\{ f \in L^1_{\mathrm{loc}}(\mathbb{R}^n) \,\middle|\, \|f; \mathrm{bmo}\| = \sup_{|Q|<1} \frac{1}{|Q|} \int_Q |f - f_Q| dx + \sup_{|Q|=1} \int_Q |f(x)| dx < \infty \right\}.$$

Similarly, one can define BMO_p and bmo_p for $0 < p < \infty$ by replacing the L^1-norm with the L^p-norm in the integral in the above definition.

Remark 1.1.9.

(i) By the Fefferman–Stein theorem, it follows that

$$h_1' = \mathrm{bmo}_1 = \mathrm{bmo}, \quad \mathcal{H}_1' = \mathrm{BMO}_1 = \mathrm{BMO}.$$

(ii) For $0 < p < \infty$, it follows from the Littlewood–Paley characterization that

$$h_p = F^0_{p,2}, \quad \mathrm{bmo}_p = F^0_{p',2}.$$

(X) The Morrey spaces (also called as the Morrey–Campanato spaces)
For $1 \le q, p < \infty$, the Morrey–Campanato space M_q^p is defined by

$$M_q^p = \left\{ f \in L^q_{\text{loc}}(\mathbb{R}^n) \;\middle|\; \|f\|_{M_q^p} = \sup_{\substack{x_0 \in \mathbb{R}^n \\ 0<R\le 1}} R^{-n(\frac{1}{q}-\frac{1}{p})} \Big(\int_{B_R(x_0)} |f|^q dx \Big)^{\frac{1}{q}} < \infty \right\},$$

where $B_R(x_0)$ denotes the ball of centered at x_0, radius R. The associated homogeneous space is defined as

$$\dot{M}_q^p = \left\{ f \in L^q_{\text{loc}}(\mathbb{R}^n) \;\middle|\; \|f\|_{\dot{M}_q^p} = \sup_{\substack{x_0 \in \mathbb{R}^n \\ 0<R<\infty}} R^{-n(\frac{1}{q}-\frac{1}{p})} \Big(\int_{B_R(x_0)} |f|^q dx \Big)^{\frac{1}{q}} < \infty \right\}.$$

Remark 1.1.10. The spaces M_q^p and $\dot{M}_q^p$ are integrable spaces (similar to the L^p space). We can define generalized Besov spaces with the Morrey space as the base space. In fact, let $(X, \|\cdot\|)$ be a Banach space consisting of distributions on $\mathbb{R}^n$ and let $\{\varphi_j\}_{j=0}^{\infty}$, $\{\psi_j(\xi)\}_{j\in\mathbb{Z}}$ be the Littlewood–Paley decomposition. With the theory of Littlewood–Paley decomposition, we can define the abstract Besov space with the base space X as

$$\begin{aligned} B^s_{X,q} = \Bigg\{ f \in \mathcal{S}'(\mathbb{R}^n) \;\Bigg|\; \|f\|_{B^s_{X,q}} &= \Big(\sum_{j=0}^{\infty} 2^{2sq} \|\check{\varphi}_j * f\|_X^q \Big)^{\frac{1}{q}} \\ &= \Big(\sum_{j=0}^{\infty} 2^{jsq} \|\varphi_j(D) f\|_X^q \Big)^{\frac{1}{q}} < \infty \Bigg\}, \\ \dot{B}^s_{X,q} = \Bigg\{ f \in \mathcal{S}'(\mathbb{R}^n) \;\Bigg|\; \|f\|_{\dot{B}^s_{X,q}} &= \Big(\sum_{j\in\mathbb{Z}} 2^{jsq} \|\check{\psi}_j * f\|_X^q \Big)^{\frac{1}{q}} \\ &= \Big(\sum_{j\in\mathbb{Z}} 2^{jsq} \|\psi_j(D) f\|_X^q \Big)^{\frac{1}{q}} < \infty \Bigg\}, \end{aligned}$$

where $1 \le q < \infty$ and $s \in \mathbb{R}$. When $q = \infty$, the above definitions remain true with obvious modification of the norms.

Finally, we consider the Gaussian semigroup characterization and the Poisson semigroup characterization for the Besov spaces. It is well known that for $f \in L^p(\mathbb{R}^n)$ with $1 < p < \infty$, the functions

$$\begin{aligned} u(x,t) &= W(t)f(x) = \mathcal{F}^{-1}(e^{-t|\xi|^2}\hat{f}), \quad x \in \mathbb{R}^n,\ t \ge 0, \\ v(x,t) &= P(t)f(x) = \mathcal{F}^{-1}(e^{-t|\xi|}\hat{f}), \quad x \in \mathbb{R}^n,\ t \ge 0, \end{aligned}$$

are solutions to the problem

$$\begin{cases} u_t = \Delta u, & (x,t) \in \mathbb{R}^n \times \mathbb{R}^+, \\ u(0) = f(x), & x \in \mathbb{R}^n, \end{cases}$$

and the problem

$$\begin{cases} v_{tt} + \Delta v = 0, & (x,t) \in \mathbb{R}^n \times \mathbb{R}^+, \\ v(x,0) = f(x), & x \in \mathbb{R}^n, \end{cases}$$

respectively. We have the following characterizations of Besov spaces: for $0 < s < \infty$, $1 < p < \infty$, $1 \le q \le \infty$ and $m, k \in \mathbb{N}$ satisfying that $m > s/2$, $k > s$, we have

$$B^s_{p,q} = \left\{ f \in L^p(\mathbb{R}^n) \;\middle|\; \|f; B^s_{p,q}\| = \|f\|_p + \left(\int_0^\infty t^{(m-\frac{s}{2})q} \left\| \frac{\partial^m W(t) f}{\partial t^m} \right\|_p^q \frac{dt}{t} \right)^{\frac{1}{q}} < \infty \right\},$$

$$B^s_{p,q} = \left\{ f \in L^p(\mathbb{R}^n) \;\middle|\; \|f; B^s_{p,q}\| = \|f\|_p + \left(\int_0^\infty t^{(k-s)q} \left\| \frac{\partial^k P(t) f}{\partial t^k} \right\|_p^q \frac{dt}{t} \right)^{\frac{1}{q}} < \infty \right\}.$$

These characterizations involve the idea of interchange between the temporal and spatial differentiation, which is very important in the study of nonlinear evolution equations such as the heat equation and the Navier–Stokes equations. We will mention this point again in Chapters 2 and 3.

Remark 1.1.11.

(i) The theory of function spaces abound. They are the foundation for the study of PDEs. The different characterizations of their norms provide a variety of methods for nonlinear estimates. Due to the space limitation, we only give simple definitions here. For more characterizations such as integer analytic function characterizations, atomic and molecular characterizations, abstract interpolation characterizations and intrinsic characterizations, one can find in the books [14, 273, 274].

(ii) The function spaces defined above can be extended to the case of a general smooth domain $\Omega \subset \mathbb{R}^n$ by means of intrinsic characterizations or restriction methods. The interested reader is referred to the books [14, 273, 274].

We now review several important results from harmonic analysis, which will be useful in the later chapters.

(I) The Riesz interpolation theorem

Let $1 \le p_j, q_j \le \infty, j = 1, 2$ and let $T : L^{p_0} \cap L^{p_1} \longrightarrow L^{q_0} \cap L^{q_1}$ be a linear operator and satisfy that

$$\|Tf\|_{q_j} \le M_j \|f\|_{p_j}, \quad j = 0, 1.$$

Then

$$\|Tf\|_{q_t} \le M_0^{1-t} M_1^t \|f\|_{p_t}, \quad f \in L^{p_0} \cap L^{p_1},$$

where

$$\frac{1}{p_t} = \frac{1-t}{p_0} + \frac{t}{p_1}, \quad \frac{1}{q_t} = \frac{1-t}{q_0} + \frac{t}{q_1}, \quad 0 \le t \le 1. \tag{1.9}$$

See [241, 244] for a proof.

(II) The Stein interpolation theorem

To state the Stein interpolation theorem, it is necessary to introduce some definitions. Let

$$D = \{z \in \mathbb{C}, 0 < \mathrm{Re}(z) < 1\}$$

be a strip region in the complex plane. We call $\{T_z\}_{z \in D}$ a family of analytic operators if:

(i) for any simple function f, $T_z f$ is a measurable function;

(ii) for any simple functions f and g, the mapping

$$z \longrightarrow \int g(x)\, T_z f(x) dx$$

is a bounded continuous function in $\overline{D}$ and is analytic in D.

The Stein interpolation theorem

Let $1 \le p_0, p_1, q_0, q_1 \le \infty$ and let T_z be a family of analytic operators on D. If there exist positive constants M_0 and M_1 such that

$$\|T_{ib} f\|_{q_0} \le M_0 \|f\|_{p_0}, \quad \|T_{1+ib} f\|_{q_1} \le M_1 \|f\|_{p_1}, \quad \forall b \in \mathbb{R},$$

then for any $z = a + bi \in D$, T_z can be extended as a bounded operator from L^p to L^q and satisfies that

$$\|T_z f\|_q \le M_0^{1-a} M_1^a \|f\|_p,$$

where

$$\frac{1}{p} = \frac{1-a}{p_0} + \frac{a}{p_1}, \quad \frac{1}{q} = \frac{1-a}{q_0} + \frac{a}{q_1}.$$

A proof can be found in [244].

(III) The Marcinkiewicz and Hunt interpolation theorems

The Marcinkiewicz interpolation theorem

Assume that $1 \le p_j \le q_j \le \infty, j = 0, 1, q_0 \ne q_1$, and that T is a sublinear operator from the measurable space (X, μ) to the measurable space (Y, ν), that is,

$$|T(f+g)| \le C(|Tf| + |Tg|).$$

Assume further that T is a bounded operator from $L^{p_j}(X, d\mu)$ to $L_W^{q_j}(Y, d\nu)$, that is,

$$(Tf)_*(\alpha) \le \left(\frac{M_j \|f\|_{p_j}}{\alpha}\right)^{q_j}, \quad j = 0, 1.$$

Then T is a bounded operator from $L^{p_t}(X)$ to $L^{q_t}(Y)$ and satisfies that

$$\|Tf\|_{q_t} \le C(q_0, q_1, t) M_0^t M_1^{1-t} \|f\|_{p_t}, \tag{1.10}$$

where (p_t, q_t) satisfies (1.9) and

$$C(q_0, q_1, t) = 2\left(\frac{q_t}{q_1 - q_t} + \frac{q_t}{q_t - q_0}\right)^{\frac{1}{q_t}}.$$

The Hunt interpolation theorem

Assume that $1 \le p_0 < p_1 \le \infty, 1 \le q_0 < q_1 \le \infty$ and that T is a bounded linear operator from $L^{p_j}(X, d\mu)$ to $L^{q_j}(Y, d\nu)$ and satisfies

$$\|Tf; L^{q_j}(Y, d\nu)\| \le C_j \|f; L^{p_j}(X, d\mu)\|, \quad j = 0, 1.$$

Then, for any $t \in (0, 1)$, T can be extended as a bounded linear operator from $L_W^{p_t}(X, d\mu)$ to $L_W^{q_t}(Y, d\nu)$ satisfying that

$$\|Tf; L_W^{q_t}(Y, d\nu)\| \le C(t, q_0, q_1, p_0, p_1) \|f; L_W^{p_t}(X, d\mu)\|$$

and being bounded at the end points, where p_t and q_t satisfy (1.9).

Remark 1.1.12.

(i) For a proof of the Marcinkiewicz interpolation theorem, see [241]. This theorem implies that weak-type estimates at end points can be transferred into strong type estimates at mid points by interpolation, which is crucial in the study of PDEs.

(ii) In the Marcinkiewicz interpolation formula (1.10), the interpolation coefficient satisfies that

$$\lim_{t \to 0} C(q_0, q_1, t) = \infty, \quad \lim_{t \to 1} C(q_0, q_1, t) = \infty.$$

(iii) For a proof of the Hunt interpolation theorem, see [226, Theorem IX.19].

(IV) The Young and generalized Young inequalities

As a direct result of the Riesz interpolation theorem, we have the following Young's and generalized Young's inequalities.

Young's theorem

Let $1 \le r \le \infty$ and $K(x,y)$ be a differentiable function. If

$$\sup_x \|K(x,\cdot)\|_r \le C, \quad \sup_y \|K(\cdot,y)\|_r \le C,$$

then the operator

$$Tf(x) = \int_{\mathbb{R}^n} K(x,y)f(y)dy$$

satisfies

$$\|Tf\|_q \le C\|f\|_p,$$

where $1 \le p \le r'$ and $\frac{1}{q} = \frac{1}{p} + \frac{1}{r} - 1$. Hereafter, for a positive number s, s' is so defined that $1/s' + 1/s = 1$.

As a direct result of the Young theorem, we have the following Young inequality.

The Young inequality

Let $f \in L^p(\mathbb{R}^n)$, $g \in L^r(\mathbb{R}^n)$ and let

$$\frac{1}{q} = \frac{1}{p} + \frac{1}{r} - 1, \quad 1 \le p,\, q,\, r \le \infty.$$

Then

$$\|f * g\|_q \le C\|f\|_p\|g\|_r.$$

Using the Marcinkiewicz interpolation theorem, we get the generalized Young inequality.

The generalized Young inequality

Let $f \in L^p(\mathbb{R}^n)$, $g \in L^r_w(\mathbb{R}^n)$ and let

$$\frac{1}{q} = \frac{1}{p} + \frac{1}{r} - 1, \quad 1 < p,\, q,\, r < \infty.$$

Then $f * g \in L^q(\mathbb{R}^n)$ and

$$\|f * g\|_q \le C\|f\|_p\|g\|_{L^r_w}.$$

(V) The Hardy–Littlewood maximal function
For a measurable function f on $\mathbb{R}^n$, the function

$$(\mathcal{M}f)(x) = \sup_{x \in B} \frac{1}{|B|} \int_B |f(y)| dy$$

is called the Hardy–Littlewood maximal function, where the supremum is taking over all the balls containing x. It is easy to see that

$$\|\mathcal{M}f\|_\infty \leq \|f\|_\infty.$$

On the other hand, using the Vitali covering lemma, it is easy to show that $\mathcal{M}$ is a weak $(1,1)$-type operator, that is,

$$(\mathcal{M}f)_*(\alpha) \leq \frac{C}{\alpha} \|f\|_1.$$

Thus, from the Marcinkiewicz interpolation theorem, it is known that

$$\|\mathcal{M}f\|_p \leq C\|f\|_p, \quad 1 < p \leq \infty.$$

With the help of the Hardy–Littlewood maximal operator, it is easy to prove the well-known Hardy–Littlewood–Sobolev inequality.

(VI) The Hardy–Littlewood–Sobolev inequality and its generalizations
Let $0 < \gamma < n, 1 < p < q < \infty$ and

$$\frac{1}{q} = \frac{1}{p} + \frac{\gamma}{n} - 1. \tag{1.11}$$

Then

$$\big\||x|^{-\gamma} * f\big\|_q \leq C\|f\|_p. \tag{1.12}$$

In fact, noting that

$$|x|^{-\gamma} * f(x) = \int_{|y| \geq R} \frac{f(x-y)}{|y|^\gamma} dy + \int_{|y| < R} \frac{f(x-y)}{|y|^\gamma} dy = I_1 + I_2,$$

we easily derive by the Hölder inequality that

$$|I_1| \leq \|f\|_p \Big(\int_{|y| \geq R} |y|^{-\gamma p'} dy \Big)^{\frac{1}{p'}} \leq C R^{\frac{n}{p'} - \gamma} \|f\|_p,$$

where we have used the fact that $\gamma p' > n$ which follows from (1.11). On the other hand, it is seen that

$$|I_2| \le \sum_{k=0}^{\infty} \int_{2^{-k-1}<\frac{|y|}{R}<2^{-k}} \frac{f(x-y)}{|y|^\gamma} dy \le C \sum_{k=0}^{\infty} \frac{1}{(2^{-k}R)^\gamma} \int_{|y|\le 2^{-k}R} f(x-y)dy$$

$$\le 2^{-\gamma} C \sum_{k=0}^{\infty} (2^{-k}R)^{n-\gamma} \mathcal{M}f(x) \le CR^{n-\gamma}\mathcal{M}f,$$

where we have used the fact that $\gamma < n$. Hence,

$$\big||x|^{-\gamma} * f(x)\big| \le CR^{\frac{n}{p'}-\gamma}\|f\|_p + CR^{n-\gamma}\mathcal{M}f(x).$$

Let $R = R(x)$ satisfy that $R^{\frac{n}{p'}-\gamma}\|f\|_p = R^{n-\gamma}\mathcal{M}f(x)$, that is,

$$R(x) = \left(\frac{\|f\|_p}{\mathcal{M}f(x)}\right)^{\frac{p}{n}}.$$

Then, we have

$$\big||x|^{-\gamma} * f\big| \le C\|f\|_p^{1-\frac{p}{q}} (\mathcal{M}f(x))^{\frac{p}{q}}.$$

The inequality (1.12) follows by taking the L^q-norm of both sides and the fact that the Hardy–Littlewood maximal operator is (p,p)-type operator.

Remark 1.1.13.

(i) From Fourier transform of Riesz potential, it is found that

$$I_\alpha f = C(n,\alpha)|x|^{-n+\alpha} * f.$$

Thus, by the Hardy–Littlewood–Sobolev inequality, we get

$$\|I_\alpha f\|_q = \big\|(-\Delta)^{-\frac{\alpha}{2}} f\big\|_q \le C\|f\|_p, \quad \frac{1}{q} = \frac{1}{p} - \frac{\alpha}{n}. \tag{1.13}$$

Naturally, (1.13) can also be regarded as the embedding theorem of fractional Soboblev spaces, that is,

$$L^p \hookrightarrow \dot{W}^{-\alpha,q}, \quad \frac{1}{q} = \frac{1}{p} - \frac{\alpha}{n}.$$

(ii) From the generalized Young inequality, the equivalent bilinear form of the Hardy–Littlewood–Sobolev inequality follows easily.

The generalized Hardy–Littlewood–Sobolev inequality
Let $1 < p, r < \infty$, $0 < \lambda < n$ and

$$\frac{1}{p} + \frac{1}{r} + \frac{\lambda}{n} = 2.$$

Then

$$\iint_{\mathbb{R}^n \times \mathbb{R}^n} \frac{|f(x)||h(y)|}{|x-y|^\lambda} dxdy \leq C(p, r, \lambda, n)\|f\|_p \|h\|_r. \tag{1.14}$$

Certainly, one can prove (1.14) using the Young inequality and the Hunt interpolation inequality. In fact, for a fixed $f \in L^p(\mathbb{R}^n)$, we define the operator T_f by

$$T_f(g) = f * g.$$

It follows from the Young inequality that T_f is a bounded linear operator from L^r to L^q and satisfies

$$\|T_f g\|_q \leq C\|g\|_r, \quad 1 \leq r \leq p', \quad \frac{1}{q} = \frac{1}{p} + \frac{1}{r} - 1.$$

Taking $r = 1$ and $r = p'$, respectively, in the above inequality, we obtain by the Hunt interpolation inequality that for $1 < r < p'$ the operator T_f is bounded from $L_w^r(\mathbb{R}^n)$ to $L_w^q(\mathbb{R}^n)$ and satisfies

$$\|T_f(g); L_w^q\| = \|f * g; L_w^q\| \lesssim \|f; L_w^p\| \|g; L_w^r\| \lesssim \|f\|_p \|g; L_w^r\|.$$

For a fixed $g \in L_w^r$, this means that T_g is a bounded linear operator from L^p to L_w^q and

$$\|T_g f; L_w^q(\mathbb{R}^n)\| \leq C\|f\|_p, \quad p \leq q, \quad \frac{1}{q} = \frac{1}{p} + \frac{1}{r} - 1.$$

By the Marcinkiewicz interpolation inequality, we have $f * g \in L^q$. Thus, from the generalized Young inequality, the inequality (1.14) follows directly.

(VII) Translation invariance operators
The operator $B : L^p \longmapsto L^q$ is said to be translation invariant if:
(i) B is bounded,
(ii) B commutes with translations, that is, $\tau_h B = B\tau_h$, where $(\tau_h f)(x) = f(x-h)$.

One can easily verify that, when $q < p$, $B \equiv 0$. So, we are always restricted to the case $1 \leq p \leq q \leq \infty$.

The characterization theorem of translation invariance operators
Let $B : L^p \to L^q$ be a translation invariance operator. Then there exists a distribution $T \in \mathcal{S}'(\mathbb{R}^n)$ such that

$$Bu = T * u, \quad \forall u \in L^p(\mathbb{R}^n).$$

In other words, there is an $m(\xi) \in \mathcal{M}_p^q$ such that

$$Bu = \mathcal{F}^{-1}(m(\xi)\mathcal{F}u) \triangleq \mathcal{F}^{-1}m(\xi)\mathcal{F}u, \quad \forall u \in L^p(\mathbb{R}^n),$$

where $\mathcal{M}_p^q$ stands for the Hörmander space defined as

$$\mathcal{M}_p^q = \Big\{m(\xi) \in \mathcal{S}'(\mathbb{R}) \mid \|m(\xi)\|_{\mathcal{M}_p^q} = \sup_{\|\varphi\|_p=1,\, \varphi\in\mathcal{S}(\mathbb{R}^n)} \|\mathcal{F}^{-1}m\mathcal{F}\varphi\|_q < \infty\Big\}.$$

Remark 1.1.14.
(i) Solutions to the Cauchy problems of many classical free evolution equations can be expressed in the form of translation invariance operators. For example,

$$u(t) = \mathcal{F}^{-1}e^{-|\xi|^2 t}\mathcal{F}\varphi = \mathcal{F}^{-1}e^{-|\xi|^2 t} * \varphi \quad \text{(the heat equation)},$$
$$u(t) = \mathcal{F}^{-1}e^{\pm i|\xi|^2 t}\mathcal{F}\varphi = \mathcal{F}^{-1}e^{\pm i|\xi|^2 t} * \varphi \quad \text{(the Schrödinger equation)},$$
$$\begin{pmatrix} u \\ \partial_t u \end{pmatrix} = \mathcal{F}^{-1}e^{At}\mathcal{F}\begin{pmatrix} \varphi \\ \psi \end{pmatrix} \quad \text{(the wave equation)},$$

where

$$A = \begin{pmatrix} 0 & 1 \\ -|\xi|^2 & 0 \end{pmatrix}.$$

(ii) A direct computation shows that any convolution operator commutes with translations.
(iii) $\mathcal{M}_p^q = \mathcal{M}_{q'}^{p'}$, $\mathcal{M}_1^1 = (C_c)^*$.

(VIII) The classical Calderón–Zygmund operator
The operator

$$T(f)(x) = K * f = \text{P.V.} \int_{\mathbb{R}^n} K(x-y)f(y)dy$$

is said to be a classical Calderón–Zygmund singular integral operator if the kernel $K(x) \in L^1_{\text{loc}}(\mathbb{R}^n \setminus \{0\})$ satisfies:
(i) size condition: $|K(x)| \le C|x|^{-n}$,

(ii) smoothness condition: $|\nabla K(x)| \leq C|x|^{-n-1}$,
(iii) cancelation condition: $\int_{0<a\leq|x|\leq b<\infty} K(x)dx = 0$.

The Calderón–Zygmund theorem A
Let T be a classical Calderón–Zygmund singular integral operator. Then:
(i) For any $1 < p < \infty$, T is a (p,p)-type operator, that is, $\|T(f)\|_p \leq C\|f\|_p$.
(ii) T is a weak $(1,1)$-type operator, that is,

$$m\{x \in \mathbb{R}^n : |T(f)(x)| > \lambda\} \leq \frac{\|f\|_1}{\lambda}.$$

Remark 1.1.15.
(i) The sketch of the proof of the first Calderón–Zygmund theorem: One first proves that the Calderón–Zygmund singular integral operator T is a $(2,2)$ type operator and then shows that T is a weak $(1,1)$-type operator, employing the Calderón–Zygmund decomposition theorem [251]. Thus, by interpolation, T is a (p,p)-type operator for $1 < p \leq 2$. Finally, from the facts that T is translation invariant and $\mathcal{M}_p^p = \mathcal{M}_{p'}^{p'}$, it follows that the Calderón–Zygmund operator T is a (p,p)-type operator for $1 < p < \infty$.
(ii) The Hilbert transform and the Riesz transform are classical examples of Calderón–Zygmund singular integral operators.

The Calderón–Zygmund theorem B
The linear operator T defined in L^2 is a classical Calderón–Zygmund operator if the following three conditions are satisfied:
(i) T is a bounded operator from L^2 to L^2.
(ii) There is a measurable function K such that

$$Tf(x) = \int_{\mathbb{R}^n} K(x-y)f(y)dy, \quad x \notin \operatorname{supp}(f),$$

where the integral is absolutely convergent for $x \notin \operatorname{supp}(f)$.
(iii) There are constants $C > 1$ and $A > 0$ such that

$$\int_{|x|>C|y|} |K(x-y) - K(x)|dx < A$$

holds uniformly in y.

We now introduce the Mikhlin–Hörmander multiplier theorem, which is frequently used to verify if a convolution operator is a (p,p)-type operator.

The Mikhlin–Hörmander multiplier theorem
Let $m(\xi) \in C^l(\mathbb{R}^n \setminus \{0\})$ and $l > n/2$. If

$$|\partial_\xi^\alpha m(\xi)| \le C|\xi|^{-\alpha}, \quad |\alpha| \le l, \tag{1.15}$$

then the operator T, defined by

$$Tf = \mathcal{F}^{-1} m \mathcal{F} f = K * f,$$

is a Calderón–Zygmund operator.

It is easy to verify that the Hilbert transform and the Riesz transform have the multipliers

$$m(\xi) = -i \operatorname{sgn} \xi \quad \text{(the Hilbert transform)},$$
$$m_j(\xi) = C_n \frac{\xi_j}{|\xi|} \quad \text{(the Riesz transform)}, \quad j = 1, \dots, n,$$

respectively, both of which satisfy (1.15). More generally, let $\Omega(x)$ be a smooth function defined on the unit sphere and satisfying that

$$\int_{\Sigma^n} \Omega(x) d\sigma = 0,$$

where Σ^n represents the unit sphere in $\mathbb{R}^n$. The natural extension of $\Omega(x)$ to $\mathbb{R}^n \setminus \{0\}$ is still denoted by itself, that is,

$$\Omega(x) = \Omega\left(\frac{x}{|x|}\right), \quad \forall x \ne 0.$$

Then

$$K(x) = \frac{\Omega(x)}{|x|^n}$$

satisfies the size condition, smoothness condition and cancelation condition in the definition of Calderón–Zygmund singular integral operator. The corresponding $m(\xi) = \mathcal{F}K$ satisfies (1.15).

1.2 Elliptic boundary value problems

In this section, we use the Laplace equation as an example to illustrate the role played by harmonic analysis and especially singular integral operators in the study of elliptic boundary value problems. We only briefly introduce the research background, history and methods.

It is well known that the boundary value problems for the Laplace equation

$$\begin{cases} \Delta u = 0, & x \in \Omega \subset \mathbb{R}^n, \\ u = f, & x \in \partial\Omega \end{cases} \quad \text{(Dirichlet problem)}, \tag{1.16}$$

$$\begin{cases} \Delta u = 0, & x \in \Omega' = \mathbb{R}^n \backslash \Omega, \\ u = f, & x \in \partial\Omega' \end{cases} \quad \text{(Exterior Dirichlet problem)}, \tag{1.17}$$

$$\begin{cases} \Delta u = 0, & x \in \Omega, \\ \frac{\partial u}{\partial n} = f, & x \in \partial\Omega \end{cases} \quad \text{(Neumann problem)}, \tag{1.18}$$

$$\begin{cases} \Delta u = 0, & x \in \Omega' = \mathbb{R}^n \backslash \Omega, \\ \frac{\partial u}{\partial n} = f, & x \in \partial\Omega' \end{cases} \quad \text{(Exterior Neumann problem)}, \tag{1.19}$$

are equivalent to the following integral equations, respectively:

$$\frac{1}{2}\phi + T_K\phi = f \quad \text{(Dirichlet problem)}, \tag{1.20}$$

$$-\frac{1}{2}\phi + T_K\phi = f \quad \text{(Exterior Dirichlet problem)}, \tag{1.21}$$

$$-\frac{1}{2}\phi + T_{K^*}\phi = f \quad \text{(Neumann problem)}, \tag{1.22}$$

$$\frac{1}{2}\phi + T_{K^*}\phi = f \quad \text{(Exterior Neumann problem)}, \tag{1.23}$$

where

$$K(x,y) = \partial_{n_y} N(x,y), \quad K^*(x,y) = K(y,x), \tag{1.24}$$

$$N(x,y) = N(x-y) = \begin{cases} \frac{|x-y|^{2-n}}{(2-n)\omega_n}, & n > 2, \\ \frac{1}{2\pi}\log|x-y|, & n = 2, \end{cases} \tag{1.25}$$

$$(T_K\phi)(x) = \int_{\partial\Omega} K(x,y)\varphi(y)d\sigma(y), \quad x \in \partial\Omega, \tag{1.26}$$

$$(T_{K^*}\phi)(x) = \int_{\partial\Omega} K(y,x)\varphi(y)d\sigma(y), \quad x \in \partial\Omega, \tag{1.27}$$

here ω_n is the surface area of the unit sphere in $\mathbb{R}^n$. If $\partial\Omega \in C^2$ is bounded, $f \in L^p(\partial\Omega)$, $1 < p < \infty$ (in the case of Neumann problems f also needs some necessary conditions), then the integral equations (1.20), (1.21), (1.22) and (1.23) are solvable. Denote their solutions by ϕ_1, ϕ_2, ϕ_3 and ϕ_4, respectively. Then

$$u(x) = (\mathcal{D}\phi_1)(x) = \int_{\partial\Omega} \frac{\partial}{\partial n_y} N(x,y)\phi_1(y)d\sigma(y),$$

$$u(x) = (\mathcal{D}\phi_2)(x) = \int_{\partial\Omega} \frac{\partial}{\partial n_y} N(x,y)\phi_2(y)d\sigma(y),$$

$$u(x) = (\mathcal{S}\phi_3)(x) = \int_{\partial\Omega} N(x,y)\phi_3(y)d\sigma(y),$$

$$u(x) = (\mathcal{S}\phi_4)(x) = \int_{\partial\Omega} N(x,y)\phi_4(y)d\sigma(y),$$

are the solution to the Dirichlet problem (1.16), the exterior Dirichelt problem (1.17), the Neumann problem (1.18) and the exterior Neumann problem (1.19), respectively, where n_y denotes the unit normal vector at $y \in \partial\Omega$ directed to the outside of Ω.

Definition 1.2.1.

$$(\mathcal{D}\phi)(x) = \int_{\partial\Omega} \frac{\partial}{\partial n_y} N(x,y)\phi(y)d\sigma(y), \quad x \notin \partial\Omega, \tag{1.28}$$

$$(\mathcal{S}\phi)(x) = \int_{\partial\Omega} N(x,y)\phi(y)d\sigma(y), \quad x \notin \partial\Omega \tag{1.29}$$

are called the double- and single-layer potentials, respectively.

From the above discussion, we see that the solvability of elliptic boundary value problems is essentially reduced to that of the boundary integral equations (1.20), (1.21), (1.22) and (1.23). If one can prove that the operator T_K (or T_{K^*}) is compact, then one can use the Fredholm theory(or the more general Riesz–Schauder theory) to solve the integral equations. Substituting the solution thus obtained into the single- or double-layer potentials gives the solution to the (exterior) Dirichlet problem or the (exterior) Neumann problem.

Fortunately, when $\partial\Omega \in C^2$, T_K and T_{K^*} are compact operators in $L^p(\partial\Omega)$ and $C(\partial\Omega)$, where $1 < p < \infty$. In fact, without loss of generality, we assume that $\partial\Omega$ is the graph of a C^2-function $h(x)$. Let

$$P = (x, h(x)), \quad Q = (y, h(y)).$$

Then

$$K(P,Q) = \frac{\langle P - Q, n_Q\rangle}{\omega_n |P - Q|^n},$$

where $\langle \cdot, \cdot \rangle$ denotes the inner product of two vectors,

$$n_Q = \frac{(\nabla h(y), -1)}{\sqrt{1 + |\nabla h(y)|^2}}.$$

Note that

$$h(x) = h(y) + \langle x - y, \nabla h(y) \rangle + e(x,y)$$

with $e(x,y) \sim O(|x-y|^2)$. Hence,

$$|K(P,Q)| \le C \frac{|\langle P - Q, (\nabla h(y), -1) \rangle|}{|P - Q|^n} \le C \frac{|e(x,y)|}{|P - Q|^n} \le \frac{C}{|P - Q|^{n-2}}.$$

This implies that T_K and T_{K^*} are weakly singular integral operators and compact in $L^p(\partial\Omega)$ and $C(\partial\Omega)$, where $1 < p < \infty$. More generally, with help of the $L^p (1 < p < \infty)$ boundedness about Cauchy-type integral on C^1 curve proved by Calderón, one can prove that when $\partial\Omega \in C^1$ is bounded, the boundary value problem of Laplace equation is solvable.

Remark 1.2.1. Potential theory and variational methods are general methods to solve elliptic boundary value problems. With the generalized form of potential theory, that is, the theory of pseudodifferential operators, one can study more general elliptic boundary value problems. Of course, there are many methods suitable for cases of special domains, which will be discussed briefly below:

(i) Consider the Dirichlet problem in the upper half-space $\mathbb{R}^{n+1}_+$:

$$\begin{cases} \Delta u = 0, & \text{in } \Omega = \mathbb{R}^{n+1}_+ \\ u = f, & \text{on } \partial\Omega = \mathbb{R}^n. \end{cases} \tag{1.30}$$

Use Fourier transform to get

$$u(x,t) = \mathcal{F}^{-1}(e^{-|\xi|t}\mathcal{F}f) = P_t * f,$$

where

$$P_t = \frac{\Gamma(\frac{n+1}{2})}{\pi^{\frac{n+1}{2}}} \frac{t}{(|x|^2 + t^2)^{\frac{n+1}{2}}}$$

is the classical Poisson kernel. It is easy to see that for any $f \in L^p(\mathbb{R}^n), 1 < p < \infty$, the function $u = P_t * f$ is a harmonic function in the upper half-space and

$$\lim_{t \to 0} u(x,t) \overset{\text{a.e.}}{=} f(x).$$

In other words, (1.30) determines a mapping from $X = L^p(\mathbb{R}^n)$ to

$$Y = \left\{ u(x,t) \mid u(x,t) \text{ is harmonic in } \mathbb{R}^{n+1}_+, \sup_{t>0} \|u(x,t)\|_p \le \|f\|_p \right\}.$$

Certainly, $u(x,t) \overset{\Delta}{=} T(t)f(x) = P_t * f$ generates a C_0-semigroup with respect to $t \in \mathbb{R}^+$ in $L^p(\mathbb{R}^n), 1 \le p < \infty$, which plays an important part in the characterization of Besov-type function spaces.

(ii) If $\Omega \subset \mathbb{R}^n$ is a special domain such as sphere, cube, 1/4-space, then one can use the Newton potentials to construct the corresponding Green function, that is, to look for solutions to the problem

$$\begin{cases} \Delta G = \delta(x-y), \\ G|_{\partial\Omega} = 0. \end{cases}$$

Of course, one can also obtain $G(x,y)$ using physical or geometrical methods. Thus, the function

$$u(x) = \int_{\Omega} G(x,y)f(y)dy + \int_{\partial\Omega} \partial_{n_y} G(x,y)g(y)d\sigma(y)$$

is a solution to the following boundary value problem:

$$\begin{cases} \Delta u = f(x), & x \in \Omega, \\ u|_{\partial\Omega} = g(x). \end{cases}$$

(iii) For the Dirichlet and Neumann problem in a spherical domain or the exterior problems outside of a sphere, one can express their solutions explicitly in spherical harmonic functions and Bessel functions (see Folland's book [91] for details). It should be pointed out that spherical harmonic functions provide an orthogonal basis for many spaces of differentiable functions and have wide applications in approximation theory and computational mathematics.

(iv) With the theory of pseudodifferential operators or the generalized form of potential theory, general elliptic boundary value problems can be transformed into boundary integral equation formulations to study. Precisely, for given functions $f \in H^{s-2,p}(\Omega)$ and $g \in B_*^{s-1-\frac{1}{p},p}(\partial\Omega)$, seek a solution $u \in H^{s,p}(\Omega)$ to the elliptic boundary value problem

$$\begin{cases} Au = f, & x \in \Omega, \\ Bu = g, & x \in \partial\Omega, \end{cases} \tag{1.31}$$

where the elliptic operator A and the boundary operator B are given, respectively, by

$$Au = \sum_{i,j=1}^{n} \frac{\partial}{\partial x_i}\left(a^{ij}(x)\frac{\partial u}{\partial x_j}\right) + \sum_{i=1}^{n} b^i(x)\frac{\partial u}{\partial x_i} + c(x)u, \quad \forall u \in H^{s,p}(\Omega),$$

$$Bu = a\frac{\partial u}{\partial \nu} + bu\bigg|_{\partial\Omega} \stackrel{\Delta}{=} a\gamma_1 u + b\gamma_0 u, \quad \forall u \in H^{s,p}(\Omega),$$

$$\frac{\partial}{\partial \nu} = \sum_{j=1}^{n} a^{ij} n_j \frac{\partial}{\partial x_i}$$

with $\vec{n}$ being the exterior unit normal vector to $\partial\Omega$. For simplicity, we assume that $a^{ij}, b^i, c \in C^\infty(\Omega)$ and that $a, b \in C^\infty(\partial\Omega)$. The space $B_*^{s-1-\frac{1}{p},p}(\partial\Omega)$ is an interpolation-type Besov space, that is,

$$B_*^{s-1-\frac{1}{p},p}(\partial\Omega) = \{\varphi = a\varphi_1 + b\varphi_2 \mid \varphi_1 \in B_{p,p}^{s-1-\frac{1}{p}}(\partial\Omega),\ \varphi_2 \in B_{p,p}^{s-\frac{1}{p}}(\partial\Omega)\}$$

with the norm

$$\|\varphi\|_{B_*^{s-1-\frac{1}{p},p}} = \inf_{\varphi=a\varphi_1+b\varphi_2} \left(\|\varphi_1\|_{B_{p,p}^{s-1-\frac{1}{p}}} + \|\varphi_2\|_{B_{p,p}^{s-\frac{1}{p}}}\right).$$

The solvability of the general elliptic problem (1.31) is equivalent to prove that the operator $\mathcal{A} = (A, B)$, defined by

$$\begin{cases} \mathcal{A} : H^{s,p}(\Omega) \longrightarrow H^{s-2,p}(\Omega) \times B_*^{s-1-\frac{1}{p},p}, \\ \mathcal{A}u = \{Au, Bu\}, \quad \forall u \in D(\mathcal{A}) = H^{s,p}, \end{cases}$$

is bijective. By the theory of pseudodifferential operators, this is equivalent to show that the boundary operator $\mathcal{T}$, defined by

$$\begin{cases} \mathcal{T} : B_{p,p}^{s-\frac{1}{p}}(\partial\Omega) \longrightarrow B_{p,p}^{s-\frac{1}{p}-1}(\partial\Omega), \\ \mathcal{T}\varphi = a\Pi\varphi + b\varphi \equiv (a\frac{\partial}{\partial\nu}(P\varphi) + b\varphi)|_{\partial\Omega} \end{cases}$$

is bijective, that is, the boundary integral equation

$$T\varphi = a\Pi\varphi + b\varphi = f, \quad \forall f \in B_{p,p}^{s-\frac{1}{p}-1}(\partial\Omega) \tag{1.32}$$

is solvable, where $P : B_{p,p}^{s-\frac{1}{p}}(\partial\Omega) \longrightarrow H^{s,p}(\Omega)$ is an abstract Poisson operator. In other words, the sufficient and necessary condition for the operator $\mathcal{A}$ to be Fredholm is that $\mathcal{T}$ is Fredholm. A dense defined closed linear operator $\mathcal{T}$ from the Banach space X to the Banach space Y is said to be a Fredholm operator if

(a) $\dim N(\mathcal{T}) < \infty$,
(b) The range $R(\mathcal{T})$ is closed in Y, and
(c) $\operatorname{codim} R(\mathcal{T}) < \infty$.

If $\mathcal{T}$ is a compact operator from the Banach space X to itself, then one can make a direct use of the Riesz–Schauder theorem (or the Fredholm alternative).

The Riesz–Schauder theorem

Let $\mathcal{T}$ be a compact operator from the Banach space X (the Banach space can be replaced by a general normed linear space) to itself. Then one of the following results holds:

(a) The homogeneous equation $x - \mathcal{T}x = 0$ has a nontrivial solution $x \in X$.
(b) For each $y \in X$, the equation $x - \mathcal{T}x = y$ has a unique solution $x \in X$. In this case, the operator $(I - \mathcal{T})^{-1}$ is bounded.

In the special case when X is a Hilbert space (we may assume that $X = X^*$), we have the following more delicate result.

The Fredholm theorem

Let $\mathcal{T}$ be a compact linear operator from the Hilbert space X to itself. For any $\lambda \in \mathbb{C}$, let $V_\lambda = \{x \in X \mid \mathcal{T}x = \lambda x\}$, $W_\lambda = \{x \in X \mid \mathcal{T}^* = \lambda x\}$. Then we have
(a) There are a finite number of or an infinitely accountable number of λ such that $V_\lambda \neq \{0\}$. In the later case, only 0 is the accumulate point of V_λ. Further, for any $\lambda \neq 0$, $\dim V_\lambda < \infty$.
(b) For $\lambda \neq 0$, $\dim V_\lambda = \dim W_{\bar{\lambda}}$.
(c) For $\lambda \neq 0$, $R(\lambda I - \mathcal{T})$ is a closed set in X.

In general, how does one verify that $\mathcal{T}$ is a Fredholm operator from the Banach space X to the Banach space Y? Peetre provided an answer for this question, that is, the following.

The Peetre theorem

Let X, Y, Z be Banach spaces, $X \hookrightarrow\hookrightarrow Z$ and the linear operator $\mathcal{T} : X \mapsto Y$ be a closed operator with the domain $D(\mathcal{T})$. Then the following two conclusions are equivalent:
(a) $\dim N(\mathcal{T}) < \infty$, $R(\mathcal{T})$ is a closed set in Y.
(b) There exists a constant $C > 0$ such that

$$\|x\|_X \le C(\|\mathcal{T}x\|_Y + \|x\|_Z), \quad x \in D(\mathcal{T}).$$

It can be shown that the operator $\mathcal{T}$ in (1.32) is Fredholm. Then it remains to prove that $\mathcal{T}$ is surjective, that is,

$$\operatorname{ind}(\mathcal{T}) = \dim(N(\mathcal{T})) - \operatorname{codim}(R(\mathcal{T})) = 0.$$

From this, the solvability of (1.31) follows (see Taira's book [255] for details).

(v) The case with the boundary data $f \in C(\partial\Omega)$ can also be dealt with using the method of harmonic measures. For example, consider the Dirichlet problem:

$$\begin{cases} \Delta u = 0, & x \in \Omega, \\ u|_{\partial\Omega} = f, & f \in C(\partial\Omega). \end{cases} \tag{1.33}$$

For any $x \in \Omega$, the maximum principle gives that the mapping $f \longrightarrow u(x)$ is a continuous linear functional on $C(\partial\Omega)$. By the Riesz representation theorem, there is a unique positive Borel measure w^x such that

$$u(x) = \int_{\partial\Omega} f(Q)d\omega^x(Q),$$

where ω^x is called as a harmonic measure at x. As a direct consequence of the Harnack principle, ω^{x_1} and ω^{x_2} are absolutely continuous with each other. Therefore, it can be shown that $u(x)$ is a solution to (1.33). See Kenig's monograph [134] for details.

We now consider the role and connection of singular integral operators in the study of elliptic boundary value problems from the development of singular integral operators.

1. The first generation Calderón–Zygmund singular integral operators

Roughly speaking, the first generation Calderón–Zygmund singular integral operators is convolution-type singular integral operators, of which the Hilbert transform and the Riesz transform are classical examples. Utilizing the fact that Calderón–Zygmund operators are (p,p)-type operators, one can easily prove the regularity of solutions. For example, given that

$$\Delta u = f, \quad f(x) \in L^2(\mathbb{R}^n),$$

we want to prove that

$$\frac{\partial^2 u}{\partial x_i \partial x_j} \in L^2(\mathbb{R}^n), \quad 1 \le i,j \le n.$$

In fact, by Fourier transform, it is seen that

$$\widehat{\Delta u} = \hat{f} \quad \Longrightarrow \quad \hat{u} = -\frac{1}{|\xi|^2}\hat{f}(\xi).$$

Thus,

$$\left\| \frac{\partial^2 u}{\partial x_i \partial x_j} \right\|_2 = \left\| \frac{\xi_i \xi_j}{|\xi|^2} \hat{f}(\xi) \right\|_2 = \|R_i R_j(f)\|_2 \le \|f\|_2,$$

where

$$R_j(f) = \text{P.V.} \int_{\mathbb{R}^n} \frac{x_j - y_j}{|x-y|^{n+1}} f(y)dy, \quad j = 1, \dots, n,$$

is a Riesz transform. Further, since R_j is a (p,p)-type operator with $1 < p < \infty$, we have

$$\left\| \frac{\partial^2 u}{\partial x_i \partial x_j} \right\|_p \le C\|f\|_p, \quad f \in L^p,\ 1 < p < \infty.$$

We now give equivalent definitions of the first generation Calderón–Zygmund operators from the multipliers point view.

Remark 1.2.2. The classical Calderón–Zygmund operator, defined in the last section by

$$Tf = \text{P.V.} \int_{\mathbb{R}^n} K(x-y)f(y)dy \triangleq (K * f)(x), \tag{1.34}$$

is usually called as the first generation Calderón–Zygmund singular integral operator. With the theory of multipliers, T can be written as

$$Tf = \mathcal{F}^{-1}m(\xi) * f.$$

For the multiplier $m(\xi)$, are there corresponding conditions on the size, smoothness, and cancellation that are satisfied by the kernel $K(x)$? This is an interesting question. In fact, for the Calderón–Zygmund kernel $K(x) = \frac{\Omega(x)}{|x|^n}$ with homogeneity, the corresponding multiplier $m(\xi)$ satisfies that:

(i) $m(\xi) = m(\lambda\xi), \lambda > 0$,

(ii) $m(\xi) \in C^\infty(\mathbb{R}^n \backslash \{0\})$,

(iii) $\int_{\Sigma^n} m(\xi)d\sigma(\xi) = 0$, where Σ^n is the surface of unit sphere in $\mathbb{R}^n$.

Remark 1.2.3. Let $K(x)$ be a distribution, the corresponding Fourier transform of which is $m(\xi)$. We have the following results:

(i) Let $m(\xi)$ be bounded function satisfying $m(\xi) \in C^\infty(\mathbb{R}^n \setminus \{0\})$, and

$$|\partial_\xi^\alpha m(\xi)| \leq C_\alpha |\xi|^{-|\alpha|}, \quad \forall \alpha.$$

Then $K(x) \in C^\infty(\mathbb{R}^n \setminus \{0\})$ satisfies

$$|\partial_x^\alpha K(x)| \leq C'_\alpha |x|^{-n-|\alpha|}, \quad \forall \alpha.$$

(ii) Let $\ell > \frac{n}{2}$, $m(\xi)$ be bounded function satisfying $m(\xi) \in C^\ell(\mathbb{R}^n \setminus \{0\})$, and

$$|\partial_\xi^\alpha m(\xi)| \leq C_\alpha |\xi|^{-|\alpha|}, \quad \forall 0 \leq |\alpha| \leq \ell.$$

Then $K(x) \in L^1_{\text{loc}}(\mathbb{R}^n \setminus \{0\})$ satisfies

$$\int_{|x|\geq 2|y|} |K(x-y) - K(x)| \leq C, \quad \forall y \neq 0.$$

Remark 1.2.4. The first generation Calderón–Zygmund singular integral operator has the kernel $K(x) = \mathcal{F}^{-1}m(\xi)$, which satisfies that

$$K(x,y) \triangleq K(x-y) \sim O(|x-y|^{-n}).$$

For convenience, we now generalize the concept of Calderón–Zygmund singular integral operators. Let the kernel K corresponding to the convolution operator (1.34) satisfy that

$$K(x,y) \sim O(|x-y|^{-n+\alpha}).$$

Introduce the following concepts:

(i) When $\alpha = 0$, the operator corresponding to $K(x-y)$ is the first generation Calderón–Zygmund singular integral operator.

(ii) When $0 < \alpha < n$, the convolution operator corresponding to $K(x-y)$ is called as weakly singular integral operators.

(iii) When $\alpha < 0$, the convolution operator corresponding to $K(x-y)$ is called as strongly singular integral operators.

We find that when Ω is a bounded smooth domain in $\mathbb{R}^n$ the integral operators (cf. (1.26) and (1.27)), appeared in the integral equations arising from the Dirichlet and Neumann problems as well as the corresponding exterior problems for the Laplace equation, have a kernel K satisfying that

$$|K(x,y)| = |K^*(x,y)| \sim O(|x-y|^{-n+2}).$$

This means that T_K and T_{K^*} are weakly singular integral operators. Note that weakly singular integral operators T_K and T_{K^*}, defined on a bounded smooth boundary $\partial\Omega$, are compact. Thus, the Fredholm alternative can be used to study the well-posedness of elliptic boundary value problems. From this, it is seen that the first generation Calderón–Zygmund singular integral operators play a key part in dealing with elliptic boundary value problems with constant coefficients.

2. The second generation Calderón–Zygmund singular integral operators

When dealing with elliptic boundary value problems with variable coefficients, the first generation Calderón–Zygmund singular integral operators are not applicable. The reason consists in the fact that the solution operator of elliptic equations with variable coefficients is no longer a convolution-type operator. This induces the second generation singular integral operators.

Consider the following elliptic operator with variable coefficients in $\mathbb{R}^n$:

$$\mathcal{L}u = -\sum_{i,j=1}^{n} a_{ij}(x)\frac{\partial^2 u}{\partial x_i \partial x_j}.$$

Using the Fourier transform, $\mathcal{L}u$ can be written as

$$\begin{aligned}
\mathcal{L}u &= -\sum_{i,j=1}^{n} a_{ij}(x)\left(\left(\frac{\partial^2 u}{\partial x_i \partial x_j}\right)^{\wedge}\right)^{\vee}(x) \\
&= (2\pi)^{-\frac{n}{2}} \int_{\mathbb{R}^n} \sum_{i,j=1}^{n} a_{ij}(x)\xi_i\xi_j \hat{u}(\xi) e^{ix\cdot\xi} d\xi \\
&= (2\pi)^{-\frac{n}{2}} \int_{\mathbb{R}^n} \sum_{i,j=1}^{n} a_{ij}(x)\xi_i\xi_j \int_{\mathbb{R}^n} u(y) e^{-iy\cdot\xi} dy\, e^{ix\cdot\xi} d\xi \\
&= (2\pi)^{-\frac{n}{2}} \int_{\mathbb{R}^n}\int_{\mathbb{R}^n} \sum_{i,j=1}^{n} a_{ij}(x)\xi_i\xi_j u(y) e^{i(x-y)\cdot\xi} dy d\xi \\
&= (2\pi)^{-\frac{n}{2}} \int_{\mathbb{R}^n}\int_{\mathbb{R}^n} \sum_{i,j=1}^{n} a_{ij}(x)\xi_i\xi_j e^{i(x-y)\cdot\xi} d\xi u(y) dy \\
&\triangleq \int_{\mathbb{R}^n} L(x, x-y) u(y) dy.
\end{aligned}$$

Roughly speaking, the solvability of the linear partial differential equation $\mathcal{L}u = g$ is reduced to the boundedness and invertibility of the singular integral operator T defined in (1.35) below.

Definition 1.2.2. The operator

$$Tf = \text{P.V.} \int_{\mathbb{R}^n} L(x, x-y) f(y) dy \tag{1.35}$$

is said to be a second generation Calderón–Zygmund singular integral operator if:
(i) $L(x, \lambda z) = \lambda^{-n} L(x, z)$ for all $\lambda > 0$,
(ii) $L(x, z) \in C^{\infty}(\mathbb{R}^n \times \mathbb{R}^n \setminus \{0\})$,
(iii) $\int_{\Sigma^n} L(x, z) d\sigma(z) = 0$ for all $x \in \mathbb{R}^n$; here Σ^n is the surface of unit sphere in $\mathbb{R}^n$.

With the expansion technique in terms of spherical harmonic functions, one can prove the following.

Proposition 1.2.1. *For any* $1 < p < \infty$, *the second generation Calderón–Zygmund singular integral operator* T, *defined by* (1.35), *is of* (p, p) *type. Furthermore,* T *is a weak* $(1, 1)$*-type operator.*

Remark 1.2.5.
(i) Formally, the second generation Calderón–Zygmund singular integral operators are semiconvolution operators.
(ii) The second generation Calderón–Zygmund singular integral operator is the prototype of pseudodifferential operators. So, pseudodifferential operators are generalizations of the second generation Calderón–Zygmund singular integral operators and potentials. In general, pseudodifferential operators can be expressed as

$$Tf = \int_{\mathbb{R}^n} \sigma(x,\xi)\hat{f}(\xi)e^{ix\xi}d\xi,$$

where σ is the symbol of T. Pseudodifferential operators are clarified according to the growth size of their symbol and its derivatives. For any $m \in \mathbb{N}$, the symbol $\sigma(x,\xi)$ is said to be in S^m if

$$\left|\partial_x^\beta \partial_\xi^\alpha \sigma(x,\xi)\right| \le C_{\alpha,\beta}(1+|\xi|)^{m-|\alpha|}.$$

It can be shown that the pseudodifferential operator corresponding to $\sigma(x,\xi) \in S^m$ can be written in the form

$$Tf = \text{P.V.} \int_{\mathbb{R}^n} R(x, x-y)f(y)dy.$$

More generally, $\sigma(x,\xi) \in S^m_{\rho,\delta}$ means that

$$\left|\partial_x^\beta \partial_\xi^\alpha \sigma(x,\xi)\right| \le A_{\alpha,\beta}(1+|\xi|)^{m-\rho|\alpha|+\delta|\beta|}.$$

Pseudodifferential operators are one of the most powerful tools in dealing with general linear elliptic partial differential equations with variable coefficients. The interested reader is referred to Hörmander's monographs [120].

3. The third generation Calderón–Zygmund singular integral operators

The singular integral operators discussed earlier can all be represented as convolution of a tempered distribution. From a mathematical point of view, this special structure helps us directly use the Fourier transform to prove the boundedness on L^2 of the singular integral operator T. If we assume that the operator T is L^2-bounded, it is easy to see that the boundedness of the first and second generation singular integral operators T depends only on the following Hörmander condition:

$$\int_{|x|>2|y|} \left|K(x-y) - K(x)\right| dx \le A, \quad \forall\ |y| > 0. \tag{1.36}$$

This reminds us that for the integral operators T defined by nonconvolutional kernels, if the boundedness on L^2 of T is assumed in advance, then its L^p boundedness should be established.

In fact, denote by Δ the diagonal in $\mathbb{R}^n \times \mathbb{R}^n$, that is, $\Delta = \{(x,x) : x \in \mathbb{R}^n\}$. For the third-generation Calderón–Zygmund singular integral operator T with nonconvolutional kernels, one can directly utilize the Calderón–Zygmund decomposition theory to obtain the following preliminary results.

Theorem 1.2.1. *Assume that T is $L^2(\mathbb{R}^n)$-bounded and that $K(x,y)$ is a function defined in $\mathbb{R}^n \times \mathbb{R}^n \backslash \Delta$ satisfying that if $f \in L^2(\mathbb{R}^n)$ has compact support, then T has the following explicit expression:*

$$Tf(x) = \int_{\mathbb{R}^n} K(x,y)f(y)\,dy, \quad \forall\, x \notin \operatorname{supp} f. \tag{1.37}$$

Assume further that the kernel $K(x,y)$ satisfies the following conditions:

$$\int_{|x-y|>2|y-z|} \big|K(x,y) - K(x,z)\big|\,dx \le C, \tag{1.38}$$

$$\int_{|x-y|>2|x-w|} \big|K(x,y) - K(w,y)\big|\,dx \le C. \tag{1.39}$$

Then the operator T is (p,p)-type $(1 < p < \infty)$ and weak $(1,1)$-type.

Corollary 1.2.1. *Under the assumptions of Theorem 1.2.1, if the kernel $K(x,y)$ satisfies the conditions (also called the Calderón–Zygmund standard kernel conditions) that there exists $\delta > 0$ such that*

$$\big|K(x,y)\big| \le \frac{C}{|x-y|^n} \quad \textit{(scaling condition)}, \tag{1.40}$$

$$\big|K(x,y) - K(w,y)\big| \le C\frac{|x-w|^\delta}{|x-y|^{n+\delta}}, \quad |x-w| \le \frac{1}{2}|x-y|, \tag{1.41}$$

$$\big|K(x,y) - K(x,z)\big| \le C\frac{|y-z|^\delta}{|x-y|^{n+\delta}}, \quad |y-z| \le \frac{1}{2}|x-y|, \tag{1.42}$$

then the third generation Calderón–Zygmund singular integral operator T is (p,p)-type $(1 < p < \infty)$ and weak $(1,1)$-type.

Proof. A direct computation shows that the standard kernel conditions (1.40)–(1.42) introduced by Coifman–Meyer implies the kernel conditions (1.38)–(1.39) in Theorem 1.2.1. The required result then follows from Theorem 1.2.1. □

Motivation. In addition to the basic motivation of studying elliptic boundary value problems on nonsmooth domains, the third generation Calderón–Zygmund singular integral operators also originated from the self-development of harmonic analysis.

Cauchy integral on the Lipschitz curve

Assume that $A(t)$ is a Lipschitz function defined in $\mathbb{R}$, that is, $A' = a \in L^\infty$. Let $\Gamma = (t, A(t))$ be a planar curve. For any $f \in \mathcal{S}(\mathbb{R})$, instead of studying the Cauchy integral

$$\tilde{C}_\Gamma f(z) = \frac{1}{2\pi i}\int_\Gamma \frac{f(\zeta)}{\zeta - z}\,d\zeta, \quad \zeta = t + A(t)i,$$

we consider its alternative form

$$C_\Gamma f(z) = \frac{1}{2\pi i} \int_{-\infty}^{\infty} \frac{f(t)(1 + ia(t))}{t + iA(t) - z}\, dt, \tag{1.43}$$

which is an analytic function defined in the open set

$$\Omega_+ = \{z = x + iy \in \mathbb{C} : y > A(x)\}.$$

By splitting the integration region and using the residue theorem, we deduce that the boundary-value of such an analytic function

$$\lim_{\varepsilon \to 0} C_\Gamma f(x + i(A(x) + \varepsilon))$$

can be replaced by

$$\frac{1}{2}\left(f(x) + \frac{i}{\pi} \lim_{\varepsilon \to 0} \int_{|x-t|>\varepsilon} \frac{f(t)(1 + ia(t))}{x - t + i(A(x) - A(t))}\, dt \right).$$

Since $a \in L^\infty$, the problem is then reduced to the study of the third generation Calderón–Zygmund singular integral operator

$$Tf(x) = \lim_{\varepsilon \to 0} \int_{|x-y|>\varepsilon} \frac{f(y)}{x - y + i(A(x) - A(y))}\, dy \tag{1.44}$$

with a nonconvolutional kernel function

$$K(x,y) = \frac{1}{x - y + i(A(x) - A(y))}. \tag{1.45}$$

The kernel function (1.45) satisfies the conditions (1.40)–(1.42) with $\delta = 1$.

The Calderón commutators

If $\|A'\|_\infty < 1$, then the kernel in (1.45) can be expressed as the series

$$K(x,y) = \frac{1}{x - y} \sum_{k=1}^{\infty} \left(i \frac{A(x) - A(y)}{x - y} \right)^k (-1)^k. \tag{1.46}$$

Thus, it is natural to consider the third generation Calderón–Zygmund singular integral operators

$$T_k f(x) = \lim_{\varepsilon \to 0} \int_{|x-y|>\varepsilon} \left(\frac{A(x) - A(y)}{x - y} \right)^k \frac{f(y)}{x - y}\, dy, \quad k \geq 0 \tag{1.47}$$

with the nonconvolutional kernel

$$K_k(x,y) = \frac{1}{x-y}\left(i\frac{A(x)-A(y)}{x-y}\right)^k, \quad k \geq 0. \tag{1.48}$$

The kernel function (1.48) satisfies the conditions (1.40)–(1.42) with $\delta = 1$.

Definition 1.2.3 (Third generation Calderón–Zygmund singular integral operators). The operator T is called the third generation Calderón–Zygmund singular integral operator if:
(i) T is bounded from $L^2(\mathbb{R}^n)$ to $L^2(\mathbb{R}^n)$,
(ii) there exists a Calderón–Zygmund standard kernel function K such that for any $f \in L^2_c(\mathbb{R}^n)$

$$Tf(x) = \int_{\mathbb{R}^n} K(x,y)f(y)\,dy, \quad x \notin \operatorname{supp} f. \tag{1.49}$$

By Theorem 1.2.1, one can obtain the Calderón–Zygmund theorem.

Theorem 1.2.2 (Calderón–Zygmund theorem). *Assume that the operator T is a third generation Calderón–Zygmund singular integral operator. Then:*
(i) $\|Tf\|_{L^p} \leq C_p\|f\|_{L^p}$, $1 < p < \infty$,
(ii) $|\{x \in \mathbb{R}^d : |Tf| > \lambda\}| \leq \frac{C}{\lambda}\|f\|_{L^1}$.

Question. For the general nonconvolutional operator T, how to verify that T is $(2,2)$-type operator? In other words, is it possible to find a replacement of the vanishing condition? This is the famous David–Journé T(1) theorem, which gives an answer to the above question.

Theorem 1.2.3 (David–Journe T(1) theorem, [56]). *Suppose $K(x,y)$ is the kernel function appeared in the third generation Calderón–Zygmund singular integral operator*

$$Tf(x) = \int_{\mathbb{R}^n} K(x,y)f(y)\,dy.$$

Then T is $(2,2)$-type Calderón–Zygmund singular integral operator if and only if the following conditions are satisfied:
(i) $T(1) \in \mathrm{BMO}(\mathbb{R}^n)$,
(ii) $T^*(1) \in \mathrm{BMO}(\mathbb{R}^n)$,
(iii) $T \in$ WBO *(Weakly Bounded Operator), that is, for any $f, g \in \mathcal{D}(\mathbb{R}^n)$ satisfying that* $\operatorname{supp}(f)$, $\operatorname{supp}(g) \subset Q$, *where Q is a cube with* $\operatorname{diam} Q \leq t$, *the following inequality holds:*

$$|\langle Tf, g\rangle| \leq Ct^{n+2\eta}\|f\|_\eta\|g\|_\eta, \quad \forall\, \eta > 0.$$

The Cauchy integral on a Lipschitz curve results in a typical third-generation singular integral operator (nonconvolutional operator), so the third generation singular integral operator can be regarded as a Calderón–Zygmund singular integral operator over non-flat spaces.

1.3 Background on harmonic analysis methods for evolution equations

In this section, we give an overview of the study of nonlinear evolution equations, emphasizing on the role of harmonic analysis methods in the recent study of evolution equations. As is well known, Cauchy problems and initial boundary value problems for nonlinear evolution equations can be reduced to the following abstract Cauchy problem:

$$\begin{cases} u_t + Au = F(u), \\ u(0) = \varphi. \end{cases} \tag{1.50}$$

Its study is mainly concerned with the following problems:
(A) Well-posedness of the problem,
(B) Behavior of solutions including finite time blowup, bifurcation, propagation of singularities and the structure of singularities.
(C) Scattering theory (mainly for wave and dispersive wave equations) and asymptotic behavior of solutions (mainly for dissipative-type equations such as parabolic equations).

Here, (A) includes local and global well-posedness. In other words, for a given initial data φ belonging to a Banach space X, the question is whether or not the Cauchy problem (1.50) determines a unique continuous flow $u(t) \in C((-T,T);X)$ or $u(t) \in C((0,T);X)$. In particular, when $T = \infty$, $u(t) \in C(\mathbb{R};X)$ or $u(t) \in C(\mathbb{R}^+;X)$ becomes a global flow, that is, the problem (1.50) is globally well-posed. Here, the Banach space X guarantees that the operator A generates a C_0-semigroup (or C_0-group) in X. In other words, the corresponding free problem

$$\begin{cases} v_t + Av = 0, \\ v(0) = \varphi \end{cases} \tag{1.51}$$

has a unique solution $v(t) = e^{-At}\varphi$, where e^{-At} is the operator semigroup generated by A.

Now if the problem (1.50) does not determine a unique global flow, then when does the solution blowup? When do bifurcation phenomena occur? How do singularities of solutions propagate? How is the structure of singularities of solutions and is it possible to characterize this? If the problem (1.50) determines a continuous global flow, then what is the asymptotic behavior at $t \to \infty$ or $t \to \pm\infty$ of the solution $u(t)$? All these questions are what (C) is going to study. In particular, For wave and dispersive equations, their global well-posedness induces the well-known scattering theory. The scattering theory mainly involves in existence and asymptotic completeness of wave operators, which will be discussed in Chapters 4 and 5. The reader is also referred to the monographs [21, 184, 242, 247].

I. Comparison between classical methods and modern harmonic analysis methods

Classical methods

The abstract semigroup method as well the compactness method based on Galerkin finite approximations mainly use spaces such as $L^\infty(I;B)$ or $C(I;B)$, where $B = L^p,\ W^{m,p}$ to establish well-posedness of nonlinear evolution equations. It is easy to see that solutions of the evolution equation belong to some self-reflective Banach space with respect to spatial variables (under the condition that the associated linear equation is globally well-posed in B, that is, there exists $u(t) \in C(\mathbb{R};B)$ or $u(t) \in C(\mathbb{R}^+;B)$ to the linear equation) and to L^∞ with respect to the time variable. This method of establishing the well-posedness of nonlinear evolution equations with the help of only spaces $L^\infty(I;B)$ or $C(I;B)$ does not necessarily work for general nonlinear evolution equations. The reason consists in the fact that it does not involve in the integrability property in time and space of solutions. In general, for parabolic equations (whose solutions are smooth as long as they exist), due to the regularity property of analytic semigroup, the abstract Segal theorem can be easily verified, so it is suitable to use the solution space $C(I;B)$. However, for wave and dispersive equations, it is difficult to use only $C(I;L^2)$ or $C(I;H^s)$ to establish the local well-posedness. It is necessary to make use of time-space Banach spaces and study the well-posedness in a subspace of the time-space Banach space. This is what harmonic analysis methods provide.

Modern harmonic analysis methods

In the study of restriction estimates to smooth compact surfaces or noncompact surfaces with nonzero curvature of Fourier transform, it is found that the dual form of the restriction estimate is just the time-space estimate of solutions to some free evolution equations. For example, let

$$\tau = \{\eta \in \mathbb{R}^{n+1} | \eta = (\xi, |\xi|^2),\ \xi \in \mathbb{R}^n\},$$

which is a paraboloid of nonzero curvature. By the restriction estimates to smooth compact surfaces of Fourier transform and a scaling argument, we deduce the following restricted L^2 estimate to the noncompact paraboloid τ (see [153, 184, 251]):

$$\left(\int |\hat{f}|_\tau^2 d\tau(\eta) \right)^{1/2} \leq C\|f\|_{L^{q'}(\mathbb{R};L^{r'}(\mathbb{R}^n))}, \tag{1.52}$$

where $d\tau(\eta) = d\xi$ and (q, r) satisfies

$$\frac{2}{q} = n\left(\frac{1}{2} - \frac{1}{r}\right), \tag{1.53}$$

$$2 \le r \begin{cases} \le \frac{2n}{n-2}, & n \ge 3, \\ < \infty, & n = 2, \\ \le \infty, & n = 1, \end{cases} \tag{1.54}$$

and $q' = q/(q-1), r' = r/(r-1)$. The pair (q, r) satisfying (1.53) and (1.54) is usually called as a time-space admissible pair for the Schrödinger equation.

In essence, the dual form of the estimate (1.52) is just the Strichartz-type time-space estimate

$$\|S(t)\varphi\|_{L^q(\mathbb{R};L^r(\mathbb{R}^n))} \le C\|\varphi\|_{L^2(\mathbb{R}^n)}, \tag{1.55}$$

where

$$v(t) = S(t)\varphi = \mathcal{F}^{-1}(e^{-i|\xi|^2 t}\mathcal{F}\varphi)$$

is the solution to the Cauchy problem of the free Schrödinger equation

$$\begin{cases} iv_t = -\Delta v, & (x, t) \in \mathbb{R}^n \times \mathbb{R}, \\ v(0) = \varphi(x), & x \in \mathbb{R}^n. \end{cases} \tag{1.56}$$

Further, if the time integration interval $\mathbb{R}$ is replaced by $I \subset \mathbb{R}$ $(0 \in \bar{I})$, then the estimate (1.55) still holds. In particular, when $q = r = 2 + 4/n$ or $q = \infty$ and $r = 2$, the estimate (1.55) becomes

$$\|S(t)\varphi\|_{L^q_{t,x}(I\times\mathbb{R}^n)} \le C\|\varphi\|_{L^2(\mathbb{R}^n)}, \tag{1.57}$$

$$\|S(t)\varphi\|_{L^\infty(I;L^2(\mathbb{R}^n))} \le C\|\varphi\|_{L^2(\mathbb{R}^n)}. \tag{1.58}$$

Furthermore, as a direct result of the estimate (1.55), we have the following estimate for the inhomogeneous part of the solution of the linear equation:

$$\left\|\int_0^t S(t-\tau)g(x,\tau)d\tau\right\|_{L^{q_1}(I;L^{r_1}(\mathbb{R}^n))} \le C\|g\|_{L^{q_2'}(I;L^{r_2'}(\mathbb{R}^n))}, \tag{1.59}$$

where (q_1, r_1) and (q_2, r_2) are two arbitrary time-space admissible pairs. Thus, for the nonlinear Schrödinger equation, the solution space can be chosen as

$$\mathcal{X}(I) = C(I;L^2) \cap \bigcap_{(q,r)\in\Lambda} L^q(I;L^r(\mathbb{R}^n)) \tag{1.60}$$

or

$$\mathcal{X}(I) = C(I; H^s) \cap \bigcap_{(q,r)\in\Lambda} L^q(I; H^{s,r}(\mathbb{R}^n)), \quad s \in \mathbb{R}, \tag{1.61}$$

where Λ denotes the set of all time-space admissible pairs. Hence, for any $\varphi \in H^s$, one can study the well-posedness in $\mathcal{X}(I)$, defined by (1.61), of the Cauchy problem for the nonlinear Schrödinger equation, that is, $u(t) \in \mathcal{X}(I)$. It can be seen that the Strichartz-type estimates and the construction of the solution space $\mathcal{X}(I)$ provide a powerful method and a variety of choices for nonlinear estimates.

Further study of harmonic analysis shows that the above restriction estimates of Fourier transform can be reduced to the special case of the following Stein–Tomas theorem.

Theorem 1.3.1. *Let $\mu \in M(\Sigma)$, $\frac{d\mu}{d\sigma} \in L^2(d\sigma)$. Then $\hat{\mu} \in L^{\frac{2(n+1)}{n-1}}(\mathbb{R}^n)$ and satisfies*

$$\|\hat{\mu}\|_{L^{\frac{2(n+1)}{n-1}}(\mathbb{R}^n)} \lesssim \left\| \frac{d\mu}{d\sigma} \right\|_{L^2(\Sigma)}, \tag{1.62}$$

where Σ represents the unit sphere in $\mathbb{R}^n$ or a $n-1$-dimensional smooth compact manifold of nonzero Riemann curvature and $M(\Sigma)$ denotes the set of all measurable functions on Σ.

The solution of the free Schrödinger equation (1.56) can be expressed as

$$\begin{aligned} S(t)\varphi &= (2\pi)^{-\frac{n}{2}} \int_{\mathbb{R}^n} e^{ix\cdot\xi + i|\xi|^2 t} \hat{\varphi} d\xi \\ &= (2\pi)^{-\frac{n}{2}} \int_{\mathbb{R}^{n+1}} e^{i(x,t)\cdot(\xi,\tau)} \hat{\varphi}(\xi) \delta(\tau - |\xi|^2) d\tau d\xi \\ &= (2\pi)^{-\frac{n}{2}} \int_{\mathbb{R}^{n+1}} e^{2i\pi\bar{x}\cdot\bar{\xi}} \varphi(\xi) d\mu(\bar{\xi}) \\ &= (2\pi)^{\frac{n}{2}} \mathcal{F}^{-1}(\hat{\varphi} d\mu), \end{aligned}$$

where $d\mu(\bar{\xi}) = \delta(\tau - |\xi|^2) d\tau d\xi$, $\bar{x} = (x,t)$ and $\bar{\xi} = (\xi,\tau)$.

Let $\Sigma_0 \subset \Sigma$. Noting that $\forall f(\bar{\xi}) \in L^2(\Sigma_0)$ and $g(\bar{x}) \in L^p(\mathbb{R}^{n+1})$,

$$\langle \hat{g}(\bar{\xi})|_{\Sigma_0}, f(\bar{\xi}) \rangle_{L^2(\Sigma_0)} = \langle g(\bar{x}), (f(\bar{\xi}) d\sigma(\bar{\xi}))^\vee \rangle_{L^2(\mathbb{R}^{n+1})} \le \|g\|_p \|(f(\bar{\xi}) d\sigma(\bar{\xi}))^\vee\|_{p'},$$

it follows by the Tomas–Stein restriction estimate (Theorem 1.3.1) that

$$\left(\int_{\Sigma\cap\{\frac{1}{2}\le|\xi|\le 2\}} |\hat{\varphi}(\bar{\xi})|^2 d\mu(\bar{\xi}) \right)^{1/2} \le A_p \|\varphi\|_p, \quad 1 \le p \le \frac{2n+4}{n+4},$$

where Σ stands for a n-dimensional smooth Riemann manifold of nonzero curvature in $\mathbb{R}^{n+1}$. By the Littlewood–Paley theory, the scaling principle and the duality principle, we get

$$\|S(t)\varphi\|_{L^{\frac{2(n+2)}{n}}(\mathbb{R}^{n+1})} \leq \|\hat{\varphi}\|_2 = \|\varphi\|_2.$$

Theorem 1.3.1 is a L^2-type restriction estimate. We also have the following well-known restriction conjecture.

The Stein conjecture

Let $\mu \in M(\Sigma^n)$, $\frac{d\mu}{d\sigma} \in L^\infty(d\sigma)$. Then

$$\hat{\mu} \in L^q(\mathbb{R}^n), \quad q > \frac{2n}{n-1}, \tag{1.63}$$

where Σ^n is unite sphere in $\mathbb{R}^n$.

When $n = 2$, Σ^n is a circle and the Stein conjecture (1.63) has been completely resolved. When $n = 3$, if

$$q > 3 + \frac{1}{7},$$

it has been shown that the Stein conjecture (1.63) holds; see Wang–Wu [285]. However, the Stein conjecture remains open in the general case. This conjecture is closely connected with the study of modern harmonic analysis and partial differential equations. For example, in the case $n = 2$, with the estimate (1.63), one can get many deep results for the 2D Schrödinger equation; see Bourgain's work [19].

Remark 1.3.1.

(i) Space-time estimates provide suitable solution spaces for the study of nonlinear evolution equations. For example, for any $\varphi \in L^2(\mathbb{R}^n)$, the following Cauchy problem for the semilinear Schrödinger equation:

$$\begin{cases} iu_t - \frac{1}{2}\Delta u = -|u|^{p-1}u, & (x,t) \in \mathbb{R}^n \times \mathbb{R}, \quad 1 < p < 1 + \frac{4}{n}, \\ u(0) = \varphi, & x \in \mathbb{R}^n \end{cases} \tag{1.64}$$

determines a global flow in $L^2(\mathbb{R}^n)$, that is, $u(t) \in C(\mathbb{R}; L^2(\mathbb{R}^n))$ and satisfies

$$u(t) \in C(\mathbb{R}; L^2(\mathbb{R}^n)) \cap \bigcap_{(q,r)\in\Lambda} L^q_{\mathrm{loc}}(\mathbb{R}; L^r(\mathbb{R}^n)).$$

It is well known that it is impossible to get the local well-posedness in L^2 of the problem (1.64) if one only uses the $C(I; L^2(\mathbb{R}^n))$-type spaces. However, one may utilize the subspace

$$X = C(I; L^2(\mathbb{R}^n)) \cap \bigcap_{(q,r)\in\Lambda} L^q(I; L^r(\mathbb{R}^n))$$

to establish the local well-posedness in $L^2(\mathbb{R}^n)$ of the problem (1.64). Further, with the help of the fact that the L^2 integral of the local solution is conservative, one can prove that the problem (1.64) determines a global flow in $L^2(\mathbb{R}^n)$ (see [184, 275] for details). Similarly, a key role is also played by the space-time estimates in the study of well-posedness and scattering theory of the Cauchy problem for the wave and Schrödinger equations. Here are some typical examples: with the Strichartz spacetime estimates, Strauss established the scattering theory of small solutions in the energy space to the wave and Schrödinger equations [246], Brenner proved the scattering theory for the Klein–Gordon equation [25], and Ginibre–Velo dealt with the scattering theory for the nonlinear Schrödinger equation [103].

(ii) With the spacetime integrability, one can derive energy identities in the weaker sense. For example, let $u(t) \in C(\mathbb{R}; H^1(\mathbb{R}^n))$ be the H^1 global solution to the following Cauchy problem for the semilinear Schrödinger equation

$$\begin{cases} iu_t = -\frac{1}{2}\Delta u + f(u), \\ u(0) = \varphi(x). \end{cases}$$

If $u \in L^q_{\mathrm{loc}}(\mathbb{R}; L^r(\mathbb{R}^n))$ with $(q, r) \in \Lambda$, then

$$E(u(t)) = \int_{\mathbb{R}^n} (|\nabla u|^2 + V(u))dx = E(\varphi),$$

where

$$\frac{\partial V(z)}{\partial \bar{z}} = f(z).$$

For the Cauchy problem of the Navier–Stokes equations,

$$\begin{cases} u_t - \Delta u + (u \cdot \nabla)u + \nabla P = 0, \\ \operatorname{div} u = 0, \\ u(0) = \varphi, \quad \varphi \in L^2(\mathbb{R}^n), \end{cases}$$

the Leray–Hopf weak solution $u(t) \in L^\infty(\mathbb{R}^+; L^2(\mathbb{R}^n)) \cap L^2(\mathbb{R}^+; \dot{H}^1(\mathbb{R}^n))$ only satisfies the energy inequality

$$\|u(t)\|_2^2 + 2\int_0^t \|\nabla u\|_2^2 dt \le \|u_0(x)\|_2^2, \quad 0 \le t \le T < \infty.$$

However, if we assume that the Leray–Hopf weak solution $u(t)$ belongs to $L^q([0, T); L^r(\mathbb{R}^n))$, where

$$\frac{2}{q} = n\left(\frac{1}{r} - \frac{1}{p}\right), \quad n \le r \le p \le \infty, \quad (q, r, n) \neq (\infty, 3, 3). \tag{1.65}$$

then $u(t)$ satisfies the energy equality

$$\|u(t)\|_2^2 + 2\int_0^t \|\nabla u\|_2^2 dt = \|u_0(x)\|_2^2, \quad 0 \le t \le T < \infty.$$

Furthermore, by the regularity theory of Serrin–von Wahl, it is found that $u(t,x) \in C^\infty((0,T) \times \mathbb{R}^n)$.

(iii) Different types of evolution equations have different space-time estimates, so they have different admissible pairs (or admissible triplets) and space-time spaces. The reason is that the associated free linear equations correspond to different oscillatory integrals. In other words, the free linear equations correspond to restriction estimates to different smooth manifolds of Fourier transform. This is also a very difficult question in harmonic analysis, and is the bridge of connection between harmonic analysis and evolution equations.

(iv) The Strichartz estimate provides a standard method for studying the well-posedness of nonlinear evolution equations, and mainly solve the problems in differentiable function spaces with nonnegative order. However, from the scaling analysis, one can predict that sometimes the problems may be solved in differentiable function spaces with negative order, that is, the so-called low regularity problem. It is a meaningful work that how to study nonlinear evolution equations in low regular spaces. According to the feature of the linear part of equations, Bourgain introduced a class of function spaces with double parameters and combined the structure of nonlinear term and Fourier frequencies splitting technique to obtain a good study for the low regularity problem of nonlinear dispersive equations and nonlinear wave equations; see Bourgain [19, 21] or Chapter 4.

II. Estimates of multipliers and well-suited Banach spaces

Let A be a differential operator with constant coefficients and let us consider the abstract Cauchy problem

$$\begin{cases} u_t + Au = f(u), \\ u(0) = \varphi. \end{cases} \tag{1.66}$$

It is well known that the abstract Cauchy problem (1.66) models initial (or initial boundary) value problems for many classical PDEs such as the heat equations, the Schrödinger equation, the wave equation and general dispersive wave equations. Given $\varphi \in X$ with X being a Banach space, to prove that the problem (1.66) determines a strong continuous flow in X, that is, the problem (1.66) has a solution $u(t) \in C(I;X)$, where I is an interval in $\mathbb{R}$ or $\mathbb{R}^+$ containing the origin, it at least should be required that the following Cauchy problem for the corresponding free equation:

$$\begin{cases} v_t + Av = 0, \\ v(0) = \varphi \end{cases} \tag{1.67}$$

determines a global continuous flow $v(t) \in C(\mathbb{R}; X)$ or $v(t) \in C(\mathbb{R}^+; X)$. It is known that the solution $v(t)$ of the problem (1.67) can be written in the form

$$v(t) = \mathcal{F}^{-1}(e^{-A(i\xi)t}\mathcal{F}\varphi) = e^{-At}\varphi. \tag{1.68}$$

From this, it is seen that the problem (1.67) is well posed in X (i. e., determines a continuous flow in X) if and only if A at least generates a C_0-semigroup in X. The following question arises naturally: how to verify that the differential operator A generates a C_0-semigroup (or C_0-group) in L^p or the Sobolev space $H^{s,p}$?

Formally, the solution operator $e^{-At} = \mathcal{F}^{-1}e^{-A(i\xi)t}*$, determined by the problem (1.68), generates an "algebraic" semigroup. However, it does not necessarily generate a C_0-semigroup in L^p or $H^{s,p}$. This is because an "algebraic" semigroup is only required to have an algebraic structure, that is, to satisfy certain operation rules, while a C_0-semigroup is required to satisfy not only certain algebraic operations but also certain topological conditions such as

$$\lim_{t\to 0^+} e^{-At}\varphi \overset{L^p}{=\!=} \varphi. \tag{1.69}$$

The condition (1.69) corresponds to the continuous dependence on the initial data of the solution of the problem (1.67), which is partially required by the well-posedness of the problem (1.67).

There are many methods to characterize a semigroup [214, 293]. It is known that the solution operator $e^{-A(i\xi)t}*$ is a convolution operator and, therefore, is translation invariant. Whether or not e^{-At} is a C_0-semigroup on L^p or $H^{s,p}$ depends on if the multiplier $e^{-A(i\xi)t}$ of e^{-At} belongs to the Hörmander space $\mathcal{M}_p^p$. Precisely, we have the following multiplier characterization of semigroup.

Theorem 1.3.2. *The operator A generates a C_0-semigroup e^{-At} on L^p if and only if one of the following conditions holds:*
(i) *For all $t > 0$, $m(\xi) = \mathcal{F}^{-1}e^{-A(i\xi)t} \in \mathcal{M}_p^p$.*
(ii) *There is an $\omega \in \mathbb{R}$ such that, when $\operatorname{Re}\lambda > \omega$, we have*

$$(\lambda I + A)^{-1} = \int_0^\infty e^{-\lambda t}T(t)dt. \tag{1.70}$$

This means that the resolvent of A is the Laplace transform of the following abstract operator:

$$T(t) = e^{-At} : [0,\infty) \longrightarrow L^p,$$

where $T(t)$ satisfies

$$\|T(t)\| \le Me^{\omega t}, \quad \forall t \ge 0.$$

Example 1. The heat operator and the solution operator of general $2m$th-order parabolic equations correspond to the following multipliers, respectively:

$$e^{-|\xi|^2 t} \in \mathcal{M}_p^p, \quad t > 0,\ 1 \le p \le \infty,$$
$$e^{-P_{2m}(\xi)t} \in \mathcal{M}_p^p, \quad t > 0,\ 1 \le p \le \infty,$$

where $P_{2m}(\xi)$ is an elliptic polynomial of degree $2m$. Further,

$$e^{-|\xi|^2(t+is)},\ e^{-P_{2m}(\xi)(t+is)} \in \mathcal{M}_p^p, \quad t > 0,\ s \in \mathbb{R}.$$

Thus, it follows that Δ and $P_{2m}(D)$ generate an analytic semigroup on L^p, $1 \le p < \infty$. When $p = \infty$, the result remains true with L^∞ being replaced by C_b.

Example 2 (Navier–Stokes equations). The solution operator of the associated linear equations with the Navier–Stokes equations corresponds to the multiplier

$$e^{-|\xi|^2 t}\delta_{ij} + \frac{\xi_i}{|\xi|}\frac{\xi_j}{|\xi|}e^{-|\xi|^2 t} \in \mathcal{M}_p^p, \quad 1 < p < \infty.$$

Similar to the case of parabolic equations, $\mathcal{P}\Delta = A$ generates an analytic semigroup on $E^p = \{u \in (L^p)^n \mid \operatorname{div} u = 0\}$, $1 < p < \infty$.

Example 3. Taking $A = \pm i\Delta$ or $iQ(D)$, where Q is a rational real function, gives us the Schrödinger equation and general dispersive wave equations, respectively. The solution operator of the associated free equations corresponds to the multipliers $e^{\pm i|\xi|^2 t}$ and $e^{iQ(\xi)t}$, respectively. It has been shown that $e^{\pm|\xi|^2 t}$ and $e^{iQ(\xi)t}$ belong to $\mathcal{M}_p^p$ if and only if $p = 2$. Therefore, the operators $\pm i\Delta$ and $iQ(D)$ generate a C_0-semigroup only in the Hilbert-type space L^2 or H^s and do not generate a C_0-semigroup in L^p, $p \ne 2$, in general; see [119, 184]. This is also the reason why we can only study dispersive wave equations in $X = L^2$ or H^s.

Example 4. Consider the Cauchy problem for the wave equation (similarly for the Klein–Gordon equation):

$$\begin{cases} u_{tt} - \Delta u = 0, \\ u(0) = \varphi(x), \quad u_t(0) = \psi(x). \end{cases} \tag{1.71}$$

Let $v = u_t$. Then the problem (1.71) can be reduced to the following first-order abstract Cauchy problem:

$$\frac{d}{dt}\begin{pmatrix} u \\ v \end{pmatrix} - \begin{pmatrix} 0 & I \\ \Delta & 0 \end{pmatrix}\begin{pmatrix} u \\ v \end{pmatrix} = 0, \quad \begin{pmatrix} u \\ v \end{pmatrix}(0) = \begin{pmatrix} \varphi(x) \\ \psi(x) \end{pmatrix}.$$

The solution operator of the linear wave equation corresponds to the multiplier e^{Bt}, where

$$B = \begin{pmatrix} 0 & 1 \\ -|\xi|^2 & 0 \end{pmatrix}.$$

The sufficient and necessary condition for e^{Bt} to be in $\mathcal{M}_p^p$ is $p = 2$. Therefore, the wave equation (or the Klein–Gordon equation) must be studied in the spaces like $X = \dot{H}^s \times \dot{H}^{s-1}$ (or $X = H^s \times H^{s-1}$).

1.4 Scaling and well-suited space-time Banach spaces for evolution equations

In the previous section based on multipliers, we discussed the well-suited Banach space X for the study of the Cauchy problem for nonlinear evolution equations

$$\begin{cases} u_t + Au = F(u), & x \in \mathbb{R}^n, \ t \in \mathbb{R}^+ \ (\text{or } \mathbb{R}), \\ u(0) = \varphi. \end{cases} \tag{1.72}$$

It is required that A generates a C_0-semigroup on X. In other words, the Cauchy problem for the corresponding linear equation

$$\begin{cases} v_t + Av = 0, & x \in \mathbb{R}^n, \ t \in \mathbb{R}^+ \ (\text{or } \mathbb{R}), \\ v(0) = \varphi(x) \end{cases} \tag{1.73}$$

is required to determine a continuous flow $v(t) \in C(\mathbb{R}^+; X)$ or $v(t) \in C(\mathbb{R}; X)$. Under this condition, can one prove that the nonlinear problem (1.72) is well posed in $C(I; X)$? To achieve this goal, it is difficult (and sometimes impossible) to only use $C(I; X)$ to establish the well-posedness of (1.72). It is necessary to do so in a subspace of $C(I; X)$. The subspace is, in fact, the intersection of $C(I; X)$ with some space-time Banach space, which, as is easily seen, is natural since the solution of the linear problem (1.73) not only belongs to $C(I; X)$ but also is of temporal and spatial integrability properties in some sense, that is, the solution belongs to some space-time Banach space. The purpose of this section is to use the scaling principle to study space-time Banach spaces corresponding to different types of evolution equations and the conditions satisfied by them. In addition, from the point view of scaling, we will discuss the relationship between the well-suited Banach space X and the nonlinear term, which will then be used to verify if (1.72) is well posed in $C(I; X)$.

For simplicity, let $A(\xi)$ be the symbol of the spatial differential operator $A = A(\nabla)$ satisfying that

$$A(\lambda i\xi) = \lambda^m A(i\xi), \quad \lambda \in \mathbb{R}^+.$$

In general, the well-suited Banach space is restricted to the following form:

$$X = H^{s,p}(\mathbb{R}^n) = (I-\Delta)^{-\frac{s}{2}}L^p, \quad s \in \mathbb{R},\ 1 < p < \infty$$

or

$$\dot{X} = \dot{H}^{s,p}(\mathbb{R}^n) = (-\Delta)^{-\frac{s}{2}}L^p, \quad s \in \mathbb{R},\ 1 < p < \infty.$$

It is clear that $e^{-A(i\xi)} \in \mathcal{M}_p^p$ is the necessary condition for $X = H^{s,p}$ or $\dot{H}^{s,p}$ to be a well-suited Banach space for the study of (1.72). It is well known that the symbol $e^{-A(i\xi)}$, corresponding to the solution operator of the parabolic and Navier–Stokes equations, belongs to $\mathcal{M}_p^p$ with $1 < p < \infty$. However, for the Schrödinger, wave-type or general dispersive equations, the symbol $e^{-A(i\xi)} \in \mathcal{M}_p^p$ if and only if $p = 2$.

In general, the well-posedness of the problem (1.72) can be studied by dealing with the following integral equation:

$$u(t) = e^{-tA}\varphi + \int_0^t e^{-(t-\tau)A}F(u)(\tau)d\tau. \tag{1.74}$$

To be clear, we need some definitions.

Definition 1.4.1 (Local well-posedness). The problem (1.72) or equivalently (1.74) is said to be locally well posed in X if the following conditions are satisfied:

(i) Given $\varphi \in X$, there exists a positive constant $T = T(\|\varphi\|_X)$ (or $T = T(\varphi)$ if X is a critical space) such that the problem (1.72) or equivalently (1.74) has a unique solution $u \in \mathcal{X}(I) := C(I;X) \cap \cdots$, where "$\cdots$" denotes some well-suited space-time Banach space. Further, $T(\delta) \to \infty$ as $\delta \to 0$.

(ii) There exist $r = r(\|\varphi\|_X) > 0$ and $M(\|\varphi\|_X) > 0$ such that for any $\tilde{\varphi} \in X$ satisfying $\|\tilde{\varphi} - \varphi\|_X < r(\|\varphi\|_X)$, we have

$$\|\tilde{u}(t) - u(t)\|_{\mathcal{X}(I)} \le M(\|\varphi\|_X)\|\tilde{\varphi} - \varphi\|_X,$$

where $\tilde{u}(t)$ and $u(t)$ are the solutions of (1.74) with data $\tilde{\varphi}$ and φ, respectively, and $I = [0,T)$ is the common existence time interval. In other words, the solution operator $\tilde{\varphi} \longmapsto \tilde{u}(t)$ determines a local Lipschitz mapping, which maps $\{\tilde{\varphi} \mid \|\tilde{\varphi}-\varphi\|_X < r\}$ into $\mathcal{X}(I)$.

(iii) If $\varphi \in Y \hookrightarrow X$, then $u \in \mathcal{Y}(I) := C(I;Y) \cap \cdots$

Remark 1.4.1.

(a) The solution of (1.74) is usually called as the mild solution of (1.72), so (i) implies the existence and uniqueness of the mild solution. The conditions (ii) and (iii) essentially imply that mild solutions are strong solutions to (1.72), which can be obtained as the limit in the norm of $\mathcal{X}(I)$ of the smooth solution of (1.72). When initial data is smooth enough, the corresponding solution is the classical solution.

(b) It is difficult to solve the problem (1.72) directly in $C(I;X)$ (i. e., directly use the norm of $C(I;X)$ to construct fixed points) unless X is a Banach algebra and the nonlinear term $F(u)$ satisfies a suitable smooth condition. However, for the general nonlinear term, $u(t) \in L^\infty(I;X)$ obtained by compactness method or else methods is not necessarily unique and one can not directly obtain the continuous dependence on the data. The Strichartz estimates provide a efficient method to study the abstract Cauchy problem (1.74) in a subspace $C(I;X) \cap Y$ of $C(I;X)$ for some space-time Banach space Y. In other words, one needs to consider the time and spatial integrability property of the solution.

(c) When X is a critical space, then there exists an $\eta > 0$ such that, when $\|\varphi\|_Y < \eta$, one can obtain the global well-posedness of solution, that is, $u \in C(\mathbb{R};X) \cap \cdots$ (or $u \in C(\mathbb{R}^+;X) \cap \cdots$).

Definition 1.4.2. Let X be a Banach space consisting of distributions defined on $\mathbb{R}^n$ and let $\dot{X}$ be the corresponding homogeneous space. The highest smoothness degree of X is defined and denoted by

$$h - \deg(X) = \deg(\dot{X}) = \log_\lambda \Lambda(\lambda),$$

where $\Lambda(\lambda) = \|\varphi(\lambda x)\|_{\dot{X}}/\|\varphi(x)\|_{\dot{X}}$ and ϕ is a nonzero function in X.

It is easy to see that $\Lambda(\lambda)$ is the homogeneous function on λ and is independent of the choice of $\varphi \neq 0$. For example,

$$X = L^p(\mathbb{R}^n), \quad h - \deg(X) = -\frac{n}{p}, \quad 1 \le p \le \infty,$$

$$X = H^{s,p}(\mathbb{R}^n), \quad h - \deg(X) = s - \frac{n}{p}, \quad 1 \le p \le \infty, \ s \in \mathbb{R},$$

$$X = B^s_{p,q}(\mathbb{R}^n), \quad h - \deg(X) = s - \frac{n}{p}, \quad 1 \le p, q \le \infty, \ s \in \mathbb{R},$$

$$X = F^s_{p,q}(\mathbb{R}^n), \quad h - \deg(X) = s - \frac{n}{p}, \quad 1 \le p < \infty, \quad 1 \le q \le \infty, \ s \in \mathbb{R},$$

$$X = M^p_q, \quad h - \deg(X) = -\frac{n}{p}, \quad 1 \le q \le p < \infty.$$

Suppose now the nonlinear term $F(u)$ is of the form

$$F(u) \sim |u|^\alpha u$$

for some parameter $\alpha > 0$. Let u be the mild solution of (1.72). Then it is seen that $u_\lambda(t) = \lambda^\theta u(\lambda^{-1}x, \lambda^{-m}t)$ solves the problem

$$u_t + Au = F(u), \quad t \in \mathbb{R}^+ \text{ (or } \mathbb{R}), \quad u(0,x) = \lambda^\theta \varphi(\lambda^{-1}x), \quad x \in \mathbb{R}^n \tag{1.75}$$

if and only if $\theta = -\frac{m}{\alpha}$. In general, if the nonlinear term $F(u)$ satisfies

$$F(\lambda^\theta u(\lambda^{-1}x, \lambda^{-m}t)) = \lambda^{-m+\theta} F(u(\lambda^{-1}x, \lambda^{-m}t)), \quad \forall \lambda \in \mathbb{R}^+ \tag{1.76}$$

($F(u)$ may contain the derivatives of u), then $u_\lambda = \lambda^\theta u(\lambda^{-1}x, \lambda^{-m}t)$ is also a solution of (1.75).

Suppose A generates a C_0-semigroup in L^p with $1 \le p < \infty$ (e. g., in the case of parabolic equations). Let $X = H^{s,p}$. Then one easily sees that $v(t) = \exp(-At)\varphi \in C(I; H^{s,p})$ solves the problem (1.73). For the nonlinear term $F(u)$ satisfying (1.76), we are interested in conditions on s, p under which (1.72) generates a unique continuous flow in $C(I; H^{s,p}) \cap Y$ for some space-time Banach space Y. In particular, if $H^{s,p}$ is a critical space in the study of (1.72), then what kind of conditions should s and p satisfy? The scaling principle suggests that:

(i) If $h\text{-}\deg(\dot{H}^{s,p}) \ge \theta$, that is, $s \ge \frac{n}{p} + \theta$, then (1.72) is locally well posed in $X = H^{s,p}$;
(ii) If $h\text{-}\deg(\dot{H}^{s,p}) < \theta$, that is, $s < \frac{n}{p} + \theta$, then (1.72) is ill posed in $X = H^{s,p}$;
(iii) When $h\text{-}\deg(\dot{H}^{s,p}) = \theta$ (i. e., $s = \frac{n}{p} + \theta$), $X = H^{s,p}$ is a critical space for (1.72), and we usually write $s_p = \frac{n}{p} + \theta$.

However, for different nonlinear evolution equations and for different nonlinear terms, the well-posedness results for (1.72) may not be in agreement with the above results predicted by the scaling principle. In some cases, the results are better than predicted, such as in the Cauchy problem for the convection-diffusion equation:

$$u_t - \Delta u = \vec{a} \cdot \nabla(|u|^\alpha u), \quad t > 0,\ x \in \mathbb{R}^n, \quad u(0,x) = \varphi(x), \quad x \in \mathbb{R}^n, \tag{1.77}$$

where $\vec{a} \in \mathbb{R}^n \setminus \{0\}$; see [85] for details. It is sometimes worse than the predictive results, for example, in the case of the Cauchy problem for the following semilinear wave equation:

$$u_{tt} - \Delta u = u^2, \quad x \in \mathbb{R}^n,\ t \in \mathbb{R}, \quad u(0,x) = f(x), \quad u_t(0,x) = g(x), \quad x \in \mathbb{R}^n. \tag{1.78}$$

The scaling principle suggests that (1.78) should be well posed in $X = H^s \times H^{s-1}$, where $s \ge -1/2$. However, Lindblad proved in [176] that (1.78) is ill posed in X if $s < 0$. However, for certain special types of nonlinear terms, for example, those in the following problem:

$$u_{tt} - \Delta u = \sum_{j=1}^{3} (\partial_j u)^2 - (\partial_t u)^2, \quad u(0,x) = f(x), \quad \partial_t u(0,x) = g(x), \tag{1.79}$$

or more general nonlinear terms satisfying the null condition (see [151, 152]), the well-posedness results are in agreement with those predicted by the scaling principle. For example, when $s > 3/2$, the problem (1.79) is well posed in $X = H^s \times H^{s-1}$ as predicted by the scaling principle. In general, the scaling principle provides a basic criterion for the study of well-posedness of nonlinear evolution equations. In particular, for parabolic equations, the well-posedness results basically are in agreement with those suggested by the scaling principle. In the subsequent chapters, we will discuss in detail several classes of important equations such as the heat equation, the Navier–Stokes equations, the Schrödinger equation and the wave equations.

We now discuss what kinds of space-time Banach spaces the solution of linear evolution equations belongs to. These space-time Banach spaces will play a decisive role in the study of nonlinear evolution equations. We will consider the following four cases.

I. Parabolic equations

Suppose the mth-order homogeneous operator $-A$ generates an analytic semigroup in L^r $(1 < r < \infty)$. It can be verified directly that if $v = v(x,t)$ is a solution to the problem (1.73), then $v_\lambda(t) = v(\lambda x, \lambda^m t)$ is also a solution to the problem (1.73) with the initial data $\varphi(x)$ being replaced by $\varphi(\lambda x)$. If $v(x,t) \in L^q(\mathbb{R}^+; L^p(\mathbb{R}^n))$, $1 < p,\ q < \infty$, then what kinds of conditions do p and q satisfy? To this end, let

$$\mathcal{E}_{p,q}(\lambda) = \|v(\lambda x, \lambda^m t)\|_{L^q(\mathbb{R}^+;L^p(\mathbb{R}^n))}, \tag{1.80}$$

$$\mathcal{E}_r(\lambda) = \|\varphi(\lambda x)\|_{L^r(\mathbb{R}^n)}. \tag{1.81}$$

Then the necessary condition for $v(t)$ to be in $L^q(\mathbb{R}^+; L^p(\mathbb{R}^n))$ is

$$\operatorname{order}(\mathcal{E}_{p,q}(\lambda)) = \operatorname{order}(\mathcal{E}_r(\lambda)),$$

that is,

$$\frac{m}{q} = n\left(\frac{1}{r} - \frac{1}{p}\right). \tag{1.82}$$

We will see in Chapter 2 that when

$$1 < r \le p < \begin{cases} \frac{rn}{n-m}, & n > m, \\ \infty, & n \le m, \end{cases}$$

the sufficient and necessary condition with $v(t) \in L^q(\mathbb{R}^+; L^p(\mathbb{R}^n))$ is that (q,p) satisfies (1.82). Furthermore, if

$$1 \le r \le p < \begin{cases} \frac{rn}{n-rm}, & n > rm, \\ \infty, & n \le rm, \end{cases}$$

the (q,p) determined by (1.82) guarantees that

$$\sup_{t\geq 0} t^{\frac{1}{q}} \|v(t)\|_p < \infty,$$

or $v \in \mathcal{C}_{q(p,r)}(\mathbb{R}^+; L^p(\mathbb{R}^n))$, where, for $\sigma > 0$ and $I \subseteq \mathbb{R}$ with $\dot{I}$ being the open interval of I, we define

$$\mathcal{C}_\sigma(I; L^p(\mathbb{R}^n)) = \left\{u \in C(\dot{I}; L^p(\mathbb{R}^n)) \mid \sup_{t\in I} t^{\frac{1}{\sigma}} \|u\|_p < \infty\right\}.$$

In the study of Cauchy problems for parabolic equations and the Navier–Stokes equations, the above space plays the same role as $L^q(I; L^p(\mathbb{R}^n))$ does in the study of wave and dispersive equations.

The admissible relation (1.82) is derived from the scaling principle. We now discuss this from the $L^p - L^q$ estimate. For the solution to the parabolic-type problem (1.73), we have

$$\begin{aligned} \|v(t)\|_\ell &\leq \|\varphi\|_\ell, \quad 1 \leq \ell \leq \infty, \\ \|v(t)\|_\infty &\leq Ct^{-\frac{n}{m}} \|\varphi\|_1, \quad t > 0. \end{aligned}$$

By interpolation, it is seen that

$$\|v(t)\|_p \leq Ct^{-\frac{n}{m}(\frac{1}{r}-\frac{1}{p})} \|\varphi\|_r, \quad 1 < r \leq p.$$

From this, it is found that, in order that $v \in L^q(\mathbb{R}^+; L^p(\mathbb{R}^n))$ or $\mathcal{C}_{q(p,r)}(\mathbb{R}^+; L^p(\mathbb{R}^n))$, we need to have the relation (1.82).

Remark 1.4.2. When A is an inhomogeneous differential operator, it is not obvious that using scaling to discuss the admissible relation corresponding to the space-time Banach space $L^q(I; L^p(\mathbb{R}^n))$. For example, when the wave equation is reduced to a first-order evolution equation, the corresponding operator is inhomogeneous. In this case, we may use the $L^p - L^q$ estimate for the free equation to analyze the corresponding admissible relation.

II. Dispersive wave equations

Let the operator A generate a C_0-semigroup only in $L^2(\mathbb{R}^n)$ or $H^s(\mathbb{R}^n)$, but cannot generate a C_0-semigroup in general $L^r (r \neq 2)$. Similar to the case of parabolic equations, if A corresponds to an mth-order homogeneous differential operator and $v(t) = v(x,t)$ is a solution to the problem (1.73), then $v_\lambda(t) = v(\lambda x, \lambda^m t)$ is a solution to (1.73) with the initial data $\varphi(x)$ being replaced by $\varphi(\lambda x)$. It is easy to see that $v \in L^q(\mathbb{R}; L^p(\mathbb{R}^n))$ only if

$$\operatorname{order}(\mathcal{E}_{p,q}(\lambda)) = \operatorname{order}(\mathcal{E}_2(\lambda)),$$

where $\mathcal{E}_{p,q}(\lambda)$ and $\mathcal{E}_2(\lambda)$ are defined as in (1.80) and (1.81). The above equality gives the admissible relation

$$\frac{m}{q} = n\left(\frac{1}{2} - \frac{1}{p}\right). \tag{1.83}$$

Furthermore, as will be seen in Chapters 4 and 5, when

$$2 \le p < \begin{cases} \frac{2n}{n-m}, & n > m, \\ \infty, & n \le m, \end{cases}$$

the pair (q, p) determined by (1.83) ensures that $v \in L^q(\mathbb{R}; L^p(\mathbb{R}^n))$.

On the other hand, when $m > 2$, $iA(i\xi)$ corresponds to an mth-order positive (negative) definite polynomial, then the solution of the free problem satisfies the regularity $L^{p'} \to L^p$ estimate (see [135])

$$\|S(t)\varphi\|_{\dot{H}^{2s_p,p}} = \|\partial^{2s_p} S(t)\varphi\|_p \le C|t|^{-n(\frac{1}{2}-\frac{1}{p})}\|\varphi\|_{p'},$$

where $v(t) = e^{-At}\varphi(x)$,

$$s_p = \frac{n(m-2)}{2}\left(\frac{1}{2} - \frac{1}{p}\right)$$

and

$$2 \le p \begin{cases} < 2^* = 2n/(n-2), & n \ge 2, \\ \le \infty, & n = 1. \end{cases}$$

Thus, the necessary condition with $v(t) = S(t)\varphi \in L^q(\mathbb{R}; \dot{H}^{s_p,p}(\mathbb{R}^n))$ is that

$$\frac{2}{q} = n\left(\frac{1}{2} - \frac{1}{p}\right). \tag{1.84}$$

In particular, for the classical Schrödinger equation, the admissible relation (1.83) is equivalent to the admissible relation (1.84).

Remark 1.4.3. As will be seen, the admissible relation for general dispersive wave equations is similar to that corresponding to general parabolic equations. However, this is not case if the admissible relation is examined from the $L^p \to L^{p'}$ and $L^p \to L^r$ estimates. The admissible relation (1.82) for parabolic equations determines an admissible triplet (q, p, r), while, for dispersive wave equations, $r \equiv 2$ and the solution operator is bounded from $L^{p'}$ to L^p $(t \ne 0)$ and is not bounded from L^2 to L^p, that is,

$$\|S(t)\varphi\|_p \le C|t|^{-\frac{2n}{m}(\frac{1}{2}-\frac{1}{p})}\|\varphi\|_{p'} = C|t|^{-\frac{n}{m}(\frac{1}{p'}-\frac{1}{p})}\|\varphi\|_{p'}.$$

Therefore, by the TT^* method and the Hardy–Littlewood–Sobolev inequality, we get the admissible relation

$$\frac{2}{q} = \frac{2n}{m}\left(\frac{1}{2} - \frac{1}{p}\right) = \frac{n}{m}\left(\frac{1}{p'} - \frac{1}{p}\right).$$

This is equivalent to the relation (1.83) derived by scaling. In this way, the regular admissible relation (1.84) is determined by the regular $L^{p'} \to L^p$ estimate in conjunction with the Hardy–Littlewood–Sobolev inequality and the TT^* method.

III. Wave equations

It is well known that the Cauchy problem

$$\begin{cases} u_{tt} - \Delta u = 0, \\ u(0) = \varphi(x), \quad u_t(0) = \psi(x) \end{cases} \tag{1.85}$$

can be changed into the abstract Cauchy problem (1.73), where

$$A = \begin{pmatrix} 0 & I \\ \Delta & 0 \end{pmatrix}, \quad V = \begin{pmatrix} u \\ u_t \end{pmatrix}, \quad V(0) = \begin{pmatrix} \varphi \\ \psi \end{pmatrix}.$$

Since A is not a homogeneous differential operator of order 2, we will use the $L^p - L^{p'}$ estimate of solutions to the problem (1.85) together with the TT^* method and the Hardy–Littlewood–Sobolev inequality to examine the well-suited space-time Banach space for the study of nonlinear wave equations.

Let

$$u[0] = (\varphi, D^{-1}\psi), \quad D^{-1} = (-\Delta)^{-\frac{1}{2}}.$$

Then it is easy to verify that

$$u(t) = \mathcal{F}^{-1}(\cos|\xi|t\mathcal{F}\varphi) + \mathcal{F}^{-1}\left(\frac{\sin|\xi|t}{|\xi|}\mathcal{F}\psi\right)$$

is a solution to the problem (1.85) and satisfies the energy identity

$$E(u(t), u_t(t)) = \int_{\mathbb{R}^n} \left[|\partial_t u|^2 + \sum_{j=1}^{n} |\partial_{x_j} u|^2\right] dx = E(\varphi(x), \psi(x)).$$

For any $s \geq 0$, we have the energy inequality

$$\|\partial u(t)\|_{H^s} \leq \|\partial u(0)\|_{H^s}, \quad \partial = (\partial_t, \partial_1, \ldots, \partial_n).$$

See [221, 242] for details. On the other hand, we have the following dispersive inequality:

$$\|u(t)\|_{\mathrm{BMO}} \le C|t|^{-\frac{n-1}{2}} \|D^{\frac{n+1}{2}} u[0]\|_{\mathcal{H}_1},$$

where $\mathcal{H}_1$ denotes the Hardy space. By interpolation, we get the Strichartz–Brenner estimate

$$\|u(t)\|_r \le C|t|^{-\gamma(r)} \|\nabla^\sigma \partial u(0)\|_{r'}, \tag{1.86}$$

where

$$\gamma(r) = (n-1)\left(\frac{1}{2} - \frac{1}{r}\right), \quad \frac{1}{r} + \frac{1}{r'} = 1, \quad r \ge 2,$$
$$\sigma = 2\delta(r) - \gamma(r) - 1, \quad \delta(r) = n\left(\frac{1}{2} - \frac{1}{r}\right).$$

Therefore, for $(\varphi, \psi) \in H^\sigma \times H^{\sigma-1}$, by the TT^* method and the Strichartz–Brenner estimate (1.86), it follows that the estimate

$$\|u(t)\|_{L^q(\mathbb{R};L^r(\mathbb{R}^n))} \le C\|u[0]\|_{H^\sigma} \tag{1.87}$$

holds only if

$$\sigma = \delta(r) - \frac{1}{q} = n\left(\frac{1}{2} - \frac{1}{r}\right) - \frac{1}{q}. \tag{1.88}$$

Further, if (q, r) satisfies

$$\frac{2}{q} \le \gamma(r), \quad q \ge 2, \quad (q, r, n) \ne (2, \infty, 3),$$

then the pair (q, r), determined by the admissible relation (1.88), guarantees that (1.87) holds.

Remark 1.4.4.

(i) When $\sigma = 0$, the wave equation has the following special admissible relation:

$$\frac{1}{q} = n\left(\frac{1}{2} - \frac{1}{r}\right). \tag{1.89}$$

It is easy to see that, if $u = u(x, t)$ is a solution to the problem (1.85), then $u_\lambda = u(\lambda x, \lambda t)$ is also a solution to the problem (1.85) with the initial data $(\varphi(\lambda x), \lambda\psi(\lambda x))$. Thus, in order that $u \in L^q(\mathbb{R}; L^r(\mathbb{R}^n))$ it is necessary that

$$\mathrm{order}(\mathcal{E}_{q,r}(\lambda)) = \mathrm{order}(\mathcal{E}_2(\lambda)),$$

where

$$\mathcal{E}_{q,r}(\lambda) = \|u_\lambda\|_{L^q(\mathbb{R};L^r(\mathbb{R}^n))}, \quad \mathcal{E}_2(\lambda) = \|\varphi(\lambda x)\|_2 + \|\lambda\psi(\lambda x)\|_{\dot{H}^{-1}}.$$

This implies the admissible relation (1.89).

(ii) When $\sigma = \beta(r) = \frac{n+1}{2}(\frac{1}{2} - \frac{1}{r})$, the wave equation has the following special admissible relation (the optimal admissible relation as usually used):

$$\frac{1}{q} = \frac{n-1}{2}\left(\frac{1}{2} - \frac{1}{r}\right). \tag{1.90}$$

It is easy to find that, if $u = u(x,t)$ is a solution to the problem (1.85), then $u_\lambda = u(\lambda x, \lambda t)$ is also a solution to the problem (1.85) with the initial data $(\varphi(\lambda x), \lambda\psi(\lambda x))$. Thus, in order that $u \in L^q(\mathbb{R}; L^r(\mathbb{R}^n))$, we have to require that

$$\operatorname{order}(\mathcal{E}_{q,r}(\lambda)) = \operatorname{order}(\mathcal{E}_2^{\beta(r)}(\lambda)),$$

where

$$\mathcal{E}_2^{\beta(r)}(\lambda) = \|\varphi(\lambda x)\|_{\dot{H}^{\beta(r)}} + \|\lambda\psi(\lambda x)\|_{\dot{H}^{\beta(r)-1}}.$$

This gives exactly the admissible relation (1.90).

Remark 1.4.5. A direct calculation shows that $v(t) = \mathcal{F}^{-1}E_{1+\alpha}(-|\xi|^2 t^{1+\alpha}) * \varphi$ is a solution to the following Cauchy problem for the fractional-order evolution equation:

$$\begin{cases} \frac{\partial^{\alpha+1} v}{\partial t^{\alpha+1}} = \Delta v, \quad 0 < \alpha < 1, \ x \in \mathbb{R}^n, \ t \geq 0, \\ v(0) = \varphi, \end{cases} \tag{1.91}$$

where $E_{1+\alpha}(-|\xi|^2)$ is the Mittag-Leffler function (see [117]). In view of that $E_{1+\alpha}(-|\xi|^2) \in \mathcal{M}_p$, we know that $\mathcal{F}^{-1}E_{1+\alpha}(-|\xi|^2 t^{1+\alpha})*$ generates a global flow in L^p $(1 < p < \infty)$. By scaling, if $v(t) = v(x,t)$ is a solution to the problem (1.91), then $v_\lambda(t) = v(\lambda x, \lambda^{\frac{2}{1+\alpha}} t)$ is also a solution to the problem (1.91) with the initial data $\varphi_\lambda(x) = \varphi(\lambda x)$. Thus, in order that $v(t) \in L^q(\mathbb{R}^+; L^p(\mathbb{R}^n))$, it is needed that

$$\operatorname{order}(\mathcal{E}_{p,q}(\lambda)) = \operatorname{order}(\mathcal{E}_r(\lambda)),$$

where

$$\mathcal{E}_{p,q}(\lambda) = \|v(\lambda x, \lambda^{\frac{2}{1+\alpha}} t)\|_{L^q(\mathbb{R}^+, L^p(\mathbb{R}^n))}, \quad \mathcal{E}_r(\lambda) = \|\varphi(\lambda x)\|_r.$$

This means that

$$\frac{1}{q} = \frac{(\alpha+1)n}{2}\left(\frac{1}{r} - \frac{1}{p}\right),$$

where

$$1 < r \le p < \begin{cases} \frac{n(1+a)r}{n(1+a)-2}, & n \ge 2, \\ \infty, & n = 1. \end{cases}$$

With this, we may construct the well-suited space-time Banach space $L^q(I; L^p(\mathbb{R}^n))$ to study the nonlinear fractional-order evolution equations; see [117] for details.

1.5 Sobolev spaces and wave operator adapted to Schrödinger operator with inverse-square potential

In this section, we discuss several basic harmonic analysis questions related to the following Schrödinger operator:

$$\mathcal{L}_a = -\Delta + \frac{a}{|x|^2} \quad \text{with} \quad a \ge -\left(\frac{n-2}{2}\right)^2 \tag{1.92}$$

in dimensions $n \ge 2$. This plays an important role in the study of large-data problems at the development of dynamics of nonlinear dispersive equations with inverse square potential. A key principle in attacking such problems in that one must first study the effective equations in various limiting regimes. Given the (possibly broken, but still intrinsic) scale and space-translation invariance of such problems, one must accept that solutions may live at any possible length scale, as well as at any spatial location. Naturally, taking a scaling limit results in a scale invariant problem. Thus, the broad goal of demonstrating well-posedness of critical problems in general geometries rests on a thorough understanding of limiting problems, such as that associated to the inverse-square potential. The harmonic analysis tools developed in the paper [142] have been essential to [141], which considers global well-posedness and scattering for both the defocusing and focusing energy-critical NLS with inverse-square potential (also in Chapter 6), and to [143], in which sharp thresholds for well-posedness and scattering are established for the focusing cubic NLS with inverse-square potential.

We interpret $\mathcal{L}_a$ as the Friedrichs extension of this operator defined initially on $C_c^\infty(\mathbb{R}^n \setminus \{0\})$. More precisely, for $a \ge -(\frac{n-2}{2})^2$, the map

$$Q(f) := \int_{\mathbb{R}^n} \left(|\nabla f(x)|^2 + \frac{a}{|x|^2} |f(x)|^2 \right) dx$$

defines a positive definite quadratic form on $C_c^\infty(\mathbb{R}^n \setminus \{0\})$. Indeed, this can be verified by observing that

$$Q(f) = \int_{\mathbb{R}^n} \left| \nabla f + \frac{\sigma x}{|x|^2} f \right|^2 dx \quad \text{with} \quad \sigma = \frac{n-2}{2} - \frac{1}{2}\sqrt{(n-2)^2 + 4a}.$$

We then define the operator $\mathcal{L}_a$ as the Friedrichs extension of the quadratic form $Q(f)$; see, for example, [226, Section X.3] for the general theory of such extensions and [142, Section 1.1] or [124] for more on this particular operator. We choose the Friedrichs extension for physically motivated reasons: (i) when $a = 0$, it yields the usual Laplacian $-\Delta$ and (ii) the Friedrichs extension appears when one takes a scaling limit of more regular potentials. For example,

$$L_k := -\Delta + \frac{ak^2}{1 + k^2|x|^2} \longrightarrow \mathcal{L}_a \quad \text{as} \quad m \to \infty$$

in strong resolvent sense (where we understand L_k as having domain $H^2(\mathbb{R}^n)$).

The fact that $Q(f) \geq 0$ when $a \geq -(\frac{n-2}{2})^2$, but not smaller, is one realization of the sharp Hardy inequality. On the other hand, the sharp Hardy inequality also shows that

$$Q(f) = \|\sqrt{\mathcal{L}_a} f\|_{L_x^2}^2 \sim \|\nabla f\|_{L_x^2}^2 \quad \text{for} \quad a > -\left(\frac{n-2}{2}\right)^2. \tag{1.93}$$

Note that this is an assertion of the isomorphism of the Sobolev spaces $\dot{H}^1(\mathbb{R}^n)$ defined in terms of powers of $\mathcal{L}_a$ and via the usual gradient. The spectral theorem allows one to define functions of the operator $\mathcal{L}_a$ and provides a simple necessary and sufficient condition for their boundedness on L^2. A sufficient condition for functions of an operator to be L^p-bounded is a basic prerequisite for the modern approach to many PDE problems. In the setting of constant-coefficient differential operators, this role is played by the classical Mikhlin multiplier theorem. The analogue for the operator $\mathcal{L}_a$ is established in Killip–Miao–Visan–Zhang–Zheng [142] as follows.

Theorem 1.5.1 (Mikhlin multipliers, [142]). *Suppose $m : [0, \infty) \to \mathbb{C}$ satisfies*

$$|\partial^j m(\lambda)| \lesssim \lambda^{-j} \quad \textit{for all} \quad j \geq 0 \tag{1.94}$$

and that either
- $a \geq 0$ *and* $1 < p < \infty$, *or*
- $-(\frac{n-2}{2})^2 \leq a < 0$ *and* $r_0 < p < r_0' := \frac{n}{\sigma}$.

Then $m(\sqrt{\mathcal{L}_a})$ extends to a bounded operator on $L^p(\mathbb{R}^n)$.

In the case $a \geq 0$, the operator $\mathcal{L}_a$ obeys Gaussian heat kernel bounds and so the result follows from general results covering this class of operators; see, for example, [3, 116]. When $-(\frac{n-2}{2})^2 \leq a < 0$, the heat kernel for $\mathcal{L}_a$ no longer obeys Gaussian bounds; indeed, the heat kernel is singular at the origin, which were found by Liskevich–Sobol [177] and Milman–Semenov [200].

Theorem 1.5.2 (Heat kernel bounds). *Assume $n \geq 3$ and $a \geq -(\frac{n-2}{2})^2$. Then there exist positive constants C_1, C_2 and c_1, c_2 such that for all $t > 0$ and all $x, y \in \mathbb{R}^n \setminus \{0\}$,*

$$C_1 C\sigma(x,y,t)t^{-\frac{n}{2}}e^{-\frac{|x-y|^2}{c_1 t}} \le e^{-t\mathcal{L}_a}(x,y) \le C_2 C\sigma(x,y,t)e^{-\frac{|x-y|^2}{c_2 t}},$$

where $C\sigma(x,y,t) = (1 \vee \frac{\sqrt{t}}{|x|})^\sigma(1 \vee \frac{\sqrt{t}}{|y|})^\sigma$, *and we denote* $A \vee B := \max\{A, B\}$.

As a consequence of the multiplier theorem and the usual randomization argument, one can easily get the square function estimates of Littlewood–Paley; see Theorem 1.5.4 below. Such estimates are an invaluable tool since they allow one to analyze problems one length scale at a time and then reassemble the pieces.

It is convenient for us to consider two kinds of Littlewood–Paley projections: one defined via a $C_c^\infty((0,\infty))$ multiplier and another defined as a difference of heat kernels. The former notion more closely matches modern expositions of the classical translation-invariant theory, while the heat kernel version allows one to exploit heat-kernel bounds (and the semigroup property) to prove estimates. There is no great cost associated to either choice, since the multiplier theorem permits one to readily pass back and forth between the two notions.

One important application of the traditional Littlewood–Paley theory is the proof of Leibniz (=product) and chain rules for differential operators of noninteger order. For example, if $1 < p < \infty$ and $s > 0$, then

$$\|fg\|_{H^{s,p}(\mathbb{R}^n)} \lesssim \|f\|_{H^{s,p_1}(\mathbb{R}^n)}\|g\|_{L^{p_2}(\mathbb{R}^n)} + \|f\|_{L^{p_3}(\mathbb{R}^n)}\|g\|_{H^{s,p_4}(\mathbb{R}^n)} \tag{1.95}$$

whenever $\frac{1}{p} = \frac{1}{p_1} + \frac{1}{p_2} = \frac{1}{p_3} + \frac{1}{p_4}$. For a textbook presentation of these theorems and original references, see [270].

Rather than pursue a direct proof of such inequalities, the authors in [142] prove a result that allows one to deduce such results directly from their Euclidean counterparts as follows.

Theorem 1.5.3 (Equivalence of Sobolev norms). *Suppose* $n \ge 3$, $a \ge -(\frac{n-2}{2})^2$ *and* $0 < s < 2$. *If* $1 < p < \infty$ *satisfies* $\frac{s+\sigma}{n} < \frac{1}{p} < \min\{1, \frac{n-\sigma}{n}\}$, *then*

$$\|(-\Delta)^{\frac{s}{2}}f\|_{L^p} \lesssim_{n,p,s} \|\mathcal{L}_a^{\frac{s}{2}}f\|_{L^p} \quad \textit{for all} \quad f \in C_c^\infty(\mathbb{R}^n). \tag{1.96}$$

If $\max\{\frac{s}{n}, \frac{\sigma}{n}\} < \frac{1}{p} < \min\{1, \frac{n-\sigma}{n}\}$, *which ensures already that* $1 < p < \infty$, *then*

$$\|\mathcal{L}_a^{\frac{s}{2}}f\|_{L^p} \lesssim_{n,p,s} \|(-\Delta)^{\frac{s}{2}}f\|_{L^p} \quad \textit{for all} \quad f \in C_c^\infty(\mathbb{R}^n). \tag{1.97}$$

As a direct consequence (1.95) and Theorem 1.5.3, we get the fractional product rule for the operator $\mathcal{L}_a$.

Lemma 1.5.1 (Fractional product rule). *Fix* $a > -\frac{(n-2)^2}{4}$. *Then for all* $f, g \in C_c^\infty(\mathbb{R}^n \setminus \{0\})$, *we have*

$$\|\sqrt{\mathcal{L}_a}(fg)\|_{L^p(\mathbb{R}^n)} \lesssim \|\sqrt{\mathcal{L}_a}f\|_{L^{p_1}(\mathbb{R}^n)}\|g\|_{L^{p_2}(\mathbb{R}^n)} + \|f\|_{L^{q_1}(\mathbb{R}^n)}\|\sqrt{\mathcal{L}_a}g\|_{L^{q_2}(\mathbb{R}^n)},$$

for any exponents satisfying

$$\max\left\{\frac{1}{n}, \frac{\sigma}{n}\right\} < \frac{1}{p} < \min\left\{1, \frac{n-\sigma}{n}\right\}, \qquad \frac{s+\sigma}{n} < \frac{1}{p_1}, \frac{1}{q_2} < \min\left\{1, \frac{n-\sigma}{n}\right\},$$

and $\frac{1}{p} = \frac{1}{p_1} + \frac{1}{p_2} = \frac{1}{q_1} + \frac{1}{q_2}$.

Note that for the applications described above, both inequalities (1.96) and (1.97) are equally important. The crucial role of this result in the study of nonlinear Schrödinger equations with inverse-square potential is documented in [141, 143]. Here again, (1.96) and (1.97) are equally important. In the case of [141], it is also essential that Theorem 1.5.3 covers both the regimes where $s < 1$ and $s > 1$; see the details in Chapter 6.

As a direct consequence of the multiplier theorem, one can obtain the basic Littlewood–Paley theory, such as Bernstein and square function inequalities, adapted to the operator $\mathcal{L}_a$. Let $\phi : [0, \infty) \to [0, 1]$ be a smooth function such that

$$\phi(\lambda) = 1 \quad \text{for} \quad 0 \le \lambda \le 1 \quad \text{and} \quad \phi(\lambda) = 0 \quad \text{for} \quad \lambda \ge 2.$$

For each dyadic number $N \in 2^{\mathbb{Z}}$, we define

$$\phi_N(\lambda) := \phi(\lambda/N) \quad \text{and} \quad \psi_N(\lambda) := \phi_N(\lambda) - \phi_{N/2}(\lambda).$$

Clearly, $\{\psi_N(\lambda)\}_{N\in 2^{\mathbb{Z}}}$ forms a partition of unity for $\lambda \in (0, \infty)$. We define the Littlewood–Paley projections as follows:

$$P^a_{\le N} := \phi_N(\sqrt{\mathcal{L}_a}), \quad P^a_N := \psi_N(\sqrt{\mathcal{L}_a}) \quad \text{and} \quad P^a_{>N} := I - P^a_{\le N}.$$

We also define another family of Littlewood–Paley projections via the heat kernel as follows:

$$\tilde{P}^a_{\le N} := e^{-\mathcal{L}_a/N^2}, \quad \tilde{P}^a_N := e^{-\mathcal{L}_a/N^2} - e^{-4\mathcal{L}_a/N^2} \quad \text{and} \quad \tilde{P}^a_{>N} := I - \tilde{P}^a_{\le N}.$$

Lemma 1.5.2 (Bernstein estimates, [142]). *Let* $1 < p \le q \le \infty$ *when* $a \ge 0$ *and let* $r_0 < p \le q < r_0' = \frac{n}{\sigma}$ *when* $-(\frac{n-2}{2})^2 \le a < 0$. *Then:*

(1) *The operators* $P^a_{\le N}$, P^a_N, $\tilde{P}^a_N$ *and* $\tilde{P}^a_{\le N}$ *are bounded on* L^p;

(2) $P^a_{\le N}$, P^a_N, $\tilde{P}^a_N$ *and* $\tilde{P}^a_{\le N}$ *are bounded from* L^p *to* L^q *with norm* $O(N^{\frac{n}{p}-\frac{n}{q}})$;

(3) $N^s\|P^a_N f\|_{L^p(\mathbb{R}^n)} \sim \|(\mathcal{L}_a)^{\frac{s}{2}} P^a_N f\|_{L^p(\mathbb{R}^n)}$ *for all* $f \in C_c^\infty(\mathbb{R}^n)$ *and all* $s \in \mathbb{R}$.

Theorem 1.5.4 (Square function estimates). *Fix* $s \ge 0$. *Let* $1 < p < \infty$ *when* $a \ge 0$ *and let* $r_0 < p < r_0' = \frac{n}{\sigma}$ *when* $-(\frac{n-2}{2})^2 \le a < 0$. *Then for any* $f \in C_c^\infty(\mathbb{R}^n)$,

$$\left\|\left(\sum_{N\in 2^{\mathbb{Z}}} N^{2s}|P^a_N f|^2\right)^{\frac{1}{2}}\right\|_{L^p(\mathbb{R}^n)} \sim \|\mathcal{L}_a^{\frac{s}{2}} f\|_{L^p(\mathbb{R}^n)} \sim \left\|\left(\sum_{N\in 2^{\mathbb{Z}}} N^{2s}|(\tilde{P}^a_N)^k f|^2\right)^{\frac{1}{2}}\right\|_{L^p(\mathbb{R}^n)},$$

provided the integer $k \ge 1$ *satisfies* $2k > s$.

Note that the function $\lambda \mapsto e^{-\lambda^2/N^2} - e^{-4\lambda^2/N^2}$ used to define $\tilde{P}_N^a$ only vanishes to second order at $\lambda = 0$. The restriction $k > s/2$ ensures that $N^s(\mathcal{L}_a)^{-s}(\tilde{P}_N^a)^k$ is actually a Mikhlin multiplier.

1.5.1 The wave operator

In this part, we state the $W^{s,p}$-boundedness for stationary wave operators of the Schrödinger operator with inverse-square potential $\mathcal{L}_a$ in dimensions $n \geq 2$, which was established in Miao–Su–Zheng [190]. In the time-dependent scattering theory, for a real-valued potential V, the wave operators for the pair of self-adjoint operators $H_0 = -\Delta$ and $H = H_0 + V$ are defined as

$$W_\pm = \text{s-}\lim_{t\to\pm\infty} e^{itH}e^{-itH_0}, \tag{1.98}$$

$$W_\pm^* = \text{s-}\lim_{t\to\pm\infty} e^{itH_0}e^{-itH}, \tag{1.99}$$

where s-lim indicates the strong limit in $L^2(\mathbb{R}^d)$. In [127], Kato and Kuroda derived the following formula for the wave operators in the stationary theory:

$$W_\pm = \int_{-\infty}^{\infty} \frac{dE(\lambda)}{d\lambda}(H - \lambda)R_0(\lambda \pm i0)\, d\lambda, \tag{1.100}$$

$$W_\pm^* = \int_{-\infty}^{\infty} \frac{dE_0(\lambda)}{d\lambda}(H_0 - \lambda)R(\lambda \pm i0)\, d\lambda, \tag{1.101}$$

where $\mathrm{d}E(\lambda)$, $\mathrm{d}E_0(\lambda)$, $R(\lambda \pm i0)$, $R_0(\lambda \pm i0)$ are the spectral measures and resolvents associated to H and H_0, respectively. The authors in [127] proved that under some restrictive assumptions, the wave operators as defined in the time-dependent theory (1.98), (1.99) and stationary theory (1.100), (1.101) are identical.

For suitable potentials V, for example, short range potentials, one can prove that the $W_\pm$ and $W_\pm^*$ from the time-dependent theory exist and are complete; see Agmon [1]. This implies that they are isometries from $L^2(\mathbb{R}^n)$ to $L^2(\mathbb{R}^n)$, and that there is no singular continuous spectrum. It is natural to ask: for what $p \neq 2$ are the wave operators $W_\pm$, $W_\pm^*$ bounded in $L^p(\mathbb{R}^n)$, or bounded in $W^{s,p}(\mathbb{R}^n)$? The $W^{s,p}$-boundedness of the wave operators was first investigated by Yajima [286–290] under the assumption that zero energy is neither a eigenvalue nor a resonance. These results are established by using the perturbation method and the resolvent identity. However, the above results only apply to a restricted class of potentials. To relax the conditions, Beceanu and Schlag recently [12] established a structure formula for wave operators using an abstract version of Wiener's theorem, which implies the L^p-boundedness for more general potentials.

Recently, in [190], the authors proved the $W^{s,p}$-boundedness of the stationary wave operators associated to the operator $\lambda = -\Delta + \frac{a}{|x|^2}$ for $a \geq -\frac{(n-2)^2}{4}$ in dimensions $n \geq 2$ with suitable $p \in [1, \infty]$ and $s \in \mathbb{R}$.

Theorem 1.5.5 (L^p-boundedness). *Let $n \geq 2$ and $a \geq -\frac{(n-2)^2}{4}$. Then the stationary wave operators $W_\pm$ and $W_\pm^*$ are bounded in $L^p(\mathbb{R}^n)$ with*

$$\max\left\{0, \frac{\sigma}{n}\right\} < \frac{1}{p} < \min\left\{1, \frac{n-\sigma}{d}\right\}, \tag{1.102}$$

where $\sigma = \frac{n-2}{2} - \frac{1}{2}\sqrt{(n-2)^2 + 4a}$.

Remark 1.5.1. The condition (1.102) is equivalent to

$$\begin{cases} 1 < p < +\infty & \text{if} \quad a > 0, \\ p_0' < p < p_0 & \text{if} \quad a < 0, \end{cases} \tag{1.103}$$

where $p_0 = \frac{n}{\sigma}$.

Theorem 1.5.6 ($W^{s,p}$-boundedness). *Assume that $n \geq 2$ and $a \geq -\frac{(n-2)^2}{4}$.*

1. *Let $-n < \alpha < 2 + 2\nu_0$. The stationary wave operators $W_\pm$ are bounded in $W^{\alpha,p}(\mathbb{R}^n)$ if p satisfies*

$$\max\left\{0, \frac{\sigma}{n}, \frac{\sigma+\alpha}{n}\right\} < \frac{1}{p} < \min\left\{1, \frac{n-\sigma}{n}, \frac{n+\alpha}{n}\right\}. \tag{1.104}$$

2. *Let $-2\nu_0 - 2 < \beta < n$. The stationary wave operators $W_\pm^*$ are bounded in $W^{\beta,p}(\mathbb{R}^n)$ if p satisfies*

$$\max\left\{0, \frac{\sigma}{n}, \frac{\beta}{n}\right\} < \frac{1}{p} < \min\left\{1, \frac{n-\sigma}{n}, \frac{n-\sigma+\beta}{n}\right\}. \tag{1.105}$$

1.5.2 Applications

Since $\mathcal{L}_a$ is a positive self-adjoint operator on $L^2(\mathbb{R}^n)$, by the spectral theorem, for any bounded Borel function F, we can define the operator $F(\mathcal{L}_a)$ by the formula

$$F(\mathcal{L}_a) = \int_0^\infty F(\lambda)\, \mathrm{d}E_{\mathcal{L}_a}(\lambda).$$

By the intertwining property $W_\pm(-\Delta) = \mathcal{L}_a W_\pm$ and the spectral theorem, we have $W_\pm\, \mathrm{d}E_{-\Delta}(\lambda) = \mathrm{d}E_{\mathcal{L}_a}(\lambda) W_\pm$. Hence,

$$F(\mathcal{L}_a) = W_\pm F(H_0) W_\pm^*. \tag{1.106}$$

Application 1: Dispersive estimates

Let $e^{it\mathcal{L}_a}f$ be the solution of the free Schrödinger equation with inverse-square potential and initial data f. Using (1.106) and the L^p-$L^{p'}$ estimates for the free Schrödinger operator $e^{-it\Delta}$,

$$\|e^{-it\Delta}f\|_{L^{p'}(\mathbb{R}^n)} \lesssim |t|^{-n(\frac{1}{p}-\frac{1}{2})}\|f\|_{L^p(\mathbb{R}^n)}, \quad \forall\, t \neq 0$$

with $1 \leq p \leq 2$, and by Theorem 1.5.5, we obtain the dispersive estimates for the Schrödinger operator $e^{-it\mathcal{L}_a}$.

Corollary 1.5.1 (Dispersive estimate for Schrödinger equations). *Let $n \geq 2$, $a \geq -\frac{(n-2)^2}{4}$ and $\frac{1}{2} \leq \frac{1}{p} < \min\{1, \frac{n-\sigma}{n}\}$. Then there holds*

$$\|e^{-it\mathcal{L}_a}f\|_{L^{p'}(\mathbb{R}^n)} \lesssim |t|^{-n(\frac{1}{p}-\frac{1}{2})}\|f\|_{L^p(\mathbb{R}^n)}, \quad \forall\, t \neq 0. \tag{1.107}$$

Remark 1.5.2. Fanelli, Felli, Fontelos and Primo have proven (1.107) with $n \in \{2, 3\}$ in [88, 89].

Application 2: Strichartz estimates

Using the classical Strichartz estimates for the free Schrödinger operator $e^{-it\Delta}$ (see Keel–Tao [129] and Strichartz [250]),

$$\|e^{-it\Delta}f\|_{L_t^qL_x^r(\mathbb{R}\times\mathbb{R}^n)} \lesssim \|f\|_{L^2(\mathbb{R}^n)},$$

with $(q, r) \in \Lambda_0 = \{(q, r) : \; q, r \geq 2, \; \frac{2}{q} = n(\frac{1}{2} - \frac{1}{r}), \; (q, r, n) \neq (2, \infty, 2)\}$, one can recover the Strichartz estimates for $e^{it\mathcal{L}_a}$, which was first proven by Burq, Planchon, Stalker and Tahvildar-Zadeh [29].

Corollary 1.5.2 (Strichartz estimates for Schrödinger equations). *Let $n \geq 2$ and $a \geq -\frac{(n-2)^2}{4}$. Then for any $(q, r) \in \Lambda_0$, we have*

$$\|e^{it\mathcal{L}_a}f\|_{L_t^qL_x^r(\mathbb{R}\times\mathbb{R}^n)} \lesssim \|f\|_{L^2(\mathbb{R}^n)}.$$

Application 3: Uniform resolvent inequality

For the Laplacian $-\Delta$, the uniform Sobolev inequality has been proved by Kenig–Ruiz–Sogge [138] and is given as follows: for $z \in \mathbb{C} \setminus \mathbb{R}^+$ and $f \in C_0^\infty(\mathbb{R}^n)$,

$$\|(-\Delta - z)^{-1}f\|_{L^q(\mathbb{R}^n)} \leq C|z|^{\frac{n}{2}(\frac{1}{p}-\frac{1}{q})-1}\|f\|_{L^p(\mathbb{R}^n)}, \tag{1.108}$$

where $n \geq 3$ and (p, q) satisfies the conditions

$$\frac{2}{n+1} \leq \frac{1}{p} - \frac{1}{q} \leq \frac{2}{n}, \quad \frac{2n}{n+3} < p < \frac{2n}{n+1}, \quad \frac{2n}{n-1} < q < \frac{2n}{n-3}. \tag{1.109}$$

See also the paper of Gutiérrez [114], which proved that the condition (1.109) is sharp. More recently, Evéquoz [86] applied the method of [114] to show the uniform resolvent estimate (1.108) in dimension $n = 2$ provided that $(1/p, 1/q)$ is contained in the pentagon

$$\left\{\left(\frac{1}{p}, \frac{1}{q}\right) : \frac{2}{3} \leq \frac{1}{p} - \frac{1}{q} < 1, \frac{3}{4} < \frac{1}{p} \leq 1, 0 \leq \frac{1}{q} < \frac{1}{4}\right\}. \tag{1.110}$$

As a consequence of Theorem 1.5.5, one can extend (1.108) to the operator $\mathcal{L}_a$. Let

$$\nu_0 = \sqrt{\left(\frac{n-2}{2}\right)^2 + a}, \quad \mu_a = \min\left\{\frac{1}{2}, \nu_0\right\}.$$

Then we have the following result.

Corollary 1.5.3 (Uniform resolvent inequality). *For $n \geq 3$, we assume that*

$$\frac{2}{n+1} \leq \frac{1}{p} - \frac{1}{q} \leq \frac{2}{n}, \quad \frac{2n}{n+2(1+\mu_a)} < p < \frac{2n}{n+1}, \quad \frac{2n}{n-1} < q < \frac{2n}{n-2(1+\mu_a)}. \tag{1.111}$$

For $n = 2$, we assume that $(1/p, 1/q)$ is as in (1.110). Then there exists a positive constant C such that for $z \in \mathbb{C} \setminus \mathbb{R}^+, f \in C_0^\infty(\mathbb{R}^n)$, we have

$$\|(\mathcal{L}_a - z)^{-1} f\|_{L^q(\mathbb{R}^n)} \leq C|z|^{\frac{n}{2}(\frac{1}{p}-\frac{1}{q})-1} \|f\|_{L^p(\mathbb{R}^n)}. \tag{1.112}$$

Remark 1.5.3. For $n \geq 3$, the estimate (1.112) was first proved by Bouclet and Mizutani [16] and Mizutani [201] under (1.111) and an extra assumption $\frac{1}{p} + \frac{1}{q} = 1$. Mizutani–Zhang–Zheng [202] proved Corollary 1.5.3 for the case

$$\mu_a = \begin{cases} 1/2, & \nu_0 \geq 1/2; \\ \tilde{\nu}_0, & 0 < \nu_0 < 1/2, \end{cases}$$

with $\tilde{\nu}_0 = \frac{\nu_0^2}{1-2\nu_0^2}$. The result of Corollary 1.5.3 extends the above results to the case $\mu_a = \min\{\frac{1}{2}, \nu_0\}$ since $\tilde{\nu}_0 < \nu_0$ for $\nu_0 < 1/2$, see Figure 1.1.

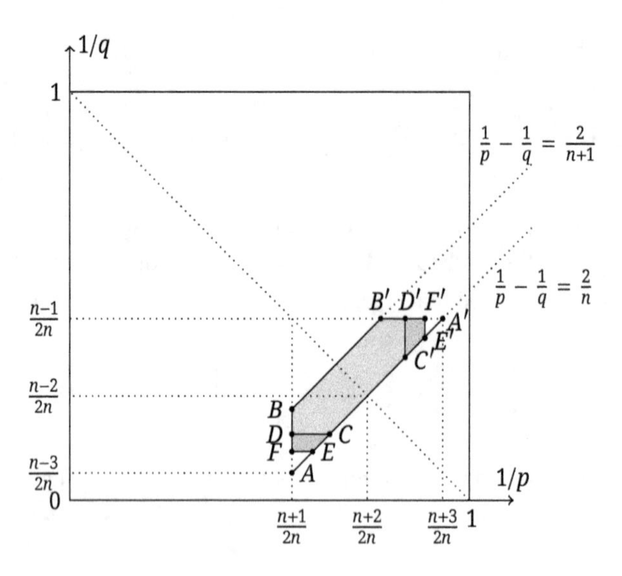

Figure 1.1: An illustration of condition (1.111) of Corollary 1.5.3. The two trapeziums $CDFE$ and $C'D'F'E'$ are new when $-\frac{(n-2)^2}{4} < a < -\frac{(n-2)^2}{4} + \frac{1}{4}$.

Application 4: Multiplier theorem

It follows from [241, Theorem 0.2.6] that $m(\sqrt{-\Delta})$ is bounded in $L^p(\mathbb{R}^n)$ provided

$$\sum_{0\le|\alpha|<k} \sup_{\lambda>0} \lambda^{-d} \| |\lambda|^{\alpha} D^{\alpha}(\chi(\cdot/\lambda)m(\cdot))\|^2_{L^2(\mathbb{R}^n)} < \infty, \tag{1.113}$$

for some integer $k > \frac{n}{2}$, whenever $\chi \in C_0^\infty(\mathbb{R}^n \setminus \{0\})$. This, together with L^p-boundedness of wave operators, yields the following multiplier theorem for $\mathcal{L}_a$.

Corollary 1.5.4. *Let $m \in L^\infty(\mathbb{R}^n)$ satisfy the condition* (1.113). *Then $m(\sqrt{\mathcal{L}_a})$ is bounded in $L^p(\mathbb{R}^n)$ for $1 < p < \infty$ when $a > 0$, and for $p_0' < p < p_0 = \frac{n}{\sigma}$ when $-\frac{(n-2)^2}{2} \le a < 0$.*

Remark 1.5.4. Noting that the condition

$$|\partial_\lambda^j m(\lambda)| \lesssim \lambda^{-j}, \quad \forall j \ge 0$$

implies that m satisfies the assumption (1.113), Corollary 1.5.4 also implies the Mikhlin-type multiplier theorem, that is, Theorem 1.5.1.

1.5.3 Local smoothing estimate

We state a local smoothing result for the linear propagator $e^{-it\mathcal{L}_a}$.

Lemma 1.5.3 (Local smoothing). *Let $n = 3$. Fix $a > -\frac{1}{4}$ and let $u = e^{-it\mathcal{L}_a}u_0$. Then*

$$\int_{\mathbb{R}}\int_{\mathbb{R}^3} \frac{|\nabla u(t,x)|^2}{R\langle R^{-1}x\rangle^3} + \frac{|u(t,x)|^2}{R|x|^2}\,dx\,dt \lesssim \|u_0\|_{L^2_x}\|\nabla u_0\|_{L^2_x} + R^{-1}\|u_0\|^2_{L^2_x}, \tag{1.114}$$

$$\int_{\mathbb{R}}\int_{|x-z|\le R} \frac{1}{R}|\nabla u(t,x)|^2\,dx\,dt \lesssim \|u_0\|_{L^2_x}\|\nabla u_0\|_{L^2_x} + R^{-1}\|u_0\|^2_{L^2_x}, \tag{1.115}$$

uniformly for $z \in \mathbb{R}^3$ and $R > 0$.

Proof. Theorem 1 of [29] shows that

$$\int_{\mathbb{R}}\int_{\mathbb{R}^3} \frac{|u(t,x)|^2}{R|x|^2}\,dx\,dt \lesssim R^{-1}\|u_0\|^2_{L^2_x}. \tag{1.116}$$

To complete the proof of (1.114), we use a weighted momentum identity analogous to the virial and Morawetz identities. To this end, let

$$F(t) := \int_{\mathbb{R}^3} 2\Im(\bar{u}\partial_j u)\, b_j(x)\,dx \quad \text{with} \quad b_j(x) = \frac{x_j/R}{\langle x/R\rangle},$$

where repeated indices are summed. Clearly,

$$|F(t)| \le 2\|u_0\|_{L^2_x}\|\nabla u_0\|_{L^2_x}, \tag{1.117}$$

while a direct computation shows

$$\partial_t F(t) = \int_{\mathbb{R}^3} -b_{jjkk}|u|^2 + 4b_{jk}\bar{u}_j u_k + 4ab_j x_j \frac{|u|^2}{|x|^4}\,dx,$$

where additional subscripts indicate differentiation and repeated indices are summed.

For our particular choice of b, we have

$$-b_{jjkk}(x) = \frac{15}{R^3\langle x/R\rangle^7} \ge 0, \quad 4ab_j(x)\frac{x_j}{|x|^4} \ge -\frac{4|a|}{R|x|^2},$$

and

$$b_{jk}(x) = \frac{\delta_{jk}}{R\langle x/R\rangle} - \frac{x_j x_k}{R^3\langle x/R\rangle^3} \ge \frac{\delta_{jk}}{R\langle x/R\rangle^3}$$

in the sense of matrices. The estimate (1.114) now follows by applying the fundamental theorem of calculus and using (1.116) to bound the potential term, which has an unfavorable sign when $a < 0$.

We now turn our attention to (1.115). When $|z| \leq 2R$, this estimate follows from (1.114). Thus, we need only consider the case $|z| \geq 2R$. We use the weighted momentum identity once more with choosing

$$b_j(x) = \frac{(x_j - z_j)/R}{\langle (x-z)/R \rangle}\phi(x/R),$$

where $\phi(y) \geq 0$ is a smooth function vanishing when $|y| \leq \frac{1}{2}$ and obeying $\phi(y) = 1$ when $|y| \geq 1$. For this choice of weight, we again have (1.117). Moreover,

$$-b_{jjkk}(x) \gtrsim_\phi -\frac{1}{R|x|^2}, \quad 4ab_j(x)\frac{x_j}{|x|^4} \geq -\frac{8|a|}{R|x|^2},$$

and

$$b_{jk}(x) \geq \frac{\delta_{jk}}{R\langle (x-z)/R\rangle^3}\phi(x/R) - \frac{1}{R}|[\nabla\phi](x/R)|.$$

The estimate (1.115) now follows by applying the fundamental theorem of calculus, using (1.114) to bound those terms with an unfavorable sign. □

The following corollary will be used to prove a Palais–Smale condition for minimizing sequences of blowup solutions in Chapter 6.

Corollary 1.5.5. *Fix $a > -\frac{1}{4} + \frac{1}{25}$ and let $w_0 \in \dot{H}^1_a(\mathbb{R}^3)$. Then*

$$\begin{aligned}
&\|\nabla e^{-it\mathcal{L}_a} w_0\|_{L^5_t L^{\frac{15}{8}}_x([\tau-T,\tau+T]\times\{|x-z|\leq R\})} \\
&\lesssim T^{\frac{29}{320}} R^{\frac{51}{160}} \|e^{-it\mathcal{L}_a} w_0\|^{\frac{1}{32}}_{L^{10}_{t,x}(\mathbb{R}\times\mathbb{R}^3)} \|w_0\|^{\frac{31}{32}}_{\dot{H}^1_x} \\
&\quad + T^{\frac{29}{280}} R^{\frac{41}{140}} \|e^{-it\mathcal{L}_a} w_0\|^{\frac{1}{28}}_{L^{10}_{t,x}(\mathbb{R}\times\mathbb{R}^3)} \|w_0\|^{\frac{27}{28}}_{\dot{H}^1_x},
\end{aligned}$$

uniformly in w_0 and the parameters $R, T > 0$, $\tau \in \mathbb{R}$ and $z \in \mathbb{R}^3$.

Proof. Replacing w_0 by $e^{-i\tau\mathcal{L}_a} w_0$, we see that it suffices to treat the case $\tau = 0$.

By Hölder's inequality, Theorem 1.5.3 and the Strichartz estimates,

$$\begin{aligned}
&\|\nabla e^{-it\mathcal{L}_a} w_0\|_{L^5_t L^{\frac{15}{8}}_x([-T,T]\times\{|x-z|\leq R\})} \\
&\lesssim R^{\frac{1}{4}} \|\nabla e^{-it\mathcal{L}_a} w_0\|^{\frac{3}{4}}_{L^{10}_t L^{\frac{30}{13}}_x([-T,T]\times\mathbb{R}^3)} \|\nabla e^{-it\mathcal{L}_a} w_0\|^{\frac{1}{4}}_{L^2_{t,x}([-T,T]\times\{|x-z|\leq R\})} \\
&\lesssim R^{\frac{1}{4}} \|w_0\|^{\frac{3}{4}}_{\dot{H}^1(\mathbb{R}^3)} \|\nabla e^{-it\mathcal{L}_a} w_0\|^{\frac{1}{4}}_{L^2_{t,x}([-T,T]\times\{|x-z|\leq R\})}.
\end{aligned}$$

To continue, we consider small and high frequencies separately. Fix $N \in 2^{\mathbb{Z}}$. Using Theorem 1.5.3, the Hölder inequality and the estimates of Bernstein and Strichartz, we can estimate at the low frequencies as follows:

$$
\begin{aligned}
&\|\nabla e^{-it\mathcal{L}_a} P^a_{<N} w_0\|_{L^2_{t,x}([-T,T]\times\{|x-z|\le R\})} \\
&\lesssim T^{\frac{29}{60}} R^{\frac{1}{5}} \|\nabla e^{-it\mathcal{L}_a} P^a_{<N} w_0\|_{L^{60}_t L^{\frac{30}{13}}_x(\mathbb{R}\times\mathbb{R}^3)} \\
&\lesssim T^{\frac{29}{60}} R^{\frac{1}{5}} N^{\frac{1}{6}} \|(\mathcal{L}_a)^{\frac{5}{12}} e^{-it\mathcal{L}_a} P^a_{<N} w_0\|_{L^{60}_t L^{\frac{30}{13}}_x(\mathbb{R}\times\mathbb{R}^3)} \\
&\lesssim T^{\frac{29}{60}} R^{\frac{1}{5}} N^{\frac{1}{6}} \|e^{-it\mathcal{L}_a} w_0\|^{\frac{1}{6}}_{L^{10}_{t,x}(\mathbb{R}\times\mathbb{R}^3)} \|(\mathcal{L}_a)^{\frac{1}{2}} e^{-it\mathcal{L}_a} w_0\|^{\frac{5}{6}}_{L^\infty_t L^2_x(\mathbb{R}\times\mathbb{R}^3)} \\
&\lesssim T^{\frac{29}{60}} R^{\frac{1}{5}} N^{\frac{1}{6}} \|e^{-it\mathcal{L}_a} w_0\|^{\frac{1}{6}}_{L^{10}_{t,x}(\mathbb{R}\times\mathbb{R}^3)} \|w_0\|^{\frac{5}{6}}_{\dot H^1(\mathbb{R}^3)}.
\end{aligned}
$$

To estimate the case at the high frequencies, we use Lemma 1.5.3 and the Bernstein estimate to get

$$
\begin{aligned}
&\|\nabla e^{-it\mathcal{L}_a} P^a_{\ge N} w_0\|^2_{L^2_{t,x}([-T,T]\times\{|x-z|\le R\})} \\
&\lesssim R\|P^a_{\ge N} w_0\|_{L^2_x} \|\nabla P^a_{\ge N} w_0\|_{L^2_x} + \|P^a_{\ge N} w_0\|^2_{L^2_x} \\
&\lesssim (RN^{-1} + N^{-2})\|w_0\|^2_{\dot H^1(\mathbb{R}^3)}.
\end{aligned}
$$

The required estimate follows by optimizing the choice of N. □

1.5.4 Linear profile decomposition in $\dot H^1_a(\mathbb{R}^3)$

In this part, we aim to establish the linear profile decomposition for the propagator $e^{-it\mathcal{L}_a}$ associated to bounded sequences in $\dot H^1_a(\mathbb{R}^3)$, which was established in Killip–Miao–Visan–Zhang–Zheng [141]. This will play an important role in finding the existence of a minimal blowup solution in Chapter 6. To do it, we first give some notation.

Definition 1.5.1. Given a sequence $\{y_n\} \subset \mathbb{R}^3$, we define

$$
\mathcal{L}^n_a := -\Delta + \frac{a}{|x+y_n|^2} \quad \text{and} \quad \mathcal{L}^\infty_a := \begin{cases} -\Delta + \frac{a}{|x+y_\infty|^2} & \text{if } y_n \to y_\infty \in \mathbb{R}^3, \\ -\Delta & \text{if } |y_n| \to \infty. \end{cases}
$$

Theorem 1.5.7 ($\dot H^1_a(\mathbb{R}^3)$ linear profile decomposition). *Let $\{f_n\}$ be a bounded sequence in $\dot H^1_a(\mathbb{R}^3)$. After passing to a subsequence, there exist $J^* \in \{0,1,2,\dots,\infty\}$, nonzero profiles $\{\phi^j\}_{j=1}^{J^*} \subset \dot H^1(\mathbb{R}^3)$, $\{\lambda^j_n\}_{j=1}^{J^*} \subset (0,\infty)$ and $\{(t^j_n, x^j_n)\}_{j=1}^{J^*} \subset \mathbb{R}\times\mathbb{R}^3$ such that for each finite $0 \le J \le J^*$, we have the decomposition*

$$
f_n = \sum_{j=1}^{J} \phi^j_n + w^J_n \quad \text{with} \quad \phi^j_n = G^j_n[e^{-it^j_n \mathcal{L}^{n_j}_a} \phi^j] \quad \text{and} \quad w^J_n \in \dot H^1_a(\mathbb{R}^3)
$$

satisfying

$$\lim_{J\to J^*}\limsup_{n\to\infty}\|e^{-it\mathcal{L}_a}w_n^J\|_{L^{10}_{t,x}(\mathbb{R}\times\mathbb{R}^3)}=0,$$

$$\lim_{n\to\infty}\left\{\|f_n\|^2_{\dot H^1_a(\mathbb{R}^3)}-\sum_{j=1}^J\|\phi_n^j\|^2_{\dot H^1_a(\mathbb{R}^3)}-\|w_n^J\|^2_{\dot H^1_a(\mathbb{R}^3)}\right\}=0,$$

$$\lim_{n\to\infty}\left\{\|f_n\|^6_{L^6_x(\mathbb{R}^3)}-\sum_{j=1}^J\|\phi_n^j\|^6_{L^6_x(\mathbb{R}^3)}-\|w_n^J\|^6_{L^6_x(\mathbb{R}^3)}\right\}=0.$$

Here, $\mathcal{L}_a^{n_j}$ is as in Definition 1.5.1 with $y_n^j=\frac{x_n^j}{\lambda_n^j}$ and $[G_n^j h](x):=(\lambda_n^j)^{-\frac12}h(\frac{x-x_n^j}{\lambda_n^j})$. Moreover, for all $j\neq k$, we have the asymptotic orthogonality property

$$\left|\log\frac{\lambda_n^j}{\lambda_n^k}\right|+\frac{|x_n^j-x_n^k|^2}{\lambda_n^j\lambda_n^k}+\frac{|t_n^j(\lambda_n^j)^2-t_n^k(\lambda_n^k)^2|}{\lambda_n^j\lambda_n^k}\to\infty\quad as\quad n\to\infty. \tag{1.118}$$

In addition, we may assume that for each j either $t_n^j\equiv 0$ or $t_n^j\to\pm\infty$.

The following typifies the manner in which the operators $\mathcal{L}_a^n$ appear: for any $y_n\in\mathbb{R}^3$ and $N_n>0$,

$$N_n^{\frac12}e^{-it\mathcal{L}_a}[\phi(N_nx-y_n)]=N_n^{\frac12}[e^{-iN_n^2t\mathcal{L}_a^n}\phi](N_nx-y_n). \tag{1.119}$$

The operator $\mathcal{L}_a^\infty$ appears as the limit of the operators $\mathcal{L}_a^n$ in various manners that we now justify.

Lemma 1.5.4 (Convergence of operators). *Fix $a>-\frac14$. Suppose we are given sequences $t_n\to t\in\mathbb{R}$ and $y_n\to y_\infty\in\mathbb{R}^3\cup\{\infty\}$. With $\mathcal{L}_a^n$ and $\mathcal{L}_a^\infty$ as in Definition 1.5.1, we have*

$$\lim_{n\to\infty}\|\mathcal{L}_a^n\psi-\mathcal{L}_a^\infty\psi\|_{\dot H^{-1}(\mathbb{R}^3)}=0\quad for\ all\quad \psi\in\dot H^1(\mathbb{R}^3), \tag{1.120}$$

$$\lim_{n\to\infty}\|(e^{it_n\mathcal{L}_a^n}-e^{it\mathcal{L}_a^\infty})\psi\|_{\dot H^{-1}(\mathbb{R}^3)}=0\quad for\ all\quad \psi\in\dot H^{-1}(\mathbb{R}^3), \tag{1.121}$$

$$\lim_{n\to\infty}\|[\sqrt{\mathcal{L}_a^n}-\sqrt{\mathcal{L}_a^\infty}]\psi\|_{L^2(\mathbb{R}^3)}=0\quad for\ all\quad \psi\in\dot H^1(\mathbb{R}^3), \tag{1.122}$$

and for all admissible pairs $\frac2q+\frac3r=\frac32$ with $2<q\le\infty$ we have

$$\lim_{n\to\infty}\|(e^{it\mathcal{L}_a^n}-e^{it\mathcal{L}_a^\infty})\psi\|_{L_t^qL_x^r(\mathbb{R}\times\mathbb{R}^3)}=0\quad for\ all\quad \psi\in L^2(\mathbb{R}^3). \tag{1.123}$$

If, in addition, $y_\infty\neq 0$, then

$$\lim_{n\to\infty}\|[e^{-\mathcal{L}_a^n}-e^{-\mathcal{L}_a^\infty}]\delta_0\|_{\dot H^{-1}(\mathbb{R}^3)}=0. \tag{1.124}$$

Proof. By Theorem 1.5.3, the operators $\mathcal{L}_a^n$ and $\mathcal{L}_a^\infty$ are isomorphisms of $\dot H_x^1$ onto $\dot H_x^{-1}$ with bounds independent of n. Consequently, it suffices to prove (1.120) for the dense subclass

$C_c^\infty(\mathbb{R}^3 \setminus \{y_\infty\})$ of $\dot{H}^1(\mathbb{R}^3)$. (In the case $y_\infty = \infty$, we may understand $\mathbb{R}^3 \setminus \{y_\infty\}$ as $\mathbb{R}^3$.) In this subclass, convergence in $L_x^{6/5} \hookrightarrow \dot{H}_x^{-1}$ is trivial.

We now prove (1.123) for the case $q = \infty$ and $r = 2$. Note that all other cases of (1.123) follow by interpolation with the end-point Strichartz inequality. Note also that the Strichartz inequality allows us to assume that $\psi \in C_c^\infty(\mathbb{R}^3)$.

Let us assume first that $y_n \to \infty$ so that $\mathcal{L}_a^\infty = -\Delta$. By the Duhamel formula and the Strichartz inequality, we obtain that

$$\|(e^{it\mathcal{L}_a^n} - e^{-it\Delta})\psi\|_{L_t^\infty L_x^2(\mathbb{R}\times\mathbb{R}^3)} \lesssim \left\| \frac{a}{|x+y_n|^2} e^{-it\Delta}\psi \right\|_{L_{t,x}^{\frac{10}{7}}(\mathbb{R}\times\mathbb{R}^3)}. \tag{1.125}$$

On the other hand, as $\psi \in C_c^\infty(\mathbb{R}^3)$, we have

$$|[e^{-it\Delta}\psi](x)| \lesssim_\psi \langle t\rangle^{-3/2}\left(1 + \frac{|x|}{\langle t\rangle}\right)^{-m}$$

for any $m > 0$. That RHS of (1.125) converges to zero as $n \to \infty$ follows easily from this and the fact that $y_n \to \infty$.

Let us assume now that $y_n \to y_\infty \in \mathbb{R}^3$. By translating ψ if necessary, we may assume that $y_\infty = 0$. This implies that $\mathcal{L}_a^\infty = \mathcal{L}_a$. Adopting the notation

$$V_n(x) = \frac{a}{|x+y_n|^2} - \frac{a}{|x|^2} = -a\frac{y_n \cdot (2x + y_n)}{|x|^2|x+y_n|^2}$$

and proceeding as above with using the Duhamel formula and the Strichartz inequality, we obtain that

$$\|(e^{it\mathcal{L}_a^n} - e^{it\mathcal{L}_a})\psi\|_{L_t^\infty L_x^2(\mathbb{R}\times\mathbb{R}^3)} \lesssim \|V_n e^{it\mathcal{L}_a}\psi\|_{L_t^2 L_x^{\frac{6}{5}}(\mathbb{R}\times\mathbb{R}^3)}. \tag{1.126}$$

For arbitrarily given $\epsilon > 0$ and $|y_n| < \frac{\epsilon}{2}$, we have

$$\begin{aligned} \text{the RHS of (1.126)} &\lesssim \|V_n\|_{L_x^{\frac{15}{11}}(\{|x|\le\epsilon\})} \|e^{it\mathcal{L}_a}\psi\|_{L_t^2 L_x^{10}} + \|V_n\|_{L_x^{\frac{3}{2}}(\{|x|\ge\epsilon\})} \|e^{it\mathcal{L}_a}\psi\|_{L_t^2 L_x^6} \\ &\lesssim \epsilon^{\frac{1}{5}} \|(\mathcal{L}_a)^{1/10}\psi|_{L_x^2} + |y_n|\epsilon^{-1}\|\psi\|_{L_x^2} \\ &\lesssim_\psi \epsilon^{\frac{1}{5}} + o(1) \quad \text{as} \quad n \to \infty. \end{aligned}$$

As $\epsilon > 0$ is arbitrary, we deduce that RHS of (1.126) $\to 0$ as $n \to \infty$, thereby completing the proof of (1.123).

We now prove (1.124). Note that $y_\infty \neq 0$ is necessary when $\sigma > 0$; otherwise, $e^{-\mathcal{L}_a^\infty}\delta_0 \equiv \infty$ as seen from Theorem 1.5.2. Let us consider first the case $y_n \to \infty$, for which $\mathcal{L}_a^\infty = -\Delta$. From Theorem 1.5.3, we see that $e^{-s\mathcal{L}_a^n}$ is bounded on $\dot{H}_x^{-1}$ uniformly in n and $s \ge 0$. From the Duhamel formula and the Sobolev embedding theorem, we may then deduce that

$$\|[e^{-\mathcal{L}_a^n} - e^{-\mathcal{L}_a^\infty}]\delta_0\|_{\dot{H}_x^{-1}} \lesssim \int_0^1 \left\| e^{-(1-s)\mathcal{L}_a^n} \frac{a}{|x+y_n|^2} e^{s\Delta}\delta_0 \right\|_{\dot{H}_x^{-1}} ds$$
$$\lesssim \left\| \frac{a}{|x+y_n|^2} e^{s\Delta}\delta_0 \right\|_{L_s^1 L_x^{6/5}([0,1]\times\mathbb{R}^3)}.$$

To estimate this inequality, we first observe that for $|y_n| \geq 2R > 0$ we have

$$\left\| \frac{a}{|x+y_n|^2} e^{s\Delta}\delta_0 \right\|_{L_x^{6/5}(\{|x+y_n|<R\})}$$
$$\lesssim \left\| \frac{a}{|x+y_n|^2} \right\|_{L_x^{6/5}(\{|x+y_n|<R\})} \|e^{s\Delta}\delta_0\|_{L_x^\infty(\{|x+y_n|<R\})}$$
$$\lesssim R^{1/2} s^{-3/2} e^{-c\frac{|y_n|^2}{s}},$$

which converges to zero in $L_s^1([0,1])$ as $n \to \infty$, independently of $R > 0$. On the other hand,

$$\left\| \frac{a}{|x+y_n|^2} e^{s\Delta}\delta_0 \right\|_{L_x^{6/5}(\{|x+y_n|>R\})} \lesssim \left\| \frac{a}{|x+y_n|^2} \right\|_{L_x^3(\{|x+y_n|>R\})} \|e^{s\Delta}\delta_0\|_{L_x^2(\mathbb{R}^3)}$$
$$\lesssim R^{-1} s^{-\frac{3}{4}},$$

which may be made arbitrarily small in $L_s^1([0,1])$ by choosing R large. This completes the proof of (1.124) in the case $y_n \to \infty$.

Suppose now that $y_n \to y_\infty \in \mathbb{R}^3 \setminus \{0\}$. Proceeding as in the previous case, the problem reduces to showing that

$$\|V_n e^{-s\mathcal{L}_a^\infty}\delta_0\|_{L_s^1 L_x^{6/5}([0,1]\times\mathbb{R}^3)} \to 0 \quad \text{as} \quad n \to \infty,$$

where

$$V_n(x) = \frac{a}{|x+y_n|^2} - \frac{a}{|x+y_\infty|^2} = a\frac{(y_\infty - y_n)\cdot(2x+y_n+y_\infty)}{|x+y_\infty|^2|x+y_n|^2}.$$

This can be shown in a way similar to that used in the previous case. Suppose $0 < \epsilon < \frac{1}{3}|y_\infty|$ and $|y_n - y_\infty| < \frac{1}{3}\epsilon$. Then

$$\|V_n e^{-s\mathcal{L}_a^\infty}\delta_0\|_{L_x^{6/5}(\{|x+y_\infty|>\epsilon\})} \lesssim \|V_n\|_{L_x^\infty(\{|x+y_\infty|>\epsilon\})} \|e^{-s\mathcal{L}_a^\infty}\delta_0\|_{L_x^{6/5}(\mathbb{R}^3)}$$
$$\lesssim |y_n - y_\infty| \epsilon^{-3} s^{-1/4},$$

which converges to zero in $L_s^1([0,1])$ as $n \to \infty$, independently of $\epsilon > 0$, while

$$
\begin{aligned}
&\|V_n e^{-s\mathcal{L}_a^\infty}\delta_0\|_{L_x^{6/5}(\{|x+y_\infty|<\epsilon\})} \\
&\quad \lesssim \|V_n\|_{L_x^{6/5}(\{|x+y_\infty|<\epsilon\})}\|e^{-s\mathcal{L}_a^\infty}\delta_0\|_{L_x^\infty(\{|x+y_\infty|<\epsilon\})} \\
&\quad \lesssim \epsilon^{1/2}s^{-3/2}\Big(1+\frac{\sqrt{s}}{\epsilon}\Big)^{2(\sigma\vee 0)} e^{-\frac{c}{s}|y_\infty|^2},
\end{aligned}
$$

which converges to zero in $L_s^1([0,1])$ as $\epsilon \to 0$. This completes the proof of (1.124).

We turn now to the proof of (1.122). Again it suffices to consider $\psi \in C_c^\infty(\mathbb{R}^3 \setminus \{y_\infty\})$. Given $\epsilon > 0$, let us define

$$
f_\epsilon(E) := \begin{cases} \sqrt{|E|} & \text{if} \quad |E| < \epsilon^{-2}, \\ \epsilon^{-1} & \text{if} \quad |E| \geq \epsilon^{-2}. \end{cases}
$$

By mollification and then Fourier inversion, there is a finite (signed) measure $d\mu_\epsilon$ (with $\mu_\epsilon(\{0\}) = \epsilon^{-1}$) so that

$$
\sup_{E\in\mathbb{R}}\Big|f_\epsilon(E) - \int e^{itE}\, d\mu_\epsilon(t)\Big| \leq \epsilon.
$$

Combining this with (1.123) yields

$$
\limsup_{n\to\infty}\|f_\epsilon(\mathcal{L}_a^\infty)\psi - f_\epsilon(\mathcal{L}_a^n)\psi\|_{L_x^2} \leq 2\epsilon\|\psi\|_{L_x^2}.
$$

On the other hand, using the fact that $|f_\epsilon(E) - \sqrt{|E|}| \leq \epsilon|E|$, we get

$$
\begin{aligned}
&\|\sqrt{\mathcal{L}_a^n}\psi - f_\epsilon(\mathcal{L}_a^n)\psi\|_{L_x^2} + \|\sqrt{\mathcal{L}_a^\infty}\psi - f_\epsilon(\mathcal{L}_a^\infty)\psi\|_{L_x^2} \\
&\quad \leq \epsilon\|\mathcal{L}_a^\infty\psi\|_{L_x^2} + \epsilon\|\mathcal{L}_a^n\psi\|_{L_x^2} \to 2\epsilon\|\mathcal{L}_a^\infty\psi\|_{L_x^2}
\end{aligned}
$$

as $n \to \infty$. As $\epsilon > 0$ is arbitrary, this completes the proof of (1.122).

Note that we also have the following analogue of (1.122):

$$
\lim_{n\to\infty}\|[\sqrt{\mathcal{L}_a^n} - \sqrt{\mathcal{L}_a^\infty}]\phi\|_{\dot{H}^{-1}(\mathbb{R}^3)} = 0 \quad \text{for all} \quad \phi \in L^2(\mathbb{R}^3). \tag{1.127}
$$

Indeed, any such $\phi \in L_x^2$ can be written as $\sqrt{\mathcal{L}_a^\infty}\psi$ for some $\psi \in \dot{H}_x^1$. The claim then follows by expanding

$$
[\sqrt{\mathcal{L}_a^n} - \sqrt{\mathcal{L}_a^\infty}]\phi = \sqrt{\mathcal{L}_a^n}[\sqrt{\mathcal{L}_a^\infty} - \sqrt{\mathcal{L}_a^n}]\psi + [\mathcal{L}_a^n - \mathcal{L}_a^\infty]\psi
$$

and then applying (1.122) and (1.120).

Now only one part of the lemma still requires justification, namely (1.121). Theorem 1.5.3 guarantees that any $\psi \in \dot{H}_x^{-1}$ can be written as $\sqrt{\mathcal{L}_a^\infty}\phi$ for some $\phi \in L_x^2$. Decomposing

$$(e^{it_n\mathcal{L}_a^n} - e^{it\mathcal{L}_a^\infty})\sqrt{\mathcal{L}_a^\infty}\phi = e^{it_n\mathcal{L}_a^n}[\sqrt{\mathcal{L}_a^\infty} - \sqrt{\mathcal{L}_a^n}]\phi + \sqrt{\mathcal{L}_a^n}[e^{it_n\mathcal{L}_a^n} - e^{it_n\mathcal{L}_a^\infty}]\phi$$
$$+ \sqrt{\mathcal{L}_a^n}[e^{it_n\mathcal{L}_a^\infty} - e^{it\mathcal{L}_a^\infty}]\phi + [\sqrt{\mathcal{L}_a^n} - \sqrt{\mathcal{L}_a^\infty}]e^{it\mathcal{L}_a^\infty}\phi$$

and applying (1.127), (1.123) and Theorem 1.5.3, we deduce that

$$\limsup_{n\to\infty}\|(e^{it_n\mathcal{L}_a^n} - e^{it\mathcal{L}_a^\infty})\psi\|_{\dot{H}_x^{-1}} \leq \limsup_{n\to\infty}\|(e^{it_n\mathcal{L}_a^\infty} - e^{it\mathcal{L}_a^\infty})\phi\|_{L_x^2}.$$

The last limit is easily seen to vanish by applying the spectral theorem. This completes the proof of (1.121), and thus that of Lemma 1.5.4. □

Our next result is essential for proving the decoupling of the potential energy in Theorem 1.5.7.

Corollary 1.5.6. *Fix $a > -\frac{1}{4}$. Given $\psi \in \dot{H}_x^1$, $t_n \to \pm\infty$, and any sequence $\{y_n\} \subset \mathbb{R}^3$, we have*

$$\|e^{it_n\mathcal{L}_a^n}\psi\|_{L_x^6} \to 0 \quad \textit{as} \quad n \to \infty,$$

where $\mathcal{L}_a^n$ is as in Definition 1.5.1.

Proof. Without loss of generality, we may assume that $y_n \to y_\infty \in \mathbb{R}^3 \cup \{\infty\}$. Let $\mathcal{L}_a^\infty$ be as in Definition 1.5.1.

Writing

$$\sqrt{\mathcal{L}_a^\infty}[e^{it_n\mathcal{L}_a^n}\psi - e^{it_n\mathcal{L}_a^\infty}\psi] = [\sqrt{\mathcal{L}_a^\infty} - \sqrt{\mathcal{L}_a^n}]e^{it_n\mathcal{L}_a^n}\psi + e^{it_n\mathcal{L}_a^n}[\sqrt{\mathcal{L}_a^n} - \sqrt{\mathcal{L}_a^\infty}]\psi$$
$$+ [e^{it_n\mathcal{L}_a^n} - e^{it_n\mathcal{L}_a^\infty}]\sqrt{\mathcal{L}_a^\infty}\psi$$

and using (1.122), and (1.123), we see that $e^{it_n\mathcal{L}_a^n}\psi - e^{it_n\mathcal{L}_a^\infty}\psi \to 0$ in $\dot{H}_x^1$. Thus, by the Sobolev embedding theorem it suffices to show that

$$\|e^{it_n\mathcal{L}_a^\infty}\psi\|_{L_x^6} \to 0 \quad \text{as} \quad n \to \infty; \tag{1.128}$$

moreover, by density, we may assume $\psi \in C_c^\infty(\mathbb{R}^3 \setminus \{y_\infty\})$.

Let $F(t) := \|e^{it\mathcal{L}_a^\infty}\psi\|_{L_x^6}$. By the Strichartz inequality, $F \in L_t^2(\mathbb{R})$. Moreover, F is Lipschitz; indeed, by the Sobolev embedding,

$$|\partial_t F(t)| \leq \|\partial_t[e^{it\mathcal{L}_a^\infty}\psi]\|_{L_x^6} \lesssim \|\mathcal{L}_a^\infty\psi\|_{\dot{H}_x^1} \lesssim_\psi 1.$$

Therefore, $F(t_n) \to 0$ as $n \to \infty$. This completes the proof of the corollary. □

Since we are dealing with an energy-critical problem, we also need convergence of propagators in suitable energy-critical spaces. For this, we need the assumption $a > -\frac{1}{4} + \frac{1}{25}$.

Corollary 1.5.7. *Fix $a > -\frac{1}{4} + \frac{1}{25}$. Suppose we are given $y_n \to y_\infty \in \mathbb{R}^3 \cup \{\infty\}$ and let $\mathcal{L}_a^n$ and $\mathcal{L}_a^\infty$ be as in Definition 1.5.1. Then*

$$\lim_{n\to\infty} \|(e^{it\mathcal{L}_a^n} - e^{it\mathcal{L}_a^\infty})\psi\|_{L^{10}_{t,x}(\mathbb{R}\times\mathbb{R}^3)} = 0 \quad \text{for all} \quad \psi \in \dot{H}^1(\mathbb{R}^3).$$

Proof. By the Strichartz inequality, it suffices to prove the claim for $\psi \in C_c^\infty(\mathbb{R}^3 \setminus \{y_\infty\})$. By the Hölder inequality,

$$\|(e^{it\mathcal{L}_a^n} - e^{it\mathcal{L}_a^\infty})\psi\|_{L^{10}_{t,x}} \leq \|(e^{it\mathcal{L}_a^n} - e^{it\mathcal{L}_a^\infty})\psi\|^{\theta}_{L^{\frac{10}{3}}_{t,x}} \|(e^{it\mathcal{L}_a^n} - e^{it\mathcal{L}_a^\infty})\psi\|^{1-\theta}_{L^q_{t,x}}$$

for any $q > 10$ and $\theta \in (0,1)$ given by $\frac{1}{10} = \frac{3\theta}{10} + \frac{1-\theta}{q}$. By (1.123), the first term on the RHS above converges to zero as $n \to \infty$. On the other hand, by the Sobolev embedding, Theorem 1.5.3 and the Strichartz inequality,

$$\|(e^{it\mathcal{L}_a^n} - e^{it\mathcal{L}_a^\infty})\psi\|_{L^q_{t,x}(\mathbb{R}\times\mathbb{R}^3)} \lesssim \||\nabla|^{\frac{3}{2}-\frac{5}{q}}\psi\|_{L^2_x} \lesssim 1,$$

provided q is chosen sufficiently close to 10 so that Theorem 1.5.3 may be applied. □

To prove Theorem 1.5.7, we follow the general method outlined in [148]. For the problems treated in those notes, all defects of compactness in the Strichartz inequality are associated with symmetries of the equation. The presence of a potential in (6.1) breaks the translation symmetry. The new difficulties that this introduces are reminiscent of those overcome in [150], which derived a linear profile decomposition in the exterior of a convex obstacle. This earlier work provides inspiration (but little technical overlap) for much of what follows.

The bulk of the proof of Theorem 1.5.7 is affected by the inductive application of an inverse Strichartz inequality; see Proposition 1.5.1 below. This in turn relies on a refinement of the classical Strichartz inequality.

Lemma 1.5.5 (Refined Strichartz inequality). *Fix $a > -\frac{1}{4} + \frac{1}{25}$. For $f \in \dot{H}_a^1(\mathbb{R}^3)$, we have*

$$\|e^{-it\mathcal{L}_a}f\|_{L^{10}_{t,x}(\mathbb{R}\times\mathbb{R}^3)} \lesssim \|f\|^{\frac{1}{5}}_{\dot{H}_a^1(\mathbb{R}^3)} \sup_{N\in 2^{\mathbb{Z}}} \|e^{-it\mathcal{L}_a}f_N\|^{\frac{4}{5}}_{L^{10}_{t,x}(\mathbb{R}\times\mathbb{R}^3)}.$$

Proof. As $a > -\frac{1}{4} + \frac{1}{25}$, we have $\sigma < \frac{3}{10}$ and so we may choose $0 < \epsilon \leq 1$ so that $\sigma < \frac{3}{10}(1-\epsilon)$. Using the square function estimate Lemma 1.5.4, Lemma 1.5.2 and the Strichartz inequality, we obtain

$$\begin{aligned}
\|e^{-it\mathcal{L}_a}f\|^{10}_{L^{10}_{t,x}} &\lesssim \iint_{\mathbb{R}\times\mathbb{R}^3} \Big(\sum_{N\in 2^{\mathbb{Z}}} |e^{-it\mathcal{L}_a}f_N|^2\Big)^5 \,dx\,dt \\
&\lesssim \sum_{N_1\leq\cdots\leq N_5} \iint_{\mathbb{R}\times\mathbb{R}^3} |e^{-it\mathcal{L}_a}f_{N_1}|^2 \cdots |e^{-it\mathcal{L}_a}f_{N_5}|^2 \,dx\,dt
\end{aligned}$$

$$
\begin{aligned}
&\lesssim \sum_{N_1\le\cdots\le N_5} \|e^{-it\mathcal{L}_a} f_{N_1}\|_{L_t^{10} L_x^{\frac{10}{1-\epsilon}}} \|e^{-it\mathcal{L}_a} f_{N_1}\|_{L_{t,x}^{10}} \prod_{j=2}^{4} \|e^{-it\mathcal{L}_a} f_{N_j}\|_{L_{t,x}^{10}}^2 \\
&\quad \cdot \|e^{-it\mathcal{L}_a} f_{N_5}\|_{L_{t,x}^{10}} \|e^{-it\mathcal{L}_a} f_{N_5}\|_{L_t^{10} L_x^{\frac{10}{1+\epsilon}}} \\
&\lesssim \sup_{N\in 2^{\mathbb{Z}}} \|e^{-it\mathcal{L}_a} f_N\|_{L_{t,x}^{10}}^8 \sum_{N_1\le N_5} \Big[1+\log\Big(\frac{N_5}{N_1}\Big)\Big]^3 N_1^{1+\frac{3\epsilon}{10}} \|e^{-it\mathcal{L}_a} f_{N_1}\|_{L_t^{10} L_x^{\frac{30}{13}}} \\
&\quad \cdot N_5^{1-\frac{3\epsilon}{10}} \|e^{-it\mathcal{L}_a} f_{N_5}\|_{L_t^{10} L_x^{\frac{30}{13}}} \\
&\lesssim \sup_{N\in 2^{\mathbb{Z}}} \|e^{-it\mathcal{L}_a} f_N\|_{L_{t,x}^{10}}^8 \sum_{N_1\le N_5} \Big[1+\log\Big(\frac{N_5}{N_1}\Big)\Big]^3 \Big(\frac{N_1}{N_5}\Big)^{\frac{3\epsilon}{10}} \|f_{N_1}\|_{\dot H_x^1} \|f_{N_5}\|_{\dot H_x^1} \\
&\lesssim \sup_{N\in 2^{\mathbb{Z}}} \|e^{-it\mathcal{L}_a} f_N\|_{L_{t,x}^{10}}^8 \|f\|_{\dot H_x^1}^2,
\end{aligned}
$$

where all space-time norms are over $\mathbb{R}\times\mathbb{R}^3$. □

Proposition 1.5.1 (Inverse Strichartz inequality). *Assume $a > -\frac{1}{4}+\frac{1}{25}$. Let $\{f_n\}\subset \dot H_a^1(\mathbb{R}^3)$ be such that*

$$
\lim_{n\to\infty} \|f_n\|_{\dot H_a^1(\mathbb{R}^3)} = A < \infty \quad \textit{and} \quad \lim_{n\to\infty} \|e^{-it\mathcal{L}_a} f_n\|_{L_{t,x}^{10}} = \epsilon > 0.
$$

Then there exist a subsequence in n, $\phi\in \dot H^1(\mathbb{R}^3)$, $\{N_n\}\subset 2^{\mathbb{Z}}$ and $\{(t_n,x_n)\}\subset \mathbb{R}\times\mathbb{R}^3$ such that

$$
g_n(x) := N_n^{-\frac{1}{2}} (e^{-it_n\mathcal{L}_a} f_n)\Big(x_n + \frac{x}{N_n}\Big) \rightharpoonup \phi(x) \quad \textit{weakly in} \quad \dot H^1(\mathbb{R}^3), \tag{1.129}
$$

$$
\lim_{n\to\infty} \{\|f_n\|_{\dot H_a^1(\mathbb{R}^3)}^2 - \|f_n-\phi_n\|_{\dot H_a^1(\mathbb{R}^3)}^2\} \gtrsim_{\epsilon,A} 1, \tag{1.130}
$$

$$
\lim_{n\to\infty} \{\|f_n\|_{L_x^6}^6 - \|f_n-\phi_n\|_{L_x^6}^6 - \|\phi_n\|_{L_x^6}^6\} = 0, \tag{1.131}
$$

where

$$
\phi_n(x) = N_n^{\frac{1}{2}} e^{it_n\mathcal{L}_a}[\phi(N_n(x-x_n))] = N_n^{\frac{1}{2}} (e^{iN_n^2 t_n \mathcal{L}_a^n}\phi)(N_n(x-x_n)),
$$

with $\mathcal{L}_a^n$ as in Definition 1.5.1 with $y_n = N_n x_n$. Moreover, we may assume that either $N_n^2 t_n \to \pm\infty$ or $t_n\equiv 0$ and that either $N_n|x_n|\to\infty$ or $x_n\equiv 0$.

Proof. By Lemma 1.5.5, for n sufficiently large there exists $N_n\in 2^{\mathbb{Z}}$ such that

$$
\|e^{-it\mathcal{L}_a} P_{N_n}^a f_n\|_{L_{t,x}^{10}} \gtrsim \epsilon^{\frac{5}{4}} A^{-\frac{1}{4}}. \tag{1.132}
$$

By the Mikhlin multiplier theorem for $\mathcal{L}_a$, there is an L_x^{10}-bounded multiplier m so that $P_N^a = m(\mathcal{L}_a)\tilde P_N^a$. Thus, (1.132) implies

$$\|e^{-it\mathcal{L}_a}\tilde{P}^a_{N_n}f_n\|_{L^{10}_{t,x}} \gtrsim \epsilon^{\frac{5}{4}}A^{-\frac{1}{4}}. \tag{1.133}$$

Using the heat kernel bounds provided by Theorem 1.5.2, it is easy to check that

$$\|\tilde{P}^a_{N_n}f\|_{L^{10}_x(\{|x|\le \alpha N_n^{-1}\})} \lesssim \alpha^{\frac{3}{10}-\sigma}\|f\|_{L^{10}_x}.$$

Therefore, (1.133) guarantees the existence of an f_n-independent $\eta > 0$ so that

$$\|e^{-it\mathcal{L}_a}\tilde{P}^a_{N_n}f_n\|_{L^{10}_{t,x}(\mathbb{R}\times\{|x|\ge\eta(\epsilon/A)^c N_n^{-1}\})} \gtrsim \epsilon^{\frac{5}{4}}A^{-\frac{1}{4}}, \tag{1.134}$$

where $c = \frac{5}{2(3-10\sigma)}$.

On the other hand, by the Strichartz estimate and Bernstein inequality

$$\|e^{-it\mathcal{L}_a}\tilde{P}^a_{N_n}f_n\|_{L^{\frac{10}{3}}_{t,x}(\mathbb{R}\times\mathbb{R}^3)} \lesssim \|\tilde{P}^a_{N_n}f_n\|_{L^2_x(\mathbb{R}^3)} \lesssim N_n^{-1}A.$$

This, together with (1.134) and Hölder's inequality, yields

$$\begin{aligned} A^{-\frac{1}{4}}\epsilon^{\frac{5}{4}} &\lesssim \|e^{-it\mathcal{L}_a}\tilde{P}^a_{N_n}f_n\|_{L^{10}_{t,x}(\mathbb{R}\times\{|x|\ge\eta(\epsilon/A)^c N_n^{-1}\})} \\ &\lesssim \|e^{-it\mathcal{L}_a}\tilde{P}^a_{N_n}f_n\|^{\frac{1}{3}}_{L^{\frac{10}{3}}_{t,x}(\mathbb{R}\times\mathbb{R}^3)} \|e^{-it\mathcal{L}_a}\tilde{P}^a_{N_n}f_n\|^{\frac{2}{3}}_{L^{\infty}_{t,x}(\mathbb{R}\times\{|x|\ge\eta(\epsilon/A)^c N_n^{-1}\})} \\ &\lesssim N_n^{-\frac{1}{3}}A^{\frac{1}{3}}\|e^{-it\mathcal{L}_a}\tilde{P}^a_{N_n}f_n\|^{\frac{2}{3}}_{L^{\infty}_{t,x}(\mathbb{R}\times\{|x|\ge\eta(\epsilon/A)^c N_n^{-1}\})}. \end{aligned}$$

Therefore, there exist $\tau_n \in \mathbb{R}$ and $x_n \in \mathbb{R}^3$ with $N_n|x_n| \gtrsim (\epsilon/A)^c$ such that

$$N_n^{-\frac{1}{2}}|(\tilde{P}^a_{N_n}e^{-i\tau_n\mathcal{L}_a}f_n)(x_n)| \gtrsim \epsilon\left(\frac{\epsilon}{A}\right)^{\frac{7}{8}}. \tag{1.135}$$

Passing to a subsequence, we may assume $N_n^2\tau_n \to \tau_\infty \in [-\infty,\infty]$. When τ_∞ is finite, we set $t_n := 0$; otherwise, we define $t_n := \tau_n$.

Having chosen the parameters t_n, x_n and N_n, we now proceed to the construction of the profile ϕ. The inequality (1.135) will underlie the demonstration that ϕ carries nontrivial energy.

With g_n defined as in (1.129),

$$\|g_n\|_{\dot{H}^1_x} = \|\nabla e^{-it_n\mathcal{L}_a}f_n\|_{L^2_x} \sim \|\sqrt{\mathcal{L}_a}e^{-it_n\mathcal{L}_a}f_n\|_{L^2_x} = \|f_n\|_{\dot{H}^1_a} \lesssim A,$$

and so, passing to a subsequence, there exists ϕ so that $g_n \rightharpoonup \phi$ weakly in $\dot{H}^1_x$. This proves (1.129).

Next, we turn to proving (1.130). Changing variables, then using Lemma 1.5.4 and the fact that $g_n \rightharpoonup \phi$ weakly in $\dot{H}^1_x$, we get

$$\|f_n\|^2_{\dot{H}^1_a} - \|f_n - \phi_n\|^2_{\dot{H}^1_a} = 2\mathbb{R}\langle g_n, \mathcal{L}^n_a\phi\rangle - \langle\phi, \mathcal{L}^n_a\phi\rangle \to \langle\phi, \mathcal{L}^\infty_a\phi\rangle \quad \text{as} \quad n\to\infty,$$

where $\mathcal{L}_a^\infty$ is as in Definition 1.5.1. Thus, it suffices to prove a lower bound on the $\dot{H}_x^1$-norm of ϕ.

Setting

$$h_n(x) = \begin{cases} N_n^{-3}[e^{-i\tau_n \mathcal{L}_a} \tilde{P}_{N_n}^a \delta_{x_n}](x_n + \frac{x}{N_n}) = [e^{-iN_n^2 \tau_n \mathcal{L}_a^n} \tilde{P}_1^n \delta_0](x) & \text{if} \quad \tau_\infty \in \mathbb{R}, \\ N_n^{-3}[\tilde{P}_{N_n}^a \delta_{x_n}](x_n + \frac{x}{N_n}) = [\tilde{P}_1^n \delta_0](x) & \text{if} \quad \tau_\infty = \pm\infty \end{cases}$$

with $\tilde{P}_1^n := e^{-\mathcal{L}_a^n} - e^{-4\mathcal{L}_a^n}$, and performing a change of variables, (1.135) becomes

$$|\langle h_n, g_n \rangle| \gtrsim \epsilon \left(\frac{\epsilon}{A}\right)^{\frac{7}{8}}.$$

By construction, $g_n \rightharpoonup \phi$ in $\dot{H}_x^1$. We set $\tilde{P}_1^\infty := e^{-\mathcal{L}_a^\infty} - e^{-4\mathcal{L}_a^\infty}$. By Lemma 1.5.4, $h_n \to \tilde{P}_1^\infty \delta_0$ in $\dot{H}_x^{-1}$ when $\tau_\infty = \pm\infty$ and $h_n \to e^{-i\tau_\infty \mathcal{L}_a^\infty} \tilde{P}_1^\infty \delta_0$ in $\dot{H}_x^{-1}$ when τ_∞ is finite. Thus, when τ_∞ is finite, we deduce that

$$\epsilon \left(\frac{\epsilon}{A}\right)^{\frac{7}{8}} \lesssim |\langle \tilde{P}_1^\infty \delta_0, e^{+i\tau_\infty \mathcal{L}_a^\infty} \phi \rangle| \lesssim \|\phi\|_{\dot{H}_x^1} \|\tilde{P}_1^\infty \delta_0\|_{L_x^{6/5}}.$$

In view of Theorem 1.5.2 and the fact that $N_n|x_n| \gtrsim (\epsilon/A)^c$, we have

$$\|\tilde{P}_1^\infty \delta_0\|_{L_x^{6/5}} \sim \left[1 + \left(\frac{\epsilon}{A}\right)^{-c}\right]^\sigma,$$

which completes the proof of (1.130) when τ_∞ is finite. Trivial modifications of the last steps handle the case $\tau_\infty = \pm\infty$.

We now turn to (1.131). If $t_n \equiv 0$, then by Rellich–Kondrashov (which yields $g_n \to \phi$ a. e.) and the Fatou lemma, we have

$$\|g_n\|_{L_x^6}^6 - \|g_n - \phi\|_{L_x^6}^6 - \|\phi\|_{L_x^6}^6 \to 0 \quad \text{as} \quad n \to \infty.$$

A change of variables then yields (1.131) in this case. If $t_n = \tau_n$, then it suffices to observe that $\phi_n \to 0$ in L_x^6 by Corollary 1.5.6.

Finally, passing to another subsequence if necessary, we may assume that either $N_n|x_n| \to \infty$ or $N_n x_n \to y_\infty \in \mathbb{R}^3$. In the latter case, we may take $x_n \equiv 0$ by replacing the profile ϕ found previously by $\phi(x - y_\infty)$. □

With Proposition 1.5.1 in place, the greater part of the proof of Theorem 1.5.7 amounts to careful bookkeeping. We omit the details; they can be found, for example, in [150]. However, as one sees there, two extra facts about the free propagator are needed to verify (1.118). For the problem at hand, these two additional inputs are provided by Lemmas 1.5.6 and 1.5.7 below; they are the analogues of Lemmas 5.5 and 5.4 in [150] and play the same roles as these earlier results in the proof of (1.118).

Lemma 1.5.6 (Weak convergence). *Let $f_n \in \dot H^1(\mathbb{R}^3)$ be such that $f_n \rightharpoonup 0$ weakly in $\dot H^1(\mathbb{R}^3)$ and let $t_n \to t_\infty \in \mathbb{R}$. Then for any $y_n \in \mathbb{R}^3$,*

$$e^{-it_n\mathcal{L}_a^n} f_n \rightharpoonup 0 \quad \textit{weakly in} \quad \dot H^1(\mathbb{R}^3),$$

where $\mathcal{L}_a^n$ is as in Definition 1.5.1.

Proof. Without loss of generality, we may assume that $y_n \to y_\infty \in \mathbb{R}^3 \cup \{\infty\}$. Let $\mathcal{L}_a^\infty$ be as in Definition 1.5.1. For any $\psi \in C_c^\infty(\mathbb{R}^3 \setminus \{y_\infty\})$, we have

$$\begin{aligned} |\langle [e^{-it_n\mathcal{L}_a^n} - e^{-it_\infty \mathcal{L}_a^n}] f_n, \psi\rangle_{\dot H_x^1}| &\lesssim \|[e^{-it_n\mathcal{L}_a^n} - e^{-it_\infty \mathcal{L}_a^n}] f_n\|_{L_x^2} \|\Delta\psi\|_{L_x^2} \\ &\lesssim |t_n - t_\infty|^{1/2} \|\sqrt{\mathcal{L}_a^n} f_n\|_{L_x^2} \|\Delta\psi\|_{L_x^2}, \end{aligned}$$

which converges to zero as $n \to \infty$. To see the last inequality above, we used the spectral theorem together with the elementary inequality

$$|e^{-it_n\lambda} - e^{-it_\infty\lambda}| \lesssim |t_n - t_\infty|^{1/2}\lambda^{1/2} \quad \text{for} \quad \lambda \ge 0.$$

Thus, to prove the lemma it suffices to prove that $e^{-it_\infty\mathcal{L}_a^n} f_n \rightharpoonup 0$ in $\dot H_x^1$. For $\psi \in C_c^\infty(\mathbb{R}^3 \setminus \{y_\infty\})$,

$$\langle e^{-it_\infty\mathcal{L}_a^n} f_n, \psi\rangle_{\dot H_x^1} = \langle f_n, [e^{it_\infty\mathcal{L}_a^n} - e^{it_\infty\mathcal{L}_a^\infty}](-\Delta\psi)\rangle_{L_x^2} + \langle f_n, e^{it_\infty\mathcal{L}_a^\infty}(-\Delta\psi)\rangle_{L_x^2}.$$

The first inner product on the RHS above converges to zero by Lemma 1.5.4. As $f_n \rightharpoonup 0$ in $\dot H^1(\mathbb{R}^3)$ by assumption and $e^{it_\infty\mathcal{L}_a^\infty}\Delta\psi \in \dot H^{-1}(\mathbb{R}^3)$, the second inner product also converges to zero. This completes the proof of the lemma. □

Lemma 1.5.7 (Weak convergence). *Let $f \in \dot H^1(\mathbb{R}^3)$ and let $\{(t_n, x_n)\}_{n\ge1} \subset \mathbb{R}\times\mathbb{R}^3$ and $\{y_n\} \subset \mathbb{R}^3$. Then*

$$[e^{-it_n\mathcal{L}_a^n} f](x + x_n) \rightharpoonup 0 \quad \textit{weakly in} \quad \dot H^1(\mathbb{R}^3) \quad \textit{as} \quad n \to \infty, \tag{1.136}$$

whenever $|t_n| \to \infty$ or $|x_n| \to \infty$. Here, $\mathcal{L}_a^n$ is as in Definition 1.5.1.

Proof. Naturally, it suffices to prove (1.136) along subsequences where $y_n \to y_\infty \in \mathbb{R}^3 \cup \{\infty\}$. Let $\mathcal{L}_a^\infty$ be as in Definition 1.5.1.

We first prove (1.136) when $t_n \to \infty$; the proof when $t_n \to -\infty$ follows symmetrically. Let $\psi \in C_c^\infty(\mathbb{R}^3 \setminus \{y_\infty\})$ and let

$$F_n(t) := \langle e^{-it\mathcal{L}_a^n} f(x + x_n), \psi\rangle_{\dot H^1(\mathbb{R}^3)}.$$

To establish (1.136), we need to show

$$F_n(t_n) \to 0 \quad \text{as} \quad n \to \infty. \tag{1.137}$$

A computation yields

$$|\partial_t F_n(t)| = |\langle \mathcal{L}_a^n e^{-it\mathcal{L}_a^n} f(x + x_n), \Delta\psi\rangle_{L^2(\mathbb{R}^3)}| \lesssim \|f\|_{\dot{H}_x^1} \|\Delta\psi\|_{\dot{H}_x^1} \lesssim_{f,\psi} 1.$$

On the other hand,

$$\begin{aligned} \|F\|_{L_t^{\frac{10}{3}}([t_n,\infty))} &\lesssim \|e^{-it\mathcal{L}_a^n} f\|_{L_{t,x}^{\frac{10}{3}}([t_n,\infty)\times\mathbb{R}^3)} \|\Delta\psi\|_{L_x^{\frac{10}{7}}(\mathbb{R}^3)} \\ &\lesssim_\psi \|[e^{-it\mathcal{L}_a^n} - e^{-it\mathcal{L}_a^\infty}]f\|_{L_{t,x}^{\frac{10}{3}}([t_n,\infty)\times\mathbb{R}^3)} + \|e^{-it\mathcal{L}_a^\infty} f\|_{L_{t,x}^{\frac{10}{3}}([t_n,\infty)\times\mathbb{R}^3)}. \end{aligned}$$

The first term on the RHS above converges to zero by Lemma 1.5.4. The second term converges to zero by the Strichartz inequality combined with the monotone convergence theorem. Putting everything together, we derive (1.137) and so (1.136) when $t_n \to \infty$.

Now assume $\{t_n\}_{n\geq 1}$ is bounded, but $|x_n| \to \infty$ as $n \to \infty$. Without loss of generality, we may assume $t_n \to t_\infty \in \mathbb{R}$ as $n \to \infty$. As

$$[e^{-it_n\mathcal{L}_a^n} f](x + x_n) = [e^{-it_n(-\Delta + \frac{a}{|\cdot + y_n + x_n|^2})} f(\cdot + x_n)](x),$$

the claim now follows from Lemma 1.5.6. □

2 Parabolic equations

This chapter is devoted to the study of the following Cauchy problem for the abstract parabolic equation:

$$u_t + Au = F(u), \quad t \in \mathbb{R}^+, \quad u(0) = \varphi, \tag{2.1}$$

where $A = -P_{2m}(D)$ is an elliptic operator of order $2m$ and generates an analytic semigroup in $L^p(\Omega)$ with $1 < p < \infty$, and $F(u)$ is a nonlinear function (containing at most $(2m-1)$th-order derivatives of u). It is easy to see that, in the case when $\Omega = \mathbb{R}^n$, (2.1) corresponds to the classical Cauchy problem, while in the case when $\Omega \subset \mathbb{R}^n$ is a bounded smooth domain and $D(A) = W^{2m,p}(\Omega) \cap W_0^{m,p}(\Omega)$, (2.1) corresponds to the homogeneous Dirichlet problem. This chapter only considers Cauchy problems. For initial-boundary value problems, the corresponding results remain true, which will not be discussed in detail here due to page limitation. However, some remarks will be given to illustrate these results if necessary.

Similar to the case of wave and dispersive wave equations, we can also introduce space-time admissible triplets and space-time Banach spaces to study nonlinear parabolic equations. The reader will see that for parabolic equations we will select space-time Banach spaces, which are different from those for the wave and dispersive wave equations so that the successive continuation method can be used. Of course, we can also select the space-time Banach spaces similar to the wave equations to study the nonlinear parabolic equations. For parabolic equations, the corresponding elliptic operator generates an analytic semigroup in any L^p space with $1 < p < \infty$, thus a much wider choice exists for the spaces for initial data and admissible triplets. This is important in establishing nonlinear estimates and illustrates sufficiently that parabolic operators are of better regularity properties. However, wave operators or dispersive wave operators generate C_0-semigroups only in the Hilbert spaces H^s ($s \in \mathbb{R}$), so the initial data only belong to the Hilbert spaces H^s with $s \in \mathbb{R}$. The selection of the admissible pairs and the corresponding time-space spaces is thus fixed.

In this chapter, we will give a unified method to deal with semilinear parabolic equations. The main idea is to choose, based on the scaling argument, well-suited time-space admissible triplets for different types of nonlinear terms and then establish, in well-suited solution spaces, the well-posedness of the Cauchy problem for the nonlinear parabolic equations by utilizing time-space estimates for the solution of the linear parabolic equations. In addition, we will give some methods of dealing with parabolic equations with inhomogeneous nonlinear terms and in particular, techniques and methods of studying small global well-posedness in suitable working spaces as well as both local well-posedness and small global well-posedness in critical Banach spaces of the Cauchy problem for parabolic equations with the exponential growing nonlinear term.

https://doi.org/10.1515/9783111384726-002

2.1 Space-time estimates for linear parabolic equations

In this section, we consider the following Cauchy problem for the linear parabolic equation:

$$u_t + Au = 0, \quad t \in \mathbb{R}^+,\ x \in \mathbb{R}^n, \quad u(0,x) = \varphi(x), \quad x \in \mathbb{R}^n \tag{2.2}$$

with $\varphi \in L^r(\mathbb{R}^n)$ $(1 < r < \infty)$, where $A = -P_{2m}(D)$ is an elliptic operator of order $2m$, $D_x = i^{-1}\partial_x$ and $P_{2m}(\xi)$ is the symbol of the differential operator, which is an elliptic polynomial in ξ of degree $2m$ and satisfies that

$$\operatorname{Re} P_{2m}(\xi) < 0, \quad \forall \xi \in \mathbb{R}^n \backslash \{0\}. \tag{2.3}$$

Note that initial-boundary value problems can be formulated as the following abstract Cauchy problem:

$$u_t + Au = 0, \quad t \in \mathbb{R}^+,\ x \in \Omega, \quad u(0) = \varphi \tag{2.4}$$

with $\varphi \in D(A)$, where $D(A)$ denotes the domain of A. Since A generates an analytic semigroup e^{-At} in $L^r(\Omega)$ with $1 < r < \infty$, then the solution $u = e^{-At}\varphi$ of (2.4) satisfies that

$$\|A^{\gamma} e^{-At}\varphi\|_p \le Ct^{-\gamma}\|\varphi\|_p, \quad t > 0,\ 1 < p < \infty, \tag{2.5}$$

where $\gamma \ge 0$.

We will establish the space-time estimates for the solution

$$u = \mathcal{F}^{-1} e^{P_{2m}(\xi)t} \mathcal{F}\varphi = e^{-At}\varphi$$

of the Cauchy problem (2.2). To do this, we need the following definitions of admissible triplets and generalized admissible triplets.

Definition 2.1.1. The triplet (q, p, r) is called an admissible triplet for $2m$th-order parabolic operators if

$$\frac{1}{q} = \frac{n}{2m}\left(\frac{1}{r} - \frac{1}{p}\right), \tag{2.6}$$

where

$$1 < r \le p < \begin{cases} \frac{nr}{n-2m}, & \text{for} \quad n > 2m, \\ \infty, & \text{for} \quad n \le 2m. \end{cases}$$

Definition 2.1.2. The triplet (q, p, r) is called a generalized admissible triplet for $2m$th-order parabolic operators if

$$\frac{1}{q} = \frac{n}{2m}\left(\frac{1}{r} - \frac{1}{p}\right), \tag{2.7}$$

where

$$1 < r \le p < \begin{cases} \frac{nr}{n-2mr}, & \text{for} \quad n > 2mr, \\ \infty, & \text{for} \quad n \le 2mr. \end{cases}$$

Remark 2.1.1.
(i) The case $m = 1$ corresponds to the classical heat equation and the Navier–Stokes equation.
(ii) It is easy to see that if (q, p, r) is an admissible triplet or a generalized admissible triplet then q is uniquely determined by p and r, so we may write $q = q(p, r)$.
(iii) Clearly, $r < q \le \infty$ if (q, p, r) is an admissible triplet, and $1 < q \le \infty$ if (q, p, r) is a generalized admissible triplet.

Definition 2.1.3. Let B be a Banach space and for $\sigma > 0$, $I = [0, T)$ and $\dot{I} = (0, T)$ with $T \le \infty$. Let us define $C_\sigma(I; B)$ and its homogeneous space $\dot{C}_\sigma(I; B)$ as follows:

$$C_\sigma(I; B) = \left\{f \in C(\dot{I}; B),\ \|f; C_\sigma(I; B)\| = \sup_{t\in I} t^{\frac{1}{\sigma}} \|f\|_B < \infty\right\}, \tag{2.8}$$

$$\dot{C}_\sigma(I; B) = \left\{f \in C_\sigma(I; B),\ \lim_{t\to 0^+} t^{\frac{1}{\sigma}} \|f\|_B = 0\right\}. \tag{2.9}$$

Remark 2.1.2. It is easy to see that:
(i) $f \in C_\sigma(I; L^p)$ if and only if $t^{\frac{1}{\sigma}} f \in C_b(I; L^p)$, here $B = L^p$, $1 < p < \infty$;
(ii) if (q, p, r) is a generalized admissible triplet or admissible triplet, then

$$C_{q(p,r)}(I; L^p) = \left\{f \in C(\dot{I}; L^p),\ \|f\|_{C_{q(p,r)}(I;L^p)} = \sup_{t\in I} t^{\frac{1}{q}} \|f\|_p < \infty\right\}, \tag{2.10}$$

$$\dot{C}_{q(p,r)}(I; L^p) = \left\{f \in C_{q(p,r)}(I; L^p),\ \lim_{t\to 0^+} t^{\frac{1}{q}} \|f\|_p = 0\right\}; \tag{2.11}$$

in particular, $C_{q(p,r)}(I; L^p) = C_b(I; L^p)$ if $p = r$.

Lemma 2.1.1 ($L^p - L^r$ estimate). *Let $P_{2m}(\xi)$ satisfy (2.3) and let $p \ge r \ge 1$ and $\varphi \in L^r(\mathbb{R}^n)$. Then the solution $u(t) = e^{-At}\varphi$ of (2.2) satisfies that*

$$\|e^{-At}\varphi\|_p = \|\mathcal{F}^{-1}(e^{P_{2m}(\xi)t}\mathcal{F}\varphi)\|_p \le Ct^{-\frac{n}{2m}(\frac{1}{r}-\frac{1}{p})}\|\varphi\|_r, \quad \forall t > 0. \tag{2.12}$$

Proof. Without loss of generality, we may assume that $\operatorname{Re} P_{2m}(\xi) = P_{2m}(\xi)$. We denote the kernel of the operator e^{-At} by $K_t(x)$. Then it is easy to see that

$$K_t(x) = \mathcal{F}^{-1}\{e^{tP_{2m}(\xi)}\}(x).$$

We claim that the kernel $K_t(x)$ satisfies the following pointwise estimate:

$$|K_t(x)| \le Ct^{-\frac{n}{2m}}\left(1 + t^{-\frac{1}{2m}}|x|\right)^{-n-2m}, \quad \forall\, x \in \mathbb{R}^n,\ t > 0. \tag{2.13}$$

With this claim in hand, we easily obtain (2.12) by Young's inequality.

Now, we turn to prove the claim (2.13). By (2.3), we know that there is an $\alpha_0 > 0$ such that

$$P_{2m}(\xi) \le -\alpha_0|\xi|^{2m}. \tag{2.14}$$

This implies

$$|K_t(x)| \le \int_{\mathbb{R}^n} e^{-t\alpha_0|\xi|^{2m}}\, d\xi \le Ct^{-\frac{n}{2m}}.$$

It remains to prove

$$|K_t(x)| \le Ct|x|^{-n-2m}, \quad |x| \ge t^{\frac{1}{2m}},\ t > 0. \tag{2.15}$$

This will be shown by stationary-phase argument as in [57, Lemma A.1]. Define the invariant derivative operator $L := \frac{x\cdot\nabla_\xi}{i|x|^2}$. Then we have

$$Le^{ix\cdot\xi} = e^{ix\cdot\xi}, \quad L^* = -\frac{x\cdot\nabla_\xi}{i|x|^2}.$$

We split the kernel $K_t(x)$ into two parts

$$\begin{aligned} K_t(x) &= \int_{\mathbb{R}^n} e^{ix\cdot\xi}L^*\left(e^{tP_{2m}(\xi)}\right) d\xi \\ &= \int_{\mathbb{R}^n} e^{ix\cdot\xi}\chi(\xi/\lambda)L^*\left(e^{tP_{2m}(\xi)}\right) d\xi + \int_{\mathbb{R}^n} e^{ix\cdot\xi}(1-\chi(\xi/\lambda))L^*\left(e^{tP_{2m}(\xi)}\right) d\xi \\ &\triangleq K_{t,1}(x) + K_{t,2}(x), \end{aligned}$$

where we will take $\lambda = |x|^{-1}$ and $\chi(\xi)$ is a smooth function satisfying

$$\chi(\xi) = \begin{cases} 1, & |\xi| \le 1, \\ 0, & |\xi| > 2. \end{cases}$$

The estimate of the term $K_{t,1}(x)$: We have

$$|K_{t,1}(x)| \leq C\frac{t}{|x|}\int_{|\xi|\leq 2\lambda}|P_{2m-1}(\xi)|\,d\xi \qquad (2.16)$$
$$\leq C\frac{t}{|x|}\int_{|\xi|\leq 2\lambda}|\xi|^{2m-1}\,d\xi \leq Ct|x|^{-2m-n},$$

by taking $\lambda = |x|^{-1}$.

The estimate of the term $K_{t,2}(x)$: Taking $N > 2m+n$ and integrating by parts, we get

$$\begin{aligned}|K_{t,2}(x)| &\leq \int_{\mathbb{R}^n}|e^{ix\cdot\xi}(L^*)^{N-1}(1-\chi(\xi/\lambda))L^*(e^{tP_{2m}(\xi)})|\,d\xi\\
&\leq C|x|^{-N}\int_{|\xi|\geq\lambda}\sum_{k=1}^{N}t^k|\xi|^{2km-N}e^{tP_{2m}(\xi)}\,dx\\
&\quad + C|x|^{-N}\sum_{k=1}^{N-1}C_k\lambda^{-k}\int_{\lambda\leq|\xi|\leq 2\lambda}\sum_{\ell=1}^{N-k}C_\ell t^\ell t^\ell|\xi|^{2\ell m-(N-k)}e^{tP_{2m}(\xi)}\,d\xi\\
&\leq C|x|^{-N}\int_{|\xi|\geq\lambda}t|\xi|^{2m-N}e^{tP_{2m}(\xi)}\,d\xi\\
&\quad + C|x|^{-N}\int_{|\xi|\geq\lambda}t|\xi|^{2m-N}t^{N-1}|\xi|^{2m(N-1)}e^{tP_{2m}(\xi)}\,d\xi\\
&\quad + C|x|^{-N}\sum_{k=1}^{N-1}\int_{\lambda\leq|\xi|\leq 2\lambda}(t|\xi|^{2m-n}e^{tP_{2m}(\xi)} + t^{N-k}|\xi|^{2m(N-k)-N}e^{tP_{2m}(\xi)})\,d\xi\\
&\leq Ct|x|^{-N}\Big(\int_{|\xi|\geq\lambda}|\xi|^{2m-N}\,d\xi + \int_{\lambda\leq|\xi|\leq 2\lambda}\lambda^{2m-N}\,d\xi\Big)\\
&\leq Ct|x|^{-N}\lambda^{2m-N+n} \leq Ct|x|^{-n-2m},\end{aligned}$$

where we have used that facts that

$$t^{N-1}|\xi|^{2m(N-1)}e^{tP_{2m}(\xi)} \leq t^{N-1}|\xi|^{2m(N_1)}e^{-ta_0|\xi|^{2m}} \leq C,$$
$$t^{N-k-1}|\xi|^{2m(N-k-1)}e^{tP_{2m}(\xi)} \leq t^{N-k-1}|\xi|^{2m(N-k-1)}e^{-ta_0|\xi|^{2m}} \leq C,$$

with $1 \leq k \leq N-1$. □

Lemma 2.1.2. *Let $u(t) = e^{-At}\varphi$ be the solution of the abstract Cauchy problem (2.4) and let us assume that the 2mth-order elliptic operator A generates an analytic semigroup in $L^r(\Omega)$ with $1 < r < \infty$ and $0 \in \rho(A)$. Then, for $p \geq r > 1$,*

$$\|e^{-At}\varphi\|_p \leq Ct^{-\frac{n}{2m}(\frac{1}{r}-\frac{1}{p})}\|\varphi\|_r, \quad \forall t > 0.$$

Proof. Noting that $0 \in \rho(A)$, we can define A^γ and have

$$\|A^\gamma f\|_p \sim \|f\|_{\dot{H}^{2m\gamma,p}(\omega)}, \quad 1 < p < \infty.$$

For any $p \geq r > 1$, we take $\gamma = \frac{n}{2m}(\frac{1}{r} - \frac{1}{p})$. Then $\dot{H}^{2m\gamma,r} \hookrightarrow L^p$, so by (2.5), we have

$$\begin{aligned}\|e^{-tA}\varphi\|_p &\leq C\|e^{-tA}\varphi\|_{\dot{H}^{2m\gamma,r}} \leq C\|A^\gamma e^{-tA}\varphi\|_r \\ &\leq Ct^{-\gamma}\|\varphi\|_r = Ct^{-\frac{n}{2m}(\frac{1}{r}-\frac{1}{p})}\|\varphi\|_r.\end{aligned}$$

This completes the proof of the lemma. □

As a direct consequence of Lemmas 2.1.1 and 2.1.2, we have the following results.

Theorem 2.1.1. *Let $I = [0, T)$ with $0 < T \leq \infty$. Let (q, p, r) be any generalized admissible triplet and let $\varphi \in L^r$. Then $u = e^{-At}\varphi \in C_{q(p,r)}(I; L^p)$ and*

$$\|e^{-At}\varphi\|_{C_{q(p,r)}(I;L^p)} \leq C\|\varphi\|_r, \tag{2.17}$$

where C is a constant independent of φ and T. Moreover, if $p > r$, then $u = e^{-At}\varphi \in \dot{C}_{q(p,r)}(I; L^p)$.

Proof. The estimate (2.17) follows directly from Lemmas 2.1.1 or 2.1.2.

We now prove that if $p > r$, then

$$\lim_{t\to 0} t^{\frac{1}{q}}\|e^{-At}\varphi\|_p = 0. \tag{2.18}$$

In fact, take $j(x) \in C_c^\infty(\mathbb{R}^n)$ such that

$$\int_{\mathbb{R}^n} j(x)dx = 1.$$

Let $j_\delta(x) = \delta^{-n}j(\delta^{-1}x)$, $\varphi_\delta(x) = j_\delta(x) * \varphi$. Then

$$\begin{aligned}t^{\frac{1}{q}}\|e^{-tA}\varphi\|_p &\leq t^{\frac{1}{q}}\|e^{-tA}(\varphi_\delta - \varphi)\|_p + t^{\frac{1}{q}}\|e^{-tA}\varphi_\delta\|_p \\ &\leq C\|\varphi_\delta - \varphi\|_r + Ct^{\frac{1}{q}}\|j_\delta(x)\|_\ell\|\varphi\|_r \\ &\leq C\|\varphi_\delta - \varphi\|_r + C_\delta t^{\frac{1}{q}}\|\varphi\|_r,\end{aligned}$$

where $1 + \frac{1}{p} = \frac{1}{\ell} + \frac{1}{r}$. Thus, for any $\varepsilon > 0$, there is a sufficiently small $\delta > 0$ such that $C\|\varphi_\delta - \varphi\|_r < \varepsilon/2$. Now fix such a $\delta > 0$ and then let $t \to 0$ in the above inequality to obtain that

$$C_\delta t^{\frac{1}{q}}\|\varphi\|_r < \varepsilon/2.$$

Since ε is arbitrarily small, (2.18) then follows. The theorem is thus proved. □

Theorem 2.1.2. *Let $I = [0, T)$ with $0 < T \leq \infty$. Let (q, p, r) be any admissible triplet and let $\varphi \in L^r$. Then $u = e^{-At}\varphi \in L^q(I; L^p) \cap C_b(I; L^r)$ and*

$$\|e^{-At}\varphi\|_{L^q(I;L^p)} \leq C\|\varphi\|_r, \tag{2.19}$$

where C is a constant independent of φ and T.

Proof. Clearly, when $p = r$ and $q = \infty$, (2.19) holds. We now consider the case $p > r$.

Let $U(t)\varphi = \|e^{-tA}\varphi\|_p$. Then, by Lemmas 2.1.1 or 2.1.2, we have

$$U(t)\varphi = \|e^{-tA}\varphi\|_p \leq Ct^{-\frac{1}{q}}\|\varphi\|_r.$$

Thus, for $\tau > 0$,

$$\begin{aligned} m\{t : |U(t)\varphi| \geq \tau\} &\leq m\{t : Ct^{-\frac{1}{q}}\|\varphi\|_r > \tau\} \\ &= m\left\{t : t < \left(\frac{C\|\varphi\|_r}{\tau}\right)^q\right\} \leq C^q\left(\frac{\|\varphi\|_r}{\tau}\right)^q. \end{aligned}$$

This means that $U(t)$ is of weak-type (r, q). On the other hand, $U(t)$ is a sub-additive operator and

$$\|U(t)\varphi\|_{L^\infty(I)} = \sup_{t\in I}\|e^{-At}\varphi\|_p \leq \|\varphi\|_p,$$

which implies that $U(t)$ is of weak-type (p, ∞).

Now for any admissible triplet (q, p, r), we can find an admissible triplet (q_1, p_1, r_1) such that $q_1 < q < \infty$ and $r_1 < r < p$. Noting that $U(t)$ is of both weak-type (r_1, q_1) and weak-type (p, ∞), we have by Marcinkiewicz's interpolation theorem that $U(t)$ is of strong-type (r, q) and satisfies the estimate (2.19). The proof is thus completed. □

Next, we consider the Cauchy problem for the inhomogeneous linear parabolic equation:

$$u_t + Au = f(x, t), \quad t \in \mathbb{R}^+, \; x \in \mathbb{R}^n, \quad u(0) = \varphi. \tag{2.20}$$

It is easy to see that

$$u(t) = e^{-At}\varphi + \int_0^t e^{-A(t-\tau)}f(x, \tau)d\tau \triangleq e^{-At}\varphi + Gf$$

is a global smooth solution to the Cauchy problem (2.20). For the inhomogeneous part Gf, we have the following space-time estimates.

Theorem 2.1.3. *For $b > 0$ and $T > 0$, let $r_c = \frac{nb}{2m}$ and $I = [0, T)$. Let $r \geq r_c$ if $r_c > 1$, $r > 1$ if $r_c \leq 1$ and let (q, p, r) be any generalized admissible triplet satisfying that $p > b + 1$.*

(i) *If* $f \in L^{\frac{q}{b+1}}(I; L^{\frac{p}{b+1}})$*, then* $Gf \in L^q(I; L^p) \cap C_b(I; L^r)$ *and*

$$\|Gf\|_{L^q(I;L^p)} + \|Gf\|_{L^\infty(I;L^r)} \le CT^{1-\frac{nb}{2mr}} \|f\|_{L^{\frac{q}{b+1}}(I;L^{\frac{p}{b+1}})}, \tag{2.21}$$

when $p < r(1+b)$ *and*

$$\begin{aligned}&\|Gf\|_{L^q(I;L^p)} + \|Gf\|_{L^\infty(I;L^r)} \\ &\quad \le CT^{1-\frac{nb}{2mr}} \left\| |f|^{\frac{1}{b+1}} \right\|^{\theta(b+1)}_{L^\infty(I;L^r)} \left\| |f|^{\frac{1}{b+1}} \right\|^{(1-\theta)(b+1)}_{L^q(I;L^p)},\end{aligned} \tag{2.22}$$

when $p \ge r(1+b)$*, and where* $\theta = \frac{p-r(b+1)}{(b+1)(p-r)}$*, and* $C > 0$ *is a constant independent of* f *and* T.

(ii) *If* $f \in C_{\frac{q}{b+1}}(I; L^{\frac{p}{b+1}})$*, then* $Gf \in C_q(I; L^p) \cap C_b(I; L^r)$ *and*

$$\|Gf\|_{C_q(I;L^p)} + \|Gf\|_{L^\infty(I;L^r)} \le CT^{1-\frac{nb}{2mr}} \|f\|_{C_{\frac{q}{b+1}}(I;L^{\frac{p}{b+1}})}, \tag{2.23}$$

when $p < r(1+b)$ *and*

$$\begin{aligned}&\|Gf\|_{C_q(I;L^p)} + \|Gf\|_{L^\infty(I;L^r)} \\ &\quad \le CT^{1-\frac{nb}{2mr}} \left\| |f|^{\frac{1}{b+1}} \right\|^{\theta(b+1)}_{L^\infty(I;L^r)} \left\| |f|^{\frac{1}{b+1}} \right\|^{(1-\theta)(b+1)}_{C_q(I;L^p)},\end{aligned} \tag{2.24}$$

when $p \ge r(1+b)$*, and where* $\theta = \frac{p-r(b+1)}{(b+1)(p-r)}$*, and* $C > 0$ *is a constant independent of* f *and* T.

Proof. First, we consider the case when $p < r(1+b)$. At this time, $q > 1+b$. Thus, it follows from Young's inequality or Hardy–Littlewood–Sobolev's inequality (in the case when $r = r_c > 1$) that

$$\begin{aligned}\|Gf\|_{L^q(I;L^p)} &\le C \left\| \int_0^t |t-s|^{-\frac{n}{2m}\left(\frac{b+1}{p}-\frac{1}{p}\right)} \|f(x,s)\|_{\frac{p}{b+1}} ds \right\|_q \\ &\le CT^{1-\frac{nb}{2mr}} \|f\|_{L^{\frac{q}{b+1}}(I;L^{\frac{p}{b+1}})},\end{aligned}$$

and

$$\begin{aligned}\|Gf\|_{C_{q(p,r)}(I;L^p)} &\le C \sup_{t\in I} t^{\frac{1}{q}} \int_0^t |t-s|^{-\frac{n}{2m}\left(\frac{b+1}{p}-\frac{1}{p}\right)} \|f(x,s)\|_{\frac{p}{b+1}} ds \\ &\le C \sup_{t\in I} t^{\frac{1}{q}} \int_0^t |t-s|^{-\frac{nb}{2mp}} s^{-\frac{b+1}{q}} ds \|f\|_{C_{\frac{q}{b+1}}(I;L^{\frac{p}{b+1}})}\end{aligned}$$

$$\leq C \sup_{t\in I} t^{1-\frac{nb}{2mr}} \int_0^1 |1-\tau|^{-\frac{nb}{2mp}} \tau^{-\frac{b+1}{q}} d\tau \|f\|_{C_{\frac{q}{b+1}}(I;L^{\frac{p}{b+1}})}$$

$$\leq CT^{1-\frac{nb}{2mr}} \|f\|_{C_{\frac{q}{b+1}}(I;L^{\frac{p}{b+1}})},$$

and

$$\|Gf\|_{L^\infty(I;L^r)} \leq C \int_0^t |t-s|^{-\frac{n}{2m}(\frac{b+1}{p}-\frac{1}{r})} \|f(x,s)\|_{\frac{p}{b+1}} ds$$

$$\leq C\left(\int_0^t |t-s|^{-\frac{n}{2m}(\frac{b+1}{p}-\frac{1}{r})\chi} ds\right)^{\frac{1}{\chi}} \|f\|_{L^{\frac{q}{1+b}}(I;L^{\frac{p}{1+b}})}$$

$$\leq CT^{1-\frac{nb}{2mr}} \|f\|_{L^{\frac{q}{b+1}}(I;L^{\frac{p}{b+1}})}, \quad \frac{1}{\chi} = 1 - \frac{b+1}{q},$$

and

$$\|Gf\|_{L^\infty(I;L^r)} \leq C \int_0^t |t-s|^{-\frac{n}{2m}(\frac{b+1}{p}-\frac{1}{r})} \|f(x,s)\|_{\frac{p}{b+1}} ds$$

$$\leq C \int_0^t |t-s|^{-\frac{n}{2m}(\frac{b+1}{p}-\frac{1}{r})} s^{-\frac{b+1}{q}} ds \|f\|_{C_{\frac{q}{b+1}}(I;L^{\frac{p}{b+1}})}$$

$$\leq CT^{1-\frac{nb}{2mr}} \|f\|_{C_{\frac{q}{b+1}}(I;L^{\frac{p}{b+1}})},$$

where $C = C(n,p,r,b)$ is independent of T. Thus, the desired estimates hold.

Next, we consider the case when $p \geq r(1+b)$. By the Riesz interpolation theorem and Hölder inequality, we derive

$$\|Gf\|_{L^\infty(I;L^r)} \leq C \int_0^t \||f(x,s)|^{\frac{1}{b+1}}\|_{r(b+1)}^{b+1} ds$$

$$= C \int_0^t \||f|^{\frac{1}{b+1}}\|_r^{(b+1)\theta} \||f|^{\frac{1}{b+1}}\|_p^{(b+1)(1-\theta)} ds$$

$$\leq CT^{1-\frac{(b+1)(1-\theta)}{q}} \||f|^{\frac{1}{b+1}}\|_{C(I;L^r)}^{(b+1)\theta} \||f|^{\frac{1}{b+1}}\|_{L^q(I;L^p)}^{(b+1)(1-\theta)}$$

$$\leq CT^{1-\frac{nb}{2mr}} \||f|^{\frac{1}{b+1}}\|_{C(I;L^r)}^{(b+1)\theta} \||f|^{\frac{1}{b+1}}\|_{L^q(I;L^p)}^{(b+1)(1-\theta)},$$

where we have used the fact that

$$\frac{1}{r(b+1)}=\frac{\theta}{r}+\frac{1-\theta}{p},\quad 1=\frac{(1+b)(1-\theta)}{q}+\frac{1}{\chi},\quad \frac{1}{\chi}=1-\frac{nb}{2mr};$$

$$\begin{aligned}\|Gf\|_{L^q(I;L^p)} &\le C\Big\|\int_0^t |t-s|^{-\frac{n}{2m}(\frac{1}{r}-\frac{1}{p})}\Big\||f|^{\frac{1}{b+1}}\|^{b+1}_{r(b+1)}ds\|_q\\ &\le C\Big\|\int_0^t |t-s|^{-\frac{n}{2m}(\frac{1}{r}-\frac{1}{p})}\Big\||f|^{\frac{1}{b+1}}\|_r^{(b+1)\theta}\||f|^{\frac{1}{b+1}}\|_p^{(b+1)(1-\theta)}ds\|_q\\ &\le C\Big(\int_0^T t^{-\frac{n}{2m}(\frac{1}{r}-\frac{1}{p})\chi}dt\Big)^{\frac{1}{\chi}}\||f|^{\frac{1}{b+1}}\|^{(b+1)\theta}_{C(I;L^r)}\||f|^{\frac{1}{b+1}}\|^{(b+1)(1-\theta)}_{L^q(I;L^p)}\\ &\le CT^{1-\frac{nb}{2mr}}\||f|^{\frac{1}{b+1}}\|^{(b+1)\theta}_{C(I;L^r)}\||f|^{\frac{1}{b+1}}\|^{(b+1)(1-\theta)}_{L^q(I;L^p)},\end{aligned}$$

where

$$\frac{1}{r(b+1)}=\frac{\theta}{r}+\frac{1-\theta}{p},\quad 1+\frac{1}{q}=\frac{(1+b)(1-\theta)}{q}+\frac{1}{\chi},$$

$$\frac{1}{\chi}-\frac{n}{2m}\Big(\frac{1}{r}-\frac{1}{p}\Big)=1-\frac{(1+b)}{q}(1-\theta)=1-\frac{nb}{2mr};$$

$$\begin{aligned}\|Gf\|_{L^\infty(I;L^r)} &\le C\int_0^t \||f(x,s)|^{\frac{1}{b+1}}\|^{b+1}_{r(b+1)}ds\\ &= C\int_0^t \||f|^{\frac{1}{b+1}}\|_r^{(b+1)\theta}\||f|^{\frac{1}{b+1}}\|_p^{(b+1)(1-\theta)}ds\\ &\le C\||f|^{\frac{1}{b+1}}\|^{(b+1)\theta}_{C(I;L^r)}\int_0^t \||f|^{\frac{1}{b+1}}\|_p^{(b+1)(1-\theta)}ds\\ &\le C\||f|^{\frac{1}{b+1}}\|^{(b+1)\theta}_{C(I;L^r)}\||f|^{\frac{1}{b+1}}\|^{(b+1)(1-\theta)}_{C_q(I;L^p)}\int_0^t s^{-\frac{1}{q}(b+1)(1-\theta)}ds\\ &\le CT^{1-\frac{nb}{2mr}}\||f|^{\frac{1}{b+1}}\|^{(b+1)\theta}_{C(I;L^r)}\||f|^{\frac{1}{b+1}}\|^{(b+1)(1-\theta)}_{C_q(I;L^p)},\end{aligned}$$

and

$$\begin{aligned}\|Gf\|_{C_q(I;L^p)} &\le C\sup_{t\in I} t^{\frac{1}{q}}\int_0^t |t-s|^{-\frac{n}{2m}(\frac{1}{r}-\frac{1}{p})}\||f|^{\frac{1}{b+1}}\|^{b+1}_{r(b+1)}ds\\ &\le C\sup_{t\in I} t^{\frac{1}{q}}\int_0^t |t-s|^{-\frac{n}{2m}(\frac{1}{r}-\frac{1}{p})}\||f|^{\frac{1}{b+1}}\|_r^{(b+1)\theta}\||f|^{\frac{1}{b+1}}\|_p^{(b+1)(1-\theta)}ds\\ &\le C\sup_{t\in I} t^{\frac{1}{q}}\int_0^t |t-s|^{-\frac{n}{2m}(\frac{1}{r}-\frac{1}{p})}s^{-\frac{(b+1)(1-\theta)}{q}}ds\end{aligned}$$

$$\times \big\| |f|^{\frac{1}{b+1}} \big\|^{(b+1)\theta}_{C(I;L^r)} \big\| |f|^{\frac{1}{b+1}} \big\|^{(b+1)(1-\theta)}_{C_q(I;L^p)}$$
$$\leq CT^{1-\frac{nb}{2mr}} \big\| |f|^{\frac{1}{b+1}} \big\|^{(b+1)\theta}_{C(I;L^r)} \big\| |f|^{\frac{1}{b+1}} \big\|^{(b+1)(1-\theta)}_{C_q(I;L^p)}.$$

Here, $C = C(n,p,r,b)$ is a constant independent of T. Thus, the theorem is proved. □

Similarly, we have the following.

Corollary 2.1.1. *Let $Q(D)$ be a homogeneous pseudodifferential operator of order $d \in [0,2m)$. For $b > 0$ and $T > 0$, we define $r_c = nb/(2m-d)$ and $I = [0,T)$. Let $r \geq r_c$ if $r_c > 1$, $r > 1$ if $r_c \leq 1$ and let (q,p,r) be any generalized admissible triplet satisfying that $p > b+1$.*

(i) *If $f \in L^{\frac{q}{b+1}}(I; L^{\frac{p}{b+1}})$, then $G(Q(D)f) \in L^q(I;L^p) \cap C_b(I;L^r)$ and*

$$\|G(Q(D)f)\|_{L^q(I;L^p)} + \|G(Q(D)f)\|_{L^\infty(I;L^r)}$$
$$\leq CT^{1-\frac{d}{2m}-\frac{nb}{2mr}} \|f\|_{L^{\frac{q}{b+1}}(I;L^{\frac{p}{b+1}})}, \quad p < r(1+b), \tag{2.25}$$

and

$$\|G(Q(D)f)\|_{L^q(I;L^p)} + \|G(Q(D)f)\|_{L^\infty(I;L^r)}$$
$$\leq CT^{1-\frac{d}{2m}-\frac{nb}{2mr}} \big\| |f|^{\frac{1}{1+b}} \big\|^{\theta(b+1)}_{L^\infty(I;L^r)} \big\| |f|^{\frac{1}{1+b}} \big\|^{(1-\theta)(b+1)}_{L^q(I;L^p)}, \quad p \geq r(1+b), \tag{2.26}$$

where $\theta = \frac{p-r(b+1)}{(b+1)(p-r)}$, $C > 0$ is a constant independent of f and T.

(ii) *If $f \in C_{\frac{q}{b+1}}(I; L^{\frac{p}{b+1}})$, then $G(Q(D)f) \in C_{q(p,r)}(I;L^p) \cap C_b(I;L^r)$ and*

$$\|G(Q(D)f)\|_{C_{q(p,r)}(I;L^p)} + \|G(Q(D)f)\|_{L^\infty(I;L^r)}$$
$$\leq CT^{1-\frac{d}{2m}-\frac{nb}{2mr}} \|f\|_{C_{\frac{q}{b+1}}(I;L^{\frac{p}{b+1}})}, \quad p < r(1+b), \tag{2.27}$$

and

$$\|G(Q(D)f)\|_{C_{q(p,r)}(I;L^p)} + \|G(Q(D)f)\|_{L^\infty(I;L^r)}$$
$$\leq CT^{1-\frac{d}{2m}-\frac{nb}{2mr}} \big\| |f|^{\frac{1}{1+b}} \big\|^{\theta(b+1)}_{L^\infty(I;L^r)} \big\| |f|^{\frac{1}{1+b}} \big\|^{(1-\theta)(b+1)}_{C_q(I;L^p)}, \quad p \geq r(1+b), \tag{2.28}$$

with $\theta = \frac{p-r(b+1)}{(b+1)(p-r)}$, $C > 0$ independent of f and T.

Remark 2.1.3.

(i) For the homogeneous part $S(t)\varphi = e^{-At}\varphi$ of the solution to the problem (2.20), the space-time estimate (2.19) holds only for an admissible triplet (q,p,r). There is no space-time estimate in $L^q(I;L^p)$ of $S(t)\varphi$ for any generalized admissible triplet (q,p,r) (cf. Theorem 2.1.2). However, for the non-homogeneous part Gf of the solution, the space-time estimates in $L^q(I;L^p)$ exist for any generalized admissible triplet (q,p,r), as seen from Theorem 2.1.3 and Corollary 2.1.1. This means that the inhomogeneous part of the solution for the linear problem is of better integrability (i. e., more regular) than its homogeneous part in some sense.

(ii) It is more suitable to choose spaces like $\mathcal{C}_{q(p,r)}(I;L^p)$ as auxiliary spaces of $C(I;L^r)$ to study nonlinear parabolic equations. It also extends the initial space to nonreflexive Banach spaces, and the solutions studied contain the self-similar solutions. On the other hand, since the parabolic equations is of better smoothness and integrability, we can study the well-posedness for nonlinear parabolic equations in spaces like

$$C([0,\infty);L^r)\cap\mathcal{C}_{q(p,r)}([0,\infty);L^p),$$

where (q,p,r) is generalized admissible triplet. Of course, similar to the wave equations and dispersive equations, we can choose $L^q(I;L^p)$ as auxiliary spaces of $C(I;L^r)$ to study the well-posedness for nonlinear parabolic equations; here (q,p,r) is admissible triplet.

(iii) Let (q,p,r) be any generalized admissible triplet and let $r_1<r<r_2$. Though (2.19) does not hold, we have

$$\|S(t)\varphi\|_{L^q(I;L^p)}\le C\|\varphi\|_{L^{r_1}\cap L^{r_2}}. \tag{2.29}$$

In fact, by the L^p-L^r estimate,

$$\|S(t)\varphi\|_p\le Ct^{-\frac{n}{2m}(\frac{1}{r_1}-\frac{1}{p})}\|\varphi\|_{r_1},\quad \|S(t)\varphi\|_p\le Ct^{-\frac{n}{2m}(\frac{1}{r_2}-\frac{1}{p})}\|\varphi\|_{r_2},$$

so

$$\|S(t)\varphi\|_p\le C\min(t^{-\frac{n}{2m}(\frac{1}{r_1}-\frac{1}{p})},t^{-\frac{n}{2m}(\frac{1}{r_2}-\frac{1}{p})})\|\varphi\|_{L^{r_1}\cap L^{r_2}}.$$

This together with Young's inequality implies the estimate (2.29). For example, when dealing with small global well-posedness (without use of Picard's iteration) we can establish, under the condition $\|\varphi\|_{L^{r_1}\cap L^{r_2}}\ll 1$, the existence of small global solutions in $C(\mathbb{R}^+;L^r)\cap L^q(\mathbb{R}^+;L^p)$ to nonlinear parabolic equations.

We conclude this section by introducing two variants of the Hölder–Sobolev inequalities and a Sobolev convex inequality, which are needed later.

Lemma 2.1.3 (Hölder–Sobolev's inequalities). *Let $k_j\ge 0$ and α_j be multi-indices on $\mathbb{R}^n$ with $|\alpha_j|\le k_j$ $(j=1,2,\dots,N)$. For $b_j\ge 0$ and $1\le p,r_j<\infty$, let*

$$d_j=b_j\Big(\frac{1}{r_j}-\frac{k_j-|\alpha_j|}{n}\Big),\quad j=1,2,\dots,N.$$

(i) *If* $d_j > 0$ *and* $\sum_{j=1}^N d_j = 1/p$, *then*

$$\left\| \prod_{j=1}^{N} |\partial^{\alpha_j} u_j|^{b_j} \right\|_p \leq C \prod_{j=1}^{N} \|u_j\|_{\dot{H}^{k_j,r_j}}^{b_j}. \tag{2.30}$$

(ii) *If*

$$\sum_{d_j>0} d_j \leq \frac{1}{p} \leq \sum_{j=1}^{N} \frac{b_j}{r_j},$$

then

$$\left\| \prod_{j=1}^{N} |\partial^{\alpha_j} u_j|^{b_j} \right\|_p \leq C \prod_{j=1}^{N} \|u_j\|_{H^{k_j,r_j}}^{b_j}. \tag{2.31}$$

Lemma 2.1.4. *Let* $1 < p_j, r_j < \infty$, $0 \leq \theta_j \leq 1$ *and* $\sigma_j, \sigma \in \mathbb{R}$ $(j = 1, 2, \ldots, N)$. *If* $\sum_{j=1}^N \theta_j = 1$ *and*

$$\sigma = \sum_{j=1}^{N} \theta_j \sigma_j, \quad \frac{1}{p} = \sum_{j=1}^{N} \frac{\theta_j}{p_j}, \quad \frac{1}{r} = \sum_{j=1}^{N} \frac{\theta_j}{r_j},$$

then $\bigcap_{j=1}^N \dot{B}^{\sigma_j}_{p_j,r_j} \subseteq \dot{B}^{\sigma}_{p,r}$ *and*

$$\|v\|_{\dot{B}^{\sigma}_{p,r}} \leq \prod_{j=1}^{N} \|v\|^{\theta_j}_{\dot{B}^{\sigma_j}_{p_j,r_j}}, \quad v \in \bigcap_{j=1}^{N} \dot{B}^{\sigma_j}_{p_j,r_j}. \tag{2.32}$$

Remark 2.1.4.

(i) Lemma 2.1.3 is an immediate consequence of the Sobolev embedding theorem and the Hölder inequality. Lemma 2.1.4 follows directly by interpolation (see [106] for details).

(ii) In Lemma 2.1.3, if $|\alpha_j| < k_j$ then (2.30) can be replaced by

$$\left\| \prod_{j=1}^{N} |\partial^{\alpha_j} u_j|^{b_j} \right\|_p \leq C \prod_{j=1}^{N} \|u_j\|^{b_j}_{\dot{B}^{k_j}_{r_j,2}}. \tag{2.33}$$

In fact, one can easily choose $\tilde{r}_j = r_j + \epsilon_j$ and $\tilde{k}_j = k_j - \delta_j$ such that for $j = 1, 2, \ldots, N$,

$$d_j = b_j\left(\frac{1}{r_j} - \frac{k_j - |\alpha_j|}{n}\right) = b_j\left(\frac{1}{\tilde{r}_j} - \frac{\tilde{k}_j - |\alpha_j|}{n}\right) \quad \text{and} \quad k_j - \delta_j \geq |\alpha_j|.$$

Then it follows from the Sobolev embedding theorem (i. e., $\dot{B}^{k_j}_{r_j,2} \hookrightarrow \dot{H}^{\tilde{k}_j,\tilde{r}_j}$) and Corollary 2.1.1 that

$$\left\| \prod_{j=1}^{N} |\partial^{a_j} u_j|^{b_j} \right\|_p \leq C \prod_{j=1}^{N} \|u_j\|_{\dot{H}^{\bar{k}_j,\bar{r}_j}}^{b_j} \leq C \prod_{j=1}^{N} \|u_j\|_{\dot{B}^{k_j}_{r_j,2}}^{b_j}.$$

Remark 2.1.5. Taking into account the regularity of the parabolic operators, we have more general space-time estimates.

(i) Let $I = [0, T)$ with $0 < T \leq \infty$ and let (q, p, s, r, η) satisfy

$$\frac{2m}{q} = \left(s - \frac{n}{p}\right) - \left(\eta - \frac{n}{r}\right), \quad s \geq \eta,$$

where

$$1 < r \leq p < \begin{cases} \frac{nr}{(s-\eta)r+(n-2m)}, & \text{for} \quad (s-\eta)r + (n-2m) > 0, \\ \infty, & \text{for} \quad (s-\eta)r + (n-2m) \leq 0. \end{cases}$$

Then

$$\|e^{-At}\varphi\|_{L^q(I;\dot{H}^{s,p})} \leq C\|\varphi\|_{\dot{H}^{\eta,r}}. \tag{2.34}$$

(ii) Assume that

$$\frac{2m}{q_1} - \frac{2m}{q_2} = 2m - \left(s_2 - \frac{n}{p_2}\right) + \left(s_1 - \frac{n}{p_1}\right),$$

where

$$1 < p_1 \leq p_2 < \begin{cases} \frac{np_1}{(s_2-s_1)p_1+(n-2p_1m)}, & \text{for} \quad (s_2 - s_1)p_1 > n - 2p_1m, \\ \infty, & \text{for} \quad (s_2 - s_1)p_1 \leq n - 2p_1m, \end{cases}$$

$$1 < q_1 \leq q_2 < \infty, \quad s_1 \leq s_2,$$

or

$$1 < p_1 \leq p_2 < \infty, \quad s_1 \leq s_2, \quad 1 < q_1 \leq p_2 < q_2 = \infty.$$

Then

$$\|Gf\|_{L^{q_2}(I;\dot{H}^{s_2,p_2})} \leq C\|f\|_{L^{q_1}(I;\dot{H}^{s_1,p_1})}, \tag{2.35}$$

$$\|Gf\|_{C_b(I;\dot{H}^{s_2,p_2})} \leq C\|f\|_{L^{q_1}(I;\dot{H}^{s_1,p_1})}, \tag{2.36}$$

where $I = [0, T)$ with $0 < T \leq \infty$.

2.2 The Cauchy problem for semilinear heat equations (I)

In this section, we consider the Cauchy problem for the semilinear heat equation

$$u_t - \Delta u = F(u), \quad (x,t) \in \mathbb{R}^n \times \mathbb{R}^+, \tag{2.37}$$

$$u(0) = \varphi(x), \quad x \in \mathbb{R}^n \tag{2.38}$$

with the nonlinear term $F(u) = f(u)$ or $Q(D)f(u)$, where $Q(D)$ is a homogeneous pseudodifferential operator of order $d \in [0,2)$ and $f(u)$ is a nonlinear function with polynomial growth. We will make use of the space-time estimates and a scaling argument to construct well-suited solution spaces and establish local and global well-posedness in spaces like L^p or $H^{s,p}$ for problem (2.37)–(2.38). Of course, the global well-posedness requires more strict conditions on the nonlinear term.

In this section, we only consider semilinear parabolic equations. Thus, if (q,p,r) is an admissible triplet, then

$$\frac{1}{q} = \frac{n}{2}\left(\frac{1}{r} - \frac{1}{p}\right), \tag{2.39}$$

where

$$1 < r \le p < \begin{cases} \frac{nr}{n-2}, & \text{for} \quad n > 2, \\ \infty, & \text{for} \quad n \le 2. \end{cases}$$

If (q,p,r) is a generalized admissible triplet, then

$$\frac{1}{q} = \frac{n}{2}\left(\frac{1}{r} - \frac{1}{p}\right), \tag{2.40}$$

where

$$1 < r \le p < \begin{cases} \frac{nr}{n-2r}, & \text{for} \quad n > 2r, \\ \infty, & \text{for} \quad n \le 2r. \end{cases}$$

For convenience, we denote by Λ the set of all admissible triplets satisfying $p > r$, and by $\tilde{\Lambda}$ the set of all generalized admissible triplets satisfying $p > r$.

To study the Cauchy problem (2.37)–(2.38), we will consider its equivalent integral equation,

$$u = e^{t\Delta}\varphi(x) + \int_0^t e^{(t-\tau)\Delta} F(u(\tau,x))d\tau. \tag{2.41}$$

The solution of the integral equation (2.41) is called a mild solution of the Cauchy problem (2.37)–(2.38). From the regularity theory of parabolic equations, it follows that a mild solution is also a regular solution ($t > 0$).

Theorem 2.2.1. *Assume that $F(u) = f(u)$ or $F(u) = Q(D)f(u)$, where $Q(D)$ is a pseudodifferential operator of order $d \in [0,2)$ and $f(u)$ is a nonlinear function satisfying that*

$$|f(u) - f(v)| \leq \sum_{j=1}^{2} C_j(|u|^{a_j} + |v|^{a_j})|u - v|, \tag{2.42}$$

for some $a_2 \geq a_1 > 0$ and generic constants $C_j > 0$ ($j = 1,2$). Set $r \geq r_2 > 1$, here $r_j = \frac{na_j}{2-d}$, $j = 1,2$. Then we have the following results:

(i) *Let $\varphi \in L^r$, $r \geq r_2$, $(q,p,r) \in \tilde{\Lambda}$. Then there exists a unique maximal solution $u(t)$ to problem* (2.37)–(2.38) *such that $u \in C([0,T^*);L^r) \cap \mathcal{C}_{q(p,r)}([0,T^*);L^p)$ with $T^* = T(\|\varphi\|_r)$ if $r > r_2$ or $u \in C([0,T^*);L^r) \cap \dot{\mathcal{C}}_{q(p,r)}([0,T^*);L^p)$ with $T^* = T(\varphi)$ if $r = r_2 > 1$.*
(ii) *Let $\varphi \in L^r$, $r \geq r_2$, $(q,p,r) \in \Lambda$. Then there exists a unique maximal solution $u(t)$ to problem* (2.37)–(2.38) *such that $u \in C([0,T^*);L^r) \cap L^q([0,T^*);L^p)$ with $T^* = T(\|\varphi\|_r)$ if $r > r_2$ or $u \in C([0,T^*);L^r) \cap L^q([0,T^*);L^p)$ with $T^* = T(\varphi)$ if $r = r_2 > 1$, where T^* is the same as in* (i).
(iii) *$u \in C((0,T^*);L^r \cap L^\infty)$.*
(iv) *If $T^* < \infty$, then*

$$\lim_{t \to T^*} \|u(t)\|_p = \infty, \quad r \leq p \leq \infty, \ p > r_2.$$

Precisely, we have

$$\|u(t)\|_p \geq C/(T^* - t)^{\frac{2-d}{2a_2} - \frac{n}{2p}}, \quad 0 \leq d < 2. \tag{2.43}$$

(v) *If, in addition, $f \in C^\infty(\mathbb{R})$, then $u \in C^\infty((0,T^*) \times \mathbb{R}^n)$.*

Corollary 2.2.1. *In Theorem 2.2.1, if $f(u) = b|u|^a u$ with $b < 0$ and $r \geq \max(\frac{na}{2}, 2)$, then $T^* = \infty$, that is, there exists a unique maximal solution $u(t)$ to problem* (2.37)–(2.38) *such that $u(t) \in C([0,\infty);L^r) \cap \mathcal{C}_{q(p,r)}([0,\infty);L^p)$, $(q,p,r) \in \tilde{\Lambda}$, $u(t) \in C([0,\infty);L^r) \cap \mathcal{C}_{q(p,r)}([0,\infty);L^p)$, $(q,p,r) \in \Lambda$.*

Corollary 2.2.2. *In Theorem 2.2.1, if $F(u) = \vec{a} \cdot \nabla(|u|^a u)$ with $\vec{a} \in \mathbb{R}^n$ and $r \geq na > 1$, then $T^* = \infty$, and the solution $u(t)$ to problem* (2.37)–(2.38) *satisfies the above two conditions.*

Theorem 2.2.2. *Assume that $F(u) = f(u)$ or $F(u) = Q(D)f(u)$, where $Q(D)$ is a homogeneous pseudodifferential operator of order $d \in [0,2)$ and $f(u)$ satisfies the same condition as in Theorem* 2.2.1. *Set $r_j = \frac{na_j}{2} > 1$ or $r_j = \frac{na_j}{2-d} > 1$, $j = 1,2$.*

(i) *Let $(q_j,p_j,r_j) \in \tilde{\Lambda}$ such that $p_j \leq r_j(1 + a_j)$ ($j = 1,2$). Then there is a $\delta > 0$ such that for $\varphi \in L^{r_1} \cap L^{r_2}$ with $\|\varphi; L^{r_1} \cap L^{r_2}\| < \delta$, there exists a unique global solution $u \in C([0,\infty);L^{r_1} \cap L^{r_2})$ to problem* (2.37)–(2.38) *satisfying that*

$$u \in \bigcap_{j=1}^{2} \dot{C}_{q_j(p_j,r)}([0,\infty); L^{p_j}) \bigcap_{j=1}^{2} L^{q_j}([0,\infty); L^{p_j}). \tag{2.44}$$

(ii) *Let $\alpha_1 = \alpha_2$ (e. g., $f(u)$ is monomial), $r = \frac{n\alpha}{2-d} > 1$. Then there exists $\delta > 0$, such that $\forall \varphi \in L^r$ with $\|\varphi; L^r\| < \delta$, there exists a unique global solution $u \in C([0,\infty); L^r)$ to problem* (2.37)–(2.38) *satisfying that*

$$u(t) \in L^q([0,\infty); L^p), \quad (q,p,r) \in \Lambda.$$

Remark 2.2.1.

(i) Our method can also be applied to the case when

$$F(u) = \sum_{j=1}^{m} Q_j(D) f_j(u),$$

where $Q_j(D)$ is a homogeneous pseudodifferential operator of order $d_j \in [0,2)$ and $f_j(u)$ behaves like $|u|^{\alpha_j}u$ or $|u|^{\alpha_j+1}$ with $\alpha_j > 0$ ($j = 1,\dots,m$). In fact, let $r_j = \frac{n\alpha_j}{n-d_j}$ ($j = 1,2,\dots,m$) and replace r_1 and r_2 in Theorems 2.2.1 and 2.2.2 by $r_s = \min_j r_j$ and $r_l = \max_j r_j$, respectively. Then similar results to Theorems 2.2.1 and 2.2.2 still hold.

(ii) Let $u(t,x)$ be a smooth solution of problem (2.37)–(2.38) and let $f(u) = b|u|^\alpha u$ with $b < 0$. Then we obtain by the maximum principle and energy estimate method, $\|u\|_\infty \le \|\varphi\|_\infty$, $\|u\|_2 \le \|\varphi\|_2$. Then, by the interpolation theorem, we find that $\|u\|_r \le \|\varphi\|_r$, $2 \le r \le \infty$. This implies the conclusion of Corollary 2.2.1 in this case. On the other hand, if $F(u) = \vec{a} \cdot \nabla(|u|^\alpha u)$ with $\vec{a} \in \mathbb{R}^n$ and $r \ge n\alpha > 1$, then it is easy to see that $\|u\|_r \le \|\varphi\|_r$, $2 \le r \le \infty$. Thus, the conclusion of Corollary 2.2.2 follows from Theorem 2.2.1 in this case.

(iii) One easily sees that the solution space corresponds to the "highest" nonlinear growth by the scaling argument. On the other hand, in Theorem 2.2.1, we can also deal with the case when $r_2 = \frac{n\alpha_2}{2-d} \le 1$.

Proof of Theorem 2.2.1. Proof of (i). Denote by Π the set of all generalized admissible triplets (q,p,r) satisfying that $1 + \alpha_2 < p$, and let $I = [0,T)$. Define

$$X(I) := \{u \in C_b(I; L^r) \cap C_{q(p,r)}(I; L^p),\ (q,p,r) \in \Pi\}$$

with the norm

$$\|u; X(I)\| := \sup_{(q,p,r)\in\Pi} \sup_{t\in I} t^{\frac{1}{q}} \|u\|_p + \sup_{t\in I} \|u\|_r.$$

Denote by $\mathcal{T}$ the mapping defined by (2.41), that is,

$$\mathcal{T}u := S(t)\varphi + \int_0^t S(t-\tau)F(u(\tau))d\tau = S(t)\varphi + GF(u), \quad S(t) = e^{t\Delta}. \tag{2.45}$$

By the space-time estimates of parabolic equations (see Theorem 2.1.1), we know

$$\|S(t)\varphi; X(I)\| \le C\|\varphi\|_r, \quad I = [0, T).$$

We study the operator $\mathcal{T}$ in metric space like

$$\mathcal{X}(I) = \{u \in X(I), \ \|u; X(I)\| \le 2C\|\varphi\|_r\},$$
$$d(u, v) = \|u - v; X(I)\|.$$

Notice that $Q(\xi) \le C(1 + |\xi|^2)^{\frac{d}{2}}$ and $Q(\xi)$ can be controlled by $Q(\xi) = |\xi|^d$ and $Q(\xi) \equiv 1$. Thus, we only consider the case $Q(\xi) = |\xi|^d$, $0 \le d < 2$ ($Q(\xi) = 1$ corresponds to the case $d \equiv 0$). Notice the fact that $d/2 + n\alpha_1/(2p) < 1$ and $q > 1 + \alpha_1$. From Corollary 2.1.1, it follows that

$$\begin{aligned}\|\mathcal{T}u; X(I)\| &\le C\|\varphi\|_r + C\sum_{j=1}^{2} T^{1-\frac{d}{2}-\frac{n\alpha_j}{2r}} \|u\|_{X(I)}^{1+\alpha_j} \\ &\le C\|\varphi\|_r + C\sum_{j=1}^{2} T^{1-\frac{d}{2}-\frac{n\alpha_j}{2r}} M^{1+\alpha_j}.\end{aligned} \tag{2.46}$$

Similarly, we can easily obtain that

$$d(\mathcal{T}u, \mathcal{T}v) \le C\sum_{j=1}^{2} T^{1-\frac{d}{2}-\frac{n\alpha_j}{2r}} M^{\alpha_j} d(u, v). \tag{2.47}$$

Case 1. $r > r_2$. It is seen from (2.46) and (2.47) that the operator $\mathcal{T}$ is a contraction mapping from $\mathcal{X}(I)$ to itself provided that T is suitably small, for example,

$$T \le \min\{(2^{\alpha_1+1}C^{\alpha_1+1}\|\varphi\|_r^{\alpha_1})^{-\frac{2r}{2r-dr-n\alpha_1}}, (2^{\alpha_2+1}C^{\alpha_2+1}\|\varphi\|_r^{\alpha_2})^{-\frac{2r}{2r-dr-n\alpha_2}}\}. \tag{2.48}$$

Thus, the Banach contraction mapping theorem implies that there exists a unique solution u in $\mathcal{X}(I)$ to problem (2.37)–(2.38). Moreover, by the Picard method, there is a maximal $T^* = T(\|\varphi\|_r)$ such that $u \in X([0, T^*))$ and either $T^* = \infty$ or $T^* < \infty$ and

$$\lim_{t \to T^*} \|u\|_r = \infty. \tag{2.49}$$

Case 2. $r = r_2 > 1$. Define

$$\dot{X}(I) := \{u \in C_b(I; L^r) \cap \dot{C}_{q(p,r)}(I; L^p), \ (q, p, r) \in \Pi\}$$

with the norm

$$\|u\|_{\dot{X}(I)} = \sup_{(q,p,r)\in\Pi} \sup_{t\in I} t^{\frac{1}{q}} \|u\|_p + \sup_{t\in I} \|u\|_r,$$

and the complete metric space

$$\dot{\mathcal{X}}(I) = \{u \in \dot{X}(I), \ \|u; \dot{X}(I)\| \le 2C\|\varphi\|_r\}$$

with the metric

$$d(u,v) = \|u - v\|_{\dot{X}(I)}, \quad u, v \in \dot{\mathcal{X}}(I).$$

We now use the contraction mapping principle to study the operator $\mathcal{T}$, defined in (2.45), in the above metric space. It is easy to derive that

$$\begin{aligned} \|\mathcal{T}u; \dot{X}(I)\| &\le C\|\varphi\|_r + CT^{1-\frac{d}{2}-\frac{n\alpha_1}{2r}} \|u\|^{\alpha_1+1}_{C_{q(p,r)}(I;L^p)} \\ &\quad + C\|u\|^{\alpha_2+1}_{C_{q(p,r)}(I;L^p)}, \quad p < r_2(1+\alpha_2), \end{aligned} \tag{2.50}$$

$$\begin{aligned} \|\mathcal{T}u; \dot{X}(I)\| &\le C\|\varphi\|_r + CT^{1-\frac{d}{2}-\frac{n\alpha_1}{2r}} \|u\|^{\theta_1(\alpha_1+1)}_{L^\infty(I;L^r)} \|u\|^{(1-\theta_1)(\alpha_1+1)}_{C_{q(p,r)}(I;L^p)} \\ &\quad + C\|u\|^{\theta_2(\alpha_2+1)}_{L^\infty(I;L^r)} \|u\|^{(1-\theta_2)(\alpha_2+1)}_{C_{q(p,r)}(I;L^p)}, \quad p \ge r_2(1+\alpha_2), \end{aligned} \tag{2.51}$$

$$\begin{aligned} t^{\frac{1}{q}} \|\mathcal{T}u\|_p &\le t^{\frac{1}{q}} \|S(t)\varphi\|_p + Ct^{1-\frac{d}{2}-\frac{n\alpha_1}{2r}} (t^{\frac{1}{q}} \|u\|_p)^{\alpha_1+1} \\ &\quad + C(t^{\frac{1}{q}} \|u\|_p)^{\alpha_2+1}, \quad p < r_2(1+\alpha_2), \end{aligned} \tag{2.52}$$

$$\begin{aligned} t^{\frac{1}{q}} \|\mathcal{T}u\|_p &\le t^{\frac{1}{q}} \|S(t)\varphi\|_p + Ct^{1-\frac{d}{2}-\frac{n\alpha_1}{2r}} \|u\|^{\theta_1(\alpha_1+1)}_{L^\infty(I;L^r)} (t^{\frac{1}{q}} \|u\|_p)^{(1-\theta_1)(\alpha_1+1)} \\ &\quad + C\|u\|^{\theta_2(\alpha_2+1)}_{L^\infty(I;L^r)} (t^{\frac{1}{q}} \|u\|_p)^{(1-\theta_2)(\alpha_2+1)}, \quad p \ge r_2(1+\alpha_2), \end{aligned} \tag{2.53}$$

$$\begin{aligned} d(\mathcal{T}u, \mathcal{T}v) &\le C\big[\|u\|^{\alpha_2}_{\dot{X}(I)} + \|v\|^{\alpha_2}_{\dot{X}(I)} \\ &\quad + T^{1-\frac{d}{2}-\frac{n\alpha_1}{2r}} (\|u\|^{\alpha_1}_{\dot{X}(I)} + \|v\|^{\alpha_1}_{\dot{X}(I)})\big] d(u,v), \end{aligned} \tag{2.54}$$

with $\theta_j = \frac{p-r(\alpha_j+1)}{(\alpha_j+1)(p-r)}$. Since $1-\theta_j > 0$ and

$$\lim_{t\to 0} t^{\frac{1}{q}} \|u\|_p = 0,$$

it then follows from (2.50)–(2.54) that $\mathcal{T}$ is a contraction mapping from $\dot{\mathcal{X}}_{(q,p,r)}(I)$ to itself provided that T is sufficient small, where $T = T(\varphi)$. Thus, by the Banach contraction mapping principle and Picard's method, there exists a maximal $T^* = T^*(\varphi)$ such that $u \in \dot{\mathcal{X}}([0, T^*))$.

Consider now the case $r \le 1 + \alpha_2$. By interpolation between $C(I; L^r)$ and any space $C_{\tilde{q}(\tilde{p},r)}(I; L^{\tilde{p}})$ with $(\tilde{q}, \tilde{p}, r) \in \Pi$, the solution u obtained above satisfies that either $u \in C_{q(p,r)}(I; L^p)$ or $u \in \dot{C}_{q(p,r)}(I; L^p)$. The proof of statement (i) is thus completed.

Proof of (ii). Denote by Π the set of all admissible triplets (q, p, r) satisfying that $1 + \alpha_2 < p$. Define

$$Y(I) := \{u \in C_b(I; L^r) \cap L^q(I; L^p),\ (q, p, r) \in \Pi,\ I = [0, T)\}$$

with the norm

$$\|u; Y(I)\| := \sup_{t \in I} \|u\|_{L^\infty(I;L^r)} + \sup_{(q,p,r)\in\Pi} \|u\|_{L^q(I;L^p)}.$$

By the space-time estimates of parabolic equations (see Theorems 2.1.1 and 2.1.2), we know

$$\|S(t)\varphi; Y(I)\| \le C\|\varphi\|_r.$$

Denoting by $\mathcal{T}$ the mapping defined by (2.41), we study the fixed point of the nonlinear mapping $\mathcal{T}$ in complete metric space

$$\mathcal{Y}(I) = \{u \in Y(I),\ \|u; Y(I)\| \le 2C\|\varphi\|_r\}, \quad d(u, v) = \|u - v; Y(I)\|,$$

where $I = [0, T)$ will be fixed. By completely similar discussions as in (i), we find that there exists a unique maximal solution u to problem (2.37)–(2.38) by using Theorem 2.1.3 and Corollary 2.1.1 (the space-time estimates of linear parabolic equations) and the Picard method. According to the alternative theorem, the solutions in (i) and (ii) have the same existing intervals. Further, by the regularity of parabolic equations, the solutions is completely same.

To prove the statement (iii), it suffices to show that for any $\epsilon > 0$, $u \in C([\epsilon, T^*), L^\infty)$. We divide the proof into two steps.

First step. We first consider the case $r > \frac{n}{2-d}$. Let $(q, p, r) \in \tilde{\Lambda}$ satisfy that

$$(1 + \alpha_2)\frac{n}{2 - d} < p < r(1 + \alpha_2). \tag{2.55}$$

By Lemma 2.1.1 and the corresponding regularity, we have for $r \neq \infty$,

$$\|(-\triangle)^{\frac{d}{2}} S(t)\varphi\|_p \le Ct^{-\frac{d}{2}-\frac{n}{2}(\frac{1}{r}-\frac{1}{p})}\|\varphi\|_r, \quad d \ge 0,\ 1 \le r \le p \le \infty. \tag{2.56}$$

It follows that

$$\begin{aligned}
&\|GF(u); C([\epsilon, T^*); L^\infty)\| \\
&\le C\sum_{j=1}^{2} \sup_{t\in[\epsilon,T^*)} \int_\epsilon^t |t - \tau|^{-\frac{n(\alpha_j+1)}{2p}} \left\|S\left(\frac{t-\tau}{2}\right)Q(D)f_j(u)\right\|_{\frac{p}{\alpha_j+1}} d\tau \\
&\le C\sum_{j=1}^{2} \int_\epsilon^{T^*} |T^* - \tau|^{-\frac{d}{2}-\frac{n(\alpha_j+1)}{2p}} \|u\|_p^{\alpha_j+1} d\tau
\end{aligned}$$

$$\leq C\sum_{j=1}^{2}(T^*)^{1-\frac{d}{2}-\frac{n(\alpha_j+1)}{2r}}\int_0^1(1-\tau)^{-\frac{d}{2}-\frac{n(\alpha_j+1)}{2p}}\tau^{-\frac{\alpha_j+1}{q}}d\tau\cdot\|u\|^{\alpha_j+1}_{C_{q(p,r)}([\epsilon;T^*);L^p)}$$

$$\leq C\sum_{j=1}^{2}(T^*)^{1-\frac{d}{2}-\frac{n(\alpha_j+1)}{2p}}\|u\|^{\alpha_j+1}_{C_{q(p,r)}([\epsilon;T^*);L^p)}, \tag{2.57}$$

where we have used the fact that

$$\frac{d}{2}+\frac{n(1+\alpha_j)}{2p}<1, \quad q>1+\alpha_j,\ j=1,2.$$

Thus, we have

$$\|u;C([\epsilon,T^*);L^\infty)\|\leq\epsilon^{-\frac{n}{2r}}\|\varphi\|_r+\|GF(u)\|_{C([\epsilon,T^*);L^\infty)}<\infty. \tag{2.58}$$

Second step. We consider the case that $r\leq\frac{n}{2-d}$. Let $p_0=r$ and $p_{N+1}=\infty$ and let us construct the generalized admissible triplets

$$(q_{j+1},p_{j+1},p_j)\in\tilde{\Lambda},\quad p_{j+1}\leq p_j(1+\alpha_2),\quad j=0,1,2,\dots,N-1$$

such that

$$p_{N-1}\leq\frac{n}{2-d},\quad p_N>\frac{n}{2-d}.$$

By the similar argument as in deriving (2.57), it is easy to obtain that

$$u\in C_{q_{j+1}(p_{j+1},p_j)}\left(\left[\frac{(j+1)\epsilon}{N+1},T^*\right);L^{p_{j+1}}\right),\quad j=0,1,\dots,N-1,$$

which implies that

$$u\in C\left(\left[\frac{(j+1)\epsilon}{N+1},T^*\right);L^{p_{j+1}}\right),\quad j=0,1,\dots,N-1.$$

In particular, $\|u(\frac{N\epsilon}{N+1})\|_{p_N}<\infty$, so by the estimate from the first step,

$$\|u;C([\epsilon,T^*);L^\infty)\|$$
$$\leq\left\|S\left(t-\frac{N\epsilon}{N+1}\right)u\left(\frac{N\epsilon}{N+1}\right)\right\|_\infty+\sup_{t\in[\epsilon,T^*)}\int_{\frac{N\epsilon}{N+1}}^{t}\|S(t-\tau)F(u)\|_\infty d\tau<\infty.$$

The statement (iii) is thus proved.

We now prove the statement (iv). If the statement were not true, then there would exist a $p\in[r,\infty]$ such that $\|u(T^*)\|_p<\infty$ with $T^*<\infty$. We show that this implies that

$$\|u(T^*)\|_r < \infty, \tag{2.59}$$

which contradicts with the definition of T^*. We distinguish three cases.

Case 1. $p < (1 + \alpha_2)r$. Since $\|u(T^*)\|_p < \infty$ and $u \in C_{q(p,r)}([0, T^*); L^p)$, it is easy to see that

$$\sup_{t\in[\frac{T^*}{2},T^*]} \|u(t)\|_p < \infty. \tag{2.60}$$

By Theorem 2.1.3 and Corollary 2.1.1, we obtain that

$$\|u(T^*)\|_r \le C\|\varphi\|_r + C\sum_{j=1}^{2} \int_{\frac{T^*}{2}}^{T^*} |T^* - \tau|^{-\frac{d}{2}-\frac{n}{2}(\frac{1+\alpha_j}{p}-\frac{1}{r})} d\tau \sup_{t\in[\frac{T^*}{2},T^*]} \|u(t)\|_p^{\alpha_j+1} < \infty. \tag{2.61}$$

Case 2. $r(1+\alpha_2) \le p < \infty$. Let $p_0 = r$ and $p_{N+1} = p$, and let us construct the generalized admissible triplets

$$(q_{j+1}, p_{j+1}, p_j) \in \Pi, \quad p_{j+1} \le p_j(1 + \alpha_2), \quad j = 0, \dots, N.$$

Then it follows that

$$\begin{aligned} \|u(T^*)\|_{p_N} &\le T^{*-\frac{n}{2}(\frac{1}{r}-\frac{1}{p_N})} \|\varphi(x)\|_r \\ &\quad + C\sum_{j=1}^{2} \int_{\frac{T^*}{2}}^{T^*} |T^* - \tau|^{-\frac{d}{2}-\frac{n}{2}(\frac{1+\alpha_j}{p}-\frac{1}{p_N})} d\tau \sup_{t\in[\frac{T^*}{2},T^*]} \|u(t)\|_p^{\alpha_j+1} < \infty. \end{aligned}$$

Thus, we can estimate $\|u(T^*)\|_{p_{j+1}}$ by $\|u(T^*)\|_{p_j}$, and after a finite number of steps to get the estimate (2.59).

Case 3. $p = \infty$. It is easy to see that

$$\begin{aligned} \sup_{t\in[\delta T^*,T^*]} \|u(t)\|_r &\le \|\varphi\|_r + C \int_{\delta T^*}^{T^*} |T^* - \tau|^{-\frac{d}{2}} d\tau \sum_{j=1}^{2} \sup_{t\in[\delta T^*,T^*]} (\|u(t)\|_\infty^{\alpha_j} \|u(t)\|_r) \\ &\le \|\varphi\|_r + C(T^*)^{1-\frac{d}{2}} \int_{\delta}^{1} |1 - \tau|^{-\frac{d}{2}} d\tau \sup_{t\in[\delta T^*,T^*]} \|u(t)\|_r, \end{aligned} \tag{2.62}$$

where we have used the fact that

$$\sup_{t\in[\delta T^*,T^*]} \|u(t)\|_\infty < \infty.$$

The estimate (2.59) follows by taking $\delta \to 1$ in (2.62). The statement (iv) is thus proved.

To prove (v), it suffices to show that for any $\epsilon > 0$,

$$u(t) = u(t,x) \in C^{\infty}([\epsilon, T^*) \times \mathbb{R}^n). \tag{2.63}$$

Case 1. We first consider the case $d = 0$. Since $\partial_t^m(S(t)\varphi) = (-\Delta)^m(S(t)\varphi)$ and in view of (2.41), it is enough to prove that

$$u \in C([\epsilon, T^*); C^{\infty}). \tag{2.64}$$

Let $p > \max(n, r)$. Then by (iii) $u \in C([\epsilon, T^*); L^p \cap L^{\infty})$. It is easy to see that for $\epsilon \le t < T^*$,

$$\begin{aligned}\|u\|_{\dot{W}^{1,p}} &\le C + C\int_{\epsilon}^{t} |t-\tau|^{-\frac{1}{2}} f'(\|u\|_{\infty})\|u\|_p d\tau \\ &\le C + C(T^*)^{\frac{1}{2}}\int_0^1 |1-\tau|^{-\frac{1}{2}} d\tau < \infty,\end{aligned} \tag{2.65}$$

where we have used the fact that $f(u) \sim f'(u)u$. This implies that $u \in C([\epsilon, T^*); W^{1,p})$. Now assume that for any $k \in \mathbb{N}$, $u \in C([\epsilon, T^*); W^{k,p})$. We then want to prove that

$$u \in C([\epsilon, T^*); W^{k+1,p}). \tag{2.66}$$

In fact, the Sobolev embedding theorem implies that $(1-\Delta)^{\frac{k-1}{2}} u \in L^{\infty}$. Without loss of generality, we may assume that $|\beta_j| = \max_{i=1,\dots,j} |\beta_i|$. Similarly as in the proof of (2.65), we have

$$\begin{aligned}\|u\|_{\dot{W}^{k+1,p}} &\le C + C\int_{\epsilon}^{t} |t-\tau|^{-\frac{1}{2}} \Bigg\| \sum_{j=1}^{k} \sum_{\substack{\beta_1+\dots+\beta_j=\gamma,\\ |\beta_j|\ge 1, |\gamma|=k}} \frac{\gamma!}{j!\prod \beta_j!} f^{(j)}(u) \prod_j \partial^{\beta_j} u \Bigg\|_p d\tau \\ &\le C + C\int_{\epsilon}^{t} |t-\tau|^{-\frac{1}{2}} \sum_{j=1}^{k} f^{(j)}(\|u\|_{\infty})\|\partial^{\beta_j} u\|_p d\tau < \infty,\end{aligned}$$

which implies (2.66). Noting that $u \in C([\epsilon, T^*); L^p \cap L^{\infty})$ and (2.66) holds for any $k \in \mathbb{N}$, the required result (2.64) follows from the Sobolev embedding theorem.

Case 2. We consider the case that $d \ne 0$. Let $0 < \delta < 2 - d$ and $p > n/\delta$. By the Leibniz rule for the fractional-order derivatives (see [21, 263] for details) and the induction method, one easily obtains that $u \in C([\epsilon, T^*); W^{m\delta,p})$ for any $m \in \mathbb{N}$. Thus, by the Sobolev embedding theorem, we have (2.64). This completes the proof of (v). □

Proof of Theorem 2.2.2. Let $I = [0, \infty)$ and let us denote by F the set of all generalized admissible triplets (q_j, p_j, r_j) satisfying that

$$1 + \alpha_j < p_j < r_j(\alpha_j + 1), \quad j = 1, 2, \tag{2.67}$$

and by F^* the set of all generalized admissible triplets (q_j, p_j, r_j) satisfying that

$$(q_j, p_j, r_j) \in F, \quad \frac{r_1}{p_1} = \frac{r_2}{p_2}, \quad j = 1, 2. \tag{2.68}$$

Define

$$Z(I) = \left\{ u \in C_b(I; L^{r_1} \cap L^{r_2}) \cap \bigcap_{j=1}^{2} \dot{C}_{q_j(p_j, r_j)}(I; L^{p_j}),\ \{(q_j, p_j, r_j)\}_{j=1,2} \in F^* \right\}$$

with the norm

$$\|u; Z(I)\| = \sum_{j=1}^{2} \sup_{(q_j, p_j r_j) \in F^*} \sup_{t \in I} t^{\frac{1}{q_j}} \|u\|_{p_j} + \sum_{j=1}^{2} \sup_{t \in I} \|u\|_{r_j}.$$

Consider the operator $\mathcal{T}$ defined by (2.45) in the complete metric space

$$\mathcal{Z}(I) = \{u \in Z(I),\ \|u; Z(I)\| \le \delta\}$$

with the metric

$$d(u, v) = \|u - v; Z(I)\|, \quad u, v \in \mathcal{Z}(I),$$

where $\delta > 0$ is a sufficiently small constant to be determined later.

First, it is easy to see from Lemma 2.1.2 and Theorem 2.1.1 that

$$\|S(\cdot)\varphi; Z(I)\| \le C(\|\varphi\|_{r_1} + \|\varphi\|_{r_2}). \tag{2.69}$$

Since $Q(D)$ is a homogeneous pseudodifferential operator of order d, it suffices to consider the case $Q(\xi) \sim |\xi|^d$ with $0 \le d < 2$. For any $(q_j, p_j, r_j) \in F^*$ $(j = 1, 2)$, it follows by (2.42), Lemma 2.1.1 and (2.56) that

$$\begin{aligned}
&\|GF(u); C_{q_1(p_1, r_1)}(I; L^{p_1})\| \\
&\le C\|u; C_{q_1(p_1, r_1)}(I; L^{p_1})\|^{\alpha_1 + 1} + \sup_{t \in I} t^{\frac{1}{q_1}} \int_0^t |t - \tau|^{-\frac{d}{2} - \frac{n\alpha_2}{2p}} \|u\|_{p_2}^{\alpha_2} \|u\|_{p_1} d\tau \\
&\le C\|u; C_{q_1(p_1, r_1)}(I; L^{p_1})\|^{\alpha_1 + 1} + C \int_0^1 |1 - \tau|^{-\frac{d}{2} - \frac{n\alpha_2}{2p_2}} \tau^{-\frac{1}{q_1} - \frac{\alpha_2}{q_2}} d\tau \\
&\quad \times \|u; C_{q_2(p_2, r_2)}(I; L^{p_2})\|^{\alpha_2} \|u; C_{q_1(p_1, r_1)}(I; L^{p_1})\| \\
&\le C \sum_{j=1}^{2} \|u; C_{q_j(p_j, r_j)}(I; L^{p_j})\|^{\alpha_j} \|u; C_{q_1(p_1, r_1)}(I; L^{p_1})\|
\end{aligned} \tag{2.70}$$

and

$$\begin{aligned}
&\|GF(u); C(I; L^{r_1})\| \\
&\quad \le C \sup_{t\in I} \int_0^t |t-\tau|^{-\frac{d}{2}-\frac{n}{2}(\frac{1+\alpha_1}{p_1}-\frac{1}{r_1})} \|u\|_{p_1}^{\alpha_1+1} d\tau \\
&\qquad + C \sup_{t\in I} \int_0^t |t-\tau|^{-\frac{d}{2}-\frac{n}{2}(\frac{\alpha_2}{p_2}+\frac{1}{p_1}-\frac{1}{r_1})} \|u\|_{p_2}^{\alpha_2} \|u\|_{p_1} d\tau \\
&\quad \le C \sup_{t\in I} \int_0^t |t-\tau|^{-\frac{d}{2}-\frac{n}{2}(\frac{\alpha_1+1}{p_1}-\frac{1}{r_1})} \tau^{-\frac{\alpha_1+1}{q_1}} d\tau \cdot \|u; \mathcal{C}_{q_1(p_1,r_1)}(I; L^{p_1})\|^{\alpha_1+1} \\
&\qquad + C \sup_{t\in I} \int_0^t |t-\tau|^{-\frac{d}{2}-\frac{n}{2}(\frac{\alpha_2}{p_2}+\frac{1}{p_1}-\frac{1}{r_1})} \tau^{-\frac{\alpha_2}{q_2}-\frac{1}{q_1}} d\tau \\
&\qquad \times \|u; \mathcal{C}_{q_2(p_2,r_2)}(I; L^{p_2})\|^{\alpha_2} \|u; \mathcal{C}_{q_1(p_1,r_1)}(I; L^{p_1})\| \\
&\quad \le C \sum_{j=1}^{2} \|u; \mathcal{C}_{q_j(p_j,r_j)}(I; L^{p_j})\|^{\alpha_j} \cdot \|u; \mathcal{C}_{q_1(p_1,r_1)}(I; L^{p_1})\|.
\end{aligned} \tag{2.71}$$

Similarly, we have

$$\begin{aligned}
&\|GF(u); \mathcal{C}_{q_2(p_2,r_2)}(I; L^{p_2})\| \\
&\quad \le C \sum_{j=1}^{2} \|u; \mathcal{C}_{q_j(p_j,r_j)}(I; L^{p_j})\|^{\alpha_j} \|u; \mathcal{C}_{q_2(p_2,r_2)}(I; L^{p_2})\|
\end{aligned} \tag{2.72}$$

and

$$\|GF(u); C(I; L^{r_2})\| \le C \sum_{j=1}^{2} \|u; \mathcal{C}_{q_j(p_j,r_j)}(I; L^{p_j})\|^{\alpha_j} \|u; \mathcal{C}_{q_2(p_2,r_2)}(I; L^{p_2})\|, \tag{2.73}$$

as well as that for $u, v \in \mathcal{Z}(I)$ and for $p_1 > r_1$ and $p_2 > r_2$,

$$t^{\frac{1}{q_1}} \|\mathcal{T}u\|_{p_1} \le t^{\frac{1}{q_1}} \|S(t)\varphi\|_{p_1} + Ct^{\frac{\alpha_1+1}{q_1}} \|u\|_{p_1}^{\alpha_1+1} + Ct^{\frac{\alpha_2}{q_2}} \|u\|_{p_2}^{\alpha_2} \cdot t^{\frac{1}{q_1}} \|u\|_{p_1}, \tag{2.74}$$

$$t^{\frac{1}{q_2}} \|\mathcal{T}u\|_{p_2} \le t^{\frac{1}{q_2}} \|S(t)\varphi\|_{p_2} + Ct^{\frac{\alpha_2+1}{q_2}} \|u\|_{p_2}^{\alpha_2+1} + Ct^{\frac{\alpha_1}{q_1}} \|u\|_{p_1}^{\alpha_1} \cdot t^{\frac{1}{q_2}} \|u\|_{p_2}, \tag{2.75}$$

and

$$\begin{aligned}
d(\mathcal{T}u, \mathcal{T}v) \le \sum_{j=1}^{2} [&\|u; \mathcal{C}_{q_j(p_j,r_j)}(I; L^{p_j})\|^{\alpha_j} \\
&+ \|v; \mathcal{C}_{q_j(p_j,r_j)}(I; L^{p_j})\|^{\alpha_j}] d(u, v).
\end{aligned} \tag{2.76}$$

Here, we use (2.80) in the next Remark 2.2.2.

Combining (2.69)–(2.76) and choosing δ sufficiently small we obtain that, if $C\|\varphi; L^{r_1} \cap L^{r_2}\| < \delta/2$, then

$$\|\mathcal{T}u; Z(I)\| \leq C\|\varphi; L^{r_1} \cap L^{r_2}\| + C\delta^{\alpha_2+1} + C\delta^{\alpha_1+1} < \delta, \tag{2.77}$$

$$d(\mathcal{T}u, \mathcal{T}v) \leq \frac{1}{2} d(u, v). \tag{2.78}$$

Furthermore, it follows from (2.74) and (2.75) that

$$\lim_{t\to 0} t^{\frac{1}{q_j}} \|\mathcal{T}u\|_{p_j} = 0, \quad j = 1, 2,\ p_j = r_j. \tag{2.79}$$

Thus, $\mathcal{T}$ is a contraction mapping from $\mathcal{Z}(I)$ to itself and, therefore, the Banach contraction mapping principle implies that there exists a unique solution $u \in \mathcal{Z}(I)$ to problem (2.37)–(2.38).

If (q_j, p_j, r_j) is generalized admissible triplet satisfying $p_j \leq 1 + \alpha_j$, then interpolating between $C(I; L^{r_j})$ and $C_{\tilde{q}_j(\tilde{p}_j, r_j)}(I; L^{\tilde{p}_j})$ with $(\tilde{q}_j, \tilde{p}_j, r_j) \in F^*$, we have $u \in C_{q_j(p_j, r_j)}(I; L^{p_j})$ ($j = 1, 2$). On the other hand, for any $(q_1, p_1, r_1) \in F$, there is (q_2, p_2, r_2) such that $(q_j, p_j, r_j) \in F^*$ ($j = 1, 2$), it suffices to prove that $u \in \dot{C}_{\hat{q}_2(\hat{p}_2, r_2)}(I; L^{\hat{p}_2})$ for any $(\hat{q}_2, \hat{p}_2, r_2) \in F^*$. In fact, we can choose $(q_j, p_j, r_j) \in F^*$ ($j = 1, 2$), which of course satisfy (2.80) in the next Remark 2.2.2. So, we obtain that

$$\begin{aligned}
\|u; C_{\hat{q}_2(\hat{p}_2, r_2)}(I; L^{\hat{p}_2})\| &\leq C\|\varphi\|_{r_2} + \sup_{t\in I} t^{\frac{1}{\hat{q}_2}} \int_0^t |t-\tau|^{-\frac{d}{2}-\frac{n\alpha_1}{2p_1}} \|u\|_{p_1}^{\alpha_1} \|u\|_{\hat{p}_2} \, d\tau \\
&\quad + \sup_{t\in I} t^{\frac{1}{\hat{q}_2}} \int_0^t |t-\tau|^{-\frac{d}{2}-\frac{n\alpha_2}{2p_2}} \|u\|_{p_2}^{\alpha_1} \|u\|_{\hat{p}_2} \, d\tau \\
&\leq C\|\varphi\|_{r_2} + C\|u; C_{q_1(p_1, r_1)}(I; L^{r_1})\|^{\alpha_1} \|u; C_{\hat{q}_2(\hat{p}_2, r_2)}(I; L^{\hat{p}_2})\| \\
&\quad + C\|u; C_{q_2(p_2, r_2)}(I; L^{r_2})\|^{\alpha_2} \|u; C_{\hat{q}_2(\hat{p}_2, r_2)}(I; L^{\hat{p}_2})\| \\
&\leq C\|\varphi\|_{r_2} + C(\delta^{\alpha_1} + \delta^{\alpha_2}) \|u; C_{\hat{q}_2(\hat{p}_2, r_2)}(I; L^{\hat{p}_2})\|.
\end{aligned}$$

A similar argument as above gives

$$\lim_{t\to 0} t^{\frac{1}{\hat{q}_2}} \|u\|_{\hat{p}_2} = 0,$$

that is, $u \in \dot{C}_{\hat{q}_2(\hat{p}_2, r_2)}(I; L^{\hat{p}_2})$. The proof of (i) is thus completed.

The statement (ii) can be proved in a similar way as in the proof of Theorem 2.2.1. Theorem 2.2.2 is thus proved. □

Remark 2.2.2. For fixed $r_j, j = 1, 2$ (same as Theorem 2.2.2), there exist $\{(q_j, p_j, r_j)\}_{j=1,2} \in F^*$. In fact, without loss of generality, we assume $\alpha_1 \leq \alpha_2$, and we deduce that $r_1 \leq r_2$ by the definition of r_j ($j = 1, 2$). Since

$$\frac{1+\alpha_1}{r_1} \le \frac{p_1}{r_1} < 1+\alpha_1, \qquad \frac{1+\alpha_2}{r_2} \le \frac{p_2}{r_2} < 1+\alpha_2$$

and

$$\left[\frac{2-d}{n\alpha_1}(1+\alpha_1), 1+\alpha_1\right) \subset \left[\frac{2-d}{n\alpha_2}(1+\alpha_2), 1+\alpha_2\right), \quad d \in [0,2),$$

it is easy to choose an infinite number of generalized admissible triplets $\{(q_j, p_j, r_j)\}_{j=1,2} \in F^*$. Moreover, since either $r_j = \frac{n\alpha_j}{2}$ or $r_j = \frac{n\alpha_j}{2-d}$ $(j = 1, 2)$, we also have

$$\frac{\alpha_1}{p_1} = \frac{\alpha_2}{p_2}, \quad \frac{\alpha_1}{q_1} = \frac{\alpha_2}{q_2}. \tag{2.80}$$

Remark 2.2.3.

(i) When $\alpha_1 = \alpha_2$, (2.37)–(2.38) correspond to the classical semilinear heat equations, the space-time estimate method provided here cannot only deal with the case with homogeneous nonlinear term, but also provide the method to deal with the nonlinear parabolic equations with inhomogeneous growth.

(ii) In fact, the method in this section can also be applied to many classical nonlinear evolution equations, such as the complex Ginzburg–Landau equation, convection diffusion equation, viscous Burgers equation, Navier–Stokes equations and so on.

2.3 The Cauchy problem for semilinear heat equations (II)

In the last section, we dealt with the Cauchy problem (2.37)–(2.38) for the case $\varphi \in L^r$. In this section, we consider the case when φ belongs to $H^{s,p}$ or $B^s_{p,2}$. In particular, we are interested in conditions on p and s under which the Cauchy problem (2.37)–(2.38) generates a strong continuous flow in $B^s_{p,2}$ or $H^{s,p}$, that is, the Cauchy problem (2.37)–(2.38) is well posed in $B^s_{p,2}$ or $H^{s,p}$. The scaling argument suggests that the Cauchy problem (2.37)–(2.38) is at least locally well posed in $B^s_{p,2}$ or $H^{s,p}$ if $p > 1$ and $s \ge s_c$, where $s_c = \frac{n}{p} - \frac{2}{\alpha}$ or $s_c = \frac{n}{p} - \frac{2-d}{\alpha}$ in the case when $F(u) = f(u)$ or $Q(D)f(u)$ with $f(u) \sim |u|^\alpha u$ and $s = s_c$ corresponds to the critical case.

On the other hand, when $F(u) = f(u)$ and f grows faster than a polynomial growth (e. g., f is of exponential growth), the suitable working space is the critical space $H^{s,p}$ or $B^s_{p,2}$ with $s = n/p$ for studying the well-posedness of the problem (2.37)–(2.38). When $s > n/p$, $H^{s,p}$ or $B^s_{p,2}$ is a Banach algebra, so in this case, local well-posedness results can be established without any restriction on the growth of the nonlinear term.

We first introduce the following linear estimates, which are immediate consequences of the interpolation theorem and the linear estimates in Section 2.1:

$$\|S(t)\varphi\|_{\dot{B}^{s+\theta}_{p,2}} \le Ct^{-\frac{\theta}{2}}\|\varphi\|_{\dot{B}^s_{p,2}}, \tag{2.81}$$

$$\|S(t)\varphi\|_{B^{s+\theta}_{p,2}} \le C(T)t^{-\frac{\theta}{2}}\|\varphi\|_{B^s_{p,2}}, \tag{2.82}$$

$$\|S(t)\varphi\|_{\dot H^{s+\theta,p}} \le Ct^{-\frac{\theta}{2}}\|\varphi\|_{\dot H^{s,p}}, \tag{2.83}$$

$$\|S(t)\varphi\|_{H^{s+\theta,p}} \le C(T)t^{-\frac{\theta}{2}}\|\varphi\|_{H^{s,p}}, \tag{2.84}$$

where $t \in [0,T)$, $1 \le p < \infty$ and $\theta \ge 0$ and

$$\|S(\cdot)\varphi; C_{\tilde q}(I;\dot B^{\tilde s}_{\tilde p,2})\| \le C\|\varphi; \dot B^s_{p,2}\|, \tag{2.85}$$

where $I = [0,\infty)$ or $I = [0,T)$, $\tilde p \ge p > 1$ and $\tilde s \ge s \ge 0$ satisfy that

$$\frac{2}{\tilde q} = \left(\tilde s - \frac{n}{\tilde p}\right) - \left(s - \frac{n}{p}\right) < 1. \tag{2.86}$$

We shall also use the following generalized Sobolev inequalities (see, e. g., [209, 212] for details):

$$\|u\|_q \le Cq^{\frac{1}{p'}}\|u; \dot H^{\frac{n}{p}-\frac{n}{q},p}\|, \quad 1 < p \le q < \infty, \tag{2.87}$$

$$\|u\|_{\dot B^0_{q,2}} \le Cq^{\frac{1}{p'}}\|u; \dot B^{\frac{n}{p}-\frac{n}{q}}_{p,2}\|, \quad 1 < p \le q < \infty, \tag{2.88}$$

$$\|u\|_{\dot B^s_{q,2}} \le Cq^{\frac{1}{p'}}\|u; \dot B^{s+\frac{n}{p}-\frac{n}{q}}_{p,2}\|, \quad 1 < p \le q < \infty, \ s \in \mathbb{R}, \tag{2.89}$$

$$\max\{\|u\|_q, \|u\|_{\dot B^0_{q,2}}\} \le Cq^{\frac{1}{p'}}\|u; \dot B^{\frac{n}{p}-\frac{n}{q}}_{p,2}\|, \quad 1 < p < q < \infty. \tag{2.90}$$

Lemma 2.3.1. *Let $p > 1$, $s \ge 0$ and $\alpha > 0$ satisfy*

$$\max\left\{0, \frac{n}{p} - \frac{n}{\alpha+1}\right\} \le s < \min\left\{\frac{n}{p}, 1 + \frac{\alpha}{\alpha+1}\frac{n}{p}\right\}. \tag{2.91}$$

Assume that $f \in C^{[\alpha]+1}(\mathbb{R})$ satisfies that

$$|f^{(j)}(u) - f^{(j)}(v)| \le C(|u|^{\alpha-j} + |v|^{\alpha-j})|u - v|, \quad j = 0, \dots, [\alpha], \tag{2.92}$$

$$|\partial^{[\alpha]+1}f(u) - \partial^{[\alpha]+1}f(v)| \le C|u - v|^{\alpha-[\alpha]}, \quad f^{[\alpha]+1}(0) = 0 \quad \text{if} \quad \alpha \notin \mathbb{N}, \tag{2.93}$$

where, for any $s \in \mathbb{R}$, $[s]$ denotes the largest integer $\le s$. Note that if $s \le \frac{n\alpha}{p(\alpha+1)}$, then, it is sufficient to assume only that

$$|f(u) - f(v)| \le C(|u|^\alpha + |v|^\alpha)|u - v|. \tag{2.94}$$

Then there holds

$$\|f(u)\|_{B^{s_\alpha}_{p,2}} \le C\|u\|^{\alpha+1}_{B^s_{p,2}}, \quad s_\alpha = s - \alpha\left(\frac{n}{p} - s\right). \tag{2.95}$$

Remark 2.3.1. If (2.91) is replaced by

$$\max\left(0, \frac{n}{p} - \frac{n}{\alpha+1}\right) \leq s < \min\left(\frac{n}{p}, \frac{\alpha}{\alpha+1}\left(1 + \frac{n}{p}\right)\right), \tag{2.96}$$

then

$$\|f(u)\|_{H^{s_\alpha,p}} \leq C\|u\|_{H^{s,p}}^{\alpha+1}, \quad s_\alpha = s - \alpha\left(\frac{n}{p} - s\right). \tag{2.97}$$

The estimate (2.97) was proved under the condition (2.96) by Ribaud in [227] using the para-composition technique based on the Littlewood–Paley theory [183]. It is clear that the estimate (2.95) can have wider applications. So, we only prove (2.95) below.

Proof of Lemma 2.3.1. First, if $0 \leq s \leq \frac{n\alpha}{p(\alpha+1)}$, then $s_\alpha \leq 0$ and

$$\frac{1}{\gamma} = \frac{1}{p} - \frac{s_\alpha}{n} = (\alpha+1)\left(\frac{1}{p} - \frac{s}{n}\right) < 1. \tag{2.98}$$

Thus, it follows by (2.94) and the Sobolev embedding theorem (i. e., $L^\gamma \hookrightarrow B^{s_\alpha}_{p,2}$, $B^s_{p,2} \hookrightarrow L^{\gamma(\alpha+1)}$) that

$$\|f(u)\|_{B^{s_\alpha}_{p,2}} \leq C\|u\|_{(\alpha+1)\gamma}^{\alpha+1} \leq C\|u\|_{B^s_{p,2}}^{\alpha+1}, \tag{2.99}$$

which proves (2.95).

Now, if $s > \frac{n\alpha}{p(\alpha+1)}$, then we need the following equivalent norm for Besov spaces $\dot{B}^s_{r,m}$ and $B^s_{r,m}$:

$$\begin{aligned}\|v\|_{\dot{B}^s_{r,m}} &\simeq \sum_{|\beta|=[s]}\left(\int_0^\infty t^{-m(s-[s])}\sup_{|y|\leq t}\|\triangle_y\partial^\beta v\|_r^m\frac{dt}{t}\right)^{\frac{1}{m}} \\ &\simeq \sum_{|\beta|=N}\left(\int_0^\infty t^{-m\sigma}\sup_{|y|\leq t}\|\triangle_y^2\partial^\beta v\|_r^m\frac{dt}{t}\right)^{\frac{1}{m}},\end{aligned} \tag{2.100}$$

$$\|v\|_{B^s_{r,m}} \simeq \|v\|_r + \|v\|_{\dot{B}^s_{r,m}}, \tag{2.101}$$

where $[s]$ denotes the largest integer such that $[s] \leq s$, $s = N + \sigma$ with a nonnegative integer N and $0 < \sigma < 2$, and

$$\triangle_{\pm y}v(\cdot) = v(\cdot \pm y) - v(\cdot) = \tau_{\pm y}v(\cdot) - v(\cdot), \tag{2.102}$$

$$\triangle_y^2 v \triangleq \tau_y v + \tau_{-y}v - 2v. \tag{2.103}$$

We distinguish between the following two cases.

Case 1. $s_\alpha < 2$. Since

$$\triangle_y^2 f(u) = f'(u) \triangle_y^2 u + \sum_{\pm} \triangle_{\pm y} u \int_0^1 [f'(\lambda \tau_{\pm y} u + (1-\lambda)u) - f'(u)] d\lambda, \tag{2.104}$$

it follows that

$$|\triangle_y^2 f(u)| \le \begin{cases} |f'(u)| \cdot |\triangle_y^2 u| + C \sum_\pm |\triangle_{\pm y} u|^2 (|\tau_{\pm y} u|^{\alpha-1} + |u|^{\alpha-1}), & \text{for} \quad \alpha \ge 1, \\ |f'(u)| \cdot |\triangle_y^2 u| + C \sum_\pm |\triangle_{\pm y} u|^{\alpha+1}, & \text{for} \quad \alpha < 1. \end{cases} \tag{2.105}$$

A direct calculation gives that

$$\begin{aligned} &\frac{1}{p} = \alpha\Big(\frac{1}{p} - \frac{s}{n}\Big) + \frac{1}{\chi_1}, \quad \frac{1}{\chi_1} = \frac{s\alpha}{n} - \frac{\alpha-1}{p}, \\ &\frac{1}{p} = (\alpha-1)\Big(\frac{1}{p} - \frac{s}{n}\Big) + \frac{2}{\chi_2}, \quad \frac{1}{\chi_2} = \frac{(\alpha-1)s}{2n} - \frac{\alpha-2}{2p}, \quad \alpha \ge 1, \\ &\frac{1}{p} = \frac{\alpha+1}{\chi_3}, \quad \alpha < 1, \end{aligned}$$

and by (2.91) $0 < 1/\chi_j < 1$ $(j = 1, 2, 3)$. Thus, by (2.100), Lemma 2.1.3 and the Sobolev embedding theorem

$$\Big(\text{i. e., } B^s_{p,2} \hookrightarrow L^{p(\alpha+1)}, \quad \dot B^s_{p,2} \hookrightarrow \dot B^{s_\alpha}_{\chi_1,2}, \quad \dot B^s_{p,2} \hookrightarrow \dot B^{\frac{s_\alpha}{2}}_{\chi_2,4}, \quad \dot B^s_{p,2} \hookrightarrow \dot B^{\frac{s_\alpha}{\alpha+1}}_{\chi_3,2(\alpha+1)}\Big)$$

we get that

$$\begin{aligned} \|f(u)\|_{B^{s_\alpha}_{p,2}} &= \|f(u)\|_p + \bigg(\int_0^\infty t^{-2s_\alpha} \sup_{|y|\le t} \|\triangle_y^2 f(u)\|_p^2 \frac{dt}{t}\bigg)^{\frac{1}{2}} \\ &\le \|u\|^{\alpha+1}_{p(\alpha+1)} + C\|u\|^\alpha_{\dot B^s_{p,2}} \|u\|_{\dot B^{s_\alpha}_{\chi_1,2}} + C \begin{cases} \|u\|^{\alpha-1}_{\dot B^s_{p,2}} \|u\|^2_{\dot B^{\frac{s_\alpha}{2}}_{\chi_2,4}}, & \text{for} \quad \alpha \ge 1 \\ \|u\|^{\alpha+1}_{\dot B^{\frac{s_\alpha}{\alpha+1}}_{\chi_3,2(\alpha+1)}}, & \text{for} \quad \alpha < 1 \end{cases} \\ &\le C\|u\|^{\alpha+1}_{B^s_{p,2}} \end{aligned}$$

and (2.95) is proved.

Case 2. $s_\alpha \ge 2$. Let $s_\alpha = [s_\alpha] - 1 + \sigma$. Then $1 \le \sigma < 2$ and $[s_\alpha] - 1 \ge 1$. A direct calculation using the Leibniz rule of derivatives gives that for $|\gamma| = [s_\alpha] - 1$,

$$\partial^\gamma f(u) = \sum_{k=1}^{[s_\alpha]-1} \sum_{\substack{\beta_1+\cdots+\beta_k=\gamma, \\ |\beta_j|\ge 1}} \frac{\gamma!}{k! \prod_{j=1}^k \beta_j} f^{(k)}(u) \prod_{j=1}^k \partial^{\beta_j} u. \tag{2.106}$$

For simplicity, we set

$$\tilde{f}(u) = f^{(k)}(u), \quad v = \tau_y u, \quad w = \tau_{-y} u,$$
$$u_j = \partial^{\beta_j} u, \quad v_j = \partial^{\beta_j} v, \quad w_j = \partial^{\beta_j} w.$$

Then the second-order difference of the term of $f^{(k)}(u) \prod_{j=1}^{k} \partial^{\beta_j} u$ can be computed as

$$\begin{aligned}
&\triangle_y^2 \left(f^{(k)}(u) \prod_{j=1}^{k} \partial^{\beta_j} u \right) \\
&= \tilde{f}(u) \sum_j \triangle_y^2 u_j \cdot \prod_{a<j} v_a \prod_{b>j} u_b \\
&\quad - \tilde{f}(u) \sum_j \sum_{i<j} (\triangle_{-y} u_j \cdot \triangle_y u_i) \prod_{a<j} v_a \prod_{\substack{b<j \\ b \neq i}} u_b \\
&\quad + \tilde{f}(u) \sum_j \sum_{i>j} (\triangle_{-y} u_j \cdot \triangle_{-y} u_i) \prod_{\substack{a<i \\ a \neq j}} u_a \prod_{b>i} w_b \\
&\quad + (\tilde{f}(v) - \tilde{f}(u)) \sum_j \triangle_y u_j \prod_{a<j} v_a \prod_{b>j} u_b \\
&\quad - (\tilde{f}(u) - \tilde{f}(w)) \sum_j \triangle_{-y} u_j \prod_{a<j} u_a \prod_{b>j} w_b + \tilde{f}'(u) \triangle_y^2 u \prod_j u_j \\
&\quad + \int_0^1 (\tilde{f}'(\theta v + (1-\theta) u) - \tilde{f}'(u)) d\theta (v - u) \prod_j u_j \\
&\quad + \int_0^1 (\tilde{f}'(\theta w + (1-\theta) u) - \tilde{f}'(u)) d\theta (w - u) \prod_j u_j = \sum_{k=1}^{8} I_k.
\end{aligned} \tag{2.107}$$

We now evaluate each I_k.

Estimation of I_1 and I_6. Let

$$\begin{aligned}
d_0 &= (\alpha + 1 - k)\left(\frac{1}{p} - \frac{s}{n} \right), \\
d_a &= \frac{1}{p} - \frac{s - |\beta_a|}{n}, \quad 1 \leq a \leq k \text{ and } a \neq j, \\
d_j &= \frac{1}{p} - \frac{s - (|\beta_j| + \sigma)}{n} \triangleq \frac{1}{\chi_1}.
\end{aligned}$$

Since

$$\frac{1}{p} - \frac{s}{n} > 0, \tag{2.108}$$

it follows that $d_a > 0$ for $a = 0, 1, \ldots, k$. It is easy to verify that

$$d_0 + \sum_{a=1}^{k} d_a = 1.$$

Noting that the Besov spaces are translation invariant, we obtain by Lemma 2.1.3 and the Sobolev embedding theorem that

$$\|I_1\|_p \le \|u\|_{\dot{B}^s_{p,2}}^{\alpha} \|\triangle_y^2 u_j\|_{\chi_1}. \tag{2.109}$$

Similarly, and letting

$$\begin{aligned} d_0 &= (\alpha - k)\left(\frac{1}{p} - \frac{s}{n}\right), \\ d_a &= \frac{1}{p} - \frac{s - |\beta_a|}{n}, \quad a = 1, \dots, k, \\ d_{k+1} &= \frac{1}{p} - \frac{s - \sigma}{n} \triangleq \frac{1}{\chi_6}, \end{aligned}$$

we have

$$\|I_6\|_p \le \|u\|_{\dot{B}^s_{p,2}}^{\alpha} \|\triangle_y^2 u_j\|_{\chi_6}. \tag{2.110}$$

We may estimate I_2 and I_3 similarly. In fact, let

$$\begin{aligned} d_0 &= (\alpha + 1 - k)\left(\frac{1}{p} - \frac{s}{n}\right), \\ d_a &= \frac{1}{p} - \frac{s - |\beta_a|}{n}, \quad a \neq i, j, \\ d_i &= \frac{1}{p} - \frac{s - (|\beta_i| + \frac{\sigma}{2})}{n} \triangleq \frac{1}{\chi_2}, \\ d_j &= \frac{1}{p} - \frac{s - (|\beta_j| + \frac{\sigma}{2})}{n} \triangleq \frac{1}{\chi_3}. \end{aligned}$$

Then $d_j > 0$ for $j = 0, 1, \dots, k+1$ and $\sum_{j=0}^{k+1} d_j = 1$, so it follows from Lemma 2.1.3 and the Sobolev embedding theorem that

$$\|I_2\|_p, \|I_3\|_p \le \|u\|_{\dot{B}^s_{p,2}}^{\alpha-1} \| \triangle_{\pm y} u_i\|_{\chi_2} \| \triangle_{\pm y} u_j\|_{\chi_3}.$$

To estimate I_4 and I_5, let

$$\begin{aligned} d_0 &= (\alpha - k)\left(\frac{1}{p} - \frac{s}{n}\right), \\ d_a &= \frac{1}{p} - \frac{s - |\beta_a|}{n}, \quad 1 \le a \le k \text{ and } a \neq j, \end{aligned}$$

$$d_j = \frac{1}{p} - \frac{s - (|\beta_j| + \frac{\sigma}{2})}{n} \triangleq \frac{1}{\chi_4},$$
$$d_{k+1} = \frac{1}{p} - \frac{s - \frac{\sigma}{2}}{n} \triangleq \frac{1}{\chi_5}.$$

Then, $d_j > 0$ for $j = 0, 1, \ldots, k+1$ and $\sum_{j=0}^{k+1} d_j = 1$, so it follows from Lemma 2.1.3 and the Sobolev embedding theorem that

$$\|I_4\|_p, \|I_5\|_p \leq \|u\|_{\dot{B}^s_{p,2}}^{\alpha-1} \| \triangle_{\pm y} u_j\|_{\chi_4} \| \triangle_{\pm y} u\|_{\chi_5}. \tag{2.111}$$

Finally, I_7 and I_8 can be estimated similarly as above by letting

$$d_a = \frac{1}{p} - \frac{s - |\beta_a|}{n}, \quad a = 1, \ldots, k,$$
$$d_{k+1} = \frac{1}{p} - \frac{s - \frac{\sigma}{\alpha+1-k}}{n} \triangleq \frac{1}{\chi_7} \equiv \frac{1}{\chi_8}$$

and noting that $d_a > 0$ for $a = 1, \ldots, k+1$ and $\sum_{a=1}^{k+1} d_a = 1$, and we have by Lemma 2.1.3 and the Sobolev embedding theorem

$$\|I_7\|_p, \|I_8\|_p \leq \|u\|_{\dot{B}^s_{p,2}}^{k} \cdot \| \triangle_{\pm y} u\|_{\chi_7}^{\alpha+1-k}. \tag{2.112}$$

Then the embedding relation

$$\dot{B}^s_{p,2} \hookrightarrow \dot{B}^{\sigma+|\beta_j|}_{\chi_{1,2}}, \quad \dot{B}^s_{p,2} \hookrightarrow \dot{B}^{\sigma}_{\chi_{6,2}}, \tag{2.113}$$
$$\dot{B}^s_{p,2} \hookrightarrow \dot{B}^{|\beta_j|+\frac{\sigma}{2}}_{\chi_{2,4}}, B^{|\beta_j|+\frac{\sigma}{2}}_{\chi_{4,4}}, \tag{2.114}$$
$$\dot{B}^s_{p,2} \hookrightarrow \dot{B}^{|\beta_j|+\frac{\sigma}{2}}_{\chi_{3,2}}, \dot{B}^{\frac{\sigma}{2}}_{\chi_{5,4}}, \tag{2.115}$$
$$\dot{B}^s_{p,2} \hookrightarrow B^{\frac{\sigma}{\alpha+1-k}}_{\chi_{j,2(\alpha+1-k)}}, \quad j = 7, 8 \tag{2.116}$$

along with (2.100) and (2.101) implies that

$$\|I_j\|_{\dot{B}^{s_\alpha}_{p,2}} \leq C\|u\|_{\dot{B}^s_{p,2}}^{\alpha+1}, \quad j = 1, 2, \ldots, 8, \tag{2.117}$$
$$\|f(u)\|_p \leq c\|u\|_{p(\alpha+1)}^{\alpha+1} \leq C\|u\|_{B^s_{p,2}}^{\alpha+1}. \tag{2.118}$$

These lead to the required estimate (2.95) in this case. The lemma is thus proved. □

Theorem 2.3.1.

(a) *Let $F(u) = f(u)$ or $Q(D)f(u)$, where $Q(D)$ is a homogeneous pseudodifferential operator of order $d \in [0, 2)$. Let $f(u)$, p, s satisfy the conditions of Lemma 2.3.1. Define $s_c = \frac{n}{p} - \frac{2}{\alpha}$ or $s_c = \frac{n}{p} - \frac{2-d}{\alpha}$. Then for $\varphi \in B^s_{p,2}$ with $s > s_c$, there exists a unique maximal* *solution $u \in C([0,T^*);B^s_{p,2})$ to problem (2.37)–(2.38), where either $T^* = \infty$ or $T^* < \infty$ satisfying that*

$$T^* \geq C\|\varphi\|^{-1/\gamma}_{B^s_{p,2}} \quad \text{and} \quad \lim_{t\to T^*}\|u(t)\|_{B^s_{p,2}} = \infty \tag{2.119}$$

with $\gamma = \frac{s-s_c}{2}$.

(b) *Let $p > 1$ and let us assume that one of the following conditions holds:*
   (i) *$sp = n$ and $F(z) = f(z) \in C^1(\mathbb{R}) \cap C^{[\frac{n}{p}]}(\mathbb{R})$ satisfying that*

$$f(0) = 0, \quad |f'(z)| \leq C|z|e^{\lambda|z|}, \quad |f^{(k)}(z)| \leq Ce^{\lambda|z|} \tag{2.120}$$

   for some $\lambda > 0$ and $2 \leq k \leq [\frac{n}{p}]$. Note that if $[\frac{n}{p}] \leq 2$ then the last inequality in (2.120) is superfluous.
   (ii) *$sp > n$, $F(u) = f(u)$ or $Q(D)f(u)$, where $Q(D)$ is a homogeneous pseudodifferential operator of order $d \in [0,2)$, and $f \in C^{[s]+1}(\mathbb{R})$ with $f(0) = 0$.*

Then, for $\varphi \in B^s_{p,2}$, there exists a unique maximal solution $u(t)$ to problem (2.37)–(2.38) such that $u \in C([0,T^*);B^s_{p,2})$ satisfying that either $T^* = \infty$ or $T^* < \infty$ and $\lim_{t\to T^*}\|u(t)\|_{B^s_{p,2}} = \infty$.

(c) *The solution $u(t)$, obtained in* (a) *and* (b), *has the following smoothing effects:*
   (i) *If $f \in C^\infty(\mathbb{R})$, then $u \in C^\infty((0,T^*) \times \mathbb{R}^n)$.*
   (ii) *Let $\theta < 2\alpha\nu$ if $s < n/p$ and $\theta < 2 - d$ if $s \geq n/p$. Then $u - S(\cdot)\varphi \in C([0,T^*);B^{s+\theta}_{p,2})$, where $S(t) = e^{t\Delta}$.*
   (iii) *Let $\tilde{s} \geq s$,*

$$1 < p \leq \tilde{p} < \begin{cases} \frac{np}{n-sp}, & \text{for} \quad sp < n, \\ \infty, & \text{for} \quad sp \geq n \end{cases}$$

   and

$$\frac{2}{\tilde{q}} = \left(\tilde{s} - \frac{n}{\tilde{p}}\right) - \left(s - \frac{n}{p}\right) < 1.$$

   Then $u \in C_{\tilde{q}}([0,T^*);B^{\tilde{s}}_{\tilde{p},2})$.

The following theorem establishes the global well-posedness of problem (2.37)–(2.38).

Theorem 2.3.2.

(a) *Let $p_c = n\alpha/2$ or $p_c = n\alpha/(2-d)$. Let $\max(p_c/(\alpha+1),1) < p \leq p_c$ and $s = s_c$ with s_c defined as in Theorem 2.3.1 and let us assume that $F(u) = f(u)$ or $Q(D)f(u)$, where $Q(D)$ is a homogeneous pseudodifferential operator of order $d \in [0,2)$, $f(u) \in C(\mathbb{R})$ satisfies (2.120). Then there exists an $\epsilon > 0$ such that the problem (2.37)–(2.38) has a unique global solution $u \in C([0,\infty);B^s_{p,2})$ provided that $\|\varphi\|_{B^s_{p,2}} < \epsilon$.*

(b) *Let $p = 2$, $d = 0$ and $s = n/2$. Assume that $f \in C^1(\mathbb{R}) \cap C^{[\frac{n}{2}]}(\mathbb{R})$ satisfies that for some $\lambda > 0$,*

$$|f(z_1) - f(z_2)| \le C[e^{\lambda|z_1|}|z_1|^4 + e^{\lambda|z_2|}|z_2|^4]|z_1 - z_2|, \quad n = 1, \tag{2.121}$$

$$|f'(z)| \le C|z|^2 e^{\lambda|z|}, \quad f(0) = 0, \quad n = 2, 3, \tag{2.122}$$

$$|f'(z)| \le C|z|e^{\lambda|z|}, \quad f(0) = 0, \quad n \ge 4. \tag{2.123}$$

Then there exists an $\epsilon > 0$ such that the problem (2.37)–(2.38) *has a unique global solution $u \in C([0,\infty); H^s)$ provided that $\|\varphi\|_{H^s} < \epsilon$.*

Remark 2.3.2.
(i) Theorem 2.3.1 (a) also holds with $T^* = T(\varphi)$ when $s = s_c$.
(ii) There exists a gap between condition (2.91) and the optimal condition (2.96) in the case when $n > (\alpha + 1)p$.
(iii) Theorem 2.3.2 deals with existence of small global solutions to problem (2.37)–(2.38) in the critical case $s = n/2$ only under the condition that both $d = 0$ and $p = 2$. It seems that our method does not work for the case when either $p \ne 2$ or $d \ne 0$.

Proof of Theorem 2.3.1. We first prove (a). For any $\varphi \in B^s_{p,2}$ with $s > s_c$, we have by the Sobolev embedding theorem that $\varphi \in L^{\tilde r}$, where

$$\frac{1}{\tilde r} = \frac{1}{p} - \frac{s}{n} < \frac{1}{p_c}$$

with $p_c = \frac{n\alpha}{2-d}$. By Theorem 2.2.1, there exists a unique maximal solution $u(t)$ to problem (2.37)–(2.38) such that

$$u \in C([0,\tilde T^*); L^{\tilde r}) \cap C_{\tilde q(\tilde p,\tilde r)}([0,\tilde T^*); L^{\tilde p}),$$

where $(\tilde q,\tilde p,\tilde r) \in \Pi$, $[0,\tilde T^*)$ is the maximal existence interval and

$$\tilde T^* \le C\|\varphi\|_{B^s_{p,2}}^{-\frac{2\tilde r}{2\tilde r - d\tilde r - n\alpha}}.$$

Now for $0 < T \le \tilde T^*$ and $I = [0,T)$, we consider the operator $\mathcal{T}$ defined by (2.45) in the complete metric space:

$$X(I) = \left\{u \in C(I; B^s_{p,2}), \|u\|_X = \sup_{t\in I}\|u\|_{B^s_{p,2}} \le M := 2\|\varphi\|_{B^s_{p,2}}\right\}$$

with the metric

$$d(u,v) = \sup_{t\in I}\|u - v\|_{\tilde r}, \quad u, v \in X(I).$$

By Lemma 2.1.1, (2.56) and Lemma 2.3.1 together with the Sobolev embedding theorem, it can be estimated easily that for $u, v \in X(I)$,

$$\sup_{t\in I} \|\mathcal{T}u\|_{B^s_{p,2}} \leq C\|\varphi\|_{B^s_{p,2}} + C\sup_{t\in I}\int_0^t |t-\tau|^{-\frac{d+s-s_\alpha}{2}} \|f(u)\|_{B^{s_\alpha}_{p,2}}\,d\tau$$

$$\leq C\|\varphi\|_{B^s_{p,2}} + C(\tilde{T}^*)T^{1-\frac{d+s-s_\alpha}{2}} \sup_{t\in I}\|u\|^{\alpha+1}_{B^s_{p,2}}, \tag{2.124}$$

and

$$d(\mathcal{T}u,\mathcal{T}v) \leq C\sup_{t\in I}\int_0^t |t-\tau|^{-\frac{d}{2}}\left\|S\left(\frac{t-\tau}{2}\right)(f(u)-f(v))\right\|_{\tilde{r}}\,d\tau \tag{2.125}$$

$$\leq C\sup_{t\in I}\int_0^t |t-\tau|^{-\frac{d}{2}-\frac{n\alpha}{2\tilde{r}}}\left[\|u\|^\alpha_{\tilde{r}} + \|v\|^\alpha_{\tilde{r}}\right]\|u-v\|_{\tilde{r}}d\tau$$

$$\leq CT^{1-\frac{d}{2}-\frac{n\alpha}{2\tilde{r}}}\left[\sup_{t\in I}\|u\|^\alpha_{B^s_{p,2}} + \sup_{t\in I}\|v\|^\alpha_{B^s_{p,2}}\right]d(u,v). \tag{2.126}$$

Thus, $\mathcal{T}$ is a contraction mapping from $X(I)$ to itself if T is suitably small, for example,

$$T \leq \tilde{C}\|\varphi\|_{B^s_{p,2}}^{-\frac{2\tilde{r}}{2\tilde{r}-d\tilde{r}-n\alpha}}.$$

Then the Banach contraction mapping principle implies that there exists a unique solution $u \in X(I)$ to problem (2.37)–(2.38). Moreover, by the Picard method, there is a maximal $T^* = T(\|\varphi\|_{B^s_{p,2}}) \leq \tilde{T}^*$ such that $u \in C([0,T^*;B^S_{p,2})$ and satisfies (2.119), where we have used the fact that

$$\nu = \frac{2\tilde{r}-\tilde{r}d-n\alpha}{2\tilde{r}} = 1 - \frac{d+s-s_\alpha}{2}.$$

We now prove (b). We first consider the case $sp = n$. From the assumptions on f, it follows that

$$|f(z)| \leq C|z|^2e^{\lambda|z|}, \tag{2.127}$$

$$|f(z_1)-f(z_2)| \leq C\sum_{i,j=1}^2 |z_i|e^{\lambda|z_j|}|z_1-z_2| \tag{2.128}$$

and further in the case when $[\frac{n}{p}] \geq 2$ that

$$|f^{(k-1)}(z_1) - f^{(k-1)}(z_2)| \leq C(e^{\lambda|z_1|}+e^{\lambda|z_2|})|z_1-z_2|, \quad 2\leq k\leq\left[\frac{n}{p}\right]. \tag{2.129}$$

Now for $T > 0$ and $I = [0,T)$, let us introduce the complete metric space

$$Y(I) = \left\{u \in C(I; B^{\frac{n}{p}}_{p,2}), \|u\|_{Y(I)} = \sup_{t\in I} \|u\|_{B^{\frac{n}{p}}_{p,2}} \le M := 2\|\varphi\|_{B^{\frac{n}{p}}_{p,2}}\right\}$$

with the metric

$$d(u,v) = \sup_{t\in I} \|u-v\|_p, \quad u,v \in Y(I),$$

and let us consider, in $Y(I)$, the operator $\mathcal{T}$ defined by (2.45). It is enough to deal with the case $s = n/p \notin \mathbb{N}$. The case $s = n/p \in \mathbb{N}$ can be shown similarly by replacing $B^s_{p,2}$ with $H^{s,p}$ in $Y(I)$.

Choose $1 < r < p$ such that (q,p,r) is a generalized admissible triplet satisfying that

$$\frac{1}{q} + \frac{d}{2} = \frac{n}{2}\left(\frac{1}{r} - \frac{1}{p}\right) + \frac{d}{2} < 1, \quad p < 2r. \tag{2.130}$$

Then, by a direct computation, it follows that for $u \in Y(I)$,

$$\begin{aligned}\|f(u)\|_r &\le \sum_{\ell=0}^{\infty} \||u|^{2+\ell}; L^r\| \le \sum_{\ell=0}^{\infty} \frac{\lambda^\ell}{\ell!}\|u; L^{r(\ell+2)}\|^{2+\ell} \\ &\le C\sum_{\ell=0}^{\infty} \frac{\lambda^\ell}{\ell!}\{r(2+\ell)\}^{\frac{2+\ell}{p'}} \|u; L^\infty(I; B^{\frac{n}{p}}_{p,2})\|^{2+\ell} \le CC_1M^2,\end{aligned} \tag{2.131}$$

where

$$C_1 = \sum_{\ell=0}^{\infty} a_\ell = \sum_{\ell=0}^{\infty} \lambda^\ell \{r(2+\ell)\}^{\frac{2+\ell}{p'}} \|u; L^\infty(I; B^{\frac{n}{p}}_{p,2})\|^\ell/\ell! < \infty,$$

and we have used (2.87)–(2.90) and Sobolev embedding theorem. On the other hand, by (2.128), it is seen that for $u, v \in Y(I)$,

$$\begin{aligned}\|f(u) - f(v)\|_r \le{}& \||u|e^{\lambda|u|}|u-v|\|_r + \||u|e^{\lambda|v|}|u-v|\|_r \\ &+ \||v|e^{\lambda|u|}|u-v|\|_r + \||v|e^{\lambda|v|}|u-v|\|_r \\ :={}& I_1 + I_2 + I_3 + I_4.\end{aligned} \tag{2.132}$$

For $\ell \ge 0$, let $\chi(\ell) = (1+\ell)(\frac{1}{r} - \frac{1}{p})^{-1}$. Then, by (2.130), $\chi(\ell) > (1+\ell)p \ge p$, $\forall \ell \ge 0$. Hence, we get

$$\begin{aligned}I_1 &\le \sum_{\ell=0}^{\infty} \frac{\lambda^\ell}{\ell!}\|u; L^{\chi(\ell)}\|^{1+\ell}\|u-v\|_p \\ &\le C\sum_{\ell=0}^{\infty} \frac{\lambda^\ell}{\ell!}\{\chi(\ell)\}^{\frac{1+\ell}{p'}} \|u; L^\infty(I; B^{\frac{n}{p}}_{p,2})\|^{1+\ell}\|u-v\|_p \\ &\le CC_2M\|u-v; L^\infty(I; L^p)\|,\end{aligned} \tag{2.133}$$

where

$$C_2 = \sum_{\ell=0}^{\infty} a_\ell = \sum_{\ell=0}^{\infty} \lambda^\ell \chi(\ell)^{\frac{1+\ell}{p'}} \|u; L^\infty(I; B^{\frac{n}{p}}_{p,2})\|^\ell / \ell! < \infty,$$

and we have used (2.87)–(2.90) and Sobolev embedding theorem. Similarly, we have

$$I_j \leq CC_2M\|u - v; L^\infty(I; L^p)\|, \quad j = 2, 3, 4. \tag{2.134}$$

Combining (2.132), (2.133) and (2.134), it is easy to see that for $u, v \in Y(I)$,

$$\|f(u) - f(v)\|_r \leq CC_2M\|u - v; L^\infty(I; L^p)\|. \tag{2.135}$$

We now estimate $\|f(u); L^\infty(I; \dot{B}^{\frac{n}{p}}_{r,2})\|$. First, if $[\frac{n}{p}] < 1$, then since by (2.135),

$$\|f(\tau_y u) - f(u)\|_r \leq CC_2M\|\tau_y u - u; L^p\|,$$

it follows that

$$\|f(u); L^\infty(I; \dot{B}^{\frac{n}{p}}_{r,2})\| \leq CC_2M\|u; L^\infty(I; B^{\frac{n}{p}}_{p,2})\|. \tag{2.136}$$

Now if $[\frac{n}{p}] = N \leq 1$, then by (2.106) and a direct calculation we obtain that for $|\alpha| = [\frac{n}{p}]$,

$$\begin{aligned}
\triangle_y \partial^\alpha f(u) &= (f'(\tau_y u) - f'(u))\partial^\alpha u + f'(\tau_y u)(\partial^\alpha \tau_y u - \partial^\alpha u) \\
&\quad + \sum_{k=2}^{N} \sum_{\substack{|\beta_1|+\cdots+|\beta_k|=\alpha,\\ |\beta_j|\geq 1}} C(\alpha, k, \beta)\Bigg\{(f^{(k)}(\tau_y u) - f^{(k)}(u)) \prod_{j=1}^{k} \partial^{\beta_j} u \\
&\quad + f^{(k)}(\tau_y u) \sum_{j=1}^{k} (\partial^{\beta_j} \tau_y u - \partial^{\beta_j} u)\Bigg\} \prod_{\ell=1}^{j-1} \partial^{\beta_\ell} \tau_y u \prod_{\ell=j+1}^{k} \partial^{\beta_\ell} u \\
&:= J_1 + J_2 + J_3 + J_4.
\end{aligned} \tag{2.137}$$

First, let $\sigma = \frac{n}{p} - [\frac{n}{p}]$. Then, arguing similarly as in deriving (2.133) and noting (2.100) and the fact that L^p $(1 \leq p < \infty)$ is translation invariant, we get

$$\left(\int_0^\infty t^{-2\sigma} \sup_{|y|\leq t} \|J_2\|_r^2 \frac{dt}{t}\right)^{\frac{1}{2}} \leq CC_2M\|u; L^\infty(I; \dot{B}^{\frac{n}{p}}_{p,2})\| \leq CC_2M^2. \tag{2.138}$$

To estimate J_1, let for $\ell \geq 0$, $\chi(\ell) = (1 + \ell)(\frac{1}{r} - \frac{1}{p})^{-1} \geq (1 + \ell)p \geq p$ and let

$$\frac{1}{r} = \frac{\ell}{\chi(\ell)} + \frac{1}{p_1} + \frac{1}{p_2},$$

$$\frac{1}{p_1} = \frac{1-\mu_1}{\chi(\ell)} + \frac{\mu_1}{p}, \quad \mu_1 = \frac{p}{n}\left[\frac{n}{p}\right],$$
$$\frac{1}{p_2} = \frac{1-\mu_2}{\chi(\ell)} + \frac{\mu_2}{p}, \quad \mu_2 = 1-\mu_1.$$

Then, by interpolation and the Sobolev embedding theorem, we obtain that

$$\begin{aligned}
&\left(\int_0^\infty t^{-2\sigma} \sup_{|y|\le t} \|J_1\|_r^2 \frac{dt}{t}\right)^{\frac{1}{2}} \\
&\le \sum_{\ell=0}^{\infty} \frac{\lambda^\ell}{\ell!} \|u; L^{\chi(\ell)}\|^\ell \|u; \dot{H}^{|\alpha|,p_1}\| \cdot \|u; \dot{B}^\sigma_{p_2,2}\| \\
&\le C\sum_{\ell=0}^{\infty} \frac{\lambda^\ell}{\ell!} \|u; L^{\chi(\ell)}\|^{\ell+1} \|u; \dot{B}^{\frac{n}{p}}_{p,2}\| \\
&\le C\sum_{\ell=0}^{\infty} \frac{\lambda^\ell}{\ell!} \{\chi(\ell)\}^{\frac{1+\ell}{p'}} \|u; L^\infty(I; B^{\frac{n}{p}}_{p,2})\|^{2+\ell} \\
&\le CC_2M^2.
\end{aligned} \tag{2.139}$$

Similarly, and letting

$$\begin{aligned}
\frac{1}{r} &= \frac{\ell}{\chi(\ell)} + \sum_{j=0}^{k} \frac{1}{p_j}, \\
\frac{1}{p_j} &= \frac{1-\mu_j}{\chi(\ell)} + \frac{\mu_j}{p}, \quad \mu_0 = \frac{p\sigma}{n}, \quad \mu_j = \frac{p|\beta_j|}{n} < 1, \\
\frac{\ell+k-1}{\chi(\ell)} &= \frac{1}{r} - \frac{1}{p},
\end{aligned}$$

we get by interpolation and the Sobolev embedding theorem

$$\begin{aligned}
\left(\int_0^\infty t^{-2\sigma} \sup_{|y|\le t} \|J_3\|_r^2 \frac{dt}{t}\right)^{\frac{1}{2}} &\le \sum_{\ell=0}^{\infty} \frac{\lambda^\ell}{\ell!} \|u; L^{\chi(\ell)}\|^\ell \prod_{j=1}^{k} \|u; \dot{H}^{|\beta_j|,p_j}\| \cdot \|u; \dot{B}^\sigma_{p_0,2}\| \\
&\le C\sum_{\ell=0}^{\infty} \frac{\lambda^\ell}{\ell!} \|u; L^{\chi(\ell)}\|^{\ell+k-1} \|u; \dot{B}^{\frac{n}{p}}_{p,2}\| \\
&\le C\sum_{\ell=0}^{\infty} \frac{\lambda^\ell}{\ell!} \{\chi(\ell)\}^{\frac{\ell+k-1}{p'}} \|u; L^\infty(I; B^{\frac{n}{p}}_{p,2})\|^{k+\ell} \\
&\le CC_3M^k, \quad 2 \le k \le \left[\frac{n}{p}\right],
\end{aligned} \tag{2.140}$$

where

$$C_3 = \sum_{\ell=0}^{\infty} a_\ell = \sum_{\ell=0}^{\infty} \lambda^\ell \chi(\ell)^{\frac{k+\ell-1}{p'}} \|u; L^\infty(I; B_{p,2}^{\frac{n}{p}})\|^\ell / \ell! < \infty.$$

Finally, J_4 can be estimated similarly by letting

$$\frac{1}{r} = \frac{\ell}{\chi(\ell)} + \sum_{i=1}^{k} \frac{1}{p_i},$$
$$\frac{1}{p_i} = \frac{1-\mu_i}{\chi(\ell)} + \frac{\mu_i}{p}, \quad \mu_j = \frac{p(|\beta_j| + \sigma)}{n}, \quad \mu_i = \frac{p|\beta_i|}{n}$$

for $i = 1, \dots, j-1, j+1, \dots, k$ with $\chi(\ell)$ defined as above, and we have

$$\left(\int_0^\infty t^{-2\sigma} \sup_{|y| \le t} \|J_4\|_r^2 \frac{dt}{t} \right)^{\frac{1}{2}} \le C \sum_{\ell=0}^{\infty} \frac{\lambda^\ell}{\ell!} \{\chi(\ell)\}^{\frac{\ell+k-1}{p'}} \|u; L^\infty(I; B_{p,2}^{\frac{n}{p}})\|^{k+\ell}$$
$$\le C C_3 M^k, \quad 2 \le k \le \left[\frac{n}{p}\right]. \tag{2.141}$$

Combining (2.85) with (2.131), (2.136)–(2.141) and noting (2.101) lead to the result

$$\|\mathcal{T}u\|_{Y(I)} \le C\|\varphi\|_{B_{p,2}^{\frac{n}{p}}} + C \sup_{t \in I} \int_0^t |t-\tau|^{-\frac{d}{2} - \frac{n}{2}(\frac{1}{r} - \frac{1}{p})} \|f(u)\|_{B_{r,2}^{\frac{n}{p}}} d\tau$$
$$\le C\|\varphi\|_{B_{p,2}^{\frac{n}{p}}} + CT^{1-\frac{d}{2}-\frac{n}{2}(\frac{1}{r}-\frac{1}{p})} M^2. \tag{2.142}$$

Similarly, we obtain

$$d(\mathcal{T}u, \mathcal{T}v) \le CT^{1-\frac{d}{2}-\frac{n}{2}(\frac{1}{r}-\frac{1}{p})} M d(u, v). \tag{2.143}$$

Thus, by (2.142) and (2.143), $\mathcal{T}$ is a contraction mapping from $Y(I)$ into itself if T is sufficiently small. And so by the Banach contraction mapping principle and the Picard method, we conclude the required result in the case $sp = n$.

Now, we consider the case that $sp > n$. First, it follows, on noting that $d \in [0, 2)$ and $B_{p,2}^s$ is a Banach algebra, that

$$\|f(u)\|_{B_{p,2}^s} \le C(\|u\|_\infty)\|u\|_{B_{p,2}^s} \le C(\|u\|_{B_{p,2}^s})\|u\|_{B_{p,2}^s}. \tag{2.144}$$

Then the statement (b) can be proved by substituting (2.144) for (2.95) and arguing in a similar way as in the proof of (a).

We now prove (c). We only prove (ii); the statements (i) and (iii) are an immediate consequence of (2.85), Theorem 2.2.1 and the Sobolev embedding theorem.

First, if $s < n/p$, then one easily verifies that

$$\frac{d}{2} + \frac{s - s_\alpha}{2} = 1 - \frac{\alpha(s - s_c)}{2}.$$

Since $\theta < 2\alpha\nu$, and by (2.56) and (2.95) and the Sobolev embedding theorem, it is found that

$$\begin{aligned}\|u(t) - S(t)\varphi\|_{B^{s+\theta}_{p,2}} &\leq C \int_0^t |t - \tau|^{-\frac{d}{2} - \frac{\theta+s-s_\alpha}{2}} \|u\|^{\alpha+1}_{B^s_{p,2}} d\tau \\ &\leq C \cdot (T^*)^{-\frac{\theta}{2} + \frac{\alpha(s-s_c)}{2}} \|u\|^{\alpha+1}_{B^s_{p,2}} < \infty\end{aligned}$$

for $t \in [0, T^*)$.

Now if $sp > n$, then we have again by (2.56), (2.84) and (2.95) and the Sobolev embedding theorem that

$$\begin{aligned}\|u(t) - S(t)\varphi\|_{B^{s+\theta}_{p,2}} &\leq C \int_0^t |t - \tau|^{-\frac{d}{2} - \frac{\theta}{2}} \|u\|^{\alpha+1}_{B^s_{p,2}} d\tau \\ &\leq C \cdot (T^*)^{1-\frac{d}{2}-\frac{\theta}{2}} C(\|u; B^s_{p,2}\|) \|u\|_{B^s_{p,2}} < \infty\end{aligned}$$

for $t \in [0, T^*)$. On the other hand, if $sp = n$, it is derived by the similar argument as in the proof of (2.142) that

$$\begin{aligned}\|u(t) - S(t)\varphi\|_{B^{s+\theta}_{p,2}} &\leq C \int_0^t |t - \tau|^{-\frac{d+\theta}{2} - \frac{n}{2}(\frac{1}{r} - \frac{1}{p})} \|f(u)\|_{B^s_{r,2}} d\tau \\ &\leq C \cdot (T^*)^{1-\frac{\theta+d}{2}-\frac{n}{2}(\frac{1}{r}-\frac{1}{p})} C(\|u; B^s_{p,2}\|) < \infty\end{aligned}$$

for $t \in [0, T^*)$. The statement (ii) of (c), and hence Theorem 2.3.1 are thus proved. □

Proof of Theorem 2.3.2. We first prove (a). Since $\varphi \in B^{s_c}_{p,2}$, then $\varphi \in L^{p_c} \cap \dot{B}^0_{p_c,2}$ and, by the Sobolev embedding theorem,

$$\|\varphi\|_{p_c}, \|\varphi\|_{\dot{B}^0_{p_c,2}} \leq C\|\varphi\|_{B^s_{p,2}}. \tag{2.145}$$

By Theorem 2.2.2, there exists a $\delta > 0$ such that for any $\varphi \in L^{p_c}$ with $\|\varphi\|_{p_c} < \delta$, there is a unique global solution $u \in C_b([0, \infty); L^{p_c})$ to problem (2.37)–(2.38). We now prove that $u \in C_b([0, \infty); B^s_{p,2})$. In fact, in view of that $\varphi \in L^{p_c} \cap L^p \cap \dot{B}^0_{p_c,2}$, we obtain by Ribaud's result and method in [227] that $u \in C_b([0, \infty); L^p \cap L^{p_c} \cap \dot{B}^0_{p,2})$ and

$$\|u; C_{\tilde{q}(\tilde{p},p_c)}(I; L^{\tilde{p}})\| < \infty, \quad \|u; C_{\tilde{q}(\tilde{p},p_c)}(I; \dot{B}^0_{\tilde{p},2})\| < \infty \tag{2.146}$$

for any generalized admissible triplet $(\tilde{q},\tilde{p},p_c)$ with $\max(1+\alpha,p_c)<\tilde{p}<p_c(\alpha+1)$. Since $\|u;B^s_{p,2}\| = \|u\|_p + \|u;\dot{B}^s_{p,2}\|$, it suffices to prove that $\|u(t);\dot{B}^s_{p,2}\| < \infty$ for $t>0$. In view of that $p_c < p(\alpha+1)$, one can choose $\tilde{p}$ such that

$$\frac{d}{2}+\frac{n}{2}\left(\frac{\alpha+1}{\tilde{p}}-\frac{1}{p_c}\right)=\frac{d+s_c}{2}+\frac{n}{2}\left(\frac{\alpha+1}{\tilde{p}}-\frac{1}{p}\right)<1,$$

where we have used $s_c<2-d$. Making use of (2.56) and (2.84) along with the Sobolev embedding theorem and noting that

$$\frac{\alpha+1}{\tilde{q}}+\frac{d+s_c}{2}+\frac{n}{2}\left(\frac{\alpha+1}{\tilde{p}}-\frac{1}{p}\right)=1,$$

we obtain that for $t>0$,

$$\begin{aligned}
\|u(t);\dot{B}^{s_c}_{p,2}\| &\le C\|\varphi;\dot{B}^{s_c}_{p,2}\| + C\int_0^t |t-\tau|^{-\frac{d+s_c}{2}}\left\|S\left(\frac{t-\tau}{2}\right)f(u);\dot{B}^0_{p,2}\right\|d\tau\\
&\le C\|\varphi;\dot{B}^{s_c}_{p,2}\| + C\int_0^t |t-\tau|^{-\frac{d+s_c}{2}-\frac{n}{2}(\frac{\alpha+1}{\tilde{p}}-\frac{1}{p})}\|f(u);\dot{B}^0_{\frac{\tilde{p}}{\alpha+1},2}\|d\tau\\
&\le C\|\varphi;\dot{B}^{s_c}_{p,2}\| + C\int_0^t |t-\tau|^{-\frac{d+s_c}{2}-\frac{n}{2}(\frac{\alpha+1}{\tilde{p}}-\frac{1}{p})}\|u;\dot{B}^0_{\tilde{p},2(\alpha+1)}\|^{\alpha+1}d\tau\\
&\le C\|\varphi;\dot{B}^{s_c}_{p,2}\| + C\int_0^t |t-\tau|^{-\frac{d+s_c}{2}-\frac{n}{2}(\frac{\alpha+1}{\tilde{p}}-\frac{1}{p})}\tau^{-\frac{\alpha+1}{\tilde{q}}}d\tau\\
&\quad\times\|u;C_{\tilde{q}(\tilde{p},r_c)}([0,\infty);\dot{B}^0_{\tilde{p},2}\|^{\alpha+1}<\infty.
\end{aligned}\tag{2.147}$$

This proves (a).

We now prove (b). First, as a consequence of (2.87), (2.88) and the interpolation theorem, it follows that for $p\ge r$,

$$\|u\|_p \le Cp^{\frac{1}{2}+\frac{r-2}{2p}}\|u;\dot{H}^{\frac{n}{2}}\|^{1-r/p}\|u\|_r^{r/p},\tag{2.148}$$

$$\|u\|_{\dot{B}^0_{p,2}} \le Cp^{\frac{1}{2}+\frac{r-2}{2p}}\|u;\dot{H}^{\frac{n}{2}}\|^{1-r/p}\|u\|_{\dot{B}^0_{r,2}}^{r/p}.\tag{2.149}$$

Similar to the case of Schrödinger equation, we say $(q,r)\in\Lambda$ if

$$\frac{2}{q}=n\left(\frac{1}{2}-\frac{1}{r}\right),\quad 2\le r<\begin{cases}\frac{2n}{n-2}, & \text{for } n\ge 3,\\ \infty, & \text{for } n\le 2.\end{cases}$$

Then it can be shown that for any $s\in\mathbb{R}$,

$$\|S(\cdot)\varphi;L^q(I;H^{s,r})\| \le C\|\varphi\|_{H^s},\quad \forall(q,r)\in\Lambda,\tag{2.150}$$

$$\left\| \int_0^t S(t-\tau)f(x,\tau)d\tau; L^{q_2}(I; H^{s,r_2}) \right\| \le C\|f; L^{q_1'}(I; H^{s,r_1'})\| \tag{2.151}$$

for any $(q_j, r_j) \in \Lambda$ $(j = 1, 2)$, where $I = [0, \infty)$ or $I = [0, T)$ with $T > 0$. In fact, (2.150) is an immediate result of Lemma 2.1.1 and Theorem 2.1.1. To prove (2.151), note first that for $(q_j, r_j) \in \Lambda$ $(j = 1, 2)$,

$$\frac{n}{2}\left(\frac{1}{r_1'} - \frac{1}{r_2}\right) < 1,$$
$$\frac{1}{q_2} = \frac{1}{q_1'} - 1 + \frac{n}{2}\left(\frac{1}{r_2'} - \frac{1}{r_1}\right).$$

Then, by the Hardy–Littlewood–Sobolev inequality, we get

$$\begin{aligned}\left\| \int_0^t S(t-\tau)fd\tau; L^{q_2}(I; H^{s,r_2}) \right\| &\le \left\| \int_0^t |t-\tau|^{-\frac{n}{2}(\frac{1}{r_1'}-\frac{1}{r_2})} \|f; H^{s,r_1'}\|d\tau \right\|_{q_2} \\ &\le C\|f; L^{q_1'}(I; H^{s,r_1'})\|.\end{aligned}$$

Let q_0, r_0, δ_0 be such that

$$\begin{aligned}&\max\left(1, \frac{n}{2}\right) < q_0 < \infty, \\ &\max\left(1, \frac{n}{2}\right) < r_0 < \begin{cases}\min(\frac{n+2}{2}, \frac{n^2}{2(n-2)}), & \text{for } n \ge 3, \\ \frac{n+2}{2}, & \text{for } n \le 2\end{cases}\end{aligned} \tag{2.152}$$

and $n/(2r_0) < \delta_0 < 1$. Take $(q_1, r_1), (q_2, r_2) \in \Lambda$ such that

$$\frac{1}{q_2'} = \frac{1}{q_0} - \frac{1}{q_1}, \quad \frac{1}{r_2'} = \frac{1}{r_0} - \frac{1}{r_1}.$$

Then it is easy to see that

$$\frac{2}{q_0} + \frac{n}{r_0} = 2,$$

which together with (2.152) implies that $(\frac{4q_0}{n}, \frac{4r_0}{n}) \in \Lambda$. For any $0 < \delta \le \delta_0$, we define

$$X_\delta = C([0,\infty); H^{\frac{n}{2}}) \cap \bigcap_{(q,r)\in\Lambda, \delta(r)\le\delta} L^q([0,\infty); H^{\frac{n}{2},r})$$

with the norm

$$\|u; X_\delta\| := \sup_{\frac{2}{q}=\delta(r)\le\delta} \|u; L^q([0,\infty); H^{\frac{n}{2},r})\|$$

in the case when n is even, and

$$X_\delta = C([0,\infty); H^{\frac{n}{2}}) \cap \bigcap_{(q,r)\in\Lambda,\delta(r)\leq\delta} L^q([0,\infty); \dot{B}^0_{r,2} \cap B^{\frac{n}{2}}_{r,2})$$

with the norm

$$\|u; X_\delta\| = \sup_{\frac{2}{q}=\delta(r)\leq\delta} \|u; L^q([0,\infty); \dot{B}^0_{r,2} \cap B^{\frac{n}{2}}_{r,2})\|$$

in the case when n is odd. We claim that for any $\rho > 0$ and $u, v \in X_\delta$ with $\|u; X_\delta\| < \rho$ and $\|v; X_\delta\| < \rho$, the following estimates hold:

$$\|f(u) - f(v); L^{q_2'}([0,\infty); L^{r_2'})\| \leq G(\rho)\|u - v; L^{q_1}([0,\infty); L^{r_1})\|, \tag{2.153}$$

$$\|f(u) - f(v); L^{q_2'}([0,\infty); \dot{B}^0_{r_2',2})\| \leq G(\rho)\|u - v; L^{q_1}([0,\infty); \dot{B}^0_{r_1,2})\|, \tag{2.154}$$

and when n is even

$$\|f(u); L^{q_2'}([0,\infty); H^{\frac{n}{2},r_2'})\| \leq G(\rho)\|u; L^{q_1}([0,\infty); H^{\frac{n}{2},r_1})\|, \tag{2.155}$$

and when n is odd

$$\|f(u); L^{q_2'}([0,\infty); B^{\frac{n}{2}}_{r_2',2})\| \leq G(\rho)\|u; L^{q_1}([0,\infty); B^{\frac{n}{2}}_{r_1,2})\|, \tag{2.156}$$

where $G(\rho)$ satisfies that

$$\begin{aligned} G(\rho) &= O(\rho^4), \quad n = 1,\\ G(\rho) &= O(\rho^2), \quad n = 2, 3,\\ G(\rho) &= O(\rho), \quad n \geq 4. \end{aligned}$$

The estimates (2.153)–(2.156) can be proved in exactly the same way as in [209]. We only consider the case $n = 1$; the others can be shown similarly. Using (2.148) and (2.149), we obtain

$$\begin{aligned} \|u^{4+\ell}; L^{q_0}([0,\infty); L^{r_0})\| &\leq \big\| \|u; L^{(4+\ell)r_0}\|^{4+\ell} \big\|_{q_0}\\ &\leq C^{4+\ell}((4+\ell)r_0)^{(\frac{1}{2}+\frac{\frac{4r_0}{n}-2}{2(4+\ell)r_0})(4+\ell)} \big\| \|u; \dot{H}^{\frac{n}{2}}\|^{4+\ell-\frac{4}{n}} \cdot \|u\|^{\frac{4}{n}}_{\frac{4r_0}{n}} \big\|_{q_0}\\ &\leq C^{4+\ell}((4+\ell)r_0)^{\frac{4+\ell}{2}+\frac{2}{n}-\frac{1}{r_0}} \|u; L^\infty([0,\infty); \dot{H}^{\frac{n}{2}})\|^{4+\ell-\frac{4}{n}}\\ &\quad \cdot \|u; L^{\frac{4q_0}{n}}([0,\infty); L^{\frac{4r_0}{n}})\|^{\frac{4}{n}}\\ &\leq C^{4+\ell}((4+\ell)r_0)^{(\frac{4+\ell}{2}+\frac{2}{n}-\frac{1}{r_0})} \|u; X_\delta\|^{4+\ell}\\ &\leq C^{4+\ell}((4+\ell)r_0)^{\frac{4+\ell}{2}+\frac{2}{nq_0}} \rho^{4+\ell}, \end{aligned} \tag{2.157}$$

where we have used the fact that $(\frac{4q_0}{n}, \frac{4r_0}{n}) \in \Lambda$ with $\delta(\frac{4r_0}{n}) < \delta$. Since

$$\sum_{\ell=0}^{\infty} \frac{\lambda^\ell}{\ell!} C^\ell((4+\ell)r_0)^{\frac{4+\ell}{2}+\frac{2}{q_0 n}} \rho^\ell < \infty, \quad \forall \rho > 0,$$

and in view of (2.121), it is seen by (2.157) that

$$\begin{aligned}
&\|f(u)-f(v); L^{q_2'}([0,\infty); L^{r_2'})\| \\
&\quad \le \sum_{\ell=0}^{\infty} \frac{\lambda^\ell}{\ell!} (\||u|^{4+\ell}; L^{q_0}([0,\infty); L^{r_0})\| \\
&\quad + \||v|^{4+\ell}; L^{q_0}([0,\infty); L^{r_0})\|)\||u-v|; L^{q_1}([0,\infty); L^{r_1})\| \\
&\quad \le G(\rho)\||u-v|; L^{q_1}([0,\infty); L^{r_1})\|, \quad n = 1,
\end{aligned}$$

that is, (2.153) holds. Since $\dot{B}^0_{r,2} \subset L^r$ and $L^{r'} \subset \dot{B}^0_{r',2}$ for $r \ge 2$, we obtain (2.154) by (2.153). Moreover, using (2.157), (2.153) and the equivalent norm (2.100) of Besov spaces, we obtain that

$$\begin{aligned}
&\|f(u); L^{q_2'}([0,\infty); B^{\frac{1}{2}}_{r_2',2})\| \\
&\quad = \|f(u); L^{q_2'}([0,\infty); L^{r_2'})\| + \left\| \int_0^\infty s^{-1} \sup_{|y|\le s} \|f(\tau_y u) - f(u); L^{r_2'}\| \frac{ds}{s} \right\|_{q_2'} \\
&\quad \le G(\rho)\|u; L^{q_1}([0,\infty); L^{r_1})\| + G(\rho)\|u; L^{q_1}([0,\infty); \dot{B}^{\frac{1}{2}}_{r_1,2})\| \\
&\quad \le G(\rho)\|u; L^{q_1}([0,\infty); \dot{B}^{\frac{1}{2}}_{r_1,2})\|.
\end{aligned}$$

Now we consider the operator $\mathcal{T}$ defined by (2.45) in X_δ, From (2.153)–(2.156), it follows that for sufficiently small δ, the operator $\mathcal{T}$ is a contraction mapping from X_δ into itself. And so the Banach contraction mapping principle implies the required result. Theorem 2.3.2 is thus proved. □

2.4 Abstract equations of parabolic type

We have utilized the space-time estimates method to study the Cauchy problem for semilinear parabolic equations. Similarly, this method can be applied to initial-boundary value problems. Of course, the method can also be applied to Cauchy problems or initial boundary value problems for general high-order parabolic equations. We will not discuss this in details due to the limitation of space. The aim of this section is to discuss the relationship between the abstract semigroup method and the space-time estimates method and to present the abstract Segal existence theorem and its various generalizations. It is necessary to point out that these abstract existence results can be applied to study other evolution equations in these cases where the operator A generates C_0-semigroups not analytic semigroups as in the case of parabolic equations.

Generally speaking, initial-boundary value problems and Cauchy problems for nonlinear evolution equations can all be written as the following abstract Cauchy problem:

$$u_t + Au = f(u), \quad u(0) = \varphi. \tag{2.158}$$

For the case of parabolic equations, $-A$ generates an analytic semigroup e^{-At} in any $L^p(\Omega)$ or $L^p(\mathbb{R}^n)$ with $1 < p < \infty$. For other evolution equations such as the Schrödinger and wave equations, $-A$ generates a C_0-semigroup only in Hilbert spaces such as L^2 or $H^1 \times L^2$. In general, the study of the problem (2.158) can be reduced to that of the equivalent integral equation

$$u(t) = e^{-tA}\varphi + \int_0^t e^{-(t-\tau)A} f(u(\tau))d\tau. \tag{2.159}$$

The solution of (2.159) is usually called as the mild solution of (2.158). Suppose $-A$ generates a C_0-semigroup (or analytic semigroup) in the Banach space X. We are interested in proving whether (2.158) or equivalently (2.159) determines a strong continuous flow in X, that is, whether (2.158) or equivalently (2.159) is well posed in X. The theoretical foundation of this is the following abstract Segal theorem (see [214]).

Theorem 2.4.1. *Let X be a Banach space and let $f : X \to X$ be locally Lipschitz continuous. Suppose $-A$ generates a C_0-semigroup in X. Then, for any $\varphi \in X$, there are a $T^* > 0$ and a unique continuous function $u : [0, T^*) \to X$ such that u satisfies* (2.159) *and either $T^* = \infty$ or $T^* < \infty$ and $\lim_{t \to T^*} \|u(t)\|_X = \infty$.*

When applying Segal's theorem to study concrete nonlinear evolution equations, the main difficulty arises in verifying that f is locally Lipschitz continuous in X. For example, for the simplest nonlinear function $f(u) = |u|^\alpha u$ with $\alpha > 0$, it is impossible to prove that f is Lipschitz continuous in L^r for $1 < r < \infty$. However, the space-time estimates method makes use of the space-time integrability of solutions to evolution equations to establish, in a subspace $X \cap Y$ (where Y denotes some suitable time-space Banach space), the well-posedness of the Cauchy problem for nonlinear evolution equations. Thus, in the space-time estimates method, one uses the Lipschitz continuity in $X \cap Y$ of the nonlinear mapping

$$Gf(u) = \int_0^t e^{-(t-\tau)A} f(u(\tau))d\tau$$

to replace the Lipschitz continuity in X of f. In doing this, a decisive role is played by the Strichartz time-space estimates of solutions to the linear evolution equations, as seen in the study of the heat equation in this chapter. In the later chapters on the study of the

Schrödinger equation, the wave equation and the Navier–Stokes equations, the reader will deeply understand further the essential role played by the space-time estimates method.

Of course, the well-posedness for the evolution equations, in particular, the parabolic equations, can be studied in the space $Y \subseteq X$ of better regularity by the regularity of e^{-tA}, and the conditions on the nonlinear function are more loose. But it is worse than the space-time estimate method, which sufficiently makes use of the integrability of the linear equation and provides an essential method to study the well-posedness and scattering theory of nonlinear equations such as dispersive equations and wave equations in weaker function spaces; see [19, 242, 247] for details.

We now state the Segal theorem for studying the Cauchy problem for abstract evolution equations for the reader's convenience. Let A generate a C_0-semigroup in X with the domain $D(A) \subset X$. Equipped with the norm

$$\|u\|_{D(A)} = \|u\|_X + \|Au\|_X, \quad \forall u \in D(A),$$

$D(A) \subset X$ is a complete Banach space. For many nonlinear evolution equations, the basis space X is always a self-reflective Banach space and A is a densely defined, closed operator in X. In particular, for parabolic operators, we may assume without loss of generality that $0 \in \rho(A)$ and

$$\|(\lambda + A)^{-1}\| \leq \frac{M}{1 + |\lambda|}, \quad \operatorname{Re}\lambda \geq 0. \tag{2.160}$$

With the help of (2.160), we can define the fractional-order operator A^α, $0 < \alpha \leq 1$ (see [214] for details). The following result is a generalized Segal theorem, which in some sense is easier to verify compared with Theorem 2.4.1.

Theorem 2.4.2.

(i) *Suppose $-A$ generates an analytic semigroup in the Banach space X and f is Lipschitz or locally Lipschitz continuous in $D(A^{1-\alpha}) \hookrightarrow X$ $(0 < \alpha < 1)$, that is,*

$$\|f(u) - f(v)\|_X \leq C[\|A^{1-\alpha}u\|_X + \|A^{1-\alpha}v\|_X]\|A^{1-\alpha}(u - v)\|_X. \tag{2.161}$$

Let $\varphi \in D(A)$. Then there exist a $T^* = T(\varphi) > 0$ and a unique $u \in C([0, T^*); D(A)) \cap C^1([0, T^*); X)$ such that u satisfies (2.159) *and*

$$\textit{either} \quad T^* = \infty \quad \textit{or} \quad T^* < \infty \quad \textit{and} \quad \lim_{t \to T^*} \|Au(t)\|_X = \infty. \tag{2.162}$$

(ii) *Suppose $-A$ generates a C_0-semigroup in X and $f : D(A) \to D(A)$ is Lipschitz or locally Lipschitz continuous:*

$$\|f(u) - f(v)\|_{D(A)} \leq C([\|Au\|_X + \|Av\|_X)\|u - v\|_{D(A)}. \tag{2.163}$$

Let $\varphi \in D(A)$. Then there are a $T^* = T(\varphi) > 0$ and a unique $u \in C([0,T^*);D(A)) \cap C^1([0,T^*);X)$ such that u satisfies (2.159) *and* (2.162) *holds.*

Remark 2.4.1.

(i) In Theorem 2.4.2(i), if $\varphi \in X$, then there are a $T^* = T(\varphi) > 0$ and a unique

$$u \in C([0,T^*);D(A)) \cap C^1((0,T^*);X) \cap C([0,T^*);X)$$

such that u satisfies (2.159) and (2.162) holds.

(ii) It is easy to see that one can use Theorem 2.4.2(i) to study the well-posedness of nonlinear parabolic equations. For some nonlinear functions $f(u)$, condition (2.161) is weaker and easier to verify than the Lipschitz or local Lipschitz continuity of f in X. For parabolic equations, since parabolic operators are sub-elliptic operators, which are of good regularity as seen from the regularity estimates of analytic semigroups, so the Lipschitz or local Lipschitz continuity of f in X can be relaxed to condition (2.161). For other nonlinear evolution equations, such as the Schrödinger and wave equations, one can only apply Theorem 2.4.2(ii) so as to obtain only the well-posedness in $D(A) \subset X$ of the evolution equations. This can be overcome with the help of the space-time estimates; see, for example, [19, 198, 241, 247, 275] for results on the Schrödinger and wave equations.

Using the abstract integration by parts, Von Wahl generalized Theorem 2.4.2 by relaxing conditions (2.161) and (2.163) (see [283] for details).

Theorem 2.4.3. *Let X be a self-reflective Banach space.*

(i) *Suppose that $-A$ is a densely defined closed operator in X satisfying* (2.160), *that is, $-A$ generates an analytic semigroup e^{-At} in X. Let f satisfy that for $0 < \alpha < 1$ and $u, v \in D(A)$,*

$$\|f(u) - f(v)\|_X \le K(\|Au\|_X + \|Av\|_X)\|A^{1-\alpha}(u - v)\|_X, \tag{2.164}$$

$$\|f(u)\|_X, \ \|f(v)\|_X \le K(\|Au\|_X + \|Av\|_X), \tag{2.165}$$

where $K(\cdot)$ is a positive nondecreasing function. If $\varphi \in D(A)$, then there are a $T^* = T(\varphi) > 0$ and a unique $u \in \bigcap_{0<T<T^*} C^1([0,T];X)$ such that u satisfies (2.159) *with the property*

$$u(t) \in D(A), \quad Au \in \bigcap_{0<T<T^*} C^1([0,T];X), \tag{2.166}$$

and (2.162) *holds.*

(ii) *Assume that $-A$ generates a C_0-semigroup in X satisfying that*

$$\{\lambda : Re\lambda \ge 0\} \subseteq \rho(A). \tag{2.167}$$

Let f satisfy that for $u, v \in D(A)$,

$$\|f(u) - f(v)\|_X \leq K(\|Au\|_X + \|Av\|_X)\|u - v\|_X, \tag{2.168}$$

$$\|f(u)\|_X, \|f(v)\|_X \leq K(\|Au\|_X + \|Av\|_X), \tag{2.169}$$

where $K(\cdot)$ is a positive increasing function. If $\varphi \in D(A)$, then there are a $T^* = T(\varphi) > 0$ and a unique continuous function $u \in C^1([0, T^*); X)$ such that u satisfies (2.159), (2.166) *and* (2.162).

(iii) *(Regularity) If, in addition, $-A$ is the infinitesimal generator of an analytic semigroup, then the solution u also has the regularity property that for $0 < \alpha < 1$ and $0 < t < T^*$,*

$$u'(t) \in D(A^{1-\alpha}),$$

and

$$t^{1-\alpha}A^{1-\alpha}u'(t) \in \bigcap_{0<\varepsilon<T<T^*} C([\varepsilon, T]; X) \cap \bigcap_{0<T<T^*} L^\infty((0,T); X).$$

Sketch of proof. Theorem 2.4.3 relies on the abstract integrating by parts formula

$$\int_0^t e^{-(t-s)A} f(u)ds = -\int_0^t e^{-(t-s)A}A^{-1}f(u)'(s)ds + A^{-1}f(u(t)) - e^{-At}A^{-1}f(\varphi). \tag{2.170}$$

In the closed subset of $C^1([0, T); X) \cap C([0, T); D(A))$,

$$Z(I) = \{u \in C^1([0, T); X) \cap C([0, T); D(A)), \|u'(t)\|_X + \|u\|_{D(A)} \leq M = 2\|\varphi\|_{D(A)}\} \tag{2.171}$$

by introducing the metric

$$d(u, v) = \|u - v\|_X, \quad u, v \in Z(I) \tag{2.172}$$

and making use of the fixed theorem, one can prove Theorem 2.4.3. □

Remark 2.4.2.

(a) In the abstract integrating by parts formula (2.170), $f(u)'(s)$ denotes the weakly continuous differential, which has the following result.

Proposition 2.4.1.

(i) *Let $f(u)$ satisfy* (4.14), *and $\forall T > 0$, $u(t) \in C^1([0, T); X)$, $u(t) \in C([0, T]; D(A))$. Then $f(u)(s)$ is a weakly continuous differentiable, that is,*

$$\int_0^T \langle f(u)'(s), \psi(s)\rangle ds = -\int_0^T \langle f(u)(s), \psi'(s)\rangle ds,$$
$$\forall \psi \in C^1([0,T];X^*) \text{ with compact support,} \tag{2.173}$$

and has the estimate

$$\sup_{0\le s\le T} \|f(u)'(s)\|_X \le K\Big(2\sup_{0\le s\le T}\|Au(s)\|_X\Big) \sup_{0\le s\le T}\|u'(s)\|_X. \tag{2.174}$$

(ii) *Let* $f(u)$ *satisfy* (4.10), $u(t) \in C^1([0,T];X) \cap C([0,T];D(A))$ *and*

$$u'(t) \in D(A^{1-\rho}), \quad A^{1-\rho}u'(t) \in C([0,T];X), \quad t \in [0,T],\ 0<\rho<1, \tag{2.175}$$

then $f(u(s))$ *is weakly continuous differentiable and satisfies* (4.19). *In addition,* $f(u(s))$ *satisfies* (4.20) *and*

$$\sup_{0\le s\le T} \|f(u)'(s)\|_X \le K\Big(2\sup_{0\le s\le T}\|Au(s)\|_X\Big) \sup_{0\le s\le T}\|A^{1-\rho}u'(s)\|_X, \tag{2.176}$$

where $K(\sigma)$ *is the same as Theorem* 2.4.3.

(b) For studying the well-posedness for nonlinear evolution directly in the domain $D(A) \subseteq X$ of an infinitesimal generator, one can use Theorem 2.4.2 or Theorem 2.4.3. The disadvantage of this method is that it is very hard to verify the locally Lipschitz continuity, especially for the wave equations and dispersive equations in the high dimension spaces.

In the above, we have compared the space-time estimate and the abstract semigroup method. From this, we can see that the space-time estimate method is much better than using the abstract semigroup method. We know that the space-time admissible pair (triplet) is different for different evolution equations, and depends completely on the type of evolution equations. Now, the problem is how to choose a suitable work space to study the well-posedness for certain nonlinear evolution equations with different nonlinear growth. We take the Cauchy problem of parabolic equation with homogeneous nonlinear growth

$$u_t - \Delta u = \mu|u|^\alpha u, \quad u(0) = \varphi \tag{2.177}$$

or $u_t - \Delta u = \mu Q(D)(|u|^\alpha u)$, $u(0) = \varphi$

$$Q(D) \text{ is a homogeneous pseudodifferential operator of order } d \in [0,2) \tag{2.178}$$

as an example to illustrate the relation between nonlinear growth and work spaces.

We know that, for $\forall 1 < p < \infty$, $e^{-|\xi|^2} \in \mathcal{M}_p$ (L^p multiplier). $e^{-t\Delta}$ is the analytic semigroup on L^p, hence the Cauchy problem of free heat equation

$$u(t) - \Delta u = 0, \quad (x,t) \in \mathbb{R}^n \times \mathbb{R}^+, \quad u(0) = \varphi, \quad x \in \mathbb{R}^n, \tag{2.179}$$

is well posed in spaces like $L^p, H^{s,p}, B^s_{p,2}$, $s \in \mathbb{R}$. What p or s, r can ensure that the nonlinear problem (2.177) or (2.178) is well posed? Is there any judging method? Next, we give a basic judge by the scaling method, which can give a direction on the study of nonlinear equations, and simply illustrate in what case the nonlinear problem is well posed and in what case it is ill posed. Though for the different evolution equation, the accord degree of this judge and the real case is different, it is very useful, especially in the parabolic case, with almost complete accord.

In general, $f(x) \in H^{s,p}(\mathbb{R}^n)$ or $B^s_{p,2}(\mathbb{R}^n)$. Then $f(\lambda x) \in H^{s,p}(\mathbb{R}^n)$ or $B^s_{p,2}(\mathbb{R}^n)$ and

$$\begin{aligned} &\|f(\lambda x)\|_{\dot{H}^{s,p}} = \lambda^{s-\frac{n}{p}} \|f(x)\|_{\dot{H}^{s,p}}, \quad \|f(\lambda x)\|_p = \lambda^{-\frac{n}{p}} \|f(x)\|_p, \\ &\|f(\lambda x)\|_{\dot{B}^s_{p,2}} = \lambda^{s-\frac{n}{p}} \|f(x)\|_{\dot{B}^s_{p,2}}. \end{aligned} \tag{2.180}$$

Thus, it is easy to see that the sufficient and necessary condition of $H^{s_1,p_1} \hookrightarrow H^{s_2,p_2}$ or $B^{s_1}_{p_1,2} \hookrightarrow B^{s_2}_{p_2,2}$ is

$$s_1 - \frac{n}{p_1} \geq s_2 - \frac{n}{p_2}, \quad -\frac{n}{p_1} \leq -\frac{n}{p_2}. \tag{2.181}$$

This is based on $H^{s,p} = \dot{H}^{s,p} \cap L^p$, $B^s_{p,2} = \dot{B}^s_{p,2} \cap L^p$ and the comparison of the singularity between $\|f(\lambda x)\|_{\dot{H}^{s_1,p_1}}$ and $\|f(\lambda x)\|_{\dot{H}^{s_2,p_2}}$ at $\lambda = \infty$, $\|f(\lambda x)\|_{L^{p_1}}$ and $\|f(\lambda x)\|_{L^{p_2}}$ at $\lambda = 0$. In particular, the sufficient condition for $L^{p_1}(\mathbb{R}^n) \hookrightarrow L^{p_2}(\mathbb{R}^n)$ is $p_1 = p_2$. Denote by $\dot{X}$ the corresponding homogeneous spaces to a general Sobolev (Besov) space. According to the definition of space degree introduced in Chapter 1, we can directly verify that

$$\deg(\dot{B}^s_{p,2}) = \deg(\dot{H}^{s,p}) = s - \frac{n}{p}, \quad \deg(L^p) = -\frac{n}{p}. \tag{2.182}$$

On the other hand, let $u(x,t)$ be the solution to (4.23) or (4.24). Then

$$u_\lambda(x,t) = \lambda^\theta u(\lambda^{-1}x, \lambda^{-2}t) \tag{2.183}$$

is the solution to (2.177) or (2.178) with initial function $\lambda^\theta \varphi(\lambda^{-1}x)$, thus

$$\theta = -\frac{2}{a} \quad \text{or} \quad \theta = -\frac{2-d}{a}. \tag{2.184}$$

Basic conclusion: (a) X is the suitable Banach space for (2.177) or (2.178), that is, for any $\varphi(x) \in X$, the condition for ensuring the existence of unique local solution

$$u(t) \in C(I;X) \cap \cdots$$

is

$$\deg(\dot{X}) \geq -\frac{2}{\alpha} \quad \text{or} \quad \deg(\dot{X}) \geq -\frac{2-d}{\alpha}. \tag{2.185}$$

(b) When

$$\deg(\dot{X}) < -\frac{2}{\alpha} \quad \text{or} \quad \deg(\dot{X}) < -\frac{2-d}{\alpha}, \tag{2.186}$$

X is not the suitable Banach space for (2.177) or (2.178), that is, for $\varphi \in X$ the problem (2.177) or (2.178) is ill posed in X.

Remark 2.4.3.

(i) Take $X = L^r(\mathbb{R}^n)$, which is the suitable work space for (2.177) or (2.178), if

$$r \geq r_c = \frac{n\alpha}{2} > 1 \quad \text{or} \quad r \geq r_c = \frac{n\alpha}{2-d} > 1, \tag{2.187}$$

the equality $r = r_c$ corresponds to the critical space.

(ii) Take $X = H^{s,p}(B^s_{p,2})$, $\alpha \geq 1$ fixed, then when

$$s \geq s_c = \frac{n}{p} - \frac{2}{\alpha} \quad \text{or} \quad s \geq s_c = \frac{n}{p} - \frac{2-d}{\alpha}, \quad p \geq 1, \tag{2.188}$$

X the suitable Banach space for (2.177) or (2.178). When $s = s_c$, it corresponds to the critical space.

(iii) On the contrary side, that is,

$$r < r_c = \frac{n\alpha}{2} \quad \text{or} \quad r < r_c = \frac{n\alpha}{2-d}, \tag{2.189}$$

$$s < s_c = \frac{n}{p} - \frac{2}{\alpha} \quad \text{or} \quad s < s_c = \frac{n}{p} - \frac{2-d}{\alpha}, \tag{2.190}$$

$L^r, H^{s,p}$ (or $B^s_{p,2}$) are not the suitable Banach spaces for (2.177) or (2.178).

Many studies on the parabolic equations show that the above conclusion is correct; see [58, 92, 98, 99, 198, 220]. In some special case, such as for $\alpha > 0$,

$$\partial_t u - \Delta u = \vec{a} \cdot \nabla(|u|^\alpha u), \quad t > 0,\ x \in \mathbb{R}^n,\ \vec{a} \in \mathbb{R}^n \setminus \{0\}, \quad u(0) = \varphi. \tag{2.191}$$

The space, which the scaling suggests is ill posed is probably well posed; see [85, 220] for details. There are a lot of open problems unsolved on the parabolic equations; the interested readers can refer to the paper by Ponce and Sideris [221] and the references therein.

3 Navier–Stokes equations

It is well known that the Navier–Stokes equations are the fundamental equations describing the motion of fluids and their global well-posedness is one of the most famous open problems in the mathematical and physical fields nowadays. In this chapter, we will give an overall discussion on the essential content and research methods of the problem so as to compare the difference and intrinsic connection between the classical methods and the modern harmonic analysis methods.

In general, the initial-boundary (or initial) value problem for the Navier–Stokes equations can be expressed as

$$u_t - \nu\Delta u + (u\cdot\nabla)u + \nabla P = 0, \quad (x,t) \in \Omega\times(0,T), \tag{3.1}$$

$$\operatorname{div} u = 0, \quad (x,t) \in \Omega\times[0,T), \tag{3.2}$$

$$u(x,0) = u_0(x), \quad x \in \Omega, \tag{3.3}$$

$$u(x,t) = 0, \quad x \in \partial\Omega,\ t \in [0,T), \tag{3.4}$$

where $0 < T \le \infty$ and $\Omega \subset \mathbb{R}^n$ is a bounded smooth domain or $\Omega = \mathbb{R}^n$. In the case when $\Omega = \mathbb{R}^n$, no boundary condition is required in the above problem. However, the decay condition at infinity of the following type can be imposed on the initial data $u_0(x)$:

$$|\partial_x^\alpha u_0(x)| \le C_\alpha(1+|x|)^{-k}, \quad \forall \alpha \in (\mathbb{Z}^+\cup\{0\})^n,\ \forall k \in \mathbb{Z}^+. \tag{3.5}$$

The remaining open problem for the Navier–Stokes equations can be stated as follows:

(i) Let $n = 3$. Denote by u_0 a smooth function satisfying (3.5) and the condition $\operatorname{div} u_0 = 0$. Does there exist a global smooth solution $(u(x,t), P(x,t))$

$$(u(x,t), P(x,t)) \in C^2(\mathbb{R}^3\times\mathbb{R}^+)\times C^1(\mathbb{R}^3\times\mathbb{R}^+) \tag{3.6}$$

to the Cauchy problem (3.1)–(3.3)?

(ii) Let $n = 3$. Is it possible to find a function $u_0 \in C^\infty(\mathbb{R}^3)$ with

$$\operatorname{div} u_0(x) = 0, \quad x \in \mathbb{R}^3 \tag{3.7}$$

such that the Cauchy problem (3.1)–(3.3) has no global solution?

As mentioned above, we consider the Cauchy problem in $\mathbb{R}^3$ as an example to express the open problems for the Navier–Stokes equations. In fact, from the mathematical point of view, we can also consider, in the case with $n > 3$, the Cauchy problem, the initial boundary value problem and the periodic boundary value problem (with respect to the spatial variables) for the Navier–Stokes equations. Similar to the Cauchy problem in $\mathbb{R}^3$ for the Navier–Stokes equations, these problems are among the most important open questions in community nowadays.

https://doi.org/10.1515/9783111384726-003

However, when $n = 2$, the corresponding problems for the Navier–Stokes equations have been completely resolved. The reason is that the Leray–Hopf weak solution is regular. It can be seen clearly from the Serrin–Wahl regularity theory; see [234, 283, 284] or Ladyzenskaya's monograph [159]. From mathematical point of view, a lower-dimensional space is of higher degree of smoothness (for fixed differentiability and integrability indices), so it is easy to be embedded in spaces of smooth functions. On the other hand, from physical point of view, no vortex can be generated in planar fluids. And so planar fluids are simpler than the three-dimensional ones. In fact, by the change of variables, the two-dimensional Navier–Stokes equations can be transformed into heat-conduction equations with convolution terms, which is impossible for the high-dimensional case ($n \geq 3$). For example, consider the following Cauchy problem:

$$u_t + (u \cdot \nabla)u - \Delta u + \nabla P = 0, \quad x \in \mathbb{R}^2,\ t \in \mathbb{R}^+, \tag{3.8}$$

$$\operatorname{div} u = 0, \tag{3.9}$$

$$u(x, 0) = u_0(x), \quad x \in \mathbb{R}^2, \tag{3.10}$$

$$|u_0(x)| \to 0, \quad |x| \to \infty. \tag{3.11}$$

Let

$$\xi = \frac{\partial u_2}{\partial x_1} - \frac{\partial u_1}{\partial x_2}$$

with $u = (u_1, u_2)$. Then the Cauchy problem (3.8)–(3.11) becomes

$$\partial_t \xi + (u \cdot \nabla)\xi = \Delta \xi, \tag{3.12}$$

$$u = \int_{\mathbb{R}^2} K(x - y)\xi(y, t)dy, \tag{3.13}$$

$$\xi(x, 0) = \xi_0(x), \tag{3.14}$$

where $K(x) = \frac{1}{2\pi}\frac{(-x_2, x_1)}{|x|^2}$ and $|\xi_0(x)| \to 0$ as $|x| \to \infty$. It is easy to see that $\nabla \cdot K = \operatorname{div} K = 0$, and so $\nabla \cdot u = \operatorname{div} u = 0$. Thus, we can define the operators

$$\xi = S\xi_0, \tag{3.15}$$

$$u = v\xi_0 = K * S\xi_0. \tag{3.16}$$

We now define the following iteration scheme:

$$u^{(0)} = \xi^{(0)} = 0, \tag{3.17}$$

$$\partial_t \xi^{(k)} - \Delta \xi^{(k)} = -(u^{(k-1)} \cdot \nabla)\xi^{(k)}, \tag{3.18}$$

$$u^{(k)} = K * \xi^{(k)}, \tag{3.19}$$

$$\xi^{(k)}(x, 0) = \xi_0(x), \quad x \in \mathbb{R}^2,\ k = 1, 2, \ldots. \tag{3.20}$$

Assume that $\xi_0 \in L^p(\mathbb{R}^2)$ with $1 < p < \infty$. Then, from the above iteration scheme, we can obtain the well-posedness of the global smooth solution of the Cauchy problem (3.8)–(3.11), that is, we can find functions $u,\ P \in C^\infty(\mathbb{R}^2 \times \mathbb{R}^+)$ satisfying (3.8)–(3.11).

Remark 3.0.1.

(i) The smoothness condition imposed on the initial data u_0, that is,

$$\xi_0(x) = \frac{\partial u_{0,2}}{\partial x_1} - \frac{\partial u_{0,1}}{\partial x_2} \in L^p(\mathbb{R}^2), \quad 1 < p < \infty,$$

is not essential. In fact, in the open question stated above, it has already been required that $u_0 \in C^\infty(\mathbb{R}^2)$ and decays fast at infinity.

(ii) $K(x)$ is the kernel of a weak-type Calderón–Zygmund singular integral operator, so $u(x,t)$, which is of the same regularity as $K(x)$, can be obtained through $\xi(x,t)$.

(iii) The scheme (3.17)–(3.20) is essentially the iteration scheme for the Cauchy problem for the Navier–Stokes equations under the mapping

$$\mathcal{P}_p : (L^p)^2 \to E_p := \{f \in (L^p)^2 \mid \operatorname{div} f = 0\}.$$

Thus,

$$P(x,t) = -\sum_{j=1}^{2}\sum_{k=1}^{2} R_j R_k(u_j u_k),$$

where R_j is a classical Riesz operator.

(iv) In general, the two-dimensional Navier–Stokes equations can be essentially reduced to a heat-conduction equation to deal with. However, it is impossible for the case $n \geq 3$, since even through change of variables the transformed equations remain coupled.

Although there have been extensive study and application of Navier–Stokes equations, there are still many excellent mathematicians who are devoted to completely resolving this difficult mathematical problem and many interesting results and methods have indeed appeared. This chapter aims to discuss these research methods. Due to the limitation of the authors' knowledge, we emphasize on some harmonic analysis methods. This chapter is organized as follows. Section 3.1 presents an overview of the progress in the theory of the Navier–Stokes equations, focusing on classical research methods. Additionally, some remarks are included to provide further insight. In Section 3.2, utilizing the technique of space-time estimates, we give a unified method to deal with the classical results for the Navier–Stokes equations. Section 3.3 mainly introduces the work conducted by Meyer and Cannone. Employing the Littlewood–Paley decomposition technique, they introduced the so-called suitable Banach working spaces and developed a unified approach to establish local well-posedness in the subcritical space of the Cauchy problem for the Navier–Stokes equations. In Section 3.4, we focus on Koch and Tataru's work, which gives the local well-posedness and small global well-posedness in the critical space for the Navier–Stokes equations via the semigroup characterization of BMO^{-1} (which is a inhomogeneous Banach space).

3.1 Classical theory of the Navier–Stokes equations

The mathematical study of the Navier–Stokes equations was initiated by Leray [163], who established the existence of weak solutions to the Navier–Stokes equations in the 1930s. Subsequently, by making use of the technique of functional analysis, Hopf [119] gave a beautiful method to deal with this weak solution on bounded domain and extended Leray's result to the higher-dimensional case. Hence, the weak solution obtained by Leray and Hopf is usually called the Leray–Hopf weak solution. However, the question of uniqueness and regularity of Leray–Hopf weak solutions of the 3D Navier–Stokes equations remains open. In the study of the Navier–Stokes equations, there are primarily two approaches. The first method is to establish the existence of a unique, globally smooth solution (smooth solutions are always unique) by studying the regularity of the Leray–Hopf weak solution. The second manner involves directly examining the well-posedness for the Navier–Stokes equations in stronger function spaces, such as the energy working space and $L^q(I;L^r)(r \geq n)$-type spaces. Essentially in the second situation, the existence of solutions implies the existence and uniqueness of smooth solutions. In this section, we shall consider the historic progress and the main results obtained by these two methods.

Consider the following Navier–Stokes equations

$$\begin{cases} u_t - \Delta u + (u \cdot \nabla)u + \nabla P = 0, & (x,t) \in \Omega \times (0,T), \\ \operatorname{div} u = 0, & x \in \Omega, \\ u|_{\partial\Omega} = 0, & \\ u(0) = u_0(x), \quad \operatorname{div} u_0(x) = 0, & x \in \Omega, \end{cases} \tag{3.21}$$

where $\Omega \subseteq \mathbb{R}^n$ is a smooth domain. In the case when $\Omega = \mathbb{R}^n$, the Dirichlet condition $u|_{\partial\Omega} = 0$ has to be removed in (3.21).

Definition 3.1.1 (Divergence-free vector space). For $0 < T < \infty$, let

$$\begin{aligned} V(T) = \{u(x,t) \mid u \in C_c^\infty(\Omega \times [0,T)),\ \operatorname{div} u = 0, \\ \|u(t)\|_{L^2(\Omega)} \text{ is uniformly bounded with respect to } t \in [0,T)\}. \end{aligned}$$

Then the completion space of $V(T)$ under the norm

$$\|u\|_{V(T)} = \left(\int_0^T (\|u\|_{L^2}^2 + \|\nabla u\|_{L^2}^2) dt \right)^{\frac{1}{2}}$$

is called as the divergence-free vector space. In particular, when T can be chosen as any positive number, $V(T)$ is denoted by $V(\infty)$.

Definition 3.1.2 (Free divergence-free vector space). Let

$$W = \{\varphi(x) \mid \varphi \in C_c^\infty(\Omega), \ \operatorname{div} \varphi = 0\}$$

and define

$$\|\varphi\|_W = \|\varphi\|_{L^2(\Omega)} + \|\nabla \varphi\|_{L^2(\Omega)}.$$

We call the completion space of W under the above norm as the free divergence-free vector space.

Definition 3.1.3 (Leray–Hopf weak solutions). Let $u_0 \in W$. The function $u(x,t)$ is called as a Leray–Hopf weak solution to (3.21) for the Navier–Stokes equations if $u \in V(T)$ and satisfies that

$$\int_0^\infty (u, \varphi_t + \Delta\varphi + u \cdot \nabla\varphi) dt = -(u_0(x), \varphi_0(x)), \quad \forall \varphi \in C_c^\infty(\Omega \times [0,T)), \tag{3.22}$$

where $\varphi_0(x) \equiv \varphi(x,0)$ and $(\cdot,\cdot)$ denotes the L^2-inner product.

Leray first proved the existence of weak solutions to the Navier–Stokes equations in [163], which still has big impact on the research for this problem so far. Hopf not only extended Leray's result, but the method introduced has also provided new momentum for the study of the Navier–Stokes equations. Their results can be expressed as follows.

Theorem 3.1.1 (Leray–Hopf theorem). *Let $u_0 \in W$. Then the problem* (3.21) *has at least one Leray–Hopf weak solution $u \in V(\infty)$ satisfying that*

$$\|u(t)\|_{L^2}^2 + 2\int_0^t \|\nabla u\|_{L^2}^2 d\tau \le \|u_0\|_{L^2}^2, \tag{3.23}$$

$$\lim_{t \downarrow 0} \|u(\cdot,t) - u_0(\cdot)\|_{L^2} = 0.$$

The proof of this result was based on the Galerkin approximation combined with a compactness argument (see [176]). In 1957, utilizing the technique of differentiating with respect to the time variable t and with the help of Galerkin approximations, energy integral estimates and compactness arguments, the Russian mathematicians Kiselev and Ladyzhenskaya proved the global well-posedness of smooth solutions to the Navier–Stokes equations for the case $n = 2$ and the global well-posedness of small smooth solutions for the case $n = 3$ [159] (this can be seen from the regularity theory of $L^q(I, L^p)$ solutions below).

Theorem 3.1.2. *Let $n = 2$ or $n = 3$ and $u_0 \in W^2$. Then* (3.21) *has a Leray–Hopf weak solution $u \in V(T)$ satisfying that*

$$\sup_{0\le t<T} \|\nabla u\|_{L^2}, \quad \sup_{0\le t<T} \|u_t\|_{L^2} < \infty,$$

where $T = T(u_0)$. Further, $T = \infty$ in the case of $n = 2$, or $T = \infty$ in the case of $n = 3$ by requiring $\|u_0\|_{W^2} \ll 1$.

We refer to [159, 176] for details of the proof of this result.

Remark 3.1.1.

(i) W^2 stands for second-order free divergence-free space, that is, W^2 is the completion of the set $\{\varphi \mid \varphi \in C_c^\infty(\Omega),\ \operatorname{div}\varphi = 0\}$ under the norm

$$\|\varphi\|_{W^2} = \left(\|\varphi\|_{L^2}^2 + \sum_{k=1}^n \|\partial_{x_k}^2 \varphi\|_{L^2}^2\right)^{1/2}.$$

(ii) The proof of Theorem 3.1.2 essentially can be reduced to the following estimates. Differentiate the first equation of (3.21) with respect to t and then take the L^2-inner product of both sides of the equation with u_t, respectively. Noting the Gagliardo–Nirenberg inequality or interpolation inequality,

$$\|f\|_{L^4}^2 \le \begin{cases} \|f\|_{L^2}\|f\|_{\dot H^1}, & n = 2,\\ \|f\|_{L^2}^{1/2}\|f\|_{\dot H^1}^{3/2}, & n = 3,\\ \|f\|_{\dot H^1}^2, & n = 4 \end{cases} \tag{3.24}$$

and the Young inequality

$$ab \le \frac{\varepsilon^p a^p}{p} + \frac{b^q}{\varepsilon^q q}, \quad \frac{1}{p} + \frac{1}{q} = 1, \quad \forall \varepsilon > 0,$$

we obtain that

$$\begin{aligned}\frac{d\|u_t\|_{L^2}^2}{dt} &= -2(u_t, u_t \cdot \nabla u) - 2\|u_t\|_{\dot H^1}^2 \le C\|u_t\|_{L^4}^2\|\nabla u\|_{L^2} - 2\|u_t\|_{\dot H^1}^2\\ &\le \begin{cases} C\|\nabla u\|_{L^2}^2\|u_t\|_{L^2}^2, & n = 2,\\ C\|\nabla u\|_{L^2}^4\|u_t\|_{L^2}^2, & n = 3. \end{cases}\end{aligned} \tag{3.25}$$

From the formal energy equality and the associated inequality,

$$\|u\|_2^2 + 2\int_0^t \|\nabla u\|_2^2 dt = \|u_0\|_2^2, \tag{3.26}$$

$$\|\nabla u\|_2^2 = -\frac{1}{2}\frac{d}{dt}\|u\|_2^2 \le \|u\|_2\|u_t\|_2, \tag{3.27}$$

it follows that for $n = 2$,

$$\|u_t\|_2 \le C\|u_t(\cdot,0)\|_2 \exp\left(C\int_0^t \|\nabla u\|_2^2 d\tau \right) < \infty,$$

$$\|u\|_{\dot{H}^1} < \infty,$$

where we have used

$$\|u_t(\cdot,0)\|_2 \le \|\Delta u_0\|_2 + \|u_0\|_4\|\nabla u_0\|_4 < \infty.$$

For $n = 3$, we only use (3.27) to deal with (3.25) and obtain that

$$\frac{d\|u_t\|_2^2}{dt} \le C\|u\|_2^2\|u_t\|_2^4.$$

By solving this inequality, it is seen that there is a $T > 0$ such that for $0 < t < T$,

$$\|u_t\|_2 < \infty, \quad \|\nabla u\|_2 < \infty.$$

(iii) Utilizing (3.24) and the inequality,

$$\frac{d\|u_t\|_2^2}{dt} + 2(1 - C\|\nabla u\|_2)\|\nabla u_t\|_2^2 \le 0,$$

Serrin proved in [234] that in the case when $n = 4$, the statement of Theorem 3.1.2 remains true provided that $\|u_0\|_{\dot{H}^1}$ is suitably small. When $n \ge 5$, the embedding relation $H^1 \hookrightarrow L^4$ does not hold, so it seems difficult to get the similar statement in Theorem 3.1.2 in this case.

As for the study of the regularity of Leray–Hopf weak solutions to the Navier–Stokes equations, Prodi and Serrin have made many contributions (see [234]). Von Wahl extended the regularity result of Prodi and Serrin to the critical case (see [283, 284]).

Theorem 3.1.3 (Prodi–Serrin theorem). *Let $u \in V(\infty)$ be a Leray-Hopf weak solution of the problem (3.21). Then, after modifying a set of zero measure in the direction t, the following equation holds:*

$$\int_0^T (u, \varphi_t + \Delta\varphi + u\cdot\nabla\varphi)dt = (u(T),\varphi(T)) - (u_0,\varphi_0), \quad \forall T > 0, \tag{3.28}$$

where $\varphi_0(x) = \varphi(x,0)$, $\varphi(x,t) \in \mathcal{D}(\Omega\times[0,\infty)) = C_c^\infty(\Omega\times[0,\infty))$ and $\operatorname{div}\varphi = 0$.

Outline of proof. Let $\theta(t) \in C_c^\infty([0,\infty))$ be such that

$$\theta(t) = \begin{cases} 1, & 0 \le t \le T, \\ 0, & T+h \le t < \infty, \end{cases}$$

and

$$\int_0^\infty \theta_t dt = -1, \quad |\theta_t| \le \frac{C}{h}.$$

Choosing $\Phi(x,t) = \theta(t)\varphi(x,t)$ in Definition 3.1.3, we have

$$\int_0^\infty \{(u,\varphi_t) + (u,\Delta\varphi) + (u, u\cdot\nabla\varphi)\}\theta(t)dt = -\int_0^\infty (u,\varphi)\theta_t\, dt - (u_0(x), \varphi_0(x)).$$

Letting $h \to 0$, it is easy to see that

$$\int_0^\infty \{(u,\varphi_t) + (u,\Delta\varphi) + (u, u\cdot\nabla\varphi)\}dt = (u(T),\varphi(T)) - (u_0(x), \varphi_0(x)),$$

for all T belonging to the Lebesgue set $\mathcal{L}$ of $u(t)$. This shows that (3.28) holds for a.e. $T > 0$.

Noting that $\|u(t)\|_{L_x^2}$ is uniformly bound and a standard argument, one can modify the set of zero measure in time direction t such that $u(t)$ satisfies (3.28) holds for all $T > 0$. □

It should be pointed out that the Prodi–Serrin theorem plays an important role in the search of sufficient conditions for a Leray–Hopf weak solution to be a regular solution. It can be seen as a bridge between the classical compactness method (functional analysis method) and the modern harmonic analysis method.

Theorem 3.1.4 (Serrin–von Wahl theorem). *Let (q,r) satisfy that*

$$\frac{2}{q} + \frac{n}{r} \le 1, \quad n < r \le \infty.$$

Let $u(x,t)$ be a Leray–Hopf weak solution of the problem (3.21). Then the following results hold:

(i) *If $u \in L^q((0,T); L^r(\Omega))$, then*

$$\|u\|_2^2 + 2\int_0^t \|\nabla u\|_2^2 d\tau = \|u_0\|_2^2, \quad 0 \le t \le T.$$

(ii) *If $u \in L^q((0,T); L^r(\Omega))$, then $u \in C^\infty((0,T)\times\Omega)$.*

(iii) *If $u \in L^q((0,T);L^r(\Omega))$ and v is also a Leray–Hopf weak solution of the problem* (3.21) *satisfying the energy inequality,*

$$\|v\|_2^2 + 2\int_0^t \|\nabla v\|_2^2 dt \le \|v_0\|_2^2, \quad 0 \le t \le T,$$

then $u \equiv v$ for all $t \in [0,T)$.

(iv) *If $u \in C([0,T);L^n(\Omega))$, then $u \in C^\infty((0,T)\times\Omega)$.*

Outline of proof. By choosing a suitable test function $\varphi(x,t)$ and making use of Theorem 3.1.3 and some limiting processes, we can get statements (i) and (iii) (see [234] for details). For the proof of (ii), we only need to use

$$u(x,t) = S(t)u_0(x) - \int_0^t S(t-\tau)\vec{f}(u(x,\tau))d\tau,$$

where $S(t) = e^{-At}$, $A = -\mathcal{P}\Delta$ and $\vec{f}(u) = (f_1(u),\dots,f_n(u))$ with $f_i(u) = \sum_{j=1}^n \partial_j(u_j u_i)$ and $\mathcal{P} : (L^r)^n \to E_r$. Note that $S(t)$ is an analytic semigroup on E_r and $u \in L^q((0,T);L^r)$, where $1 < r < \infty$. With the help of the regularity of the analytic semigroup, it is easy to deduce that $u \in C^\infty((0,T)\times\Omega)$. See the method of proving regularity of solutions for parabolic equations in Chapter 2. By the fact that $u(t) \in C([0,T), L^n(\Omega))$ can be approximated by simple function $v(t)$ on t, we can simply give out the proof of (iv); see [161, 162, 282, 283]. □

The following Solonnikov estimate also plays an important role in the study of the Navier–Stokes equations (see [283]).

Theorem 3.1.5 (Solonnikov estimate). *Let $p > 1$, $p \ne \frac{3}{2}$ (for the case $p = \frac{3}{2}$, a suitable modification can be made and the corresponding results remain true), $\Omega \subseteq \mathbb{R}^3$, $u_0 \in \widetilde{W}^{2-2/p,p}(\Omega) \cap E_p(\Omega)$, $f \in L^p(I;L^p(\Omega))$. Then the initial boundary value problem*

$$\begin{cases} u_t - \Delta u + \nabla P = f, \\ \operatorname{div} u = 0, \\ u(x,0) = u_0(x), \quad u|_{\partial\Omega} = 0 \end{cases}$$

has a unique solution (u,P) satisfying that

$$u \in L^p((0,T);H^{2,p}(\Omega) \cap \dot{H}^{1,p}(\Omega)), \quad P \in L^p((0,T);L^p(\Omega) \cap \dot{H}^{1,p}(\Omega))$$

with

$$u'(t) \in L^p((0,T);L^p(\Omega)), \quad u(t) \in C([0,T);\widetilde{W}^{2-\frac{2}{p},p}(\Omega)).$$

Moreover, the following Solonnikov estimate holds:

$$\int_0^T [\|u'(t)\|_p^p + \|u(t)\|_{H^{2,p}}^p + \|\nabla P(t)\|_p^p]dt + \sup_{0\le t\le T} \|u(t)\|^p_{\widetilde{W}^{2-\frac{2}{p},p}}$$

$$\le C(T)\left(\int_0^T \|f(t)\|_p^p dt + \|u_0\|_{\widetilde{W}^{2-\frac{2}{p},p}}\right). \tag{3.29}$$

Here,

$$\widetilde{W}^{2-\frac{2}{p},p}(\Omega) = \begin{cases} W^{2-\frac{2}{p},p}(\Omega), & 1 < p \le \frac{3}{2}, \\ \dot{W}^{2-\frac{2}{p},p}(\Omega), & p > \frac{3}{2}. \end{cases}$$

Remark 3.1.2.

(i) In Theorem 3.1.5, P is uniquely determined up to a function being in $L^p((0,T)\times\Omega)$.

(ii) $C(T)$ in the Solonnikov estimate can be replaced by a constant, which does not depend on T. In fact, this is done in the following way: for $k = 1, 2, \dots$, let

$$\xi_0(t) = \begin{cases} 1, & 0 \le t \le 1, \\ -3t+4, & 1 \le t < 1+\frac{1}{3}, \\ 0, & t > 1+\frac{1}{3}, \end{cases}$$

$$\xi_k(t) = \begin{cases} 0, & t < k-\frac{1}{3}, \\ 3t-(3k-1), & k-\frac{1}{3} \le t \le k, \\ 1, & k < t < k+1, \\ -3t+3(k+1)+1, & k+1 \le t \le k+1+\frac{1}{3}, \\ 0, & t > k+1+\frac{1}{3}. \end{cases}$$

It is easy to see that $\sum_{k=0}^{\infty} \xi_k(t) \le 2$ and an arbitrary $t > 0$ is contained in the support of at most two $\xi_k(t)$. Note that

$$(\xi_k u)' - \Delta(\xi_k u) + \nabla(\xi_k P) = \xi_k f + \xi_k'(t)u.$$

Therefore, by using the inequality (3.29) iteratively and then adding all these inequalities, we get the exact Solonnokov inequality with some constant C being independent of T.

(iii) When $s \le 1/p$, $\dot{W}^{s,p} = W^{s,p}$ and, therefore, we introduced the notation $\widetilde{W}^{s,p}$ in Theorem 3.1.5.

(iv) With the help of the Solonnikov estimate, we can solve the problem (3.21) locally in the space $L^p((0,T); H^{2,p}(\Omega))$.

(v) It is interesting to ask if the Solonnikov estimate can be used to study the regularity of Leray–Hopf weak solutions. Here are some thoughts in this direction:

Step 1. Let $u(t)$ be a Leray–Hopf weak solution of the problem (3.21). Then

$$u(t) \in L^\infty((0,T);L^2(\Omega)) \cap L^2((0,T);\dot{H}^1(\Omega)).$$

By the Hölder inequality, Gagliardo–Nirenberg inequality and

$$\left|(u\cdot\nabla)u\right| \lesssim |\nabla u|^{\frac{2(n+1)}{n+2}} + |u|^{\frac{2(n+1)}{n}},$$

it is seen that

$$(u\cdot\nabla)u \in L^{\frac{n+2}{n+1}}((0,T);L^{\frac{n+2}{n+1}}(\Omega)).$$

Step 2. Let $p = \frac{n+2}{n+1}$. Consider

$$\begin{cases} w' - \Delta w + \nabla\widetilde{P} = -(u\cdot\nabla)u, \\ \nabla\cdot w = 0, \end{cases} \tag{3.30}$$

with initial data

$$w(x,0) = u_0(x) \in E_2 \cap \widetilde{W}^{2-\frac{2(n+1)}{n+2},\frac{n+2}{n+1}} \subset E_{\frac{n+2}{n+1}} \cap \widetilde{W}^{2-\frac{2(n+1)}{n+2},\frac{n+2}{n+1}}, \quad \mathrm{meas}(\Omega) < \infty.$$

Here, u in the right side of the first equation denotes the Leray–Hopf weak solution. From Solonnikov's estimate, it follows that the problem (3.30) has a unique solution $(w,\widetilde{P})$ with

$$w(t) \in L^{\frac{n+2}{n+1}}((0,T);H^{2,\frac{n+2}{n+1}}), \quad \widetilde{P} \in L^{\frac{n+2}{n+1}}((0,T);H^{1,\frac{n+2}{n+1}}).$$

Step 3. In the sense of weak solutions, $w-u$ satisfies

$$(w-u)_t - \Delta(w-u) + \nabla(\widetilde{P}-P) = 0.$$

Thus, it follows from Theorem 3.1.5 that $u \equiv w$, which implies that

$$u \in L^{\frac{n+2}{n+1}}((0,T);H^{2,\frac{n+2}{n+1}}).$$

However, since the integrable index is very small, it seems impossible to get the smoothness of u from this.

When $\Omega = \mathbb{R}^n$, it corresponds to the following Cauchy problem for the Navier–Stokes:

$$\begin{cases} u_t - \Delta u + (u\cdot\nabla)u + \nabla P = 0, & (x,t) \in \mathbb{R}^n \times (0,T), \\ \operatorname{div} u = 0, & \\ u(x,0) = u_0(x), & u = (u_1,\dots,u_n). \end{cases} \tag{3.31}$$

In this case, we can make use of the tools for the Fourier analysis and the methods are more specific. In [87], Fabes, Jones and Riviere utilized the method of Fourier transforms to study the following system of heat equations with the divergence-free constraint:

$$\begin{cases} \partial_t E_{ij} - \Delta E_{ij} = 0, \\ \operatorname{div}(E_i) = \sum_{j=1}^n \partial_{x_j} E_{ij}(x,t) = 0, \quad E_i = (E_{i1}, \dots, E_{in}), \\ \int_{\mathbb{R}^n} E(x-y,t)g(y)dy \xrightarrow{L^p} g(x), \quad t \to 0^+, \ 1 < p < \infty, \end{cases} \tag{3.32}$$

where $g \in E_p$. They constructed the following fundamental solution for the problem (3.32):

$$E_{ij}(x,t) = \delta_{ij}\Gamma(x,t) - R_i R_j \Gamma(x,t),$$

where

$$\Gamma(x,t) = \frac{\exp(-|x|^2/4t)}{(4\pi t)^{n/2}},$$

and R_j is the j-th Riesz transform, which is a classical Calderón–Zygmund singular integral operator on $L^p(\mathbb{R}^n)$ $(1 < p < \infty)$ defined by

$$R_j(f) = \text{P.V.} \lim_{\varepsilon \to 0} C_j \int_{|x-y|>\epsilon} \frac{x_j - y_j}{|x-y|^{n+1}} f(y)dy$$

(see [244, 251] for details). They also derived the well-known Oseen–Fabes–Jones–Riviere formula:

$$E_{ij}(x,t) = \delta_{ij}\Gamma(x,t) + \int_0^{\frac{1}{t}} \partial^2_{x_i x_j} \Omega(xs^{\frac{1}{2}}) s^{\frac{n}{2}-1} ds, \tag{3.33}$$

where

$$\Omega(x) = (4\pi)^{-\frac{n}{2}} \exp\left(-\frac{|x|^2}{4}\right), \quad \Gamma(x,t) = \Omega_{\sqrt{t}}(x) = \sqrt{t}^{-n} \Omega(x/\sqrt{t}).$$

With the fundamental solution E_{ij}, the problem (3.31) can be formally reduced to the integral equation

$$\begin{aligned} u(x,t) &= \Gamma(\cdot,t) * u_0(x) - \int_0^t \int_{\mathbb{R}^n} \langle u(y,s), \nabla E(x-y,t-s) \rangle (u(y,s)) dyds \\ &= \Gamma(\cdot,t) * u_0(x) + B(u,u). \end{aligned} \tag{3.34}$$

From (3.33), it is seen that $\partial_{x_k} E_{ij}(x,t) \in L^1(\mathbb{R}^n \times [0,T))$, $k = 1,\dots,n$. Thus, if $u, v \in L^q((0,T); L^p(\mathbb{R}^n))$, then

$$B(u,v) \in L^{\frac{q}{2}}((0,T); L^{\frac{p}{2}}(\mathbb{R}^n)), \quad p,q \geq 2,$$

so the integral equation (3.34) can be solved in $L^q((0,T);L^p)$. Fabes–Jones–Riviere [87] established a bridge between the integral equation (3.34) and $L^{p,q}$-type weak solutions (not Leray–Hopf weak solutions) of the problem (3.31).

Definition 3.1.4. The function $u(x,t) = (u_1,\dots,u_n)$ is called an $L^{p,q}$-type weak solution of the problem (3.31) if the following conditions are satisfied:

(i) $u \in L^q((0,T); L^p(\mathbb{R}^n))$, $p, q \geq 2$.

(ii) For any $\varphi \in \mathcal{D}_T$, there holds

$$\int_0^T \int_{\mathbb{R}^n} \langle u, \partial_t \varphi + \Delta\varphi + u \cdot \nabla\varphi \rangle dxdt = -\int_{\mathbb{R}^n} \langle u_0, \varphi_0 \rangle dx,$$

where $\varphi_0(x) = \varphi(x,0)$, $\langle \cdot,\cdot \rangle$ denotes the Euclidean inner product, $\nabla\varphi$ is a matrix with the entry $\partial\varphi_i/\partial x_j$, and

$$\mathcal{D}_T = \{\varphi \in S(\mathbb{R}^{n+1}) \mid \varphi_j(x,t) = 0,\ t \geq T,\ \operatorname{div}\varphi = 0,\ \forall t \in [0,T)\}.$$

(iii) For almost all $t \in [0,T)$, $\operatorname{div} u(\cdot,t) = 0$ holds in the weak sense.

Theorem 3.1.6 (Fabes–Jones–Riviere [87]). *Let $u_0 \in E_r(\mathbb{R}^n)$, $1 \leq r < \infty$. Then $u \in L^q((0,T); L^p(\mathbb{R}^n))$ ($2 \leq p < \infty$, $2 \leq q \leq \infty$) is an $L^{p,q}$-type weak solution to the Cauchy problem* (3.31) *if and only if u is a solution to the integral equation* (3.34).

With the help of the integral equation (3.34) and the Banach fixed-point theorem, Fabes–Jones–Riviere established the existence of a unique $L^{p,q}$-type weak solution to Navier–Stokes equations for the first time.

Theorem 3.1.7. *Let $\frac{n}{p} + \frac{2}{q} \leq 1$, $n < p < \infty$. Then the following results hold:*

(i) *Let $u_0 \in E_r(\mathbb{R}^n)$ with r satisfying that $\frac{n}{p} + \frac{2}{q} > \frac{n}{r} > 0$. Then there are $T_0 = T_0(u_0) > 0$ and weak solution $u \in L^q((0,T); L^p(\mathbb{R}^n))$ to the problem* (3.31) *satisfying the integral equation* (3.34), *where $0 < T \leq T_0$.*

(ii) *Let u, v be two $L^{p,q}$-type weak solutions to the problem* (3.31). *Then $u \equiv v$.*

(iii) *Let $\frac{n}{p} + \frac{2}{q} = 1$, $u_0 \in E_{r_1} \cap E_{r_2}$ and $\frac{n}{p} + \frac{2}{q} - \frac{n}{r_1} < 0 < \frac{n}{p} + \frac{2}{q} - \frac{n}{r_2}$. Then there is an $\varepsilon > 0$ such that, when*

$$\|u_0\|_{E_{r_1} \cap E_{r_2}} = \|u_0\|_{r_1} + \|u_0\|_{r_2} < \varepsilon,$$

the problem (3.31) *has a unique $L^{p,q}$-type weak solution $u \in L^q(\mathbb{R}^+; L^p(\mathbb{R}^n))$ satisfying* (3.34).

Remark 3.1.3.

(i) Theorem 3.1.7 was proved by Fabes–Jones–Riviere in 1972 [87], and can be improved to include the endpoint case, that is, the same result as in Theorem 3.1.7 can still be obtained under the conditions

$$\frac{n}{p}+\frac{2}{q}\leq 1, \quad n\leq p<\infty \tag{3.35}$$

and

$$\frac{n}{p}+\frac{2}{q}\geq\frac{n}{r}>0.$$

In particular, for (iii), if $\|u_0\|_{E_{r_1}\cap E_{r_2}} \ll 1$. Then, when $\|u_0\|_n \ll 1$, the problem (3.31) or the integral equation (3.34) has a small global solution $u \in L^q(\mathbb{R}^+;L^p(\mathbb{R}^n))$; see [31, 33, 97, 126, 128] or Section 3.2 below.

(ii) Similar to initial boundary value problems, all $L^{p,q}$-type weak solutions with p, q satisfying (3.35) are regular solutions. It follows that if $u_0 \in E_2\cap E_p$ $(n\leq p<\infty)$ and $u \in L^q((0,T);L^p(\mathbb{R}^n))$ are a solution of the integral equation (3.34), where $\frac{2}{q}+\frac{n}{p}=1$, then u must be a Leray–Hopf weak solution.

(iii) As a direct result of (ii) in Theorem 3.1.7, it can be shown that if $u_0 \in E_2\cap E_p$ $(n\leq p<\infty)$ and u, v are two solutions satisfying the energy inequality (3.23), then there is $T_0>0$ such that $u\equiv v$ for $0\leq t\leq T_0$.

It is known that when $u_0 \in L^2(\mathbb{R}^n)$ the problem (3.31) has a Leray–Hopf weak solution

$$u(x,t)\in L^\infty((0,T);L^2(\mathbb{R}^n))\cap L^\infty((0,T);\dot{H}^1)$$

for any $T<\infty$, and when $u_0 \in L^r(\mathbb{R}^n)$ with $r\geq n$ the problem (3.31) has an $L^{p,q}$-type weak solution $u(x,t)\in L^q((0,T_0);L^p(\mathbb{R}^n))$, where

$$T_0=\begin{cases} T(\|u_0\|_r), & r>n,\\ T(u_0), & r=n.\end{cases}$$

Now, for $u_0 \in L^\ell$ with $2\leq \ell<n$, the question is whether the problem (3.31) has a global Leray–Hopf or $L^{p,q}$-type weak solution. By making use of the technique of decomposing the initial data, together with harmonic analysis tools, Calderón [31] proved the existence of $L^{p,q}$-type weak solutions for this case, which filled the gap. More importantly, the result of Calderón breaks down the constraint $r\geq n$ suggested by scaling and gives the existence of both $L^{p,q}$-type weak solutions and small global solutions in the critical case, under the general condition

$$\frac{2}{q}=n\left(\frac{1}{r}-\frac{1}{p}\right), \quad 2\leq r<\infty.$$

This also means that not all $L^{p,q}$-type weak solutions are smooth solutions and that only under the condition $\frac{2}{q} + \frac{n}{p} \leq 1$ can the regularity of $L^{p,q}$-type weak solutions be guaranteed; see [31, 234, 282] for details.

The next question is how different the $L^{p,q}$-type weak solution is from the regular solution when $\frac{2}{q} + \frac{n}{p} > 1$. Caffarelli, Kohn and Nirenberg gave a partial regularity result of suitable weak solutions in [30], which we will state by using a scaling argument for simplicity.

Let $k = -1 + \frac{n}{p} + \frac{2}{q}$. If (u, P) is a solution to the Cauchy problem (3.31), then by scaling, it is known that

$$u_\lambda = \lambda^{-1} u(\lambda^{-1} x, \lambda^{-2} t), \quad P_\lambda = \lambda^{-2} P(\lambda^{-1} x, \lambda^{-2} t)$$

are also a solution to the problem (3.31) with the initial data $u_0(x)$ being replaced with $\lambda^{-1} u_0(\lambda^{-1} x)$. It is easy to verify that

$$\|u_\lambda\|_{L^q(\mathbb{R}^+; L^p(\mathbb{R}^n))} = \lambda^k \|u\|_{L^q(\mathbb{R}^+; L^p(\mathbb{R}^n))}.$$

The number $k = -1 + \frac{n}{p} + \frac{2}{q}$ is regarded as the scaling factor for the characterization of solutions to the Navier–Stokes equations. With this number k, we can state the partial regularity result of weak solutions to the Navier–Stokes equations obtained by Caffarelli, Kohn and Nirenberg [30].

Theorem 3.1.8. *Let u be a Leray–Hopf weak solution to the Cauchy problem* (3.31) *and $u \in L^q((0, T); L^p(\mathbb{R}^n))$. Then there is a set $\Sigma \subset [0, T)$ such that* $\mathrm{mes}_{\frac{k}{2}}(\Sigma) = 0$ *and*

$$u \in C^\infty(\mathbb{R}^n \times ([0, T) \setminus \Sigma)),$$

where $\mathrm{mes}_{\frac{k}{2}}(\Sigma)$ *denotes the $\frac{k}{2}$-th-order Hausdoff measure of Σ.*

Corollary 3.1.1. *Let the conditions of Theorem* 3.1.8 *be fulfilled. If $k = -1 + \frac{n}{p} + \frac{2}{q} \leq 0$, then $u \in C^\infty(\mathbb{R}^n \times (0, T))$; in particular, when $u \in L^\infty((0, T); L^n(\mathbb{R}^n))$, $n \neq 3$, $u \in C^\infty((0, T) \times \mathbb{R}^n)$.*

Remark 3.1.4.

(i) From Corollary 3.1.1, we know that, when $n = 2$, the Leray–Hopf weak solution u to the Navier–Stokes equations satisfies that $u \in L^\infty((0, T); L^2(\mathbb{R}^2))$. This means that $u \in C^\infty([0, T) \times \mathbb{R}^2)$.

(ii) For $2/q + n/p \leq 1$ with $p > n$, all $L^{p,q}$-type weak solutions are regular solutions. When $n \geq 4$, the condition $p > n$ can be removed.

In recent years, numerous researches have been dedicated to the study of the Navier–Stokes equations. Nevertheless, it appears that significant progress has been limited. In dealing with small solutions in different types of critical Sobolev spaces, Kato gave an effective method; see [33, 35, 98, 126, 128, 153] or the following Sections 3.2–3.4 of this chapter.

We conclude this section by recalling space-time estimates for solutions of the linear Navier–Stokes equations. From the Helmholtz decomposition [93],

$$(L^r)^n = E_r \oplus G_r,$$

where

$$E_r = \{h(x) \in (L^r)^n,\ \operatorname{div} h(x) = 0 \text{ in the weak sense}\},$$
$$G_r = \{\nabla g,\ g \in W^{1,r}\}.$$

Let $\mathcal{P}$ denote the projection operator from $(L^r)^n$ to E_r and let B_r denote the Laplace operator $-\Delta$ with homogeneous boundary conditions. Define

$$A = -\mathcal{P}\Delta, \quad D(A) = E_r \cap D(B_r).$$

Then it is easy to prove that A generates an analytic semigroup $S(t) = e^{-At}$ on E_r $(1 < r < \infty)$ and that A has bounded inverse operator (see [93, 97]). Thus, the fractional-order operator A^α of A can be defined and satisfies that

$$\|A^\alpha e^{-At}\|_{\mathcal{L}(L^r,L^r)} \le C_\alpha t^{-\alpha}, \quad \forall \alpha \ge 0,\ t > 0.$$

With the help of $\mathcal{P}$, the problem (3.21) for the Navier–Stokes equations can be transformed into the following Cauchy problem for the abstract parabolic equation:

$$\begin{cases} u_t + Au = F(u), & (x,t) \in \mathbb{R}^n \times \mathbb{R}^+, \\ u(x,0) = u_0(x), \end{cases} \tag{3.36}$$

where $F(u) = -\mathcal{P}\nabla \cdot (u \otimes u)$, $u \otimes u$ stands for a function matrix with the elements $(u_k u_j)$. Clearly, $F_j(u) = \partial_k(u_k u_j)$, and the sum is represented by repeated indices. When $\Omega = \mathbb{R}^n$, the analytic semigroup $S(t)$ can be written explicitly as

$$S(t)\varphi = \mathcal{F}^{-1}\left(\delta_{ij} e^{-|\xi|^2 t} - \frac{\xi_i \xi_j}{|\xi|^2} e^{-|\xi|^2 t}\right)\mathcal{F}\varphi.$$

In particular, for $\varphi \in E_r$, we have

$$S(t)\varphi = \mathcal{F}^{-1}(\delta_{ij} e^{-|\xi|^2 t}\mathcal{F}(\varphi)) = \mathcal{F}^{-1}(e^{-|\xi|^2 t}\mathcal{F}\varphi),$$

where

$$\delta_{ij} = \begin{cases} 1, & i = j, \\ 0, & i \ne j. \end{cases}$$

The space-time estimates for parabolic equations in Section 2.1 remain valid for the Navier–Stokes equations, which will be listed as follows.

Definition 3.1.5. The triplet (q, p, r) is admissible for the Navier–Stokes equations if

$$\frac{1}{q} = \frac{n}{2}\left(\frac{1}{r} - \frac{1}{p}\right), \tag{3.37}$$

where

$$1 < r \le p < \begin{cases} \frac{rn}{n-2}, & n > 2, \\ \infty, & n \le 2. \end{cases}$$

(q, p, r) is called a generalized admissible triplet for the Navier–Stokes equations if it satisfies (3.37) and

$$1 < r \le p < \begin{cases} \frac{nr}{n-2r}, & n > 2r, \\ \infty, & n \le 2r. \end{cases}$$

From the estimate of the analytic semigroup $S(t) = e^{-At}$, it is easy to see that

$$\|e^{-At}\varphi\|_p \le Ct^{-\frac{n}{2}(\frac{1}{r}-\frac{1}{p})}\|\varphi\|_r, \quad \forall t > 0,\ p \ge r > 1.$$

See the proof of Lemmas 2.1.1 and 2.1.2 in Chapter 2 for details. A direct verification gives the following space-time estimates.

Theorem 3.1.9. *Let $I = [0, T]$ or $[0, \infty)$. Let (q, p, r) be any generalized admissible triplet and let $\varphi \in E_r$. Then $e^{-At}\varphi \in C_{q(p,r)}(I; E_p)$ and*

$$\|e^{-At}\varphi; C_{q(p,r)}(I; E_p)\| \le C\|\varphi\|_r.$$

Furthermore, if (q, p, r) is a generalized admissible triplet with $p > r$. Then $S(t)\varphi \in \dot{C}_{q(p,r)}(I, E_p)$, that is,

$$\lim_{t\to 0} t^{\frac{1}{q}}\|S(t)\varphi\|_p = 0,$$

where $C_{q(p,r)}(I, X)$ and $\dot{C}_{q(p,r)}(I, X)$ are the same as in Definition 2.1.3 of Chapter 2.

Theorem 3.1.10. *Let (q, p, r) be any admissible triplet and let $\varphi \in E_r$. Then $e^{-tA}\varphi \in L^q(I; E_p) \cap C_b(I; E_r)$ and*

$$\|e^{-tA}\varphi\|_{L^q(I;E_p)} \le C\|\varphi\|_r, \quad I = [0, T),\ 0 < T \le \infty.$$

When $g \in E_r$, we have $\|g\|_{E_r} = \|g\|_r$. Thus, Theorems 3.1.9 and 3.1.10 are completely similar to Theorems 2.1.1 and 2.1.3 in Chapter 2. As for the space-time estimates of the inhomogeneous part, we have the following.

Theorem 3.1.11. *For $T > 0$, let $I = [0, T)$. Let $Q(D)$ be a homogeneous pseudo-differential operator of order d with $0 \le d < 2$ and let $r \ge r_c = bn/(2-d) > 1$, (q, p, r) be any generalized admissible triplet satisfying that $p > b + 1$. Then the following space-time estimates hold:*

(i) *If $f \in L^{\frac{q}{b+1}}([0,T); E_{\frac{p}{b+1}})$, then*

$$GQ(D)f = \int_0^t S(t-\tau)Q(D)f d\tau \in L^q([0,T); E_p) \cap C_b([0,T); E_r)$$

and satisfies the estimate

$$\begin{aligned} &\|GQ(D)f\|_{L^q([0,T);E_p)} + \|GQ(D)f\|_{L^\infty([0,T);E_r)} \\ &\lesssim T^{1-\frac{d}{2}-\frac{nb}{2r}} \|f\|_{L^{\frac{q}{b+1}}([0,T);E_{\frac{p}{b+1}})}, \quad p < r(1+b), \end{aligned} \tag{3.38}$$

and

$$\begin{aligned} &\|GQ(D)f\|_{L^q([0,T);E_p)} + \|GQ(D)f\|_{L^\infty([0,T);E_r)} \\ &\lesssim T^{1-\frac{d}{2}-\frac{nb}{2r}} \big\||f|^{\frac{1}{1+b}}\big\|^{\theta(b+1)}_{L^\infty([0,T);E_r)} \big\||f|^{\frac{1}{1+b}}\big\|^{(1-\theta)(b+1)}_{L^q([0,T);E_p)}, \quad p \ge r(1+b), \end{aligned} \tag{3.39}$$

where $\theta = \frac{p-r(b+1)}{(b+1)(p-r)}$.

(ii) *If $f \in C_{\frac{q}{b+1}}(I; E_{\frac{p}{b+1}})$, then $GQ(D)f \in C_{q(p,r)}(I; E_p) \cap C_b([0,T); E_r)$ and satisfies that*

$$\begin{aligned} &\|GQ(D)f\|_{C_q([0,T);E_p)} + \|Gf\|_{L^\infty([0,T);E_r)} \\ &\lesssim T^{1-\frac{d}{2}-\frac{nb}{2r}} \|f\|_{C_{\frac{q}{b+1}}([0,T);E_{\frac{p}{b+1}})}, \quad p < r(1+b), \end{aligned} \tag{3.40}$$

and

$$\begin{aligned} &\|GQ(D)f\|_{C_q([0,T);E_p)} + \|Gf\|_{L^\infty([0,T);E_r)} \\ &\lesssim T^{1-\frac{d}{2}-\frac{nb}{2r}} \big\||f|^{\frac{1}{1+b}}\big\|^{\theta(b+1)}_{L^\infty([0,T);E_r)} \big\||f|^{\frac{1}{1+b}}\big\|^{(1-\theta)(b+1)}_{C_q([0,T);E_p)}, \quad p \ge r(1+b), \end{aligned} \tag{3.41}$$

where $\theta = \frac{p-r(b+1)}{(b+1)(p-r)}$.

Remark 3.1.5. It is easy to see that (3.36) corresponds to the integral equation

$$u(t) = S(t)u_0 + \int_0^t S(t-\tau)F(u)d\tau \equiv S(t)u_0 + G\nabla f, \tag{3.42}$$

where $f(u) = (f_{ij}(u))_{ij}$, $f_{ij} = -u_i u_j$. Specially when $\Omega = \mathbb{R}^n$, (3.42) becomes the integral equation (3.34). In this case, we have

$$G\nabla f = B(u,u) = -\int_0^t \int_{\mathbb{R}^n} \langle u(y,s), \nabla E(x-y,t-s)\rangle (u(y,s))dyds, \tag{3.43}$$

which satisfies the estimates in Theorem 3.1.11. Let $b = 1$, $d = 1$, $r_c = n$ and let (q,p,r) be any generalized admissible triplet with $r \geq r_c$. Then

$$\|G\nabla f\|_{L^q(I;E_p)} = \|B(u,u)\|_{L^q(I;E_p)} \leq Ct^{\frac{1}{2}-\frac{n}{2r}} \|u\|^2_{L^q(I;E_p)}, \tag{3.44}$$

$$\|G\nabla f\|_{C_{q(p,r)}(I;E_p)} \leq Ct^{\frac{1}{2}-\frac{n}{2r}} \|u\|^2_{C_{q(p,r)}(I;E_p)}. \tag{3.45}$$

On the other hand, it follows from (3.33) and the equality

$$\partial_{x_k} E_{ij} = \delta_{ij}\partial_{x_k}\Gamma(x,t) + \int_0^{\frac{1}{t}} \partial^3_{x_i x_j x_k} \Omega(xs^{\frac{1}{2}}) s^{\frac{n}{2}-\frac{1}{2}} ds$$

that

$$|E_{ij}| \leq \frac{C}{(|x| + t^{\frac{1}{2}})^n} \leq \frac{C_\theta}{|x|^{n\theta} t^{\frac{n}{2}(1-\theta)}}, \quad 0 \leq \theta \leq 1, \tag{3.46}$$

$$|\partial_{x_k} E_{ij}| \leq \frac{C}{(|x| + t^{\frac{1}{2}})^{n+1}} \leq \frac{C_\theta}{|x|^{(n+1)\theta} t^{\frac{n+1}{2}(1-\theta)}}, \quad 0 \leq \theta \leq 1. \tag{3.47}$$

With the help of the above two inequalities and the expression (3.34) (or (3.43)) of $B(u,u)$, the nonlinear estimates (3.44) and (3.45) also follow from the Young inequality or the Hardy–Littlewood–Sobolev inequality.

3.2 Methods of space-time estimates

In this section, we will make use of the space-time estimates method to study well-posedness of the Cauchy problem

$$\begin{cases} u_t - \Delta u + (u \cdot \nabla)u + \nabla P = 0, & (x,t) \in \mathbb{R}^n \times [0,T), \\ \operatorname{div} u = 0, \\ u(x,0) = u_0(x), \quad \operatorname{div} u_0(x) = 0, & x \in \mathbb{R}^n, \end{cases} \tag{3.48}$$

or the initial boundary value problem

$$\begin{cases} u_t - \Delta u + (u \cdot \nabla)u + \nabla P = 0, & (x,t) \in \Omega \times [0,T), \\ \operatorname{div} u = 0, \\ u(x,0) = u_0(x), \quad \operatorname{div} u_0(x) = 0, & x \in \Omega, \\ u|_{\partial\Omega=0}. \end{cases} \tag{3.49}$$

By applying the orthogonal projection operator,

$$\mathcal{P}: (L^r(\mathbb{R}^n))^n \to E_r(\mathbb{R}^n) \quad (1 < r < \infty)$$

to (3.48) or (3.49), the problem (3.48) or (3.49) can be changed into the following abstract Cauchy problem:

$$\begin{cases} u_t + Au = F(u), \\ u(x,0) = u_0(x), \end{cases} \tag{3.50}$$

where $A = -\mathcal{P}\Delta$, $F(u) = -\mathcal{P}\nabla \cdot (u \otimes u)$ and $u \otimes u = (u_j u_k)$ stands for an $n \times n$ matrix with the element $u_j u_k$. Thus, $F_j(u) = -\partial_k(u_j u_k)$ with the Einstein convention of summation of repeated indices. The domain of the operator A is $D(A) = \{u | u \in W^{2,r}(\mathbb{R}^n) \cap E_r\}$ or $D(A) = \{u \mid u \in \dot{W}^{2,r}(\Omega) \cap E_r\}$, $1 < r < \infty$. We denote $S(t) = e^{-At}$ by the analytic semigroup generated by A on E_r. Then the problem (3.50) can be reduced equivalently to the integral equation

$$\begin{aligned} u(t) &= e^{-At}u_0(x) - \int_0^t e^{-A(t-\tau)}\mathcal{P}\nabla \cdot (u \otimes u)d\tau \\ &\equiv e^{-At}u_0(x) + \int_0^t e^{-A(t-\tau)}F(u)d\tau. \end{aligned} \tag{3.51}$$

By scaling (see Chapter 2 or the previous section of this chapter), it follows that $r_c = n$. Therefore, we will prove, for arbitrary initial data $u_0 \in E_r$ with $r \geq r_c$, that the problem (3.51) has a unique solution $u \in C([0,T^*); E_r)$ belonging to a suitable space-time Banach space. By the regularity theory of the Navier–Stokes equations, this solution u satisfies that $u \in C^\infty((0,T^*)\times\mathbb{R}^n)$ or $u \in C^\infty((0,T^*)\times\Omega)$. Thus, to study well-posedness of smooth solutions to the problem (3.48) or (3.49), it is enough to study the well-posedness of mild solutions to the problem (3.51). In particular, when $r = r_c$, we have global well-posedness for the problem (3.48) or (3.49) provided that $\|u_0\|_{E_{r_c}} \ll 1$. When $n = 2$, it has been proved that the Leray–Hopf weak solution of the Navier–Stokes equations is a regular solution, and the problem has thus been completely solved. Thus, in the remaining sections of this chapter we assume that $n \geq 3$ unless it is otherwise indicated.

Theorem 3.2.1.

(i) *Let $r \geq r_c = n$, $u_0 \in E_r$ and (q,p,r) be any generalized admissible triplet. Then the problem* (3.51) *has a unique solution u such that*

$$\begin{cases} u \in C[0,T^*); E_r) \cap C_{q(p,r)}([0,T^*); E_p), & T^* = T(\|u_0\|_r), \quad r > r_c, \\ u \in C([0,T^*); E_r) \cap \dot{C}_{q(p,r)}([0,T^*); E_p), & T^* = T(u_0), \quad r = r_c, \end{cases}$$

where $[0,T^*)$ is the maximal existence interval for $u(t)$.

(ii) *If $T^* < \infty$, then*

$$\lim_{t\to T^*} \|u(t)\|_p = \infty, \quad r \le p \le \infty,\ p > r_c \tag{3.52}$$

and

$$\|u(t)\|_p \ge \frac{C}{(T^* - t)^{\frac{1}{2}-\frac{n}{2p}}}. \tag{3.53}$$

(iii) *Let (q, p, r) be any admissible triplet. Then the solution u obtained in* (i) *satisfies the following integrability:*

$$u(t) \in L^q((0, T^*); E_p).$$

(iv) *In the case when $r = r_c = n$, if $\|u_0\|_r \ll 1$, then the solution u obtained in* (i) *is a global one, that is, $T^* = \infty$.*

Proof. (i) To see the role played by the solution space $C([0, T); E_r)$, denote by Γ the set of all generalized admissible triplets (q, p, r) with $p \ne r$ and by Ξ the set of all admissible triplets (q, p, r) with $p \ne r$, and write $I = [0, T)$.

We first consider the case $r > r_c$. Define the solution space

$$X(I) = \{u \mid u \in C_b(I; E_r) \cap C_{q(p,r)}(I; E_p),\ (q, p, r) \in \Gamma\}$$

with the norm

$$\|u\|_{X(I)} = \max_{(q,p,r)\in\Gamma} \sup_{t\in I} t^{\frac{1}{q}} \|u(t)\|_p + \sup_{t\in I} \|u(t)\|_r.$$

From Theorem 3.1.9, it is seen that

$$\|S(t)u_0\|_{X(I)} \le C\|u_0\|_r, \quad u_0 \in E_r. \tag{3.54}$$

Now, we define

$$\mathcal{X}(I) = \{u(t) \in X(I) \mid \|u\|_{X(I)} \le 2C\|u_0\|_r = M\},$$

where M is to be determined later, and introduce the metric $d(u, v) = \|u - v\|_{X(I)}$ on $\mathcal{X}(I)$, so $\mathcal{X}(I)$ is a complete metric space. Consider the nonlinear operator $\mathcal{T}$ defined by the RHS of the integral equation (3.51):

$$\mathcal{T}u = e^{-At}u_0(x) - \int_0^t e^{-A(t-\tau)}\mathcal{P}\nabla(u \otimes u)d\tau \equiv S(t)u_0(x) + Gu. \tag{3.55}$$

From the estimates of the analytic semigroup $S(t)$, it follows that

$$\begin{aligned}\|Gu; C_{q(p,r)}(I;E_p)\| &\le \sup_{t\in I} t^{\frac{1}{q}} \int_0^t |t-\tau|^{-\frac{1}{2}-\frac{n}{2}(\frac{2}{p}-\frac{1}{p})} \|u(\tau)\|_p^2 d\tau \\ &\le CT^{\frac{1}{2}-\frac{n}{2r}} \int_0^1 (1-\tau)^{-\frac{1}{2}-\frac{n}{2p}} \tau^{-\frac{2}{q}} d\tau \|u\|^2_{C_{q(p,r)}(I;E_p)} \\ &\le CT^{\frac{1}{2}-\frac{n}{2r}} \|u\|^2_{C_{q(p,r)}(I;E_p)},\end{aligned} \tag{3.56}$$

$$\begin{aligned}\|Gu; C_b(I;E_r)\| &\le \sup_{t\in I} \int_0^t |t-\tau|^{-\frac{1}{2}-\frac{n}{2}(\frac{2}{r}-\frac{1}{r})} \|u\|_r^2 d\tau \\ &\le CT^{\frac{1}{2}-\frac{n}{2r}} \sup_{t\in I} \|u\|_r^2.\end{aligned} \tag{3.57}$$

Therefore,

$$\|\mathcal{T}u\|_{X(I)} \le C\|u_0\|_r + CT^{\frac{1}{2}-\frac{n}{2r}} M^2. \tag{3.58}$$

Likewise, for any $u, v \in \mathcal{X}(I)$, we have

$$d(\mathcal{T}u, \mathcal{T}v) \le CT^{\frac{1}{2}-\frac{n}{2r}} M d(u,v). \tag{3.59}$$

Thus, if we let

$$T \le (4C^2\|u_0\|_r)^{-\frac{2r}{r-n}}, \tag{3.60}$$

then $\mathcal{T}$ is a contractive mapping from $\mathcal{X}(I)$ into itself. From the Banach contraction mapping principle, it follows that (3.51) has a unique solution $u \in \mathcal{X}(I)$ for T satisfying (3.60). Further, by the Picard iteration, it is known that there is a maximal $T^* = T(\|u_0\|_r)$ such that $u \in X([0,T^*))$ and either $T^* = \infty$ or $T^* < \infty$ and

$$\lim_{t\uparrow T^*} \|u(t)\|_r = \infty.$$

We now consider the case $r = r_c = n$. Take $(q,p,r) = (4, 2n, n) \in \Gamma$ and define

$$\dot{X}(I) = \{u \in C_b(I;E_r) \cap \dot{C}_{q(p,r)}(I;L^p)\}$$

with the norm

$$\|u(t)\|_{\dot{X}(I)} = \|u\|_{C_{q(p,r)}(I;E_p)} + \sup_{t\in I} \|u\|_r.$$

Introduce the complete metric space

$$\dot{\mathcal{X}} = \{u(t) \in \dot{X}(I) \mid \|u(t)\|_{\dot{X}(I)} \le M = 2C\|u_0\|_r\}$$

with the metric

$$d(u,v) = \|u - v\|_{C_{q(p,r)}(I;E_p)}, \quad u,v \in \dot{\mathcal{X}}$$

and consider the nonlinear operator $\mathcal{T}$ on $\dot{\mathcal{X}}$. A similar argument as in deriving (3.56) gives that

$$\|Gu\|_{C_{q(p,r)}(I;E_p)} \le C\|u\|^2_{C_{q(p,r)}(I;E_p)},$$

$$\begin{aligned}\|Gu\|_{C_b(I;E_r)} &\le \sup_{t\in I} \int_0^t |t-\tau|^{-\frac{1}{2}-\frac{n}{2}(\frac{2}{p}-\frac{1}{r})} \|u(\tau)\|_p^2 d\tau \\ &\le \sup_{t\in I} \int_0^t |t-\tau|^{-\frac{1}{2}-\frac{n}{2}(\frac{2}{p}-\frac{1}{r})} \tau^{-\frac{2}{q}} d\tau \cdot \|u\|^2_{C_{q(p,r)}(I,E_p)} \\ &\le C\|u\|^2_{C_{q(p,r)}(I,E_p)}.\end{aligned}$$

From the above two inequalities and Theorem 3.1.9, it follows that

$$\|\mathcal{T}u\|_{\dot{X}(I)} \le C\|u_0\|_r + C\|u\|^2_{C_{q(p,r)}(I,E_p)}. \tag{3.61}$$

Similarly, we have

$$t^{\frac{1}{q}}\|\mathcal{T}u\|_p \le t^{\frac{1}{q}}\|S(t)u_0\|_p + C(t^{\frac{1}{q}}\|u\|_p)^2, \tag{3.62}$$

$$d(\mathcal{T}u, \mathcal{T}v) \le C[\|u\|_{C_{q(p,r)}(I,E_p)} + \|v\|_{C_{q(p,r)}(I,E_p)}]d(u,v). \tag{3.63}$$

From (3.61)–(3.63) and noting that

$$\lim_{t\downarrow 0} t^{\frac{1}{q}}\|u(t)\|_p = 0, \quad \lim_{t\downarrow 0} t^{\frac{1}{q}}\|S(t)u_0\|_p = 0$$

it follows that, for sufficiently small T, the operator $\mathcal{T}$ is contractive from $\dot{\mathcal{X}}(I)$ into itself. In this case, T depending on the initial data u_0 itself. Further, by the Picard iteration, it is deduced that the problem (3.51) has a unique solution $u \in \dot{\mathcal{X}}([0,T^*))$, where $T^* = T(u_0)$ is maximal.

For any other $(\hat{q},\hat{p},r) = (\hat{q},\hat{p},n) \in \Gamma$, let $(q,p,r) = (4,2n,n)$. Then we can verify directly that for $I = [0,T^*)$,

$$\begin{aligned}\|Gu\|_{C_{\hat{q}(\hat{p},r)}(I;E_{\hat{p}})} &\le \sup_{t\in I} t^{\frac{1}{\hat{q}}} \int_0^t |t-\tau|^{-\frac{1}{2}-\frac{n}{2}(\frac{2}{p}-\frac{1}{\hat{p}})} \|u(\tau)\|_p^2 d\tau \\ &\le \sup_{t\in I} t^{\frac{1}{\hat{q}}} \int_0^t |t-\tau|^{-\frac{1}{2}-\frac{n}{2}(\frac{2}{p}-\frac{1}{\hat{p}})} \tau^{-\frac{2}{q}} d\tau \cdot \|u\|^2_{C_{q(p,r)}(I;E_p)} < \infty,\end{aligned}$$

where we use the fact that

$$-\frac{1}{2}-\frac{n}{2}\left(\frac{2}{p}-\frac{1}{\hat{p}}\right)<1, \quad \frac{2}{q}<1.$$

(ii) We first prove that

$$u \in C((0,T^*), E_r \cap L^\infty). \tag{3.64}$$

It is enough to show that for any $\varepsilon > 0$, we have $u \in C([\varepsilon, T^*); L^\infty)$.

In fact, let $(q,p,r) \in \Gamma$ satisfy that $2n < p < 2r$ with $r > n$. Then we have

$$\begin{aligned}
&\|Gu, C([\varepsilon,T^*); L^\infty)\| \\
&\le C \sup_{t\in[\varepsilon,T^*)} \int_\varepsilon^t |t-\tau|^{-\frac{n}{p}} \left\| S\left(\frac{t-\tau}{2}\right)\nabla\cdot(u\otimes u)\right\|_{p/2} d\tau \\
&\le C \int_\varepsilon^{T^*} |T^*-\tau|^{-\frac{1}{2}-\frac{n}{p}} \|u\|_p^2 d\tau \\
&\le C{T^*}^{\frac{1}{2}-\frac{n}{p}} \|u\|^2_{C_{q(p,r)}([\varepsilon,T^*),E_p)} < \infty.
\end{aligned}$$

Thus, it follows that

$$\|u; C([\varepsilon,T^*); L^\infty)\| \le \varepsilon^{-\frac{n}{2r}}\|u_0\|_r + \|Gu\|_{C([\varepsilon,T^*);L^\infty)} < \infty,$$

which implies (3.64) and

$$u \in C((0,T^*); E_p) \tag{3.65}$$

for $r \le p \le \infty$.

By the alternative property, it is easy to see that the condition $p = r > r_c = n$, $T < \infty$ implies (3.52). Further, we have (3.53) by (3.60). For $p > n$, it is easy to deduce by (3.65) that if $\limsup_{t\to T^*} \|u(t)\|_p < \infty$, then

$$\|u(T^*)\|_r < \infty.$$

See the proof of Theorem 2.2.1 in Chapter 2 for details. This contradicts with the fact that T^* is maximal. Meanwhile, taking $u(t) \in E_p$ as the initial value ($t < T^*$ and very close to T^*) to solve the integral equation (3.51), we get by arguing similarly as in the proof of (i), that

$$T^* - t \le \left(4C^2\|u(t)\|_p\right)^{-\frac{2p}{p-n}}.$$

This means that (3.53) holds.

(iii) For $I = [0, T)$ with $T > 0$, define

$$Y(I) = \{u \mid u \in C_b(I; E_r) \cap \mathcal{C}_{q(p,r)}(I; E_p) \cap L^q(I; L^p),\ (q,p,r) \in \Xi\},$$

$$\|u\|_{Y(I)} = \max_{(q,p,r)\in\Xi} \sup t^{\frac{1}{q}} \|u(t)\|_p + \max_{(q,p,r)\in\Xi} \|u\|_{L^q(I;L^p)} + \sup_{t\in I} \|u\|_r.$$

By replacing $X(I)$ (or $\dot{X}(I)$) in (i) by $Y(I)$ (or the corresponding $\dot{Y}(I)$) and noting the fact that

$$\|Gu\|_{L^q(I;L^p)} \le CT^{\frac{1}{2}-\frac{n}{2r}} \|u\|^2_{L^q(I;L^p)},$$

(iii) can be shown similarly as in proving (i).

(iv) For $r = r_c = n$, if $\|u_0\|_{E_1} \ll 1$, then it can be derived by using (3.61)–(3.63) that $u \in C([0, \infty); E_r) \cap \dot{\mathcal{C}}_{q(p,r)}([0, \infty); E_p)$. The theorem is thus proved. □

Remark 3.2.1.

(i) For the Navier–Stokes equations and for any $r > r_c = n$, the fixed-point theorem can be used directly to solve the integral equation (3.51) on $C([0, T); E_r)$. However, when $r = r_c = n$, it is necessary to solve the integral equation (3.51) on some subspace of $C([0, T); E_r)$ (i. e., the intersection of $C([0, T); E_r)$ with some space-time Banach space) other than $C([0, T); E_r)$ itself. Theorem 3.2.1(iv) was first established by Kato, so it is usually called as the Kato theorem (see [126]).

(ii) For any $r > r_c = n$, one can also solve the integral equation (3.51) directly on $C(I; E_r) \cap L^q(I; E_p)$, instead of $C(I; E_r) \cap \mathcal{C}_{q(p,r)}(I; E_p)$.

We know that the smooth degree of the critical space L^n corresponding to the Navier–Stokes equations is $\deg(L^n) = -1$. In this case, provided that $u_0 \in L^n$ and $\|u_0\|_n \ll 1$, the problem (3.48) or (3.49) is solvable globally. It is natural to ask whether or not the condition $\|u_0\|_n \ll 1$ can be relaxed, and whether or not there is a larger space B of the smooth degree -1 with $\deg(B) = -1$ and $L^n \hookrightarrow B$ such that, when $u_0 \in B$ and $\|u_0\|_B \ll 1$, the Navier–Stokes equations are of similar global well-posedness for small solutions. In what follows, we take the Cauchy problem (3.48) as an example to illustrate that this is indeed the case. To this end, we recall the Littlewood–Paley decomposition, which is of many advantages in the characterization of differentiable functions.

Let $\hat{\varphi}(\xi) \in C_c^\infty(\mathbb{R}^n)$ be a radial bump function satisfying that $0 \le \hat{\varphi}(\xi) \le 1$ and

$$\begin{cases} \hat{\varphi}(\xi) = 1, & |\xi| \le 1, \\ \hat{\varphi}(\xi) = 0, & |\xi| \ge 2. \end{cases} \tag{3.66}$$

It is easy to see that the inverse Fourier transform $\varphi(x)$ of $\hat{\varphi}(\xi)$ belongs to $\mathcal{S}(\mathbb{R}^n)$. Let

$$\begin{cases} \psi(x) = 2^n\varphi(2x) - \varphi(x), & \\ \varphi_j(x) = 2^{nj}\varphi(2^j x), & j \in \mathbb{N}, \\ \psi_j(x) = 2^{nj}\psi(2^j x), & j \in \mathbb{Z} \end{cases}$$

and define

$$S_j f = \varphi_j(x) * f, \quad \Delta_j f = \psi_j(x) * f. \tag{3.67}$$

Then $\{S_j, \Delta_j\}_{j\in\mathbb{Z}}$ is the classical Littlewood–Paley decomposition, and

$$I = S_0 + \sum_{j\geq 0} \Delta_j, \quad S_0 f = \varphi * f, \tag{3.68}$$

$$I = \sum_{j\in\mathbb{Z}} \Delta_j \tag{3.69}$$

can be used to characterize inhomogeneous and homogeneous differentiable functions, respectively. It is easy to verify that $\dot{B}^{\frac{n}{p}-1}_{p,q}$ with $n < p < \infty$, $1 \leq q \leq \infty$, $\dot{H}^{\frac{n}{2}-1}$ and the homogeneous Morrey–Campanato space $\dot{M}^n_2$ are spaces including L^n with smooth degree -1. We may establish the global well-posedness of small solutions to the problem (3.48) on these spaces instead of L^n. Here, as an example, we consider the Besov space $\dot{B}^s_{p,q}$. From different characterizations of Besov spaces in Chapter 1, we have the following.

Lemma 3.2.1. *Let $1 \leq p \leq \infty$, $\alpha > 0$. Then $\dot{B}^{-\alpha}_{p,\infty}$ has the following equivalent norm:*

$$\begin{aligned} \sup_{j\in\mathbb{Z}} 2^{-j\alpha}\|\Delta_j v\|_p &\sim \sup_{j\in\mathbb{Z}} 2^{-j\alpha}\|S_j v\|_p \sim \sup_{t\geq 0} t^{\frac{\alpha}{2}}\|S(t)v\|_p \\ &\sim \sup_{t\geq 0}\|S(t)v\|_{\dot{B}^{-\alpha}_{p,\infty}}, \quad v \in \dot{B}^{-\alpha}_{p,\infty}, \end{aligned} \tag{3.70}$$

where S_j and Δ_j are defined as in (3.67) *and $S(t)$ stands for the Gaussian semigroup.*

For a proof, see [32, 273, 274].

By using the Littlewood–Paley characterization of Besov spaces,

$$\|v\|_{\dot{B}^{\alpha}_{p,\infty}} = \sup_{j\in\mathbb{Z}} 2^{-j\alpha}\|\Delta_j v\|_p, \quad v \in \dot{B}^{\alpha}_{p,\infty},$$

it is easy to see that

$$L^n(\mathbb{R}^n) \hookrightarrow \dot{B}^{-\alpha_1}_{p_1,\infty}(\mathbb{R}^n) \hookrightarrow \dot{B}^{-\alpha_2}_{p_2,\infty}(\mathbb{R}^n), \tag{3.71}$$

where $\alpha_j = 1 - n/p_j$, $j = 1, 2$, $n \leq p_1 \leq p_2 \leq \infty$. In fact, the relation (3.71) follows directly from Bernstein's inequality,

$$\|\Delta_j v\|_{p_2} \leq 2^{jn(\frac{1}{p_1}-\frac{1}{p_2})} C\|\Delta_j v\|_{p_1}.$$

It should be pointed out that the inclusion relation in (3.71) is strict. For example,

$$|x|^{-1} \in \dot{B}^{-\alpha}_{p,\infty}, \quad \alpha = 1 - \frac{n}{p},$$

which can be deduced by the equivalent norm (3.70). But $|x|^{-1} \notin L^n$.

We are now in a position to prove the global well-posedness of small solutions to the problem (3.48) in the Besov space $\dot{B}^{-\alpha}_{p,\infty}$ with $\alpha = 1 - n/p$ instead of L^n (cf. (iv) of Theorem 3.2.1). This result is called the generalized Kato theorem, which was proved by Cannone in [32] for the case $n = 3$. Here, we prove the general n-dimensional case.

Theorem 3.2.2. *Let (q, p, n) be any generalized admissible triplet with $n < p \le 2n$ and write $\alpha = \alpha(p) = n(1/n - 1/p) = 2/q$. Then there is a $\delta > 0$ such that, if $u_0 \in E_n$ and*

$$\|u_0\|_{\dot{B}^{-\alpha}_{p,\infty}} < \delta,$$

then the integral equation (3.51) (*or equivalently* (3.48)) *has a unique global mild solution*

$$u(t) \in C([0, \infty); E_n) \cap \dot{C}_{q(p,n)}([0, \infty); E_p),$$

that is,

$$t^{\frac{1}{q}} u(t) \in C([0, \infty); E_p), \quad \text{and} \quad \lim_{t \to 0} t^{\frac{1}{q}} \|u(t)\|_p = 0.$$

Proof. Denote by Π the set of all generalized admissible triplets (q, p, n) satisfying that $n < p \le 2n$. Let

$$X = \{u \in C([0, \infty); E_n) \cap \dot{C}_{q(p,n)}([0, \infty); E_p),\ (q, p, n) \in \Pi\}$$

with the norm

$$\|u\|_X = \sup_{t>0} \|u(t)\|_{\dot{B}^{-\alpha}_{p,\infty}} + \sup_{t>0} t^{\frac{1}{q}} \|u(t)\|_p,$$

where $\alpha = 1 - n/p = 2/q$ and $(q, p, n) \in \Pi$. Define the complete metric space

$$X_\delta = \{u \in X \mid \|u\|_X \le 2C\delta\}$$

with the metric

$$d(u, v) = \|u - v\|_X, \quad u, v \in X_\delta.$$

Consider the nonlinear operator $\mathcal{T}$ on X_δ (cf. (3.55)). First, by using the equivalent characterization (3.70) of Besov spaces, it can be shown easily that

$$\begin{aligned} \|S(t)u_0\|_X &\le \sup_{t>0} \|S(t)u_0\|_{\dot{B}^{-\alpha}_{p,\infty}} + \sup_{t>0} t^{\frac{1}{q}} \|S(t)u_0\|_p \\ &\le 2\|u_0\|_{\dot{B}^{-\alpha}_{p,\infty}} \le 2\delta. \end{aligned} \tag{3.72}$$

Now, by the Sobolev embedding $L^n \hookrightarrow \dot{B}^{-a}_{p,\infty}$, we derive

$$\begin{aligned}
\|Gu\|_{\dot{B}^{-a}_{p,\infty}} &\le C\|Gu\|_n \le C \sup_{t\in[0,\infty)} \int_0^t |t-\tau|^{-\frac{1}{2}-\frac{n}{2}(\frac{2}{p}-\frac{1}{n})} \|u(\tau)\|_p^2 d\tau \\
&\le C \int_0^1 (1-\tau)^{-\frac{n}{p}} \tau^{-\frac{2}{q}} d\tau \cdot \|u\|^2_{C_{q(p,n)}([0,\infty);E_p)} \\
&\le C \big\| u; C_{q(p,n)}([0,\infty);E_p) \big\|^2, \qquad (3.73)\\
\|Gu\|_{C_{q(p,r)}([0,\infty);E_p)} &\le \sup_{t\in[0,\infty)} t^{\frac{1}{q}} \int_0^t (t-\tau)^{-\frac{1}{2}-\frac{n}{2}(\frac{2}{p}-\frac{1}{p})} \|u(\tau)\|_p^2 d\tau \\
&\le C \big\| u; C_{q(p,n)}([0,\infty);E_p) \big\|^2. \qquad (3.74)
\end{aligned}$$

Thus, from the above three inequalities, it follows that

$$\|\mathcal{T}u\|_X \le C\delta + C\|u\|_X^2. \qquad (3.75)$$

On the other hand, similar to the estimates (3.73) and (3.74), we have

$$t^{\frac{1}{q}} \|Gu\|_p \le C\Big(\sup_{0\le\tau\le t} \tau^{\frac{1}{q}} \|u(t)\|_p \Big)^2, \qquad (3.76)$$

$$\|Gu\|_n \le C\Big(\sup_{0\le\tau\le t} \tau^{\frac{1}{q}} \|u(t)\|_p \Big)^2. \qquad (3.77)$$

In view of (3.76), (3.77) and the fact that $\lim_{t\to 0} t^{1/q} \|S(t)u_0\|_p = 0$, we obtain that

$$\lim_{t\to 0} t^{\frac{1}{q}} \|\mathcal{T}u\|_p = 0, \quad \lim_{t\to 0} \|\mathcal{T}u - u_0\|_n = 0. \qquad (3.78)$$

Thus, it follows from the estimates (3.72), (3.75)–(3.78) that, if δ is sufficiently small, then $\mathcal{T}$ is a contractive mapping from X_δ into itself. Therefore, the Banach contraction mapping principle implies the conclusion of the theorem. □

Remark 3.2.2. We have the strict inclusion relation $L^n \hookrightarrow \dot{B}^{-a}_{p,\infty}$ $(a = 1 - \frac{n}{p})$, so $\|u_0\|_n$ can be sufficiently large while $\|u_0\|_{B^{-a}_{p,\infty}}$ is small arbitrarily. For example, for any $f \in E_n$, take $\omega_k(x) = \exp(ix \cdot k)$. Then it is clear that $\|\omega_k f\|_n = \|f\|_n$. However,

$$\lim_{|k|\to\infty} \|\omega_k f\|_{\dot{B}^{-a}_{p,\infty}} = 0.$$

See [32] for a detailed discussion.

We now consider the question whether, for any $u_0 \in B$ with $L^n \hookrightarrow B$ and $\deg B = -1$, the problem (3.48) or (3.51) determines a global small solution $u \in C([0,\infty);B)$. As an example, we consider the case $B = \dot{B}^{-\alpha}_{p,\infty}(\mathbb{R}^n)$ with $\alpha = 1 - n/p$. Note that the difference between $\dot{B}^{-\alpha}_{p,\infty}$ and $L^n(\mathbb{R}^n)$ consists in the fact that $B = \dot{B}^{-\alpha}_{p,\infty}(\mathbb{R}^n)$ is not separable. Thus, for $u_0 \in \dot{B}^{-\alpha}_{p,\infty}$ and $(q, p, n) \in \Pi$, the limit

$$\lim_{t\to 0} t^{\frac{1}{q}} \|S(t)u_0\|_p = 0 \tag{3.79}$$

does not necessarily hold. For example, let

$$u_0(x) = \left(0, \frac{-x_3}{|x|^2}, \frac{x_2}{|x|^2}\right).$$

Then

$$\lim_{t\to 0} t^{\frac{1}{q}} \|S(t)u_0\|_p = \|S(1)u_0\|_p \neq 0,$$

where $(q, p, n) \in \Pi$. However, (3.79) plays an important role in the proof of the uniqueness and continuous dependence part of Theorem 3.2.2.

On the other hand, $S(t)$ is not a C_0-semigroup on a nonseparable space B such as $\dot{B}^{-\alpha}_{p,\infty}$ and L^∞, so $S(t)u_0(x)$ is not a continuous abstract function from $[0, \infty)$ to B. Thus, if we want to study (3.48) or (3.51) on B such as $\dot{B}^{-\alpha}_{p,\infty}$, we need to replace $C([0, \infty); B)$ by $C_*([0, \infty); B)$, where $u \in C_*([0, \infty); B)$ means that $u \in C((0, \infty); B)$ and u is continuous at $t = 0$ in the $\sigma(B, B')$ sense, that is,

$$\lim_{t\downarrow 0}(u(x, t) - u_0(x), \psi) = 0, \quad \psi \in B',$$

where B' is the dual space to B.

Theorem 3.2.3. *Let (q, p, n) be any admissible triplet satisfying that $n < p \leq 2n$ and let $\alpha = 1 - n/p = 2/q$. Then there exists a $\delta > 0$ such that, for $u_0 \in \dot{B}^{-\alpha}_{p,\infty}$ with $\operatorname{div} u_0 = 0$ (in the weak sense), if $\|u_0\|_{\dot{B}^{-\alpha}_{p,\infty}} < \delta$, then the problem (3.48) (or equivalently (3.51)) has a unique global mild solution u satisfies that*

$$\begin{cases} u \in C_*([0, \infty); \dot{B}^{-\alpha}_{p,\infty}), \\ t^{\frac{1}{q}} u(t) \in C_*((0, \infty); L^p(\mathbb{R}^n)), \end{cases} \tag{3.80}$$

$$\sup_{t\geq 0} \|u\|_{\dot{B}^{-\alpha}_{p,\infty}} + \sup_{t\geq 0} t^{\frac{1}{q}} \|u(t)\|_p \leq R(\|u_0\|_{\dot{B}^{-\alpha}_{p,\infty}}), \tag{3.81}$$

$$u(t) - S(t)u_0(x) \in C_*((0, \infty); L^n(\mathbb{R}^n)), \quad n < p \leq 2n, \tag{3.82}$$

where $R(\|u_0\|_{\dot{B}^{-\alpha}_{p,\infty}})$ is a constant depending only on $\|u_0\|_{\dot{B}^{-\alpha}_{p,\infty}}$. In particular, when $n = 3$, we have

$$u(t) - S(t)u_0 \in C_*((0, \infty); \dot{H}^{\frac{1}{2}}(\mathbb{R}^3)), \quad 3 < p \leq 4. \tag{3.83}$$

Proof. The proof follows similarly to that of Theorem 3.2.2 where X is replaced by

$$Y = \{u \mid u \in C_*([0,\infty); \dot{B}^{-a}_{p,\infty}),\ t^{\frac{1}{q}} u(t) \in C_*([0,\infty); L^p)\}$$

with

$$\|\cdot\|_Y = \sup_{t>0} \|\cdot\|_{\dot{B}^{-a}_{p,\infty}} + \sup_{t\geq 0} t^{\frac{1}{q}} \|\cdot\|_p,$$

where $(q,p,n) \in \Pi$. We only need to prove the estimates (3.82) and (3.83). In fact,

$$\begin{aligned}\|Gu\|_{C_*(([0,\infty);L^n)} &\leq C \sup_{t\in[0,\infty)} \int_0^t |t-\tau|^{-\frac{1}{2}-\frac{n}{2}(\frac{2}{p}-\frac{2}{n})} \|u(\tau)\|_p^2 d\tau \\ &\leq C\Big(\sup_{t\in[0,\infty)} t^{\frac{1}{q}} \|u(t)\|_p\Big)^2 < \infty, \quad n < p \leq 2n, \\ \|Gu\|_{C_*([0,\infty);\dot{H}^{\frac{1}{2}}(\mathbb{R}^3))} &\leq C \sup_{t\in[0,\infty)} \int_0^t |t-\tau|^{-\frac{1}{2}-\frac{1}{4}-\frac{3}{2}(\frac{2}{p}-\frac{1}{2})} \|u(\tau)\|_p^2 d\tau \\ &\leq C\Big(\sup_{t\in[0,\infty)} t^{\frac{1}{q}} \|u\|_p\Big)^2 < \infty, \quad 3 < p \leq 4,\end{aligned}$$

where we use the facts that

$$\frac{1}{2} + \frac{1}{4} - \frac{3}{2}\Big(\frac{2}{p} - \frac{1}{2}\Big) < 1, \quad \frac{2}{q} = 3\Big(\frac{1}{2} - \frac{1}{p}\Big) < 1.$$

The proof is complete. □

Theorem 3.2.3 can be used to study existence of self-similar solutions. If $(u(x,t), P(x,t))$ is a solution to the Navier–Stokes equations,

$$\begin{cases} \frac{\partial u}{\partial t} - \Delta u + (u\cdot\nabla)u + \nabla P = 0, \\ \operatorname{div} u = 0, \end{cases} \tag{3.84}$$

then for any $\lambda > 0$, $(u_\lambda(x,t), P_\lambda(x,t))$ is also a solution to (3.84), where

$$u_\lambda(t,x) = \lambda u(\lambda^2 t, \lambda x), \quad P_\lambda(t,x) = \lambda^2 P(\lambda^2 t, \lambda x).$$

Definition 3.2.1. $u(t,x)$ is a self-similar solution to the Navier–Stokes (3.84), if for any $\lambda > 0$, we have

$$u_\lambda(t,x) = \lambda u(\lambda^2 t, \lambda x) = u(t,x). \tag{3.85}$$

Let $U(x) = u(x, 1)$. Then, by (3.85), a self-similar solution is of the following form:

$$u(t,x) = \frac{1}{\sqrt{t}} U\left(\frac{x}{\sqrt{t}}\right).$$

Note that $v(t,x) \longmapsto v_\lambda(t,x)$ is invariant under the norm of Y. Thus, as a direct result of Theorem 3.2.3, we have the following.

Theorem 3.2.4. *Let (q,p,r) be any generalized admissible triplet with $n < p \leq 2n$ and let $\alpha = 1 - n/p = 2/q$. Then there exists a $\delta > 0$ such that for $u_0(x) = \lambda u_0(\lambda x) \in \dot{B}^{-\alpha}_{p,\infty}$ satisfying that* $\operatorname{div} u_0 = 0$ *and $\|u_0\|_{\dot{B}^{-\alpha}_{p,\infty}} < \delta$, the Cauchy problem* (3.48) *has a unique mild solution $u(t,x) = \frac{1}{\sqrt{t}} U(\frac{x}{\sqrt{t}})$, where $U \in \dot{B}^{-\alpha}_{p,\infty} \cap L^p(\mathbb{R}^n)$ and satisfies that*

$$U(x) = S(1)u_0(x) + W(x),$$
$$\|U\|_{\dot{B}^{-\alpha}_{p,\infty}} + \|U\|_p \leq R(\|u_0\|_{\dot{B}^{-\alpha}_{p,\infty}}),$$

where $W \in L^n(\mathbb{R}^n)$, $3 \leq p \leq 6$ and $R(\|u_0\|_{\dot{B}^{-\alpha}_{p,\infty}})$ is a constant depending only on $\|u_0\|_{\dot{B}^{-\alpha}_{p,\infty}}$. In particular, when $n = 3$ and $3 < p \leq 4$, we have $W \in \dot{H}^{\frac{1}{2}}(\mathbb{R}^3)$.

3.3 Local well-posedness—the Littlewood–Paley method

With the Littlewood–Paley decomposition technique, Cannone and Meyer [35] introduced the so-called well-suited Banach space X and further established local well-posedness on X of the Cauchy problem for the Navier–Stokes equations. It should be pointed out that the well-suited Banach space X defined in [35] via the Littlewood–Paley decomposition includes all the noncritical Banach spaces used previously to deal with the Cauchy problem for the Navier–Stokes equations. Thus, it can be said that they gave a unified method to deal with the Cauchy problem for the Navier–Stokes equations. Certainly, one may use the scaling principle to roughly analyze these so-called well-suited Banach spaces. In fact, if $u(t,x)$ is a solution to the following Cauchy problem:

$$\begin{cases} \partial_t u(t) - \Delta u + (u \cdot \nabla)u + \nabla P = 0, & (x,t) \in \mathbb{R}^n \times (0,\infty), \\ \operatorname{div} u = 0, & (x,t) \in \mathbb{R}^n \times [0,\infty), \\ u(x,0) = u_0(x), & x \in \mathbb{R}^n, \end{cases} \tag{3.86}$$

then

$$u_\lambda(x,t) = \lambda^{-1} u(\lambda^{-1}x, \lambda^{-2}t), \quad P_\lambda(x,t) = \lambda^{-2} P(\lambda^{-1}x, \lambda^{-2}t)$$

is also a solution to the problem (3.86) with u_0 replacing by $\lambda^{-1}u_0(\lambda^{-1}x)$. By scaling (see Section 1.4 in Chapter 1), the so-called well-suited Banach space X must be such that $\deg(X) \geq -1$, so that (3.86) can guarantee to determine a local continuous flow on

$C([0,T];X)$ (the case where $T = \infty$ corresponds to a global flow). In fact, the well-suited Banach space constructed by Cannone and Meyer is just a space X of differentiable functions with $\deg(X) > -1$ (which excludes the critical case). For example, X can be any of the following classical spaces:

$$X = \begin{cases} L^p(\mathbb{R}^n), & -\frac{n}{p} > -1, \\ H^{s,p}(\mathbb{R}^n), & s - \frac{n}{p} > -1,\ p > 1, \\ M_q^p(\mathbb{R}^n), & -\frac{n}{p} > -1,\ p \geq q > 1, \\ B_{p,q}^s(\mathbb{R}^n), & s - \frac{n}{p} > -1,\ 1 < p, q \leq \infty, \\ C^s(\mathbb{R}^n), & s > -1, \\ F_{p,q}^s(\mathbb{R}^n), & s - \frac{n}{p} > -1,\ 1 \leq q \leq \infty,\ 1 \leq p < \infty, \\ L^{(n,p)}(\mathbb{R}^n), & n \leq p \leq \infty, \end{cases} \tag{3.87}$$

where M_q^p stands for the Morrey–Campanato spaces, C^s denotes the Zygmund spaces and $L^{(n,p)}$ represents the Lorentz spaces.

For simplicity, we are restricted to the case $n = 3$. For the general case $n \geq 3$, the results are completely similar. Let $D_j = -i\frac{\partial}{\partial x_j}$ $(j = 1,2,3)$. Then $R_j = D_j/(-\Delta)^{\frac{1}{2}}$ is the Riesz operator with symbol $\frac{\xi_j}{|\xi|}$. Thus, the projection operator $\mathcal{P}$ from $(L^p(\mathbb{R}^3))^3$ to $E_p(\mathbb{R}^3)$ can be expressed as

$$(\mathcal{P}v)_k = v_k(x) - \sum_{j=1}^{3} R_k R_j v_j(x), \quad 1 \leq k \leq 3,$$

so its symbol is

$$\left(\delta_{kj} - \frac{\xi_k \xi_j}{|\xi|^2}\right)_{3\times 3}.$$

We can define the Littlewood–Paley decomposition $\{S_j, \Delta_j\}$ here similar to the previous section but with the bump function $\hat{\varphi}(\xi) \in C_c^\infty$ being defined to satisfy that

$$\hat{\varphi}(\xi) = \begin{cases} 1, & |\xi| \leq \frac{3}{4}, \\ 0, & |\xi| \geq \frac{3}{2}, \end{cases} \tag{3.88}$$

instead of (3.66) (this does not make any essential difference but for simplicity). With the Littlewood–Paley decomposition $\{S_j, \Delta_j\}$, we now define the well-suited Banach space X.

Definition 3.3.1 (Well-suited Banach spaces). A Banach space X of vector distributions on $\mathbb{R}^3$ is called a well-suited Banach space for the Navier–Stokes equations if the following conditions are satisfied:

(i) X is translation invariant, that is,

$$\|v(\cdot + h)\|_X = \|v(\cdot)\|_X, \quad \forall h \in \mathbb{R}^3, \ v \in X.$$

(ii) There exists a positive sequence $\{\eta_j\}_{j\in\mathbb{Z}}$ such that

$$\sum_{j\in\mathbb{Z}} 2^{-|j|}\eta_j < \infty, \quad j \in \mathbb{Z}$$

and

$$\|\Delta_j(fg)\|_X \leq \eta_j \|f\|_X \|g\|_X, \quad \forall f, g \in X, \ j \in \mathbb{Z}. \tag{3.89}$$

Definition 3.3.2. For each well-suited Banach space X, we can also define the following abstract Besov-type space:

$$\dot{B}^0_{X,1} = \Big\{ f \in X, \ \sum_{j\in\mathbb{Z}} \|\Delta_j f\|_X < \infty \Big\}.$$

It is easy to see that, if $X = L^p(\mathbb{R}^n)$ is a well-suited Banach space, then

$$\dot{B}^0_{L^p,1} = \dot{B}^0_{p,1} = \Big\{ f \in L^p(\mathbb{R}^n) \ \Big| \ \sum_{j\in\mathbb{Z}} \|\Delta_j f\|_p < \infty \Big\}.$$

Theorem 3.3.1. *Let X be a well-suited Banach space. For any $u_0 \in X$ with $\operatorname{div} u_0 = 0$ in the distributional sense, there is a $T = T(\|u_0\|_X) > 0$ such that the Cauchy problem* (3.86) *has a unique solution $u \in C([0,T); X)$ satisfying that*

$$u(t) = S(t)u_0(x) - \int_0^t \mathcal{P}S(t-\tau)\nabla \cdot (u \otimes u) d\tau, \tag{3.90}$$

where $S(t) = \mathcal{F}^{-1} e^{-|\xi|^2 t} \mathcal{F}$ is the classical heat semigroup. Further, we have

$$\omega(t) = u(t) - S(t)u_0(x) \in \dot{B}^0_{X,1}, \quad \forall t \in [0,T). \tag{3.91}$$

Remark 3.3.1.

(i) Well-suitable Banach spaces exclude the case of limiting or critical spaces, so local well-posedness can still be handled without use of space-time estimates. This section aims to give a unified result in order to embody the universality of the harmonic analysis methods by using the Littlewood–Paley decomposition.

(ii) Since $\omega \in \dot{B}^0_{X,1}$, $\omega(t)$ can be expanded into a fast convergent series in terms of a wavelets basis. However, this is not necessarily true for $u(t)$ and $S(t)u_0(x)$.

(iii) Solutions to the integral equation (3.90) are called as *mild solutions.* It is easy to see that mild solutions belonging to the suitable Banach space X are regular solutions. The reader is referred to discussions on regularity of solutions to parabolic equations in Chapter 2. Note that the Navier–Stokes equations are essentially abstract parabolic equations.

Before proceeding the proof of Theorem 3.3.1, we introduce two lemmas.

Lemma 3.3.1. *Let X be a suitable Banach space. Then there exists a function $\lambda(t) \geq 0$ such that for any $t \in [0, T)$ (we may assume that $T \leq 4$ since we are concerned with local well-posedness here), we have $\lambda(t) \in L^1([0,4])$ and*

$$\|\mathcal{P}S(t)\nabla \cdot (u \otimes v)\|_X \leq \lambda(t)\|u\|_X\|v\|_X, \quad \forall u, v \in X. \tag{3.92}$$

Proof. Note that the vector operator

$$(\mathcal{P}\nabla\cdot)_j = \mathcal{F}^{-1}\frac{\xi_j\xi_i\xi_k}{|\xi|^2}\mathcal{F}.$$

is a pseudodifferential operator of order one (acting on matrix functions). Hence, to prove (3.92), it is enough to show the following inequality:

$$\|\Lambda S(t)(fg)\|_X \leq \lambda(t)\|f\|_X\|g\|_X, \tag{3.93}$$

where $\Lambda = (-\Delta)^{1/2} = \mathcal{F}^{-1}|\xi|\mathcal{F}$ is a Calderón operator. It is easy to see that

$$[0,4] = \bigcup_{m=0}^{\infty}[4^{-m}, 4^{-m+1}] \equiv \bigcup_{m=0}^{\infty} J_m.$$

We claim that the inequality

$$\lambda(t) \leq C\sum_{j\leq m} 2^j\eta_j + \sum_{j\geq m+1} 2^{-2j+3m}\eta_j, \quad t \in J_m \tag{3.94}$$

implies the estimate (3.93), where η_j is the sequence appeared in the definition of the suitable Banach space X. In fact, using (3.94), we get

$$\begin{aligned}
\int_0^4 \lambda(t)dt &= \sum_{m=0}^{\infty}\int_{4^{-m}}^{4^{-m+1}}\lambda(t)dt \leq 3\sum_{m=0}^{\infty}\sup_{t\in J_m}\lambda(t)\cdot 4^{-m} \\
&\leq C\sum_{m=0}^{\infty}\sum_{j\leq m} 2^{j-2m}\eta_j + C\sum_{m=0}^{\infty}\sum_{j\geq m+1} 2^{-2j+m}\eta_j \\
&\leq 2C\sum_{j\in\mathbb{Z}} 2^{-|j|}\eta_j.
\end{aligned}$$

We now construct a new Littlewood–Paley decomposition. Let

$$\widetilde{\psi}(x) = 2^6\varphi(2^2x) - 2^{-3}\varphi\left(\frac{x}{2}\right),$$
$$\widetilde{\psi}_j(x) = 2^{3j}\psi(2^jx), \quad j \in \mathbb{Z},$$

where $\varphi(x)$ is the same as defined in (3.88), and denote the corresponding Littlewood–Paley decomposition by

$$\{\widetilde{S}_j, \widetilde{\Delta}_j\}_{j\in\mathbb{Z}}.$$

It is clear that

$$\widetilde{\Delta}_j\Delta_j = \Delta_j, \quad j \in \mathbb{Z}.$$

Thus, using the equality

$$\sum_{j\in\mathbb{Z}} \Delta_j = \sum_{j\in\mathbb{Z}} \widetilde{\Delta}_j\Delta_j = I,$$

we have the following operator decomposition:

$$\Lambda S(t) = \sum_{j\in\mathbb{Z}} \widetilde{\Delta}_j\Lambda S(t)\Delta_j. \tag{3.95}$$

Now let

$$\begin{cases} 2^j W_{j,t} = \widetilde{\Delta}_j\Lambda S(t), & j \leq m, \\ 2^{-2j+3m} W_{j,t} = \widetilde{\Delta}_j\Lambda S(t), & j \geq m+1. \end{cases} \tag{3.96}$$

Then the operator $W_{j,t}$ can be expressed as

$$W_{j,t}f = \omega_{j,t} * f, \quad j \in \mathbb{Z}.$$

If $\omega_{j,t} \in L^1(\mathbb{R}^3)$ and $\|\omega_{j,t}\|_{L^1} \leq C$, then from the decomposition (3.95) and (3.96), it follows that

$$\begin{aligned} \|\Lambda S(t)(fg)\|_X &\leq \left\| \sum_{j=-\infty}^{m} 2^j\Delta_j W_{j,t}(fg) + \sum_{j=m+1}^{\infty} 2^{-2j+3m}\Delta_j W_{j,t}(fg) \right\|_X \\ &\leq \sum_{j\leq m} \|2^j\Delta_j(fg) * \omega_{j,t}\|_X + \sum_{j\geq m+1} \|2^{-2j+3m}\Delta_j(fg) * \omega_{j,t}\|_X \\ &\leq \sum_{j\leq m} \|\omega_{j,t}\|_1\|2^j\Delta_j(fg)\|_X + \sum_{j\geq m+1} \|\omega_{j,t}\|_1\|2^{-2j+3m}\Delta_j(fg)\|_X \end{aligned}$$

$$\leq \sum_{j\leq m} 2^j \eta_j \|f\|_X \|g\|_X + \sum_{j\geq m+1} 2^{-2j+3m} \eta_j \|f\|_X \|g\|_X$$
$$\leq \lambda(t) \|f\|_X \|g\|_X,$$

where we use (3.89) and the translation invariance of X (to ensure that Young's inequality holds). It remains to prove that

$$\|\omega_{j,t}\|_1 \leq C, \quad j \in \mathbb{Z}, \ t \in J_m. \tag{3.97}$$

In fact, we only need to consider the case $t \in J_0$. The general case $t \in J_m$ can be transformed to the case $t \in J_0$. Since

$$\Lambda S(t) = \sum_{j\leq 0} 2^j \Delta_j W_{j,t} + \sum_{j\geq 1} 2^{-2j} \Delta_j W_{j,t}, \quad m = 0,$$

by applying the Fourier transform to both sides of the above equation, we derive that

$$|\xi| \exp(-t|\xi|^2) = \sum_{j\leq 0} 2^j \hat{\psi}(2^{-j}\xi) \hat{\omega}_{j,t}(\xi) + \sum_{j\geq 1} 2^{-2j} \hat{\psi}(2^{-j}\xi) \hat{\omega}_{j,t}(\xi). \tag{3.98}$$

Noting that $\hat{\psi}(\xi) = \hat{\psi}(\xi)\hat{\tilde{\psi}}(\xi)$, one gets

$$|\xi| \exp(-t|\xi|^2) = \sum_{j\in\mathbb{Z}} |\xi| \exp(-t|\xi|^2) \hat{\psi}(2^{-j}\xi) \hat{\tilde{\psi}}(2^{-j}\xi). \tag{3.99}$$

Comparing (3.98) and (3.99), we get

$$\begin{cases} \hat{\omega}_{j,t}(2^j\xi) = |\xi| \hat{\tilde{\psi}}(\xi) \exp(-4^j t|\xi|^2), & j \leq 0, \\ \hat{\omega}_{j,t}(2^j\xi) = 8^j |\xi| \hat{\tilde{\psi}}(\xi) \exp(-4^j t|\xi|^2), & j \geq 1. \end{cases}$$

From the definition of $\tilde{\psi}$, it is seen that $\omega_{j,t} \in S(\mathbb{R}^n)$ and the estimate (3.97) holds for $t \in J_0$. On the other hand, for $t \in J_m$, let

$$t = 4^{-m}s, \quad \xi = 2^m \eta,$$

and institute these into

$$|\xi| \exp(-t|\xi|^2) = \sum_{j\leq m} 2^j \hat{\psi}(2^{-j}\xi) \hat{\omega}_{j,t}(\xi) + \sum_{j\geq m+1} 2^{-2j+3m} \hat{\psi}(2^{-j}\xi) \hat{\omega}_{j,t}(\xi),$$

then one can get

$$\begin{aligned}
\eta \exp(-s|\eta|^2) &= \sum_{j\le m} 2^{j-m}\hat{\psi}(2^{-(j-m)}\eta)\hat{\tilde{\psi}}(2^{-(j-m)}\eta)\hat{\omega}_{j,t}(2^m\eta) \\
&\quad + \sum_{j\ge m+1} 2^{-2(j-m)}\hat{\psi}(2^{-(j-m)}\eta)\hat{\tilde{\psi}}(2^{-(j-m)}\eta)\hat{\omega}_{j,t}(2^m\eta) \\
&= \sum_{j\le m} 2^{j-m}\hat{\psi}(2^{-(j-m)}\eta)\hat{\tilde{\psi}}(2^{-(j-m)}\eta)\hat{\omega}_{j-m,t}(\eta) \\
&\quad + \sum_{j\ge m+1} 2^{-2(j-m)}\hat{\psi}(2^{-(j-m)}\eta)\hat{\tilde{\psi}}(2^{-(j-m)}\eta)\hat{\omega}_{j-m,t}(\eta) \\
&= \sum_{j\le 0} 2^{j}\hat{\psi}(2^{-j}\xi)\hat{\omega}_{j,t}(\xi) + \sum_{j\ge 1} 2^{-2j}\hat{\psi}(2^{-j}\xi)\hat{\omega}_{j,t}(\xi).
\end{aligned}$$

It can be seen from this that the estimate at the case $t \in J_m$ can be induced to the estimate at the case $s \in J_0$. □

Remark 3.3.2. Essentially, Lemma 3.3.1 implies the bilinear estimate

$$\begin{aligned}
\sup_{0\le t\le T} \|B(u,u)\|_X &= \sup_{0\le t\le T} \left\| \int_0^t \mathcal{P}S(t-s)\nabla\cdot(u\otimes u)ds \right\|_X \\
&\le C \int_0^T \lambda(t)dt \Big(\sup_{0\le t\le T} \|u(t)\|_X \Big)^2. \qquad (3.100)
\end{aligned}$$

Lemma 3.3.2. *Let Y be an abstract Banach space and let $B : Y \times Y \to Y$ be a bilinear operator satisfying that*

$$\|B(x_1,x_2)\|_Y \le \eta \|x_1\|_Y \|x_2\|_Y, \quad \forall x_1, x_2 \in Y.$$

Then, for any $y \in Y$, if $4\eta\|y\|_Y < 1$, then $x = y + B(x,x)$ has a unique solution $x \in Y$ and

$$\|x\|_Y \le \frac{1-\sqrt{1-4\eta\|y\|_Y}}{2\eta}.$$

Proof. Let

$$R = \frac{1-\sqrt{1-4\eta\|y\|_Y}}{2\eta}$$

and let $B_R = \{x \in Y, \quad \|x\|_Y \le R\}$. A direct calculation leads to the result

$$\|y\|_Y + \eta R^2 = R, \quad R < 2\|y\|_Y.$$

Now, we consider the mapping F on B_R, which is defined by

$$F(x) = y + B(x,x).$$

Clearly,

$$\begin{aligned}\|F(x)-F(x')\|_Y &\le \|B(x-x',x)\|_Y + \|B(x',x-x')\|_Y \\ &\le 2\eta R\|x-x'\|_Y, \quad x,x' \in B_R, \\ \|F(x)\|_Y &\le \|y\|_Y + \eta R^2 = R.\end{aligned}$$

Thus, $F : B_R \to B_R$ is a contraction mapping, and so the Banach fixed-point theorem implies Lemma 3.3.2. □

Proof of Theorem 3.3.1. The existence part follows from Lemmas 3.3.1 and 3.3.2. We now prove the regularity (3.91) of the flow function $\omega(x,t)$. In other words, we want to prove that the function $\omega(x,t) = u(x,t) - S(t)u_0(x)$ can be decomposed on X as

$$\omega(x,x) = u(x,t) - S(t)u_0(x) = \sum_{j\in\mathbb{Z}} d_j(t),$$

where

$$\operatorname{supp}(\hat{d}_j) \subset \{\xi \in \mathbb{R}^3,\ \alpha 2^j \le |\xi| \le \beta 2^{j+1},\ \beta > \alpha > 0\}$$

and

$$\sum_{j\in\mathbb{Z}} \|d_j(t)\|_X < \infty, \quad \forall t \in [0,T).$$

In fact, noting the operator decomposition (3.95), we obtain

$$u(t) - S(t)u_0(x) = \sum_{j\in\mathbb{Z}} \int_0^t \widetilde{\Delta}_j \mathcal{P} S(t-\tau)\Delta_j \nabla\cdot(u\otimes u)d\tau \equiv \sum_{j\in\mathbb{Z}} d_j, \tag{3.101}$$

where $d_j(t) = \int_0^t \tilde{d}_j(t-\tau,\tau)d\tau$. Thus, for $t-\tau \in J_m$, we have

$$\begin{aligned}\|\tilde{d}_j(t-\tau,\tau)\|_X &\le C2^j\eta_j\Big(\sup_{0\le t\le T}\|u(t)\|_X\Big)^2, \quad j\le m, \\ \|\tilde{d}_j(t-\tau,\tau)\|_X &\le C2^{-2j+3m}\eta_j\Big(\sup_{0\le t\le T}\|u(t)\|_X\Big)^2, \quad j>m.\end{aligned}$$

Therefore, it follows that

$$\|d_j(t)\|_X \le C(2^{-|j|}\eta_j)\Big(\sup_{0\le t\le T}\|u(t)\|_X\Big)^2, \tag{3.102}$$

which completes the proof of the theorem. □

Remark 3.3.3.

(i) In Theorem 3.3.1, it is actually assumed that $S(t)$ is a C_0-semigroup on X. Otherwise, it would not be guaranteed that $S(t)u_0$ belongs to $C([0,T];X)$.

(ii) When X is nonseparable, it is not necessary to assume that $S(t)$ is a C_0-semigroup on X. Since X is translation invariant, then we have

$$\|S(t)u_0\|_X \le \|S(t)\|_1 \|u_0\|_X = \|u_0\|_X,$$

where $\|S(t)\|_1$ denotes the L^1 norm of the kernel function corresponding to $S(t)$. This means that $S(t)u_0$ is weakly-* continuous in the distributional sense, that is,

$$S(t)u_0 \in C_*([0,T];X).$$

Thus, it is enough to replace $C([0,T),X)$ by $C_*([0,T),X)$ in Theorem 3.3.1.

We now verify whether the usual spaces of differentiable functions X such as C^s, L^p, $H^{s,p}$, the Morrey–Campamato spaces M_q^p or more general the Besov spaces $B_{p,q}^s$ and the Trieble–Lizorkon spaces $F_{p,q}^s$ are well-suited Banach spaces provided $\deg(X) > -1$. We only consider several simple examples. The more complicated examples can be found in [33, 35].

(a) The Hölder–Zygmund spaces $C^s(\mathbb{R}^3)$, $s > 0$

When $s \in \mathbb{Z}^+$, $C^s(\mathbb{R}^3)$ corresponds to the Zygmund spaces, while for $s \in \mathbb{R}^+ \setminus \mathbb{Z}^+$, C^s corresponds to the Hölder spaces (see Chapter 1). Let us recall first the Littlewood–Paley characterization of C^s. Let $\{S_j, \Delta_j\}_{j\in\mathbb{Z}}$ be the Littlewood–Paley decomposition. Then $f \in C^s$ if and only if:

(i) $\|S_0 f\|_\infty \le C$,

(ii) $\|\Delta_j f\|_\infty \le C2^{-js}$, $j \ge 0$.

Noting that C^s is a Banach algebra, we have

$$\|\Delta_j(fg)\|_{C^s} \le C\|fg\|_{C^s} \le C\|f\|_{C^s}\|g\|_{C^s}, \quad \forall s \in \mathbb{R}^+.$$

Thus, letting

$$\eta_j = \|2^{3j}\psi(2^j x)\|_1 \equiv \|\psi(x)\|_1 = C,$$

we have

$$\sum_{j\in\mathbb{Z}} 2^{-|j|}\eta_j < \infty$$

and, therefore, C^s is a well-suited Banach space for $s > 0$, so $\omega(x,t) = u(x,t) - S(t)u_0 \in \dot{B}^0_{X,1}$. We now prove that

$$\omega(x,t) = u(x,t) - S(t)u_0 \in \mathcal{C}^{s+1}(\mathbb{R}^3).$$

According to the Littlewood–Paley characterization of $\mathcal{C}^{s+1}$, we only need to show that

$$\|S_0(u(t) - S(t)u_0)\|_\infty \le C, \tag{3.103}$$

$$\|\Delta_j(u(t,x) - S(t)u_0)\|_\infty \le 2^{-j(s+1)}C, \quad j \in \mathbb{Z}^+. \tag{3.104}$$

Since $u(t,x), S(t)u_0 \in \mathcal{C}^s$, then (3.103) is obvious. Now, we utilize the equality

$$\Delta_j(u(t) - S(t)u_0) = \sum_{k\in\mathbb{Z}} \int_0^t \widetilde{\Delta}_k \mathcal{P} S(t-\tau)\Delta_k \nabla \cdot (\Delta_j(u \times u))d\tau$$

to get that

$$\begin{aligned}\|\Delta_j(u(t) - S(t)u_0)\|_\infty &\le C \sum_{j-1\le k\le j+1} 2^{-k}\eta_k \cdot \sup_{0\le t\le T} \|\Delta_j(u\times u)\|_\infty \\ &\sim 2^{-j} \cdot 2^{-js} \sim 2^{-j(s+1)},\end{aligned}$$

which is (3.104).

From the above results, we get the following.

Theorem 3.3.2. *Let $s > 0$, $X = \mathcal{C}^s(\mathbb{R}^3)$ be a Hölder–Zgymund space. For any $u_0 \in \mathcal{C}^s(\mathbb{R}^3)$ with* $\operatorname{div} u_0(x) = 0$, *there exists a $T = T(\|u_0\|_{\mathcal{C}^s}) > 0$ such that the integral equation* (3.90) *has a unique solution $u(t,x) \in C_*([0,T); \mathcal{C}^s(\mathbb{R}^3))$ satisfying that $\omega(t,x) = u(t,x) - S(t)u_0 \in \mathcal{C}^{s+1}(\mathbb{R}^3)$.*

Remark 3.3.4. Roughly speaking, since

$$\begin{aligned}\|\partial_t^\alpha S(t)f\|_{\mathcal{C}^s} &\sim \|(-\Delta)^\alpha S(t)f\|_{\mathcal{C}^s} \sim \|S(t)f\|_{\mathcal{C}^{2\alpha+s}}, \\ \|\nabla f\|_{\mathcal{C}^s} &\sim 2^k \|f\|_{\mathcal{C}^s}, \quad \operatorname{supp}(f) \subset \{2^{k-1} \le |\xi| \le 2^{k+1}\},\end{aligned}$$

and $\nabla \sim \sum_{k\in\mathbb{Z}} 2^k \Delta_k$, we have

$$\begin{aligned}|d_k| &= \left|\int_0^t \widetilde{\Delta}_k \mathcal{P} S(t-\tau)\Delta_k \nabla \cdot (u \otimes u)d\tau\right| \\ &\sim |(\partial_t)^{-1}\widetilde{\Delta}_k \mathcal{P} S(t-\tau)\Delta_k \cdot \nabla \cdot (u\otimes u)| \\ &\sim 2^{-2k} \cdot 2^k \cdot C2^{-ks} \sim C \cdot 2^{-k(s+1)}.\end{aligned}$$

(b) The Lebesgue spaces $L^p(\mathbb{R}^3)$

From the Young inequality, it is easy to see that

$$\|\Delta_j(fg)\|_p \le \|\psi_j\|_r \|f\|_p \|g\|_p, \quad \frac{1}{p} = \frac{2}{p} + \frac{1}{r} - 1.$$

This together with

$$\|\psi_j\|_r = \Big(\int_{\mathbb{R}^3} 2^{3jr}|\psi(2^j x)|^r dx\Big)^{\frac{1}{r}} = 2^{3j-\frac{3j}{r}}\|\psi\|_r = 2^{\frac{3j}{p}}C$$

implies that

$$\|\Delta_j(fg)\|_p \le 2^{\frac{3j}{p}}C\|f\|_p\|g\|_p.$$

To ensure that L^p is a well-suited Banach space, the inequality

$$\sum_{j\in\mathbb{Z}}\eta_j 2^{-|j|} < \infty, \quad \eta_j = 2^{\frac{3j}{p}}$$

should be satisfied, which implies that $p > 3$. Thus, we have the following.

Theorem 3.3.3. *Let $p > 3$, $X = L^p(\mathbb{R}^3)$. For any $u_0 \in X$ with $\operatorname{div} u_0(x) = 0$ in the distributional sense, there exists a $T = T(\|u_0\|_p) > 0$ such that the integral equation (3.90) has a unique mild solution $u \in C([0,T); L^p(\mathbb{R}^3))$. Moreover, the following regularity property holds:*

$$\omega(x,t) = u(x,t) - S(t)u_0(x) \in B^s_{p,\infty}, \quad s = 1 - \frac{3}{p}. \tag{3.105}$$

Proof. We only need to prove (3.105). From Theorem 3.3.1, it is known that $\omega(t,x) \in \dot{B}^0_{p,1}$, so $\|S_0\omega\|_p < \infty$. By the Littlewood–Paley characterization of Besov spaces, it is found that (3.105) is equivalent to

$$\|S_0\omega\|_p + \sup_{j\ge 0} 2^{sj}\|\Delta_j\omega\|_p < \infty. \tag{3.106}$$

Thus, it is enough to show that

$$2^{sj}\|\Delta_j\omega\|_p < \infty, \quad j \ge 0,\ s = 1 - \frac{3}{p}. \tag{3.107}$$

In fact, from the expression (3.101), we derive that

$$\Delta_j\omega = \sum_{j-1\le k\le j+1}\int_0^t \widetilde{\Delta}_k \mathcal{P} S(t-\tau)\Delta_k\nabla\cdot\Delta_j(u\otimes u)d\tau.$$

A direct calculation leads to the result

$$\|\Delta_j\omega\|_p \le \sum_{j-1\le k\le j+1} C2^{-|k|}\eta_k\|u\|_p^2 \le C2^{-j}\cdot 2^{\frac{3j}{p}}\|u\|_p^2,$$

which deduces (3.107). □

(c) The Sobolev spaces $H^s(\mathbb{R}^3)$

The procedure of verifying that $H^s(\mathbb{R}^3)$ is a well-suited Banach space for $s - 3/2 > -1$ leads essentially to a general technology, that is, the well-known Bony's paraproduct decomposition (see [183]) as follows. For any two arbitrary tempered distributions f and g,

$$\begin{aligned} fg &= \sum_{k=0}^{\infty}(S_{k+1}fS_{k+1}g - S_kfS_kg) + S_0fS_0g \\ &= \sum_{k=0}^{\infty}(\Delta_kf\Delta_kg + \Delta_kgS_kf + \Delta_kfS_kg) + S_0fS_0g \\ &= \sum_{k=0}^{\infty}\Delta_kfS_{k-2}g + \sum_{k=0}^{\infty}\Delta_kgS_{k-2}f + \sum_{|k-k'|<2}\Delta_{k'}f\Delta_kg + S_0fS_0g. \end{aligned}$$

From this decomposition, we neglect some nondiagonal terms for simplicity. We deduce for $j \geq 0$,

$$\Delta_j(fg) = \Delta_jfS_{j-2}g + \Delta_jgS_{j-2}f + \Delta_j\Big(\sum_{k\geq j}\Delta_kf\Delta_kg\Big). \tag{3.108}$$

From the Littlewood–Paley characterization of $H^s(\mathbb{R}^3)$, we know that $f \in H^s(\mathbb{R}^3)$ if and only if

$$\|f\|_{H^s} = \|S_0f\|_2 + \Big(\sum_{j\geq 0}2^{2js}\|\Delta_jf\|_2^2\Big)^{1/2} < \infty. \tag{3.109}$$

We now evaluate $\|\Delta_j(fg)\|_2$ by considering each term on the RHS of (3.108). First, if $j \geq 0$ and $0 < s < 3/2$, then by Bernstein's inequality, we find

$$\begin{aligned} \|S_{j-2}f\Delta_jg\|_2 \leq \|S_{j-2}f\|_2\|\Delta_jg\|_\infty &\leq 2^{\frac{3j}{2}}\|S_{j-2}f\|_2\|\Delta_jg\|_2 \\ &\leq C2^{\frac{3j}{2}}2^{-2js}\|g\|_{H^s}\|f\|_{H^s}, \end{aligned} \tag{3.110}$$

and, if $j \geq 0$ and $s = 3/2$, the same argument shows, by noting that $\|S_{j-2}u\|_\infty \leq C(j+1)\|u\|_{H^s}$ for $j \geq 0$, that

$$\|S_{j-2}f\Delta_jg\|_2 \leq Cj\|f\|_{H^s}2^{-js}\|g\|_{H^s}. \tag{3.111}$$

Combining (3.110) and (3.111), we get

$$\|S_{j-2}f\Delta_jg\|_2 \leq \max(C2^{\frac{3j}{2}-js}, Cj)2^{-js}\|f\|_{H^s}\|g\|_{H^s}, \quad 0 < s \leq 3/2. \tag{3.112}$$

On the other hand, when $j \geq 0$ and $s \geq 0$, Young's inequality gives

$$\begin{aligned}\Big\|\Delta_j\Big(\sum_{k\geq j}\Delta_k f\Delta_k g\Big)\Big\|_2 &\leq C\cdot 2^{\frac{3j}{2}}\sum_{k\geq j}\|\Delta_k f\Delta_k g\|_1\\ &\leq 2^{\frac{3j}{2}}C\sum_{k\geq j}\|\Delta_k f\|_2\cdot\|\Delta_k g\|_2\\ &\leq 2^{\frac{3j}{2}}\cdot 2^{-2js}C\|f\|_{H^s}\|g\|_{H^s}.\end{aligned} \tag{3.113}$$

Finally, when $s > 3/2$, it is well known that H^s is a Banach algebra, so

$$\|\Delta_j(fg)\|_{H^s} \leq C\|f\|_{H^s}\|g\|_{H^s}. \tag{3.114}$$

Combining (3.109), (3.112)–(3.114), we obtain

$$\eta_j \leq \max(C, Cj2^{-js}, 2^{\frac{3j}{2}-2js}C).$$

Thus, in order that

$$\sum_{j=-\infty}^{\infty} 2^{-|j|}\eta_j < \infty,$$

one needs that $s > 1/2$ and, therefore, when $s > 1/2$, H^s is a well-suited Banach space.

Theorem 3.3.4. *Let $X = H^s(\mathbb{R}^3)$ with $s > 1/2$. Then, for any initial data $u_0 \in X$ with $\operatorname{div} u_0(x) = 0$ in the sense of distributions, there exist $T = T(\|u_0\|_{H^s}) > 0$ and a unique mild solution $u(t) \in C([0,T); H^s(\mathbb{R}^3))$ of the problem (3.90). Moreover, the fluctuation function $\omega(x,t)$ satisfies the following regularity property:*

$$\omega(x,t) \in \begin{cases} H^{2s-1/2-\varepsilon}, & 1/2 < s \leq 3/2,\\ H^{s+1-\varepsilon}, & s > 3/2\end{cases} \tag{3.115}$$

for all $\varepsilon > 0$.

Proof. By Theorem 3.3.1, it is enough to prove (3.115). By (3.101) and (3.102), we find

$$\begin{aligned}\|\Delta_j(u(t) - S(t)u_0)\|_2 &= 2^{-js}\sum_{j-1\leq k\leq j+1}\int_0^t \|\tilde{d}_k(t-\tau,\tau)\|_{H^s}d\tau\\ &\leq 2^{-js}\sum_{j-1\leq k\leq j+1} 2^{-|k|}\eta_k\cdot\Big(\sup_{0\leq t<T}\|u(t)\|_{H^s}\Big)^2\\ &\sim 2^{-j}2^{\frac{3j}{2}}\cdot 2^{-2js}\Big(\sup_{0\leq t<T}\|u(t)\|_{H^s}\Big)^2.\end{aligned}$$

Thus,

$$2^{(2s-\frac{1}{2})j}\|\Delta_j(u(t)-S(t)u_0)\|_2 \sim \Big(\sup_{0\le t\le T}\|u(t)\|_{H^s}\Big)^2, \quad j\ge 0.$$

This means that for any $\varepsilon > 0$,

$$2^{(2s-1/2-\varepsilon)j}\|\Delta_j(u(t)-S(t)u_0)\|_2 \sim 2^{-\varepsilon j}\Big(\sup_{0\le t\le T}\|u(t)\|_{H^s}\Big)^2$$

for $1/2 < s < 3/2$. Therefore, when $1/2 < s \le 3/2$, we have

$$\omega(x,t) = u(t) - S(t)u_0(x) \in H^{2s-\frac{1}{2}-\varepsilon}, \quad \forall \varepsilon > 0. \tag{3.116}$$

On the other hand, when $s > 3/2$, we see, by noting that

$$\|\Delta_j(fg)\|_2 \sim 2^{-js}\|\Delta_j(fg)\|_{H^s} \le 2^{-js}\|f\|_{H^s}\|g\|_{H^s},$$

and

$$\begin{aligned}\|\Delta_j(u(t)-S(t)u_0)\|_2 &\sim \sum_{j-1\le k\le j+1} C2^{-|k|}2^{-js}\|u(t)\|_{H^s}^2\\ &\sim 2^{-j(1+s)}\Big(\sup_{0\le t\le T}\|u(t)\|_{H^s}\Big)^2.\end{aligned}$$

This implies that for any $\varepsilon > 0$,

$$2^{j(1+s-\varepsilon)}\|\Delta_j(u(t)-S(t)u_0)\|_2 \le 2^{-\varepsilon j}\Big(\sup_{0\le t\le T}\|u(t)\|_{H^s}\Big)^2. \tag{3.117}$$

From (3.116) and (3.117), the regularity result (3.115) follows. □

(d) The Morrey–Campanato spaces $M_2^p(\mathbb{R}^3)$

The Morrey–Campanato space $M_2^p(\mathbb{R}^3)$ is characterized, in terms of the Littlewood–Paley decomposition, as follows: $f \in M_2^p$ if and only if $\|S_0 f\|_\infty \le C$ and

$$\Big(\int_{Q_m}\sum_{j\ge m}|\Delta_j f|^2 dx\Big)^{1/2} \le C2^{-3m(\frac{1}{2}-\frac{1}{p})}, \quad m \in \mathbb{Z}^+ \cup \{0\},$$

where $Q_m = Q_m(x, 2^{-m})$ is an arbitrary cube with sides of length 2^{-m}, centered at x, and x can run over the whole $\mathbb{R}^3$.

Using this characterization in conjunction with the paraproduct decomposition (3.108), it is quite easy to prove, similar to the case of Sobolev spaces, that $M_2^p(\mathbb{R}^3)$ is a well-suited Banach space when $p > 3$, and we have the following theorem (see [33, 35]).

Theorem 3.3.5. *Let $X = M_2^p(\mathbb{R}^3)$, $p > 3$. Then for any initial data $u_0 \in X$ with $\operatorname{div} u_0(x) = 0$ in the sense of distributions, there exist $T = T(\|u_0\|_X) > 0$ and a unique mild solution*

$u(t) \in C_*([0,T); M_2^p(\mathbb{R}^3))$ *of the problem* (3.90). *Moreover, the fluctuation function* $\omega(t,x)$ *satisfies the following regularity property:*

$$\begin{aligned} &\omega(t,x) \in C_*([0,T); M_2^{\frac{3p}{6-p}}(\mathbb{R}^3)), \quad 3 < p < 6, \\ &\omega(t,x) \in C_*([0,T); \mathrm{BMO}(\mathbb{R}^3)), \quad p = 6, \\ &\omega(t,x) \in C_*([0,T); C^{1-\frac{6}{p}}(\mathbb{R}^3)), \quad p > 6. \end{aligned}$$

Remark 3.3.5.

(i) In general, a translation invariant function space X is a well-suited Banach space for the study of the Navier–Stokes equations if and only if $\deg(X) > -1$.

(ii) When $\deg(X) = -1$, X corresponds to a critical (or limit) space for the study of Navier–Stokes equations.

3.4 The Navier–Stokes equations in critical spaces

In Section 3.3, we have introduced the well-suited Banach space X, by employing the Littlewood–Paley decomposition theory, and proved that the following Cauchy problem of the Navier–Stokes equations:

$$\begin{cases} u_t - \Delta u + (u \cdot \nabla) u + \nabla P = 0, \\ \operatorname{div} u = 0, \\ u(x,0) = u_0(x), \quad \operatorname{div} u_0(x) = 0 \end{cases} \tag{3.118}$$

determines a local flow on X, that is, $u(t) \in C([0,T); X)$ satisfies the following integral equation, which is equivalent to the problem (3.118):

$$u(t) = S(t)u_0(x) - \int_0^t \mathcal{P} S(t-\tau) \nabla \cdot (u \otimes u) d\tau, \tag{3.119}$$

where $S(t) = \mathcal{F}^{-1} e^{-|\xi|^2 t} \mathcal{F}$ is the classical heat semigroup and $\mathcal{P} : (L^r)^n \to E_r(\mathbb{R}^n)$ is the orthogonal Helmholtz projection. As we all know, a well-suited Banach space X is a space of differentiable functions satisfying that $\deg(X) > -1$ such as those defined in (3.87). Well-suited Banach spaces only provide a sufficient but not necessary condition for the local solvability of the problem (3.118) (or equivalently (3.119)). As discussed in Section 3.2, for those spaces X given in (3.87), when $\deg(X) = -1$, it can be shown that the problem (3.118) is still locally solvable in X. Furthermore, if $\|u_0\|_X \ll 1$, the problem (3.118) is globally solvable in X.

The purpose of this section is to look for a critical space X, which is as large as possible and in which the problem (3.118) or equivalently (3.119) generates a local strong flow.

To this end, as discussed in Section 3.2, we will use the method of space-time estimates and prove that

$$u(t) \in C([0,T);X) \cap Y, \quad T = T(u_0),$$

where Y is a Banach space to be determined later. Further, we prove that, when $\|u_0\|_X \ll 1$, the problem (3.118) or (3.119) determines a global strong flow $u(t) \in C([0,\infty);X) \cap Y$.

In fact, for some critical spaces such as $X = L^n(\mathbb{R}^n)$, $\dot{B}^{\frac{n}{p}-1}_{p,\infty}(\mathbb{R}^n)$, $M^n_q(\mathbb{R}^n)$, $L^{(n,q)}(\mathbb{R}^n)$, $n \le q \le \infty$, this fact has been proved by many authors using a method due to T. Kato [126] (see also the discussion in Section 3.2). This section mainly introduces the latest result of Koch and Tataru [153]. Precisely, utilizing the Carleson characterization of BMO, we define the so-called BMO^{-1} space and prove that the problem (3.118) or (3.119) is locally well-posed on $X = \mathrm{BMO}^{-1}$. If $\|u_0\|_{\mathrm{BMO}^{-1}} \ll 1$, then the problem (3.118) or (3.119) is globally well-posed. It should be pointed out that BMO^{-1} is larger than L^n and $\dot{B}^{\frac{n}{p}-1}_{p,\infty}(n \le p < \infty)$.

We first recall the definition of BMO. It is known that

$$\mathrm{BMO} = \left\{ f \in L^1_{\mathrm{loc}}(\mathbb{R}^n) \;\middle|\; \|f\|_{\mathrm{BMO}} = \sup_{Q \subset \mathbb{R}^n} \frac{1}{|Q|} \int_Q |f - f_Q| dx < \infty \right\}$$

is the homogeneous space associated with $\mathrm{bmo}(\mathbb{R}^n)$, where

$$\mathrm{bmo}(\mathbb{R}^n) = \left\{ f \in L^1_{\mathrm{loc}}(\mathbb{R}^n) \;\middle|\; \|f\|_{\mathrm{bmo}} = \sup_{|Q|=1} \int_Q |f| dx + \sup_{|Q| \le 1} \frac{1}{|Q|} \int_Q |f - f_Q| dx < \infty \right\}.$$

Here, $Q \subset \mathbb{R}^n$ denotes the cube in $\mathbb{R}^n$, $|Q| = \mathrm{meas}(Q)$, $f_Q = \frac{1}{|Q|}\int_Q f dx$. In the framework of Besov–Triebel spaces, we have

$$\mathrm{BMO} = \dot{F}^0_{\infty,2} = (\dot{F}^0_{1,2})^* = (\mathcal{H}_1)^*,$$

where $\mathcal{H}_1$ is the Hardy space. Since $\mathcal{H}_1 \subset L^1(\mathbb{R}^n)$, then

$$L^\infty(\mathbb{R}^n) \hookrightarrow \mathrm{BMO}(\mathbb{R}^n),$$

and this inclusion is strict. For example, $\log|x| \notin L^\infty$, but $\log|x| \in \mathrm{BMO}$.

Proposition 3.4.1 (BMO's Carleson characterization). *Let $\Phi(x) = (4\pi)^{-n/2} e^{-|x|^2/4}$. Then*

$$\mathrm{BMO} = \{ v \in \mathcal{S}'(\mathbb{R}^n) \mid \|v\|_{\mathrm{BMO}} < \infty \},$$

where $\Phi_t(x) = t^{-n}\Phi(\frac{x}{t})$ and

$$\|v\|_{\mathrm{BMO}} := \sup_{x \in \mathbb{R}^n, R > 0} \left(\frac{2}{|B(x,R)|} \int_{B(x,R)} \int_0^R t |\nabla \Phi_t * v|^2 dt dy \right)^{1/2}.$$

The proof can be seen in [246]. The above Carleson characterization uses the Gaussian kernel, which is essentially equivalent to the heat semigroup characterization. For any $v \in \mathrm{BMO} \subset S'(\mathbb{R}^n)$, the problem

$$\begin{cases} \omega_t - \Delta\omega = 0, \\ \omega(x,0) = v(x) \end{cases}$$

has a solution $\omega(x,t) = \Phi_{\sqrt{t}} * v$. Therefore, we have

$$\|v\|_{\mathrm{BMO}} = \sup_{x\in\mathbb{R}^n,\, 0<R<\infty} \left(\frac{1}{|B(x,R)|} \int_{B(x,R)} \int_0^{R^2} |\nabla\omega|^2 dtdy \right)^{1/2}. \tag{3.120}$$

In fact, let $t = \sqrt{\tau}$. Then by the Carleson characterization of BMO, we have

$$\begin{aligned} \|v\|_{\mathrm{BMO}} &= \sup_{x\in\mathbb{R}^n,\, R>0} \left(\frac{2}{|B(x,R)|} \int_{B(x,R)} \int_0^{R} t|\nabla\Phi_t * v|^2 dtdy \right)^{1/2} \\ &= \sup_{x\in\mathbb{R}^n,\, R>0} \left(\frac{1}{|B(x,R)|} \int_{B(x,R)} \int_0^{R^2} |\nabla\Phi_{\sqrt{t}} * v|^2 dtdy \right)^{1/2}. \end{aligned}$$

This gives (3.120). But unfortunately, the space BMO is not an inhomogeneous Banach space. Thus, we need to introduce BMO^{-1} based on the Carleson characterization of BMO.

Definition 3.4.1. BMO^{-1} is the completion of the set

$$\{f \in \mathcal{S}(\mathbb{R}^n) \mid \|f\|_{\mathrm{BMO}^{-1}} < \infty\},$$

under the norm $\|\cdot\|_{\mathrm{BMO}^{-1}}$, where

$$\|f\|_{\mathrm{BMO}^{-1}} = \sup_{x\in\mathbb{R}^n,\, R>0} \left(\frac{1}{|B(x,R)|} \int_{B(x,R)} \int_0^{R^2} |\Phi_{\sqrt{t}} * f|^2 dtdy \right)^{1/2}.$$

It is easy to see that BMO^{-1} is a Banach space and the following result holds true.

Proposition 3.4.2. *Let $m(x) \in C^\infty(\mathbb{R}^n \setminus \{0\})$ be a homogeneous function of order 0. Then*

$$\|m(D_x)f\|_{\mathrm{BMO}^{-1}} \le C\|f\|_{\mathrm{BMO}^{-1}}. \tag{3.121}$$

Proof. For any $f \in \mathrm{BMO}^{-1}$, let $v = \Phi_{\sqrt{t}} * f$. By Definition 3.4.1, it is sufficient to prove that for any $R > 0$,

$$\frac{1}{|B(x,R)|}\int_{B(x,R)}\int_0^{R^2}|Tv|^2 dtdy \le C\|f\|^2_{\mathrm{BMO}^{-1}}, \tag{3.122}$$

where $T = m(D_x)$ and $Tv = \Phi_{\sqrt{t}} * Tf$. By the rescaling and translation property, (3.122) is reduced to

$$\int_{B(0,1)}\int_0^1 |Tv|^2 dtdy \le C\|f\|^2_{\mathrm{BMO}^{-1}}. \tag{3.123}$$

We claim that

$$|v(x,t)| \le Ct^{-\frac{1}{2}}\|f\|_{\mathrm{BMO}^{-1}}. \tag{3.124}$$

By scaling and translation invariance, (3.124) is equivalent to the estimate

$$|v(0,1)| \le C\|f\|_{\mathrm{BMO}^{-1}}. \tag{3.125}$$

This follows directly from the fact that $\Phi(x) \in \mathcal{S}(\mathbb{R}^n) \subset (\mathrm{BMO}^{-1})^*$.

We now prove (3.123). For $0 < t \le 1$, we deduce that

$$\begin{aligned} Tv(t) &= TS(t)f = T(S(t) - S(1))f + TS(1)f \\ &= T(S(t) - S(1))f - \int_1^\infty T\Delta S(\tau) f d\tau \\ &= T(1 - S(1-t))v + \int_1^\infty T\Delta S\Big(\frac{\tau}{2}\Big)v\Big(\frac{\tau}{2}\Big)d\tau. \end{aligned} \tag{3.126}$$

Note that $T(1 - S(1-t))$ is a bounded operator on L^2 with its kernel K_t satisfying that

$$|K_t(x)| \le C|x|^{-n-2}, \quad 0 < t < 1. \tag{3.127}$$

On the other hand, we use (3.124) to get

$$\Big\|\int_1^\infty T\Delta S\Big(\frac{\tau}{2}\Big)v\Big(\frac{\tau}{2}\Big)d\tau\Big\|_\infty \le C\int_1^\infty \tau^{-1}\Big\|v\Big(\frac{\tau}{2}\Big)\Big\|_\infty d\tau \le C\|f\|_{\mathrm{BMO}^{-1}}, \tag{3.128}$$

where we used the fact that the kernel $\widetilde{K}_t(x)$ of the operator $T\Delta S(t)$ satisfies the estimate

$$\|K_t\|_{L^1(\mathbb{R}^n)} \le Ct^{-1}. \tag{3.129}$$

Combining (3.126), (3.127) and (3.128) gives (3.123), and we complete the proof of this proposition. □

Proposition 3.4.3. *A distribution* $u \in \mathrm{BMO}^{-1}(\mathbb{R}^n)$ *if and only if there exist* $f^j \in \mathrm{BMO}$, $j = 1, 2, \ldots, n$, *such that*

$$u = \sum_{j=1}^{n} \partial_j f^j = \operatorname{div} f, \tag{3.130}$$

where $f = (f^1, f^2, \ldots, f^n)$.

Proof. We first prove the sufficient condition. Let $f^j \in \mathrm{BMO}$, $j = 1, 2, \ldots, n$. Then $v^j = \Phi_{\sqrt{t}} * f^j$ satisfies that

$$\begin{aligned}
&\sup_{x\in\mathbb{R}^n,\,0<R<\infty} \frac{1}{|B(x,R)|} \int_{B(x,R)} \int_0^{R^2} \sum_{j=1}^{n} |\partial_j v^j|^2 dtdy \\
&\leq \sup_{x\in\mathbb{R}^n,\,0<R<\infty} \frac{1}{|B(x,R)|} \int_{B(x,R)} \int_0^{R^2} \sum_{j=1}^{n} |\nabla\Phi_{\sqrt{t}} * f^j|^2 dtdy \\
&\leq \sum_{j=1}^{n} \|f^j\|_{\mathrm{BMO}}^2 < \infty.
\end{aligned}$$

This implies

$$\sup_{x\in\mathbb{R}^n,\,0<R<\infty} \frac{1}{|B(x,R)|} \int_{B(x,R)} \int_0^{R^2} \sum_{j=1}^{n} |\Phi_{\sqrt{t}} * \partial_j f^j|^2 dtdy < \infty,$$

which means $u = \sum_{j=1}^n \partial_j f^j \in \mathrm{BMO}^{-1}$.

Next, we turn to show the necessary condition. Let $u \in \mathrm{BMO}^{-1}$ and $R_{kj} = \partial_k \partial_j \Delta^{-1}$, $u_{kj} = R_{kj} u$, where $\hat{R}_{kj} = \frac{\xi_k \xi_j}{|\xi|^2}$ is a homogeneous function of order 0. It follows that $u_{kj} \in \mathrm{BMO}^{-1}$. Let $f^i = \partial_i \Delta^{-1} u$. Then f^i satisfies that $\partial_j f^i = u_{ij}$, which implies $f^i \in \mathrm{BMO}$ and $\partial_k u_{ij} = \partial_i u_{kj}$. From the definition of f^i, it follows that $u = \sum_{i=1}^n \partial_i f^i = \operatorname{div} f$. □

Remark 3.4.1.

(i) We may introduce the restriction BMO_R of BMO:

$$\mathrm{BMO}_R = \{f(x) \in \mathcal{S}'(\mathbb{R}^n) \mid \|f\|_{\mathrm{BMO}_R} < \infty\},$$

where

$$\|f\|_{\mathrm{BMO}_R} = \sup_{B_\rho \subset \mathbb{R}^n,\, \rho \leq R} \left(\frac{1}{|B(x,\rho)|} \int_{B(x,\rho)} \int_0^{\rho^2} |\nabla\Phi_{\sqrt{t}} * f|^2 dtdy \right)^{1/2}.$$

(ii) We use BMO_R to define VMO space (the BMO space satisfying vanishing condition):

$$\mathrm{VMO} = \Big\{ f \in \mathrm{BMO} \mid \lim_{R\to 0} \|f\|_{\mathrm{BMO}_R} = 0 \text{ or } \lim_{R\to\infty} \|f\|_{\mathrm{BMO}_R} = 0 \Big\}.$$

(iii) We may introduce the spaces $\overline{\mathrm{VMO}}$ and $\overline{\mathrm{VMO}}^{-1}$:

$$\overline{\mathrm{VMO}} = \Big\{ f \in \mathrm{BMO}_1 \mid \lim_{\rho\to 0} \|f\|_{\mathrm{BMO}_\rho} = 0 \Big\},$$
$$\overline{\mathrm{VMO}}^{-1} = \Big\{ f \in \mathrm{BMO}_1^{-1} \mid \lim_{\rho\to 0} \|f\|_{\mathrm{BMO}_\rho^{-1}} = 0 \Big\}.$$

(iv) If $C_0(\mathbb{R}^n)$ represents the set $\{f \mid f \in C(\mathbb{R}^n),\ \lim_{|x|\to\infty} f(x) = 0\}$, then VMO is the completion of $C_0(\mathbb{R}^n)$ under the BMO norm.

(v) Let $2 \le n \le p < \infty$, $q > 2$. Then we have the following embedding:

$$\dot{H}^{\frac{n}{2}-1}(\mathbb{R}^n) \hookrightarrow L^n(\mathbb{R}^n) \hookrightarrow \dot{B}_{p,\infty}^{-1+\frac{n}{p}}(\mathbb{R}^n) \hookrightarrow \mathrm{BMO}^{-1}(\mathbb{R}^n) \hookrightarrow \dot{F}_{\infty,q}^{-1}(\mathbb{R}^n) \hookrightarrow \dot{B}_{\infty,\infty}^{-1}(\mathbb{R}^n).$$

It is an interesting question to study the Navier–Stokes equations on the larger space $\dot{F}_{\infty,q}^{-1}(\mathbb{R}^n)$ or $\dot{B}_{\infty,\infty}^{-1}(\mathbb{R}^n)$.

Before stating the main theorem of this section, we see that the BMO^{-1} space is larger than the known critical spaces $\dot{B}_{p,q}^{\frac{n}{p}-1}$ $(1 < q \le \infty)$, L^n, $\dot{H}^{\frac{n}{p}-1,p}$ and M_q^n and that $\deg(\mathrm{BMO}^{-1}) = -1$.

When $p > n$, $s > 0$, $B_{p,\infty}^{-s}$ has the following Gaussian kernel characterization:

$$B_{p,\infty}^{-s} = \Big\{ u \in \mathcal{S}'(\mathbb{R}^n) \mid v = \Phi_{\sqrt{t}} * u,\ \|u\|_{B_{p,\infty}^{-s}} = \sup_{0<t\le 1} t^{\frac{s}{2}} \|v(t)\|_p < \infty \Big\}. \tag{3.131}$$

In particular,

$$\|u\|_{B_{p,\infty}^{\frac{n}{p}-1}} = \sup_{0<t\le 1} \sqrt{t}^{1-\frac{n}{p}} \|v(t)\|_p, \quad p > n. \tag{3.132}$$

Note that for $R \le 1$, we have

$$\begin{aligned} \left(\frac{1}{|B(x,R)|} \int_{B(x,R)} \int_0^{R^2} |v|^2 dt dy \right)^{1/2} &\le |B(x,R)|^{-\frac{1}{p}} \left(\int_0^{R^2} \|v\|_p^2 dt \right)^{1/2} \\ &\le \sqrt{\frac{p}{n}} |B(x,1)|^{-\frac{1}{p}} \sup_{0<t\le R^2} t^{\frac{1-\frac{n}{p}}{2}} \|v(t)\|_p. \end{aligned} \tag{3.133}$$

On the other hand, from the $L^p \to L^n$ estimate for the heat equation, it follows that

$$C t^{\frac{1-\frac{n}{p}}{2}} \|v(t)\|_p \le C \|u\|_n. \tag{3.134}$$

From (3.131)–(3.134), we get

$$L^n(\mathbb{R}^n) \hookrightarrow B^{\frac{n}{p}-1}_{p,\infty}(\mathbb{R}^n) \hookrightarrow \mathrm{BMO}^{-1}, \quad n \le p < \infty. \tag{3.135}$$

For the Morrey–Campanato space, we have

$$M^n_q = \left\{u \in L^1_{\mathrm{loc}}(\mathbb{R}^n) \;\Big|\; \sup_{\substack{x\in\mathbb{R}^n\\ 0<R\le 1}} R\left(\frac{1}{|B(x,R)|}\int_{B(x,R)} |v|^q dy\right)^{\frac{1}{q}} \sim \|u\|_{M^n_q} < \infty\right\},$$

which implies that

$$M^n_q \hookrightarrow \mathrm{BMO}^{-1}, \quad n \ge q > 1. \tag{3.136}$$

(See [153, 255] for details.) From (3.135) and (3.136), it is seen that BMO^{-1} is indeed larger than $\dot{B}^{\frac{n}{p}-1}_{p,q}$ $(n \le p < \infty)$, L^n, $\dot{H}^{\frac{n}{p}-1,p}$ and M^n_q.

Finally, since

$$\begin{aligned}
\|\varphi(\lambda x)\|_{\mathrm{BMO}^{-1}} &= \sup_{x\in\mathbb{R}^n, R>0}\left(\frac{1}{|B(x,R)|}\int_{B(x,R)}\int_0^{R^2} |\omega(\lambda y, \lambda^2 t)|^2 dt dy\right)^{\frac{1}{2}} \\
&= \sup_{x\in\mathbb{R}^n, R>0}\left(\frac{\lambda^{-2}}{|B(x,R)|}\int_{B(x,R)}\int_0^{(\lambda R)^2} |\omega(\lambda y, \tau)|^2 d\tau dy\right)^{\frac{1}{2}} \\
&= \sup_{x\in\mathbb{R}^n, R>0}\left(\frac{\lambda^{-n-2}}{|B(x,R)|}\int_{B(\lambda x,\lambda R)}\int_0^{(\lambda R)^2} |\omega(z, \tau)|^2 d\tau dz\right)^{\frac{1}{2}} \\
&= \sup_{x\in\mathbb{R}^n, R>0}\left(\frac{\lambda^{-2}}{|B(\lambda x,\lambda R)|}\int_{B(\lambda x,\lambda R)}\int_0^{(\lambda R)^2} |\omega(z, \tau)|^2 d\tau dz\right)^{\frac{1}{2}} \\
&= \lambda^{-1}\|\varphi(x)\|_{\mathrm{BMO}^{-1}},
\end{aligned}$$

where $\omega = G_{\sqrt{t}} * \varphi$, we find $\deg(\mathrm{BMO}^{-1}) = -1$. This means that under the scaling transform

$$u_\lambda(x,t) = \lambda u(\lambda x, \lambda^2 t), \quad P_\lambda = \lambda^2 P(\lambda x, \lambda^2 t),$$

we have

$$F(\lambda) = \|u_\lambda\|_{\mathrm{BMO}^{-1}} = \|u\|_{\mathrm{BMO}^{-1}} = \text{Constant}.$$

This will be an instructive role in the construction of intermediate spaces later.

We now determine the solution space and intermediate space by analogy. From Section 3.2, we know that when $u_0 \in L^n$ or $u_0 \in \dot{B}^{\frac{n}{p}-1}_{p,\infty}$, $p > n$, the solution space is

$$X = \{u(t) \in C([0,\infty); L^n(\mathbb{R}^n)) \cap \dot{C}_{q(p,n)}([0,\infty); L^p(\mathbb{R}^n)),\ \operatorname{div} u = 0\}$$

or

$$X = \{u(t) \in C_*((0,\infty); \dot{B}^{\frac{n}{p}-1}_{p,\infty}) \mid t^{\frac{1}{q}} u(t) \in C_*([0,\infty); L^p),\ \operatorname{div} u = 0\},$$

where (q,p,n) is any generalized admissible triple with $n < p \le 2n$, and the auxiliary space satisfies that

$$\|u_\lambda(t,x); \dot{C}_{q(p,n)}(\mathbb{R}^+; L^p)\| = \|u(t,x); \dot{C}_{q(p,n)}(\mathbb{R}^+; L^p)\|$$

or

$$\|t^{\frac{1}{q}} u_\lambda; C_*(\mathbb{R}^+; L^p)\| = \|t^{\frac{1}{q}} u; C_*(\mathbb{R}^+; L^p)\|.$$

Thus, based on the scaling principle, we choose an auxiliary space associated with BMO^{-1}. In essence, BMO^{-1} corresponds to the case $p = \infty$ and the special generalized admissible triplet $(2,\infty,n)$. Motivated by this and the construction of BMO^{-1}, we need to ensure that when $u_0(x) \in \mathrm{BMO}^{-1}$, then $S(t)u_0(x) \in X$, where the solution space X is chosen as

$$X = \left\{ u \in L^2_{\mathrm{loc}}(\mathbb{R}^n \times [0,\infty)) \;\middle|\; \|u\|_X = \sup_t t^{\frac{1}{2}} \|u(t)\|_\infty \right.$$
$$\left. + \left(\sup_{x\in\mathbb{R}^n, R>0} \frac{1}{|B(x,R)|} \int_{B(x,R)} \int_0^{R^2} |u|^2 dt dy \right) < \infty,\ \operatorname{div} u = 0 \right\}.$$

The corresponding intermediate space is

$$Y = \left\{ f(x,t) \in L^1_{\mathrm{loc}}(\mathbb{R}^n \times [0,\infty)) \;\middle|\; \|f\|_Y = \sup_{t>0} t \|f\|_\infty \right.$$
$$\left. + \sup_{x\in\mathbb{R}^n, R>0} \frac{1}{|B(x,R)|} \int_{B(x,R)} \int_0^{R^2} |f| dt dy < \infty \right\}.$$

The choice of Y is natural, which ensures that the nonlinear mapping $N(u) = u \otimes u$ maps from X into Y.

We are now in a position to state the main theorem in this section.

Theorem 3.4.1. *Let* $u_0 \in \mathrm{BMO}^{-1}$, $\operatorname{div} u_0 = 0$ *in the sense of distributions, and* $\|u_0\|_{\mathrm{BMO}^{-1}} \ll 1$. *Then the problem* (3.118) *or* (3.119) *has a unique global solution* $u(t) \in X$.

Similar to the definition of BMO_R^{-1}, we may define X_R as follows:

$$X_R = \left\{ u \in L^1_{\mathrm{loc}}(\mathbb{R}^n \times [0,\infty)) \,\middle|\, \|u\|_{X_R} = \sup_{t>0} t^{\frac{1}{2}} \|u(t)\|_\infty \right.$$
$$\left. + \left(\sup_{x\in\mathbb{R}^n,\, 0<\rho\le R} \frac{1}{|B(x,\rho)|} \int_{B(x,\rho)} \int_0^{\rho^2} |u|^2 dt dy \right)^{\frac{1}{2}} < \infty \right\}.$$

Let $X(I)$ and $X_R(I)$ be X and X_R, respectively, with $[0,\infty)$ being replaced by $I = [0,T)$. Then, similar to Theorem 3.4.1, we have the following.

Proposition 3.4.4. *There exists $\varepsilon > 0$ such that for any $R > 0$, if*

$$u_0 \in \mathrm{BMO}^{-1}, \quad \operatorname{div} u_0 = 0 \quad \textit{in the distributional sense}$$

and $\|u_0\|_{\mathrm{BMO}^{-1}} < \varepsilon$, then the problem (3.118) *or* (3.119) *has a unique solution $u(t,x) \in X_R(0,R^2)$. In particular, for any $u_0 \in \overline{\mathrm{VMO}}^{-1}$ with $\operatorname{div} u_0 = 0$ in the distributional sense, the problem* (3.118) *or* (3.119) *has a unique local solution.*

To prove Theorem 3.4.1, we write $N(u) = u \otimes u$, so the integral equation (3.119) becomes

$$\begin{aligned} u(t) &= S(t)u_0(x) - \int_0^t S(t-\tau)\mathcal{P}\nabla \cdot (u \otimes u) d\tau \\ &= S(t)u_0(x) - \int_0^t S(t-\tau)\mathcal{P}\nabla \cdot N(u) d\tau \\ &= S(t)u_0(x) + G\mathcal{P}\nabla \cdot N(u). \end{aligned} \tag{3.137}$$

Noting that $\mathrm{BMO}^{-1} \hookrightarrow \dot{B}^{-1}_{\infty,\infty}$, we get

$$\begin{aligned} \|S(t)u_0(x)\|_X &= \sup_{t>0} t^{\frac{1}{2}} \|S(t)u_0(x)\|_\infty \\ &\quad + \left(\sup_{x\in\mathbb{R}^n,\, 0<R} \frac{1}{|B(x,R)|} \int_{B(x,R)} \int_0^{R^2} |S(t)u_0|^2 dt dy \right)^{1/2} \\ &\le \|u_0\|_{\dot{B}^{-1}_{\infty,\infty}} + \|u_0\|_{\mathrm{BMO}^{-1}} \le C\|u_0\|_{\mathrm{BMO}^{-1}}. \end{aligned} \tag{3.138}$$

Thus, the proof of Theorem 3.4.1 is reduced to prove the following lemmas.

Lemma 3.4.1. *N is a mapping from X into Y.*

Lemma 3.4.2. *$G\nabla\mathcal{P}$ is a mapping from Y into X.*

It is clear that

$$\|N(u)\|_Y \le 2\|u\|_X^2. \tag{3.139}$$

Thus, we only need to prove Lemma 3.4.2. First, noting the following estimates:

$$|\mathcal{P}\Phi(x)| \le C(1+|x|)^{-n}, \quad \Phi(x) = (4\pi)^{-\frac{n}{2}}\exp\left(-\frac{|x|^2}{4}\right), \tag{3.140}$$

$$|K_t(x)| \le C(\sqrt{t}+|x|)^{-n}, \quad K_t(x) = \mathcal{P}\Phi_{\sqrt{t}}(x), \tag{3.141}$$

$$|\mathcal{P}\nabla\Phi_{\sqrt{t}}(x)| \le C(\sqrt{t}+|x|)^{-n-1}, \tag{3.142}$$

$$|\mathcal{P}(\delta_0 - \Phi_{\sqrt{t}}(x))| \le Ct|x|^{-n-2}, \tag{3.143}$$

where we use the fact that the multiplier $m(\xi) = \delta_{ij} - \frac{\xi_i\xi_j}{|\xi|^2}$ of the projection $\mathcal{P}$ is a classical Calderón–Zygmund singular integral operator,

$$\int \nabla\Phi dx = 0, \quad \int (\delta_0 - \Phi_t)dx = 0, \quad \int x_i(\delta_0 - \Phi_t)dx = 0,$$

and $\mathcal{P}(\delta_0 - \Phi_{\sqrt{t}}(x))$ is the kernel of the operator $\mathcal{P}(1 - S(t))$.

Let $a(x) \in C_c^\infty(\mathbb{R}^n)$, $\operatorname{supp}(a) \subset B(0,1)$. If $\widetilde{K}_t(x)$ is the kernel of the pseudo-differential operator $S(-t)a(t^{\frac{1}{2}}D_k)$, then we have

$$\widetilde{K}_t(x) = \mathcal{F}^{-1}(e^{-t|\xi|^2}a(\sqrt{t}\xi)), \quad \hat{\varphi}(\sqrt{t}\xi) \triangleq e^{-t|\xi|^2}a(\sqrt{t}\xi).$$

In view of that $\varphi(x) \in \mathcal{S}(\mathbb{R}^n)$, we find

$$|\varphi(x)| \le (1+|x|)^{-N}, \quad \forall N \ge 0.$$

By the multiplier estimate and the scaling principle, we get

$$|\widetilde{K}_t(x)| = |\varphi_{\sqrt{t}}(x)| \le C_n t^{-\frac{n}{2}}\left(1 + \frac{|x|}{\sqrt{t}}\right)^{-N}, \quad \forall N \ge 1. \tag{3.144}$$

We now prove Lemma 3.4.2 in the following five steps.

Proof of Lemma 3.4.2. *Step* 1 (Scaling and localization). It is clear that Lemma 3.4.2 is equivalent to the estimates

$$|G\nabla\mathcal{P}f| \le Ct^{-\frac{1}{2}}\|f\|_Y, \tag{3.145}$$

$$\|G\nabla\mathcal{P}f\|^2_{L^2(B(x,R)\times[0,R^2])} \le C|B(x,R)|\|f\|_Y^2. \tag{3.146}$$

Since the estimates (3.145) and (3.146) are invariant under the translation transform and the scaling transform: $f(x,t) \to \tilde{f}_\lambda(x,t) = \lambda f(\lambda x, \lambda^2 t)$, then (3.145) and (3.146) are reduced to

$$|G\nabla\mathcal{P}f(0,1)| \leq C\|f\|_{Y_1}, \tag{3.147}$$

$$\|G\nabla\mathcal{P}f\|^2_{L^2(B(0,1)\times[0,1))} \leq C\|f\|^2_{Y_1}. \tag{3.148}$$

With the fact that

$$\begin{aligned}
\int_0^t \mathcal{P}\nabla\Phi_{\sqrt{t-s}} * f(\cdot,s)ds &= \int_0^t (\mathcal{P}\nabla\Phi)_{\sqrt{t-s}} * f(\cdot,s)\frac{ds}{\sqrt{t-s}} \\
&= \int_0^1 (\mathcal{P}\nabla\Phi)_{\sqrt{1-s}\sqrt{t}} * f(\cdot,st)\frac{ds}{\sqrt{1-s}} \cdot \sqrt{t} \\
&= \left[\int_0^1 (\mathcal{P}\nabla\Phi)_{\sqrt{1-s}} * f_{\sqrt{t}^{-1}}(\cdot,ts)\frac{ds}{\sqrt{1-s}}\right]_{\sqrt{t}} \cdot \sqrt{t} \\
&= \left[\int_0^1 \mathcal{P}\nabla\Phi_{\sqrt{1-s}} * \sqrt{t}f(\sqrt{t}\cdot,ts)ds\right]\left(\frac{x}{\sqrt{t}}\right)t^{-\frac{n}{2}}\sqrt{t}^{n-1} \cdot \sqrt{t} \\
&= \left[\int_0^1 \mathcal{P}\nabla\Phi_{\sqrt{1-s}} * \tilde{f}_{(\sqrt{t})}(\cdot,s)ds\right]\left(\frac{x}{\sqrt{t}}\right),
\end{aligned}$$

thus, we obtain

$$|G\nabla\mathcal{P}f| = \left|(G\nabla\mathcal{P}\tilde{f}_{(\sqrt{t})})\left(\frac{x}{\sqrt{t}},1\right)\right| = C\|\tilde{f}_{(\sqrt{t})}\|_{Y_1} \leq Ct^{-\frac{1}{2}}\|f\|_Y.$$

Similarly, we can prove the equivalence between (3.146) and (3.148).

Now, let $\chi(x,t)$ be the character function of $B(0,2)\times[0,1]$. For $f \in Y$, we divide f into the following two parts:

$$f = \chi f + (1-\chi)f \equiv f_1 + f_2.$$

Clearly, $f_1, f_2 \in Y$. Let $\widetilde{\mathbb{Z}}^n$ denote the integer point $x \in \mathbb{Z}^n$ satisfying $|x| > 1$. Then it is easy to see that all the unit balls of centered at these points can cover the set $\{y, |y| \leq 1\}^c$. Let $Q(0,m)$ be the cube of centered at 0 and side length m. Then we find by noting the fact that the kernel $K(x,t) = \mathcal{P}\nabla\Phi_{\sqrt{t}}(x)$ of the operator $G\nabla\mathcal{P}$ satisfies the estimate (3.142), that

$$\begin{aligned}
&\|G\nabla\mathcal{P}(1-\chi)f\|_{L^\infty(B(0,1)\times[0,1])} \\
&\quad\leq \left\|\int_0^1\int_{\mathbb{R}^n} \frac{1-\chi(s,y)}{(\sqrt{t-s}+|x-y|)^{n+1}}|f(y,s)|dyds\right\|_{L^\infty(B(0,1)\times[0,1])}
\end{aligned}$$

$$
\begin{aligned}
&\leq \int_0^1 \int_{|y|\geq 1} \frac{|f(y,s)|}{|y|^{n+1}} dyds \\
&\leq C \sum_{m=2}^{\infty} \sum_{x\in\{Q(0,m)\setminus Q(0,m-1)\}\cap\mathbb{Z}^n} \int_0^1 \int_{B(x,1)} m^{-n-1}|f(y,s)|dyds \\
&\leq C \sum_{m=2}^{\infty} \frac{1}{m^2} \sup_{x\in\mathbb{R}^n} \int_0^1 \int_{B(x,1)} |f(y,s)|dyds \\
&\leq C \sup_{x\in\mathbb{R}^n} \int_{Q(x,1)} |f|dtdy,
\end{aligned}
$$

where we used the fact that $|y| \geq 2$ and $|x| \leq 1$ in obtaining the second inequality. From this, (3.147) holds for f_2.

Step 2 (The pointwise estimate). We prove the estimate (3.147) for $f_1 = \chi f$ or for f with $\operatorname{supp}(f) \subset B(0,2) \times [0,1]$. To this end, we have

$$
\begin{aligned}
\int_0^1 \mathcal{P}\nabla\Phi_{\sqrt{s}} * f(\cdot,1-s)ds &= \int_0^1 (\mathcal{P}\nabla\Phi)_{\sqrt{s}} * f(\cdot,1-s)\frac{ds}{\sqrt{s}} \\
&= \int_0^{\frac{1}{2}} (\mathcal{P}\nabla\Phi)_{\sqrt{s}} * f(\cdot,1-s)\frac{ds}{\sqrt{s}} + \int_{\frac{1}{2}}^1 (\mathcal{P}\nabla\Phi)_{\sqrt{s}} * f(\cdot,1-s)\frac{ds}{\sqrt{s}} \\
&= \int_0^{\frac{1}{2}} \int_{\mathbb{R}^n} (\mathcal{P}\nabla\Phi)_{\sqrt{s}}(0-\cdot)f(\cdot,1-s)dy\frac{ds}{\sqrt{s}} \\
&\quad + \int_{\frac{1}{2}}^1 \int_{\mathbb{R}^n} (\mathcal{P}\nabla\Phi)_{\sqrt{s}}(0-\cdot)f(\cdot,1-s)dy\frac{ds}{\sqrt{s}} \triangleq I + II.
\end{aligned}
$$

We evaluate I and II as follows:

$$
\begin{aligned}
I &= C\int_0^{\frac{1}{2}} \int_{\mathbb{R}^n} (\mathcal{P}\nabla\Phi)_{\sqrt{s}}(0-\cdot)(1-s)^{-1}(1-s)f(\cdot,1-s)dy\frac{ds}{\sqrt{s}} \\
&\leq C\int_0^{\frac{1}{2}} \|\sqrt{s}^{-1}(\mathcal{P}\nabla\Phi)_{\sqrt{s}}(y)\|_1 ds \cdot \sup_s |(1-s)f(y,1-s)| \\
&\leq C\|f\|_{Y_1},
\end{aligned}
$$

$$II \le C\int_{\frac{1}{2}}^{1}\int_{|y|\le 2} \sqrt{s}^{-1}(\mathcal{P}\nabla\Phi)_{\sqrt{s}}(y)|f(\cdot,1-s)|dyds$$
$$\le C\int_{\frac{1}{2}}^{1}\int_{|y|\le 2} \frac{1}{(\sqrt{s}+|y|)^{n+1}}|f(y,1-s)|dyds$$
$$\le C\int_{0}^{\frac{1}{2}}\int_{|y|\le 2} |f(y,t)|dydt \le C\|f\|_{Y_1}.$$

Step 3 (Cutting off high frequencies). In fact, we can prove the stronger estimate (compared with (3.148)):

$$\int_0^1\int_{\mathbb{R}^n} |\nabla Gf|^2 dxdt \le \|f\|_Y\|f\|_{L^1(\mathbb{R}^n\times\mathbb{R}^+)}, \tag{3.149}$$

where we have dispensed with the projection operator $\mathcal{P}$ since $\mathcal{P}$ is a bounded operator in L^2 and commutes with ∇G. We further deduce that (3.149) remains true for f without the restriction on the support of f.

Let $a(\xi) \in C_c^\infty(\mathbb{R}^n)$ satisfy that

$$a(\xi) = \begin{cases} 1, & |\xi| \le 1, \\ 0, & |\xi| \ge 2, \end{cases}$$

and consider the multiplier operator

$$A_t = a(t^{\frac{1}{2}}D_x).$$

This operator cuts off the frequencies larger than $2t^{-\frac{1}{2}}$. Then, for $t \le 1$, we have

$$\|(1-A_t)g\|_{H^{-1}(\mathbb{R}^n)} \le Ct^{\frac{1}{2}}\|g\|_{L^2(\mathbb{R}^n)}$$

and

$$\begin{aligned}\|G\nabla(1-A_t)f\|^2_{L^2(\mathbb{R}^n\times(0,1))} &\le C\int_0^1 \|(1-A_t)f\|^2_{H^{-1}(\mathbb{R}^n)}dt \\ &\le C\|f\|_Y\|f\|_{L^1(\mathbb{R}^n\times(0,1))}.\end{aligned}$$

Step 4 (The key estimate: the estimate of the low frequency part). It remains to look at the L^2-estimate of A_tf. Let $\widetilde{K}_t(x)$ be the kernel of $S(-t)A_t$, which is well-defined since the range of A_t consists of functions whose Fourier transform is compactly supported.

Then, for all $N \geq 1$, we have estimate (3.144). In particular, $\|\widetilde{K}_t(x)\|_{L^1(\mathbb{R}^n)} < \infty$ uniformly in $t \in [0, \infty)$. Hence, we have

$$\|S(-t)A_t f\|_Y \leq C\|f\|_Y, \quad \|S(-t)A_t f\|_1 \leq C\|f\|_1.$$

Now, let $\omega(t) = S(-t)A_t f$. Then $v(t) = \nabla G A_t f$ can be expressed as

$$v(t) = \nabla S(t) \int_0^t \omega(\tau) d\tau. \tag{3.150}$$

To conclude, we need to prove

$$\|v\|^2_{L^2(\mathbb{R}^n \times (0,1))} \leq C\|\omega\|_Y \|\omega\|_{L^1(\mathbb{R}^n \times (0,1))}. \tag{3.151}$$

In fact, a direct calculation gives

$$\begin{aligned}
\|v\|^2_{L^2(\mathbb{R}^n \times (0,1))} &= \int_0^1 \left\| \nabla S(t) \int_0^t \omega(\tau) d\tau \right\|^2_{L^2(\mathbb{R}^n)} dt \\
&= -2 \int_0^1 \int_0^t \int_0^t \langle \Delta S(2t)\omega(\tau), \omega(\theta) \rangle_{L^2(\mathbb{R}^n)} d\theta d\tau dt \\
&= -2 \int_0^1 \int_\tau^1 \int_0^\tau \langle \Delta S(2t)\omega(\tau), \omega(\theta) \rangle_{L^2(\mathbb{R}^n)} d\theta dt d\tau \\
&= \int_0^1 \int_0^\tau \langle (S(2\tau) - S(2))\omega(\tau), \omega(\theta) \rangle_{L^2(\mathbb{R}^n)} d\theta d\tau \\
&= \int_0^1 \left\langle (S(2\tau) - S(2))\omega(\tau), \int_0^\tau \omega(\theta) d\theta \right\rangle_{L^2(\mathbb{R}^n)} d\tau \\
&\leq \int_0^1 \|\omega(\tau)\|_{L^1(\mathbb{R}^n)} \left\| (S(2\tau) - S(2)) \int_0^\tau \omega(\theta) d\theta \right\|_{L^\infty(\mathbb{R}^n)} d\tau,
\end{aligned}$$

where we have made use of (3.150) for the first, integrating by parts for the second, Fubini's theorem for the third and $\partial_t S(t) = \Delta S(t)$ for the fourth equality.

If we have

$$\left\| S(2t) \int_0^t \omega(\theta) d\theta \right\|_{L^\infty(\mathbb{R}^n)} \leq C\|\omega\|_Y, \quad \forall t > 0, \tag{3.152}$$

then we get the estimate (3.151).

Step 5. The estimate of (3.152). Note that

$$\frac{1}{|B(x,R)|}\left\|\int_0^{R^2}\omega(\theta)d\theta\right\|_{L^1(B(x,R))} \leq \frac{1}{|B(x,R)|}\int_{B(x,R)}\int_0^{R^2}|\omega(y,t)|dtdy$$
$$\leq \|\omega(x,t)\|_Y.$$

The operator $S(2t)$ has a kernel $\widetilde{K}_t(x)$, which satisfies

$$\widetilde{K}_t(x) = C_n t^{-\frac{n}{2}}\exp\left(-\frac{|x|^2}{8t}\right).$$

We compute

$$\begin{aligned}
\left\|S(2t)\int_0^t\omega(\theta)d\theta\right\|_{L^\infty(\mathbb{R}^n)} &= \left\|\int_0^t\Phi_{\sqrt{2t}}*\omega(\theta)d\theta\right\|_\infty \\
&\leq \left\|\int_0^t\int_{\mathbb{R}^n}(8\pi t)^{-\frac{n}{2}}e^{-\frac{|x-y|^2}{8t}}\omega(y,\theta)dyd\theta\right\|_\infty \\
&\leq \sum_{k\in\mathbb{Z}^n}\int_0^t\int_{y-x\in\sqrt{8t}(k+[0,1]^n)}(8\pi t)^{-\frac{n}{2}}e^{-\frac{|x-y|^2}{8t}}|\omega(y,\theta)|dyd\theta \\
&\leq \sum_{k\in\mathbb{Z}^n}\sup_{z\in(k+[0,1]^n)}e^{-\frac{|z|^2}{8}}(8\pi t)^{-\frac{n}{2}}\left(\int_0^t\int_{y-x\in\sqrt{8t}(k+[0,1]^n)}|\omega(y,\theta)|dyd\theta\right) \\
&\leq C\sup_B\frac{1}{|B|}\int_Q|\omega(y,\theta)|dyd\theta.
\end{aligned}$$

This implies (3.152) and completes the proof of the theorem. □

Remark 3.4.2. The existence of a global smooth solution to the classical incompressible Navier–Stokes equations (and Euler equations) is one of the most significant unsolved problems in mathematics and physics (one of the seven Millennium Prize Problems). So far, there are sufficient methods found to solve the problem, the famous mathematician Nirenberg thought that the method to solve this problem should be the harmonic analysis method. Meyer, Chemin and their school utilize the skill of micro localization analysis to study the fluid dynamics equations such as the incompressible Euler equations, incompressible Navier–Stokes equations and make a series of contributions. The famous Beale–Kato–Majda criterion and the geometry criterion of Fefferman–Constantin can be used to judge if the smooth solution generates singularity for the fluid dynamics equations. See [34, 38, 39] for the works about studying the well-posedness of incompressible fluid dynamics equations in the layer of frequency and the Beale–Kato–Majda criteria.

We refer to [40, 41, 43] for the compressible Navier–Stokes equations, and [42] for the well-posedness of the ideal MHD equations in the Triebel–Lizorkin spaces.

3.5 Notes

In this note, we are devoted to summary the recent history of the incompressible Navier–Stokes (NS) equations:

$$\begin{cases} u_t - \Delta u + u \cdot \nabla u + \nabla P = 0, \\ \operatorname{div} u = 0, \\ u|_{t=0} = u_0, \end{cases} \tag{NS}$$

where $u(t,x) : \mathbb{R} \times \mathbb{R}^n \to \mathbb{R}^n$ is the velocity and the scalar function P denotes the pressure.

For the Cauchy problem to the Navier–Stokes equations, the previous sections provided a detailed introduction to the results concerning mild solutions. Now we mainly introduce some results on weak solutions. For the convenience of the following statement, we first provide the definitions of several types of weak solutions mentioned in the following. Let $\Omega = \mathbb{R}^n$ or $\Omega = \mathbb{T}^n$.

- *Very weak solution*: Let $u_0 \in L^2(\Omega)$ be weakly divergence-free. A vector field $u \in L^2([0,T]\times\Omega)$ is a weak solution of (NS) with initial data u_0 if the following statements hold:
  (1) For a. e. $t \in [0,T]$, u is weakly divergence-free;
  (2) For all test functions $\varphi \in C_c^\infty([0,T) \times \Omega))$,
$$\int_\Omega u_0 \varphi(x,0) dx = -\int_0^T \int_\Omega u \cdot (\partial_t \varphi + \Delta\varphi + u \cdot \nabla\varphi) dx dt.$$
- *Leray–Hopf weak solution*: Let $u_0 \in L^2(\Omega)$. A Leray–Hopf solution on $\Omega \times (0,T)$ with initial datum u_0 is a divergence-free vector field
$$u \in L_t^\infty L_x^2 \cap L^2 \dot{H}_x^1,$$
  which:
  (i) belongs to $C_w([0,T]; L^2(\Omega))$ and attains the initial data $u(\cdot,0) = u_0$;
  (ii) solves (NS) in the sense of distributions on $\Omega \times (0,T)$ for some pressure $P \in L^1_{\mathrm{loc}}(\Omega \times (0,T))$ and
  (iii) satisfies the energy inequality
$$\|u(t)\|^2_{L^2(\Omega)} + 2\int_0^t \|\nabla u(s)\|^2_{L^2(\Omega)} ds \le \|u_0\|^2_{L^2(\Omega)}, \quad \forall 0 < t < T.$$

- *Suitable weak solutions*: Let Ω be an open set in $\mathbb{R}^n$. We say that a pair (u,P) is a suitable weak solution of (NS) on the set $\Omega \times (-T_1, T)$ if the following conditions hold:
  (a) $u \in L^\infty((-T_1,T);L^2(\Omega)) \cap u \in L^2((-T_1,T); W^{1,2}(\Omega)), P \in L^{\frac{3}{2}}(\Omega \times (-T_1,T))$;
  (b) u and P satisfy (NS) in the distribution sense;
  (c) For $t \in (-T_1, T)$ and for each smooth, compactly supported test function $\varphi \geq 0$,

$$\int_\Omega \varphi |u(x,t)|^2 \mathrm{d}x + 2\int_{-T_1}^{t}\int_\Omega \varphi |\nabla u(x,s)|^2 \mathrm{d}x\mathrm{d}s \leq \int_{-T_1}^{t}\int_\Omega |u(x,s)|^2(\varphi_s + \Delta\varphi) + (|u(x,s)|^2 + 2P)(u\cdot\nabla\varphi)\mathrm{d}x\mathrm{d}s.$$

Regularity (blow-up) criteria

The existence of global weak solutions of the initial value problem for 3D (NS) was proved in the pioneering work of Leray [163] and later it was extended to the initial-boundary value problem by Hopf [119]. So far, the uniqueness issue as well as the regularity issue of the Leray–Hopf weak solutions for 3D (NS) remains to be very challenging problems.

To figure out how far we are away from solving this remarkable problem, many researchers are dedicated to establishing the uniqueness and regularity criteria of Leray–Hopf weak solutions.

Theorem 3.5.1 ([158, 222, 234]). *Let u and u_1 be two Leray–Hopf weak solutions of the Cauchy problem for 3D* (NS). *Suppose that the velocity field u satisfies the so-called Ladyzhenskaya–Prodi–Serrin condition for some $T > 0$ such that*

$$u \in L^q((0,T); L^p(\mathbb{R}^3)), \quad \textit{with} \quad \frac{2}{q} + \frac{3}{p} = 1, \quad p > 3.$$

Then $u = u_1$ in $\mathbb{R}^3 \times (0,T)$. Moreover, u is a smooth function on $(0,T)$.

For the case $\|u\|_{L_t^q L_x^p} < \infty$ with $p, q < \infty$, u is “locally small” in this norm. For the endpoint case $\|u\|_{L_t^\infty L_x^3} < \infty$, the possible “concentration effect” was the main obstacle in the proof of regularity. To suppress the concentration, Escauriaza, Seregin and Sverák in [84] used a new method based on reduction of the regularity problem to a backward uniqueness problem, which is then solved by finding suitable Carleman-type inequalities. Their result is as follows.

Theorem 3.5.2 ([84]). *Suppose that u is a weak Leray–Hopf solution of the Cauchy problem* (NS) *in $\mathbb{R}^3 \times (0,T)$ and v satisfies the additional condition $u \in L^\infty((0,T); L^3(\mathbb{R}^3))$. Then*

$$u \in L^5(\mathbb{R}^3 \times (0,T)),$$

and hence it is smooth and unique on $\mathbb{R}^3 \times (0,T)$.

One can see [94, 131] for other proofs of Theorem 3.5.2.

Theorem 3.5.2 also implies the following qualitative blowup criterion: suppose that (u,P) is a classical solution to 3D (NS) whose maximal time of existence T_* is finite, then

$$\limsup_{t\to T_*} \|u(t)\|_{L^3_x(\mathbb{R}^3)} = +\infty.$$

In a recent novel paper [264], Tao proved a quantitative version of the above blowup criterion.

Theorem 3.5.3 ([264]). *Assume that* (u,P) *is a classical solution to 3D* (NS) *on* $[0,T_*)\times\mathbb{R}^3$, *which blows up at a finite time* $0 < T_* < \infty$. *Then*

$$\limsup_{t\to T_*^-} \frac{\|u(t)\|_{L^3_x(\mathbb{R}^3)}}{(\log\log\log \frac{1}{T_*-t})^c} = +\infty$$

for an absolute constant $c > 0$.

One can see [10, 11, 213] and references therein for recent improvement in this direction.

Partial regularity of weak solutions

In a series of recent papers [229, 230], Scheffer studied the partial regularity theory of the 3D Navier–Stokes system. For a partial regularity theorem, we mean an estimate for the dimension of the set S of possible singular points. Scheffer's principal result is the following.

Theorem 3.5.4. *Let* $\Omega \subset \mathbb{R}^3$. *There exists a Leray–Hopf weak solution of the initial boundary value problem for* (NS) *on* Ω, *whose singular set* S *satisfies*

$$\mathcal{H}^{5/3}(S) < \infty,$$
$$\mathcal{H}^1(S \cap (\Omega \times t\})) < \infty \quad \textit{uniformly in} \quad t,$$

where $\mathcal{H}^k$ *denotes Hausdorff* k*-dimensional measure.*

Caffarelli, Kohn and Nirenberg in [30] improved Scheffer's theorem. Their result is a local partial regularity theorem for a particular class of weak solutions, called suitable weak solutions. They show that, for any such weak solution, the singular set has one-dimensional Hausdorff measure zero.

Theorem 3.5.5 ([30]). *For any suitable weak solution of 3D* (NS) *on a space-time open set* Ω, *there exists a positive constant* ϵ^* *such that, if for some* $(t_0,x_0) \in \Omega$, *we have*

$$\limsup_{r\to 0} \frac{1}{r} \iint_{(t_0-r^2,t_0+r^2)\times B(x_0,r)} |\nabla u|^2 dxds < \epsilon^*,$$

then u is Hölder continuous in a neighborhood of (t_0, x_0). *Moreover, the associated singular set satisfies*

$$P^1(S) = 0.$$

The measure P^1 *on* $\mathbb{R}^3 \times \mathbb{R}$ *is analogous to one-dimensional Hausdorff measure, but defined using parabolic cylinders instead of Euclidean balls.*

Lin in [172] proposed a simplified proof of the Caffarelli–Kohn–Nirenberg theorem. For other proofs, one can see [157, 162, 277].

Classical regularity criteria

The existence result of Leray–Hopf weak solutions has been shown for nearly 100 years; however, the uniqueness and regularity of Leray–Hopf weak solutions for the 3D Navier–Stokes system remain open. To figure out how far we are from dealing with this challenging problem, many researchers are dedicated to establishing the uniqueness and regularity criteria of Leray–Hopf weak solutions.

Theorem 3.5.6 ([158, 222, 234]). *Let* u *and* u_1 *be two weak Leray–Hopf solutions of the Cauchy problem* (NS). *Suppose that the velocity field u satisfies the so-called Ladyzhenskaya–Prodi–Serrin condition for some* $T > 0$ *such that*

$$u \in L^q((0,T); L^p(\mathbb{R}^3)), \quad \text{with} \quad \frac{2}{q} + \frac{3}{p} \le 1, \quad p > 3.$$

Then $u = u_1$ *in* $\mathbb{R}^3 \times (0,T)$. *Moreover, v is a smooth function on* $[0,T]$.

Nonuniqueness of a weak solution

In the seminal work [163], Leray demonstrated the existence of global weak solutions to the Navier–Stokes equations in three dimensions. Recently, there are some noticeable advancements in the study of nonuniqueness issue of weak solutions for 3D Navier–Stokes equations. Albritton, Brué and Colombo [2] exhibited two distinct Leray solutions with zero initial velocity and identical body force.

Theorem 3.5.7 ([2]). *There exist* $T > 0$, $f \in L_t^1 L_x^2$ *and two distinct suitable Leray–Hopf solutions* u, $\bar{u}$ *to the forced Navier–Stokes equations on* $\mathbb{R}^3 \times (0,T)$

$$u_t - \Delta u + u \cdot \nabla u + \nabla P = f, \quad \operatorname{div} u = 0, \quad u|_{t=0} = u_0$$

with the same body force f and initial condition $u_0 \equiv 0$.

For periodic domain $\mathbb{T}^3$, Cheskidov and Luo [44] showed the sharpness of the Ladyzhenskaya–Prodi–Serrin criteria $\frac{2}{p} + \frac{3}{q} \leq 1$ at the endpoint $(p, q) = (2, \infty)$.

Theorem 3.5.8 ([44]). *Let* $1 \leq p < 2$.

(1) *A very weak solution* $u \in L^p((0,T); L^\infty(\mathbb{T}^3))$ *of* (NS) *is not unique in the class* $L^p((0,T); L^\infty(\mathbb{T}^3))$ *if* u *has at least one interval of regularity.*
(2) *There exist non-Leray–Hopf weak solutions* $u \in L^p((0,T); L^\infty(\mathbb{T}^3))$.

For initial datum of finite kinetic energy, Leray [163] has proven in 1934 that there exists at least one global in time finite energy weak solution of the 3D Navier–Stokes equations. In the periodic setting $\mathbb{T}^3$, Buckmaster and Vicol in [28] proved that weak solutions to the 3D Navier–Stokes equations are not unique in the class of weak solutions with finite kinetic energy.

Definition 3.5.1 (Weak solution). We say $u \in C^0(\mathbb{R}; L^2(\Omega))$ is a weak solution of 3D (NS) if for any $t \in \mathbb{R}$ the vector field $u(\cdot, t)$ is weakly divergence-free, has zero mean and satisfies (NS) in $D'(\mathbb{T}^3 \times \mathbb{R})$, that is,

$$\int_0^T \int_\Omega u \cdot (\partial_t \varphi + \Delta\varphi + u \cdot \nabla\varphi) \mathrm{d}x\mathrm{d}s = 0,$$

holds for any test function $\varphi \in C_0^\infty(\mathbb{T}^3 \times \mathbb{R})$ such that $\varphi(\cdot, t)$ is divergence-free for all t.

Theorem 3.5.9 ([28]). *There exists* $\beta > 0$, *such that for any nonnegative smooth function* $e(t) : [0, T] \to \mathbb{R}_{\geq 0}$, *there exists* $v \in C_t^0([0,T]; H_x^\beta(\mathbb{T}^3))$ *a weak solution of the 3D Navier–Stokes equations, such that* $\int_{\mathbb{T}^3} |v(x,t)|^2 \mathrm{d}t = e(t)$ *for all* $t \in [0, T]$. *Moreover, the associated vorticity* $\nabla \times v$ *lies in* $C_t^0([0,T]; L^1(\mathbb{T}^3))$.

Subsequently, Buckmaster, Colombo and Vicol [27] proved that the above wild solutions can be generated by H^3 initial data.

Finally, we mention that some nonuniqueness results of the 3D viscous and resistive MHD equations have also been established. For instance, in the framework of $H_{t,x}^\epsilon$ with ϵ sufficiently small, Li, Zeng and Zhang [170] proved the nonuniqueness of a generalized MHD system; Miao and Ye [196] constructed non-unique distributional solutions in $C_t L_x^2$ for the 3D MHD system.

4 Schrödinger equations

The nonlinear Schrödinger equation is a typical dispersive equation, which arises from the study of modern physics such as quantum mechanics and has been studied by many mathematicians in recent years. However, many problems, in particular, the well-posedness and scattering theory of the Cauchy problem remain unsolved completely. The main purpose of this chapter is to introduce the latest results for the Schrödinger equation. First, beginning with the classical Strichartz estimates (originated from the Fourier restriction estimates), we briefly discuss the classical research methods and research progress for the Schrödinger equation over the last decades. Next, we consider the nonlinear Schrödinger equation with cubic nonlinearity $|u|^2u$ and discuss Bourgain's technique of Fourier high-low frequency method and I-team's I-method for dealing with low regularity problems such as well-posedness and scattering theory in H^s $(s < 1)$ and, in particular, the characterization of Bourgain-type spaces, the classical Morawetz-type estimates and the interaction Morawetz estimates. Finally, with the localized form of the Morawetz- type estimates, we establish global well-posedness and scattering theory of critical Schrödinger equations in the case with symmetric initial data $\varphi \in H^1(\mathbb{R}^n)$. It is hoped that the reader can appreciate the essential role played by harmonic analysis methods in the study of Schrödinger equations.

In general, the Cauchy problem of nonlinear Schrödinger equations can be expressed as

$$\begin{cases} iu_t = -\frac{1}{2}\Delta u + f(u), & (x,t) \in \mathbb{R}^n \times \mathbb{R}, \\ u(0) = \varphi(x), & x \in \mathbb{R}^n, \end{cases} \tag{4.1}$$

where u is a complex-valued function, Δ is the Laplace operator in $\mathbb{R}^n$ and the nonlinear term $f(u)$ satisfies the following assumptions:

(H1) $f \in C^1(\mathbb{C};\mathbb{C}), f(0) = 0$ and for some $1 < p < \infty$,

$$|f'(z)| \le \max\left(\left|\frac{\partial f}{\partial z}\right|, \left|\frac{\partial f}{\partial \bar{z}}\right|\right) \le C(1 + |z|^{p-1}) \tag{4.2}$$

for all $z \in \mathbb{C}$.

(H2) There exists a function $V \in C^1(\mathbb{C};\mathbb{R})$ with $V(0) = 0$ and $V(z) = V(|z|)$ such that

$$\frac{\partial V}{\partial \bar{z}} = f(z). \tag{4.3}$$

In other words, there exists a function $G \in C^1(\mathbb{R}^+;\mathbb{R})$ such that

$$f(z) = zG'(|z|^2), \tag{4.4}$$

that is, $V(z) = G(|z|^2)$.

https://doi.org/10.1515/9783111384726-004

Remark 4.0.1.

(i) (H1) is a polynomial growth condition and (H2) is the conformal invariance condition, that is,

$$f(e^{i\theta}z) = e^{i\theta}f(z). \tag{4.5}$$

Under the conformal invariance condition, the smooth solution of (4.1) satisfies the mass and energy conservation

$$M(u(t)) := \|u(t)\|_2^2 = M(\varphi), \tag{4.6}$$

$$E(u(t)) := \int_{\mathbb{R}^n} \Big[\frac{1}{2}|\nabla u|^2 + V(u(t))\Big] dx = E(\varphi), \tag{4.7}$$

which can be obtained directly from the integral estimates (see [184] for details).

(ii) (H1) and (H2) are satisfied by many functions such as

$$f(z) = a_1|z|^{q-1}z + a_2|z|^{p-1}z, \quad 1 \le q < p,\ 1 < p < \infty,$$
$$f(z) = (1 - e^{-\gamma|z|^2})z, \quad f(z) = z\sin(|z|^2) \text{ or } z\cos(|z|^2).$$

For the nonlinear Schrödinger equations with the nonlinear term f not satisfying (H1) and (H2), local existence results are still similar but the global existence results are only restricted to the case of small data (see [209, 210]).

(iii) To be precise but without loss of generality, we consider in this chapter the following special case of (4.1):

$$\begin{cases} iu_t = -\frac{1}{2}\Delta u + \mu|u|^{p-1}u, & (x,t) \in \mathbb{R}^n \times \mathbb{R}, \\ u(0) = \varphi(x), & x \in \mathbb{R}^n. \end{cases} \tag{4.8}$$

In this case, $E(u(t))$ is of the following form:

$$E(u(t)) = \int_{\mathbb{R}^n} \Big(\frac{1}{2}|\nabla u|^2 + \frac{2\mu}{p+1}|u|^{p+1}\Big) dx = E(\varphi). \tag{4.9}$$

(iv) When $\mu < 0$, the problem (4.8) corresponds to the case of focusing. In this case, if $p \ge 1 + 4/n$ then the smooth solution of (4.8) may blow up in a finite time [109] (see [19] for the latest results). When $\mu > 0$, (4.8) corresponds to the case of defocusing. In this case, we mainly discuss the well-posedness and scattering theory in H^s of (4.8).

Now, we recall the scaling analysis in Section 1.4 of Chapter 1. For the Schrödinger equation, $S(t) = e^{\pm\frac{1}{2}i\Delta t} = \mathcal{F}^{-1}e^{\mp\frac{1}{2}i|\xi|^2 t}\mathcal{F}$ generates a C_0-group only in the $C(I; H^s)$-type space.

When $0 \le s \le n/2$, we have to consider the Cauchy problem of nonlinear Schrödinger equations in a subspace of $C(I; H^s)$, for example,

$$\mathcal{X}(I) = C(I; H^s) \cap L^q(I; B^s_{r,2}(\mathbb{R}^n)), \quad s \ge 0. \tag{4.10}$$

It is natural that, since for $s > n/2$, H^s is a Banach algebra, by the classical well-posedness theory, one may study the Schrödinger equation in $C(I; H^s)$ directly. For (4.8), the transform $u(x,t) \quad \to \lambda^{-\frac{2}{p-1}} u(\lambda^{-1}x, \lambda^{-2}t)$ leaves the Schrödinger equation invariant. Thus, by scaling, it is known that in order for (4.8) to be solvable in H^s, we need

$$h - \deg(H^s) = s - \frac{n}{2} \ge -\frac{2}{p-1}, \tag{4.11}$$

that is, $s \ge n/2 - 2/(p-1)$. In particular, when $s = s_c = n/2 - 2/(p-1)$, H^{s_c} is a critical space for (4.8). Conversely, for any H^s ($s \ge 0$), in order to determine a continuous flow for (4.8) in H^s, it is required that $p \le 1 + 4/(n-2s)$. In particular, $p_c = 1 + 4/(n-2s)$ corresponds to the $\dot{H}^s$-critical exponent. Precisely, we have:

(I) For fixed nonlinear growth exponent p, the corresponding critical space is $\dot{H}^{s_c}(\mathbb{R}^n)$, where $s_c = n/2 - 2/(p-1)$ is the critical exponent. If $s > s_c$, $H^s(\mathbb{R}^n)$ is a subcritical space, and if $s < s_c$, then $H^s(\mathbb{R}^n)$ is a supercritical space.

(II) For fixed $s > 0$, the admissible nonlinear growth exponent of $H^s(\mathbb{R}^n)$ is $p \le p_c$. The number $p_c = 1 + 4/(n-2s)$ is called the $\dot{H}^s$-critical exponent, $p < p_c$ is called the H^s-subcritical exponent and $p > p_c$ is called the H^s-supercritical exponent. In particular, $p_c = 1 + 4/n$ is the L^2-critical exponent, and $p_c = 1 + 4/(n-2)$ is the $\dot{H}^1$-critical exponent.

We now define the admissible pair for the Schrödinger equation (cf. Section 1.4 of Chapter 1).

Definition 4.0.1. The pair (q, r) is called admissible pair for the Schrödinger equation if

$$\frac{2}{q} = n\left(\frac{1}{2} - \frac{1}{r}\right), \tag{4.12}$$

where

$$2 \le r \begin{cases} \le \infty, & n = 1, \\ < \infty, & n = 2, \\ \le \frac{2n}{n-2} \triangleq 2^*, & n \ge 3. \end{cases} \tag{4.13}$$

We usually write $(q, r) \in \Lambda$.

Clearly, $q = q(r, n)$ is uniquely determined by r and n and satisfies

$$\begin{cases} 2 \le q \le \infty, & n \ge 3, \\ 2 < q \le \infty, & n = 2, \\ 4 \le q \le \infty, & n = 1. \end{cases} \tag{4.14}$$

When $r = 2$, $q = \infty$, which corresponds to spaces like $L^\infty(I; H^s)$, $s \ge 0$. In general, (q, r) corresponds to the space-time integrable space $L^q(I; B^s_{r,2}(\mathbb{R}^n))$, $s \ge 0$.

4.1 Space-time estimates for linear Schrödinger equations

Clearly,

$$v(t) = S(t)\varphi - i\int_0^t S(t-\tau)g(x,\tau)d\tau \triangleq S(t)\varphi + Jg(t,x) \tag{4.15}$$

is a solution to the following Cauchy problem of the Schrödinger equation:

$$\begin{cases} iv_t = -\frac{1}{2}\Delta v + g(x,t), & (x,t) \in \mathbb{R}^n \times \mathbb{R} \\ v(0) = \varphi(x), & x \in \mathbb{R}^n, \end{cases} \tag{4.16}$$

where

$$S(t)\varphi = \exp\left(i\frac{t}{2}\Delta\right)\varphi = \mathcal{F}^{-1}e^{-\frac{i}{2}|\xi|^2 t}\mathcal{F}\varphi = M(t)D(t)\mathcal{F}M(t)\varphi \tag{4.17}$$

with

$$M(t) = \exp\left(i\frac{|x|^2}{2t}\right), \quad D(t)\varphi(x) = (it)^{-\frac{n}{2}}\varphi\left(\frac{x}{t}\right). \tag{4.18}$$

In fact, since

$$\int_{\mathbb{R}^n} e^{-i\frac{t}{2}|x|^2}dx = \left(\frac{2\pi}{it}\right)^{\frac{n}{2}},$$

we have

$$\begin{aligned} S(t)\varphi &= \frac{1}{(2\pi it)^{n/2}}\int_{\mathbb{R}^n} e^{\frac{i|x-y|^2}{2t}}\varphi(y)dy \\ &= \frac{1}{(2\pi it)^{n/2}}e^{\frac{i|x|^2}{2t}}\int_{\mathbb{R}^n} e^{-i\frac{x}{t}\cdot y}(e^{\frac{i|y|^2}{2t}}\varphi(y))dy \end{aligned}$$

$$= \frac{1}{(2\pi it)^{n/2}} M(2\pi)^{\frac{n}{2}} \mathcal{F}(M\varphi)\left(\frac{x}{t}\right)$$
$$= M\frac{1}{(it)^{n/2}}(\mathcal{F}M\varphi)\left(\frac{x}{t}\right) = MD(t)\mathcal{F}M\varphi.$$

Theorem 4.1.1 (Strichartz estimates). *Let $(q,r) \in \Lambda$, $(\gamma,\rho) \in \Lambda$, $I = \mathbb{R}$ or I is an interval with $0 \in I$. Then:*

(i) *For all $\varphi \in L^2(\mathbb{R}^n)$, $S(t)\varphi \in C(I;L^2) \cap L^q(I;L^r(\mathbb{R}^n))$, one has*

$$\|S(t)\varphi; L^q(I;L^r(\mathbb{R}^n))\| \le C\|\varphi\|_2 \tag{4.19}$$

for some constant C independent of I.

(ii) *Let $g(x,t) \in L^{\gamma'}(I;L^{\rho'}(\mathbb{R}^n))$. Then*

$$Jg \in L^q(I;L^r(\mathbb{R}^n)) \cap C(I;L^2(\mathbb{R}^n))$$

and

$$\|Jg; L^q(I;L^r(\mathbb{R}^n))\| \le C\|g; L^{\gamma'}(I;L^{\rho'}(\mathbb{R}^n))\|, \tag{4.20}$$

where C is a constant independent of I and

$$\frac{1}{\gamma'} + \frac{1}{\gamma} = \frac{1}{\rho'} + \frac{1}{\rho} = 1.$$

Remark 4.1.1.

(a) The estimates in Theorem 4.1.1 are called as the Strichartz estimates, the original form of which is

$$\|v\|_{L^q(\mathbb{R}^{n+1})} \le C\big[\|\varphi\|_{L^2_x} + \|g\|_{L^p(\mathbb{R}^{n+1})}\big], \tag{4.21}$$

where

$$q = \frac{2(n+2)}{n}, \quad p = q' = \frac{2(n+2)}{n+4}. \tag{4.22}$$

See [250] for details. The estimate (4.21) is essentially a direct result of the Fourier restriction estimate to a hyper-surface of nonzero curvature. Precisely, the symmetric Strichartz estimate (4.21) is exactly the symmetric form of the restriction estimate to the paraboloid of the Fourier transform. The restriction estimate to the paraboloid of the Fourier transform can be obtained through the restriction estimate to a truncated paraboloid of Fourier transforms and scaling principle.

(b) We now give a simple proof of the symmetric Strichartz estimate using the Tomas–Stein restriction estimate. The reader can appreciate the relationship between the restriction estimate of Fourier transforms and Strichartz estimates.

Theorem 4.1.2 (Tomas–Stein's restriction estimate). *Let S be a smooth hypersurface of nonzero Gaussian curvature in $\mathbb{R}^n$. Then the following L^p restriction estimate holds:*

$$\left(\int_{S_0} |\hat{f}(\xi)|^2 d\sigma(\xi)\right)^{1/2} \leq A_p(S_0)\|f\|_p, \quad \forall f \in L^p(\mathbb{R}^n), \tag{4.23}$$

for $1 \leq p \leq p_0 = \frac{2n+2}{n+3}$, where $S_0 \subset S$ is an open set and $\bar{S}_0$ is a compact subset of S.

For the proof of the Tomas–Stein theorem, see [184, 243, 248] or Chapter 5.
Now let

$$S = \{(\xi,\tau) : R(\xi,\tau) = \tau - |\xi|^2/2 = 0\} \tag{4.24}$$

be a paraboloid in $\mathbb{R}^{n+1}$. It is easy to see that the solution $S(t)f$ of the free Schrödinger equation can be expressed as

$$S(t)f = (2\pi)^{-\frac{n}{2}} \int_{\mathbb{R}^{n+1}} e^{i\bar{x}\cdot\bar{\xi}}\hat{f}(\bar{\xi})d\mu(\bar{\xi}) = \mathcal{F}^{-1}(\hat{f}d\mu), \tag{4.25}$$

where $d\mu(\bar{\xi}) = \delta(\tau - |\xi|^2/2)d\tau d\xi$, $\bar{x} = (x,t)$ and $\bar{\xi} = (\xi,\tau)$. By the Tomas–Stein restriction estimate, we obtain

$$\left(\int_{S\cap\{\frac{1}{2}\leq|\xi|\leq 2\}} |\hat{f}(\bar{\xi})|^2 d\mu(\bar{\xi})\right)^{1/2} \leq A_p\|f\|_p, \quad 1 \leq p \leq \frac{2n+4}{n+4}. \tag{4.26}$$

By scaling, it is seen that

$$\left(\int_{S\cap\{2^{k-1}\leq|\xi|\leq 2^{k+1}\}} |\hat{f}_k(\bar{\xi})|^2 d\mu(\bar{\xi})\right)^{1/2} \leq A_p\|f_k\|_p, \quad p = \frac{2n+4}{n+4}. \tag{4.27}$$

Using the Littlewood–Paley dyadic decomposition,

$$f = \sum_{k\in\mathbb{Z}} \Delta_k f \equiv \sum_{k\in\mathbb{Z}} f_k,$$

one can easily deduce that

$$\begin{aligned}\left(\int_S |\hat{f}(\xi)|^2 d\mu\right)^{1/2} &\leq \left(\sum_{k\in\mathbb{Z}} \int_{S\cap\{2^{k-1}\leq|\xi|\leq 2^{k+1}\}} |\hat{f}_k|^2 d\mu\right)^{1/2} \\ &\leq A_p\left(\sum_{k\in\mathbb{Z}} \|f_k\|_p^2\right)^{1/2} \leq A_p\left\|\left(\sum_{k\in\mathbb{Z}} |f_k|^2\right)^{1/2}\right\|_p \\ &= A_p\|f\|_{\dot{F}^0_{p,2}} \leq A_p\|f\|_p, \quad p = \frac{2(n+2)}{n+4}.\end{aligned} \tag{4.28}$$

Thus, by Tomas's symmetry principle (see [271] or [184]), one has

$$\|S(t)f\|_q = \|\mathcal{F}^{-1}(\hat{f}d\mu)\|_q \leq C\|f\|_2, \tag{4.29}$$

where

$$q = \frac{2n+4}{n} = 2 + \frac{4}{n}.$$

Taking $f(x) = \varphi(x)$, we obtain the estimate (4.21) in the case $g = 0$. The case with $g \neq 0$ is a direct result of (4.29) and the Hardy–Littlewood–Sobolev inequality.

(c) Let S be a smooth compact hypersurface in $\mathbb{R}^n$ with nonzero curvature. If

$$p < \frac{2n}{n+1}, \quad \frac{n-1}{n+1} \cdot \frac{1}{r} + \frac{1}{p} \geq 1, \tag{4.30}$$

then we have the following restriction conjecture

$$\|\hat{f}|_S\|_{L^r(d\sigma)} \leq C\|f\|_{L^p(\mathbb{R}^n)}, \tag{4.31}$$

where $d\sigma$ is the measure of the surface S.
It is easy to see that when $r = 2$, then $p \leq 2(n+1)/(n+3)$. So, the Stein conjecture (4.31) is exactly Theorem 4.1.2. On the other hand, if the Stein conjecture is true, then it is optimal. In fact, by interpolation, the Stein conjecture is essentially equivalent to

$$\|\hat{f}|_S\|_{L^1(d\sigma)} \leq C\|f\|_{L^p(\mathbb{R}^n)}, \quad p < \frac{2n}{n+1}. \tag{4.32}$$

If one takes S to be the unit sphere Σ^n in $\mathbb{R}^n$, then it is clear that

$$\hat{\sigma}(\xi) \sim \frac{\cos(|\xi|)}{|\xi|^{\frac{n-1}{2}}}, \quad \text{as} \quad |\xi| \to \infty. \tag{4.33}$$

Thus, in order that (4.32) holds, it is necessary that $p'\frac{n-1}{2} > n$ is true, that is, $p < \frac{2n}{n+1}$. This implies that the bound $p < \frac{2n}{n+1}$ is optimal.
The Stein conjecture (4.31) or (4.32) has been shown in the case $n = 2$ (see [19] for details). When $n \geq 3$, the Stein conjecture remains open. It should be pointed out that the Stein conjecture is closely related to the Kakeya needle problem. Interested readers are referred to [257] and [19]. So far, the best result on the Stein conjecture is

$$\|\hat{f}|_S\|_{L^1(d\sigma)} \leq C\|f\|_{L^p(\mathbb{R}^3)}, \quad p < \frac{26}{19}. \tag{4.34}$$

Applying (4.34) directly to the Schrödinger equation gives

$$\left\| \int_{|\xi|\le 1} e^{i(x\cdot\xi+\frac{1}{2}|\xi|^2 t)}\varphi(\xi)d\xi \right\|_{L^q(\mathbb{R}^3)} \le C\|\varphi\|_{L^\infty(\mathbb{R}^3)}, \quad q > \frac{26}{7}. \tag{4.35}$$

See [19] for a proof of this result.

Proof of Theorem 4.1.1. With the help of the energy estimate,

$$\|S(t)\varphi\|_{L^2} = \|\varphi\|_{L^2} \tag{4.36}$$

and the decay estimate

$$\|S(t)\varphi\|_{L^\infty(\mathbb{R}^n)} \le C|t|^{-\frac{n}{2}}\|\varphi\|_{L^1}, \tag{4.37}$$

we give the proof of Theorem 4.1.1 except the endpoint $(q,r) = (2, \frac{2n}{n-2})$ with $n \ge 3$. For the endpoint case, we shall use the abstract method of Keel and Tao; see Chapter 5. For the case $(q,r) = (4,\infty)$ with $n = 1$, see [135, 184].

By the Riesz–Thorin interpolation theorem, it follows from (4.36) and (4.37) that

$$\|S(t)\varphi\|_{L^r} \le C|t|^{-\delta(r)}\|\varphi\|_{L^{r'}}, \quad 2 \le r \le \infty, \tag{4.38}$$

where

$$\delta(r) = n\left(\frac{1}{2} - \frac{1}{r}\right), \quad r' = \frac{r}{r-1}. \tag{4.39}$$

Without loss of generality, we take $I = [0,T]$. For $\psi(x,t) \in \mathcal{S}(\mathbb{R}^{n+1})$, we have

$$\begin{aligned} \langle S(t)\varphi, \psi(x,t)\rangle &= \int_0^T \left(\mathcal{F}^{-1}\exp\left(-\frac{i}{2}|\xi|^2 t\right)\mathcal{F}\varphi, \psi(x,t)\right)dt \\ &= \int_0^T \left(\varphi, \mathcal{F}^{-1}\exp\left(\frac{i}{2}|\xi|^2 t\right)\mathcal{F}\psi\right)dt \\ &\le \|\varphi\|_{L^2}\left\| \int_0^T \mathcal{F}^{-1}\exp\left(\frac{i}{2}|\xi|^2 t\right)\mathcal{F}\psi dt\right\|_{L^2}, \end{aligned} \tag{4.40}$$

where $\langle\cdot,\cdot\rangle$ denotes the inner product in $L^2(\mathbb{R}^n \times [0,T])$ and $(\cdot,\cdot)$ denotes the inner product in $L^2(\mathbb{R}^n)$. Now for $0 < t \le T$, one has

$$\begin{aligned} &\left\| \int_0^t \mathcal{F}^{-1}\exp\left(\frac{i}{2}|\xi|^2\tau\right)\mathcal{F}\psi d\tau\right\|_{L^2}^2 \\ &= \left(\int_0^t \mathcal{F}^{-1}\exp\left(\frac{i}{2}|\xi|^2\tau\right)\mathcal{F}\psi d\tau, \int_0^t \mathcal{F}^{-1}\exp\left(\frac{i}{2}|\xi|^2 s\right)\mathcal{F}\psi ds\right) \end{aligned}$$

$$
\begin{aligned}
&= \int_0^t \left(\psi(x,\tau), \int_0^t \mathcal{F}^{-1} \exp\left(-\frac{i}{2}|\xi|^2(\tau - s)\right) \mathcal{F}\psi ds \right) d\tau \\
&\leq \|\psi(x,t)\|_{L^{q'}(I;L^{r'})} \left(\int_0^t \left\| \int_0^t \mathcal{F}^{-1} e^{\frac{-i(\tau-s)}{2}|\xi|^2} \mathcal{F}\psi ds \right\|_{L^r}^q d\tau \right)^{\frac{1}{q}} \\
&\leq C\|\psi(x,t)\|_{L^{q'}(I;L^{r'})} \left(\int_0^t \left(\int_0^t |\tau - s|^{-\delta(r)} \|\psi\|_{L^{r'}} ds \right)^q d\tau \right)^{\frac{1}{q}} \\
&\leq C\|\psi(x,t)\|_{L^{q'}(I;L^{r'})}^2,
\end{aligned}
$$

where we have used (4.38) and the Hardy–Littlewood–Sobolev inequality. Inserting the above inequality into (4.40) gives

$$
\langle S(t)\varphi, \psi(x,t) \rangle \leq C\|\varphi\|_{L^2} \|\psi(t,x)\|_{L^{q'}(I;L^{r'}(\mathbb{R}^n))}, \quad \forall \psi \in S(\mathbb{R}^{n+1}), \tag{4.41}
$$

which implies that

$$
\|S(t)\varphi\|_{L^q(I;L^r(\mathbb{R}^n))} \leq C\|\varphi\|_{L^2(\mathbb{R}^n)}, \tag{4.42}
$$

where C is a constant independent of I. The above proof is also valid for the case $I = \mathbb{R}$.

We now prove (4.20). We first consider three special cases, and the general case then follows from interpolation.

Case I. $(q,r) = (\gamma,\rho) \in \Lambda$. By the Hardy–Littlewood–Sobolev inequality, it follows that

$$
\|\mathcal{J}g\|_{L^q(I;L^r)} \leq C \left\| \int_0^t |t-\tau|^{-\frac{2}{q}} \|g\|_{r'} d\tau \right\|_{L^q} \leq C\|g\|_{L^{q'}(I;L^{r'})}. \tag{4.43}
$$

Case II. $(\gamma,\rho) = (\infty, 2)$ and $(q,r) \in \Lambda$. For any $\psi(x,t) \in S(\mathbb{R}^{n+1})$ and $0 < t \leq T$, we have

$$
\begin{aligned}
\langle \mathcal{J}g, \psi \rangle &= \int_0^T \left(\int_0^t \mathcal{F}^{-1} \exp\left(-\frac{i}{2}(t-s)|\xi|^2\right) \mathcal{F}g(s) ds, \psi \right) dt \\
&= \int_0^T \int_0^t \left(\mathcal{F}^{-1} \exp\left(-\frac{i}{2}(t-s)|\xi|^2\right) \mathcal{F}g(s), \psi \right) ds dt \\
&= \int_0^T \left(g, \int_s^T \mathcal{F}^{-1} \exp\left(\frac{i}{2}(t-s)|\xi|^2\right) \mathcal{F}\psi dt \right) ds
\end{aligned}
$$

$$\leq \int_0^T \|g\|_{L^2} \|J\psi\|_{L^2} ds \leq \|g\|_{L^1(I;L^2)} \|J\psi; L^\infty(I;L^2)\|$$
$$\leq C\|g\|_{L^1(I;L^2)} \|\psi\|_{L^{q'}(I;L^{r'})},$$

where we use (4.41). Thus,

$$\|Jg\|_{L^q(I;L^r(\mathbb{R}^n))} \leq C\|g\|_{L^1(I;L^2(\mathbb{R}^n))}. \tag{4.44}$$

Case III. $(q,r) = (\infty, 2)$ and $(\gamma, \rho) \in \Lambda$. From the proof of (4.41), it follows that

$$\|Jg\|_{L^\infty(I;L^2)} = \sup_{t\in I} \|Jg\|_{L^2} \leq C\|g\|_{L^{\gamma'}(I;L^{\rho'})}. \tag{4.45}$$

The estimates (4.43)–(4.45) imply that J is a bounded linear operator from $L^{\gamma'}(I;L^{\rho'})$ to $L^\gamma(I;L^\rho)$, from $L^1(I;L^2)$ to $L^q(I;L^r)$ and from $L^{\gamma'}(I;L^{\rho'})$ to $L^\infty(I;L^2)$. Thus, for $2 \leq r \leq \rho$ there exists a $\theta \in [0,1]$ such that

$$\frac{1}{r} = \frac{\theta}{\rho} + \frac{1-\theta}{2}, \quad \frac{1}{q} = \frac{\theta}{\gamma} + \frac{1-\theta}{\infty}.$$

So, by the Ginibre–Velo interpolation theorem [105], we get

$$\begin{aligned}\|Jg\|_{L^q(I;L^r)} &\leq C\|Jg; L^\gamma(I;L^\rho)\|^\theta \|Jg; L^\infty(I;L^2)\|^{1-\theta} \\ &\leq C\|g; L^{\gamma'}(I;L^{\rho'})\|^\theta \|g; L^{\gamma'}(I;L^{\rho'})\|^{1-\theta} \\ &\leq C\|g; L^{\gamma'}(I;L^{\rho'})\|.\end{aligned} \tag{4.46}$$

For $2 \leq \rho \leq r$, we have $r' \leq \rho' \leq 2$. So, by interpolation it follows from (4.43) and (4.44) that

$$\|Jg; L^q(I;L^r)\| \leq C\|g; L^{\gamma'}(I;L^{\rho'})\|. \tag{4.47}$$

Thus, combining (4.42), (4.46) and (4.47), we conclude the proof of Theorem 4.1.1. □

As an immediate result of interpolation, characterization of equivalent norms in Besov spaces and Theorem 4.1.1, we have the following.

Theorem 4.1.3. *Let $(q,r) \in \Lambda$, $(q_1, r_1) \in \Lambda$, $s \in \mathbb{R}$. Assume that $I = \mathbb{R}$ or I is an interval with $0 \in \bar{I}$. If $\varphi \in H^s(\mathbb{R}^n)$ and $g(x,t) \in L^{q_1'}(I; B^s_{r_1',2})$. Then the problem* (4.16) *has a solution $v(x,t) = S(t)\varphi + Jg(x,t)$ satisfying that*

$$v(x,t) \in C(I; H^s(\mathbb{R}^n)) \cap \bigcap_{(q,r)\in\Lambda} L^q(I; B^s_{r,2}),$$
$$\|S(t)\varphi\|_{L^q(I;B^s_{r,2})} \leq C\|\varphi\|_{H^s}, \quad (q,r) \in \Lambda, \tag{4.48}$$

$$\|\mathcal{J}g\|_{L^q(I;B^s_{r,2})} \le C\|g(x,t)\|_{L^{q_1'}(I;B^s_{r_1',2})}. \tag{4.49}$$

As we know, there is no global smoothness for the solution of the linear Schrödinger equation except the integrability properties characterized by the Strichartz estimates in Theorems 4.1.1 or 4.1.3. Precisely, if $\varphi \in H^s$, then $S(t)\varphi \in H^s$ for all $t \neq 0$. But

$$S(t)\varphi \notin H^{s+\varepsilon}, \quad \varepsilon > 0, \ \forall t \neq 0.$$

Nevertheless, the heat semigroup $e^{t\Delta}$ is of this kind of global smoothness. However, the solution of the linear Schrödinger equation (and general dispersive equations) satisfies the local smoothness effects (Kato smoothing effects).

Theorem 4.1.4.

(i) *Let $n = 1$. Then*

$$\sup_x \|D_x^{\frac{1}{2}} S(t)\varphi\|_{L^2(dt)} \le C\|\varphi\|_{L^2},$$

$$\sup_x \left(\int_{\mathbb{R}} \left| D_x \int_0^t S(t-\tau) g d\tau \right|^2 dt \right)^{1/2} \le C\|g; L^1_x(\mathbb{R}; L^2_t)\|,$$

where $D_x = (-\Delta)^{1/2}$.

(ii) *For $n \ge 2$, let $\{Q_\alpha\}$ be the cube of side-length R with*

$$\bigcup_{\alpha \in \mathbb{Z}^n} Q_\alpha = \mathbb{R}^n.$$

Then we have the local Kato smoothing effects

$$\sup_{\alpha \in \mathbb{Z}^n} \left(\int_{Q_\alpha} \left(\int_{-\infty}^{\infty} |D_x^{\frac{1}{2}} S(t)\varphi|^2 dt \right) dx \right)^{1/2} \le CR\|\varphi\|_2,$$

$$\sup_{\alpha \in \mathbb{Z}^n} \left(\int_{Q_\alpha} \int_{\mathbb{R}} \left| \nabla_x \int_0^t S(t-\tau) g(x,\tau) d\tau \right|^2 dt dx \right)^{1/2} \le CR \sum_{\alpha \in \mathbb{Z}^n} \|g\|_{L^2(Q_\alpha; L^2_t(\mathbb{R}))}.$$

Theorem 4.1.4 can be used to study the local well-posedness and small-data global well-posedness of the following Cauchy problem for the nonlinear Schrödinger equation:

$$\begin{cases} iu_t + \frac{1}{2}\Delta u + F(u, \bar{u}, \nabla_x u, \nabla_x \bar{u}) = 0, \\ u(0) = \varphi(x). \end{cases} \tag{4.50}$$

See [135, 184] for details. On the other hand, local Kato smoothness effects can also be regarded as reverse space-time estimates.

Another type of important estimates for solutions of the linear Schrödinger equation is the so-called maximal inequality, which arises from the well-known Carleson conjecture: What is the smallest $s = s_0$ such that

$$\lim_{t\to 0} S(t)\varphi = \lim_{t\to 0} e^{\frac{i}{2}t\Delta}\varphi \overset{\text{a.e.}}{=} \varphi, \quad \forall \varphi \in H^s \tag{4.51}$$

holds?

When $n = 1$, Carleson [36] proved in 1979 that (4.51) holds if $s \geq 1/4$. Then in 1982, Dahlberg and Kenig [55] proved that $s_0 = 1/4$ is optimal.

When $n \geq 2$, Dahlberg and Kenig [55] utilized Carleson's method to prove that (4.51) holds if $s \geq n/4$. Making use of a completely new method, Sjölin proved in 1987 [239] that, if $s > 1/2$, then

$$\lim_{t\to 0} S(t)\varphi \overset{\text{a.e.}}{=} \varphi, \quad \varphi \in H^s(\mathbb{R}^n),\ n \geq 2, \tag{4.52}$$

which means that $s_0 = n/4$ is not optimal. It is easy to see that the pointwise convergence (4.51) follows by the maximal operators estimate

$$\Big\|\sup_{|t|<1}\big|e^{-it\Delta}f(x)\big|\Big\|_{L^2(B(0,1))} \leq C\|f(x)\|_{H^s(\mathbb{R}^n)}, \tag{4.53}$$

where $B(0,1) \subset \mathbb{R}^n$ is the unit ball centered at zero. The first breakthrough for $s < \frac{1}{2}$ was achieved by Bourgain [17, 18], where he proved that there exists $s < \frac{1}{2}$ such that (4.53) holds true. Thereafter, Moyua–Vargas–Vega [205] further developed Tomas–Stein $X_{p,4}$-space to obtain that (4.53) holds if $s > s_0$ for some $s_0 \in (\frac{20}{41}, \frac{40}{81})$. By making use of the bilinear restriction estimate for paraboloid, Tao–Vargas [265] and Tao [259] improved the result to $s > \frac{15}{32}$ and $s > \frac{2}{5}$, respectively. Later, observing the localization properties of Schrödinger waves, Lee [160] obtained the result for $s > \frac{3}{8}$. Shao [235] gave an alternative proof by using the method of stationary phase and wave packet decomposition. In [24], Bourgain gave an counterexample to show that $s \geq \frac{1}{3}$ is necessary for (4.53) with $n = 2$. By using polynomial partitioning and the decoupling method [113], Du–Guth–Li [73] obtained the result for $s > \frac{1}{3}$, which is sharp up to the endpoint $s = \frac{1}{3}$.

Previous to [23], the results about $n \geq 3$ remained $s > \frac{1}{2}$, and $s \geq \frac{1}{4}$ was still believed to be the correct condition for (4.53) in every dimension. The study on this problem stagnated for several years until the recent work [23], where the $\frac{1}{2}$-barrier was broken for all dimensions. More precisely, Bourgain [23] proved that (4.53) holds if $s > \frac{1}{2} - \frac{1}{4n}$. More surprisingly, Bourgain also discovered some counterexamples to disprove the widely believed assertion on the $\frac{1}{4}$-threshold. Specifically, he showed that $s \geq \frac{1}{2} - \frac{1}{n}$ is necessary for (4.53) if $n \geq 5$. These examples originated essentially from an observation on arithmetical progressions. Recently, R. Luca and M. Rogers [180] showed that $s \geq \frac{1}{2} - \frac{1}{n+2}$ is necessary for (4.53) if $n \geq 3$. More recently, Bourgain [24] gave a counterexample to see that $s \geq \frac{1}{2} - \frac{1}{n+2}$ is necessary for $n \geq 3$. Up to the endpoint, Du–Zhang [74] proved the sharp result for (4.53) in higher dimensions $n \geq 3$. We remark that their result [74] also gives improved results on the size of divergence set of Schrödinger solutions, the Falconer distance set problem and the spherical average Fourier decay rates of fractal measures.

Recently, in [135], by using the estimate of oscillatory integrals, Kenig, Ponce and Vega proved that

$$\int_{\mathbb{R}^n} |\hat{\varphi}|^2 \left| \frac{P'(\xi)}{P''(\xi)} \right|^{\frac{1}{2}} d\xi < \infty, \quad n = 1, \tag{4.54}$$

where $P(\xi)$ is a polynomial of degree ≥ 2. This estimate means that the Carleson conjecture is true in the case $n = 1$ if $s \geq s_0 = 1/4$. Similar pointwise convergence results also hold for other dispersive equations (e. g., the free term $S(t)\varphi = \mathcal{F}^{-1} e^{i\xi^3 t} \mathcal{F}\varphi$ of solutions to the KdV equation).

On the other hand, the resolution of the Carleson conjecture can be reduced to the maximal estimate of solutions to the free equations, that is,

Theorem 4.1.5.
(i) *If $n = 1$, then*

$$\left\| \sup_{t \in \mathbb{R}} |S(t)\varphi| \right\|_{L_x^4(\mathbb{R})} \leq C \|\varphi\|_{H^{\frac{1}{4}}(\mathbb{R})}. \tag{4.55}$$

(ii) *If $n \geq 2$, then*

$$\left\| \sup_{t \in \mathbb{R}} |S(t)\varphi| \right\|_{L_x^2(Q)} \leq C \|\varphi\|_{H^s(\mathbb{R}^n)}, \quad s > 1/2, \tag{4.56}$$

where Q is the unit cube in $\mathbb{R}^n$.

Remark 4.1.2. The maximal inequality (4.56) is not optimal for the higher dimensional case (e. g., $n = 2$). See [19] for the relevant studies.

4.2 Progress in classical studies of nonlinear Schrödinger equations

For simplicity, we consider the following special case of (4.1):

$$\begin{cases} iu_t + \frac{1}{2}\Delta u = \lambda |u|^{p-1} u, & (t, x) \in \mathbb{R} \times \mathbb{R}^n \\ u(0) = \varphi(x), & x \in \mathbb{R}^n \end{cases} \tag{4.57}$$

and its corresponding integral equation

$$u(t) = S(t)\varphi - i\int_0^t S(t-\tau)[\lambda|u|^{p-1}u](\tau)d\tau, \tag{4.58}$$

where $S(t) = e^{\frac{i}{2}\Delta t}$. The case $\lambda > 0$ corresponds to the defocusing case, while the case $\lambda < 0$ corresponds to the focusing case. The following mass and energy conservation laws hold:

$$M(u(t)) = \int_{\mathbb{R}^n} |u|^2 dx = M(\varphi), \tag{4.59}$$

$$E(u(t)) = \int_{\mathbb{R}^n} \left[\frac{1}{2}|\nabla u|^2 + \frac{2\lambda}{p+1}|u|^{p+1}\right] dx = E(\varphi). \tag{4.60}$$

Using the pseudoconformal transform,

$$u(t) = \mathcal{C}v(x,t) = \left(\frac{1}{it}\right)^{\frac{n}{2}} \exp\left(\frac{i|x|^2}{2t}\right)\overline{v\left(\frac{1}{t}, \frac{x}{t}\right)}, \tag{4.61}$$

we can derive pseudoconformal conservation identity

$$\|xS(-t)u\|_2^2 + \frac{2\lambda t^2}{p+1}\|u\|_{p+1}^{p+1} + \frac{(np-n-4)}{p+1}\lambda \int_0^t \tau\|u\|_{p+1}^{p+1} d\tau = \|x\varphi\|_2^2 \tag{4.62}$$

for any $\varphi \in H^1 \cap \mathcal{F}H^1 = \Sigma$. In fact, by an elementary method, it is easy to derive the following equivalent form with (4.62) (see [9]):

$$\|(x+it\nabla)u\|_2^2 + \frac{2\lambda t^2}{p+1}\|u(t)\|_{p+1}^{p+1} + \frac{np-n-4}{p+1}\lambda \int_0^t \tau\|u\|_{p+1}^{p+1} d\tau = \|x\varphi\|_2^2 \tag{4.63}$$

for any $\varphi \in H^1 \cap \mathcal{F}H^1 = \Sigma = H^1 \cap L^2(|x|^2 dx)$.

Note that the nonlinear term is conformally invariant. So, the nonlinear Schrödinger equation in (4.57) is invariant under the Galilean transform

$$u \longmapsto v = G_h u = \exp\left(ih\cdot x - \frac{ih^2 t}{2}\right)u(x-ht,t), \quad h \in \mathbb{R}^n.$$

Its generator is $iJ(t)$ with

$$J(t) = x + it\nabla \quad \text{or} \quad J_j = x_j + it\partial_{x_j}, \quad j = 1,2,\dots,n.$$

A direct calculation together with the representation (4.17) of the free Schrödinger group gives

$$J(t) = S(t)xS(-t) = S(t-\tau)J(\tau)S(\tau-t) = itM(t)\nabla M(-t), \tag{4.64}$$

where $M(t) = \exp(i\frac{|x|^2}{2t})$. From (4.64), it is easy to see that (4.62) and (4.63) are equivalent. In particular, we have

$$\|(x+it\nabla)u\|_2^2 + \frac{2\lambda}{p+1}t^2\|u(t)\|_{p+1}^{p+1} = \|x\varphi\|_2^2, \quad \forall\varphi \in \Sigma,\ p = 1+\frac{4}{n}, \tag{4.65}$$

$$\|u(t)\|_{p+1}^{p+1} \le \frac{C}{t^2}, \quad p \ge 1+\frac{4}{n},\ \lambda > 0. \tag{4.66}$$

These two inequalities play an important role in scattering theory.

Theorem 4.2.1 (Local well-posedness). *Let $s_c = n/2 - 2/(p-1)$, $s \ge \max(0, s_c)$ and let $p-1 > [s]$ in the case when $p-1 \notin 2\mathbb{Z}$. Then for any $\varphi \in H^s(\mathbb{R}^n)$, there exists a $T^* > 0$ such that the problem* (4.57) *or* (4.58) *has a unique solution*

$$u(t) \in C([0,T^*); H^s) \cap \bigcap_{(q,r)\in\Lambda} L^q([0,T^*); B^s_{r,2}) \tag{4.67}$$

satisfying the alternative that either $T^* = \infty$ or $T^* < \infty$ and $\lim_{t\to T^*}\|u(t)\|_{H^s} = \infty$, where $T^* = T(\|\varphi\|_{H^s})$ in the case with $s > s_c$ (the subcritical case), and $T^* = T(\varphi)$ in the case if $s = s_c$ (the critical case). Further, in the case when $s = s_c$, if $\|\varphi\|_{H^{s_c}} \ll 1$ then $T^* = \infty$.

Outline of proof of Theorem 4.2.1. Let $I = [0,T)$ and let us take the special admissible pair (γ, ρ) as follows:

$$(\gamma,\rho) = \left(\frac{4(p+1)}{(p-1)(n-2s)}, \frac{n(p+1)}{n+(p-1)s}\right) \in \Lambda.$$

Clearly, by the Strichartz estimate (4.49), the solution obtained in the space $X(I) = L^\gamma(I; B^s_{\rho,2}) \cap C(I; H^s)$ satisfies (4.67). Consider now the metric space

$$\mathcal{X}(I) = \{u(t) \in X(I) \mid \|u(t)\|_{X(I)} \le M = 4C\|\varphi\|_{H^s},\} \subset X(I)$$

with the metric

$$d(u,v) = \|u-v\|_{L^\gamma(I;L^\rho)},$$

where

$$\|u\|_{X(I)} = \|u; L^\gamma(I; B^s_{\rho,2})\| + \sup_{t\in I}\|u(t)\|_{H^s}.$$

It is easy to see that $\mathcal{X}(I)$ is a complete metric space. Noting that $B^s_{\rho,2} = \dot{B}^s_{\rho,2} \cap L^\rho$, we see by the Strichartz estimate that

$$\|S(t)\varphi\|_{X(I)} \le 2C\|\varphi\|_{H^s}. \tag{4.68}$$

By the technique of nonlinear estimates in Besov spaces, we get

$$\begin{aligned}\left\|\int_0^t S(t-\tau)f(u(\tau))d\tau\right\|_{L^\gamma(I;\dot{B}^s_{\rho,2})} &\le C\|f(u)\|_{L^{\gamma'}(I;\dot{B}^s_{\rho',2})}\\ &\le CT^{1-\frac{p+1}{\gamma}}\|u\|^p_{L^\gamma(I;\dot{B}^s_{\rho,2})},\end{aligned} \tag{4.69}$$

$$\begin{aligned}\left\|\int_0^t S(t-\tau)f(u(\tau))d\tau\right\|_{L^\gamma(I;L^\rho)} &\le C\|f(u)\|_{L^{\gamma'}(I;L^{\rho'})}\\ &\le CT^{1-\frac{p+1}{\gamma}}\|u\|^{p-1}_{L^\gamma(I;\dot{B}^s_{\rho,2})}\|u\|_{L^\gamma(I;L^\rho)},\end{aligned} \tag{4.70}$$

and

$$\begin{aligned}&\left\|\int_0^t S(t-\tau)(f(u(\tau))-f(v(\tau)))d\tau\right\|_{L^\gamma(I;L^\rho)}\\ &\le CT^{1-\frac{p+1}{\gamma}}\left(\|u\|^{p-1}_{L^\gamma(I;\dot{B}^s_{\rho,2})}+\|v\|^{p-1}_{L^\gamma(I;\dot{B}^s_{\rho,2})}\right)\|u-v\|_{L^\gamma(I;L^\rho)},\end{aligned} \tag{4.71}$$

$$\left\|\int_0^t S(t-\tau)f(u(\tau))d\tau\right\|_{L^\infty(I;H^s)} \le CT^{1-\frac{p+1}{\gamma}}\|u\|^{p-1}_{L^\gamma(I;\dot{B}^s_{\rho,2})}\|u\|_{L^\gamma(I;B^s_{\rho,2})}. \tag{4.72}$$

We now consider the mapping $\mathcal{T}$ in the space $\mathcal{X}(I)$ defined by the RHS of the integral equation (4.58):

$$\mathcal{T}:\ u\to\mathcal{T}u=S(t)\varphi-i\int_0^t S(t-\tau)f(u(\tau))d\tau.$$

By the estimates (4.68)–(4.72), it follows that

$$\|\mathcal{T}u\|_{X(I)}\le 2C\|\varphi\|_{H^s}+2CT^{1-\frac{p+1}{\gamma}}M^p\le\frac{M}{2}+2CT^{1-\frac{p+1}{\gamma}}M^p, \tag{4.73}$$

$$d(\mathcal{T}u,\mathcal{T}v)\le 2CM^{p-1}T^{1-\frac{p+1}{\gamma}}d(u,v). \tag{4.74}$$

Thus, in the case when $s>s_c$, $\mathcal{T}$ is a contraction mapping from $\mathcal{X}(I)$ into itself if $T>0$ is sufficiently small. Further, T depends only on $\|\varphi\|_{H^s}$. The conclusion of Theorem 4.2.1 follows from the Banach contraction principle and Picard's iteration in this case.

When $s=s_c$, the estimates (4.73) and (4.74) can be modified as

$$\|\mathcal{T}u\|_{X(I)}\le 2C\|\varphi\|_{H^s}+2C\|u\|^{p-1}_{L^\gamma(I;\dot{B}^s_{\rho,2})}\|u\|_{L^\gamma(I;B^s_{\rho,2})}, \tag{4.75}$$

$$d(\mathcal{T}u,\mathcal{T}v)\le C\left(\|u\|^{p-1}_{L^\gamma(I;\dot{B}^s_{\rho,2})}+\|v\|^{p-1}_{L^\gamma(I;\dot{B}^s_{\rho,2})}\right)d(u,v). \tag{4.76}$$

If $T > 0$ is sufficiently small, it is also guaranteed that $\mathcal{T}$ is a contraction mapping from $\mathcal{X}(I)$ into itself, so the local well-posedness follows from the Banach contraction principle. In this case, T depends on φ itself. By the Picard method, T can be extended to the maximal $T^* = T(\varphi)$ and, in particular, if $\|\varphi\|_{H^{s_c}} \ll 1$, then $T^* = \infty$. □

Remark 4.2.1.

(i) We state Theorem 4.2.1 in the case $t > 0$. It can also be applied for the case $t < 0$ or for both cases at the same time. See [37, 102–104, 184].

(ii) From the estimates (4.75) and (4.76), one finds easily that for any $s \geq s_c$, when $\|\varphi\|_{\dot{H}^{s_c}} \ll 1$, the problem (4.57) or (4.58) has a unique global solution $u(t) \in C(\mathbb{R}; H^s) \cap L^q(\mathbb{R}; B^s_{r,2})$. This can be obtained from Theorem 4.2.1 and the nonlinear estimates

$$\|f(u)\|_{\dot{B}^s_{\rho',2}} \leq C\|u\|^{p-1}_{\dot{B}^{s_c}_{\rho,2}}\|u\|_{\dot{B}^s_{\rho,2}}, \quad s \geq s_c,$$
$$\|f(u)\|_{\rho'} \leq C\|u\|^{p-1}_{\dot{B}^{s_c}_{\rho,2}}\|u\|_{\rho},$$
$$\|f(u) - f(v)\|_{\rho'} \leq C[\|u\|^{p-1}_{\dot{B}^{s_c}_{\rho,2}} + \|v\|^{p-1}_{\dot{B}^{s_c}_{\rho,2}}]\|u - v\|_{\rho},$$

where $\rho = \rho_c = \frac{2n(p+1)}{n(p-1)-4}$.

(iii) When $s = 0$ and $1 < p < 1 + 4/n$ or when $s = 1$ and $1 < p < 1 + 4/(n-2)$, the solution obtained in Theorem 4.2.1 satisfies the conservation identities (4.59) and (4.60). This can be shown as follows. Utilizing the regularization technique of double convolution-type mollifiers [102], that is, taking the non-negative radial function $h(x) = h(|x|) \in C_c^\infty(\mathbb{R}^n)$ with $\|h\|_1 = 1$ to construct the following double convolution mollifying problem of (4.57):

$$\begin{cases} i\partial_t u_j(t) + \frac{1}{2}\Delta u_j = h_j * f(h_j * u), \\ u_j(0) = h_j * \varphi, \end{cases} \tag{4.77}$$

where $h_j(x) = j^n h(jx)$. By the abstract Segal theorem, it follows that the problem (4.77) has a unique smooth solution

$$u_j(t) \in \bigcap_{l=0}^{k} C^l(\mathbb{R}; H^{2k-2l}), \quad k \in \mathbb{N}^+,$$

which satisfies

$$E_0(u_j(t)) = E_0(h_j * \varphi), \tag{4.78}$$
$$E_1(u_j(t)) = E_1(h_j * \varphi). \tag{4.79}$$

Thus, taking the limit $j \to \infty$ in above two equations and using the space-time integrability of the solution obtained in Theorem 4.2.1 lead to the conservation identities (4.59) and (4.60). See [102] for details.

As a direct result of Theorem 4.2.1, the Sobolev theorem, the energy conservation integrals and interpolation, we have the following L^2 and H^1 global well-posedness results.

Theorem 4.2.2 (L^2 global well-posedness). *Let $1 < p < 1 + 4/n$. Then for $\varphi \in L^2(\mathbb{R}^n)$, the Cauchy problem* (4.57) *or the equivalent integral equation* (4.58) *has a unique global solution $u(t)$. In other words, the Cauchy problem* (4.57) *determines a global flow $u(t) \in C(\mathbb{R}; L^2(\mathbb{R}^n))$ satisfying the integral equation* (4.58)*; furthermore, for any $T > 0$,*

$$u(t) \in C(\mathbb{R}; L^2(\mathbb{R}^n)) \cap \bigcap_{(q,r)\in\Lambda} L^q([-T,T]; L^r(\mathbb{R}^n)).$$

Theorem 4.2.3 (H^1 global well-posedness). *Let $1 < p < 1 + 4/(n-2)$ and let $\varphi \in H^1(\mathbb{R}^n)$. Then we have the following global well-posedness results:*

(i) *If $\lambda > 0$, then the problem* (4.57) *or equivalently* (4.58) *has a unique global solution $u(t) \in C(\mathbb{R}; H^1(\mathbb{R}^n))$ satisfying that for any $T > 0$,*

$$u(t) \in C(\mathbb{R}; H^1(\mathbb{R}^n)) \cap \bigcap_{(q,r)\in\Lambda} L^q([-T,T]; W^{1,r}(\mathbb{R}^n)). \tag{4.80}$$

(ii) *If $\lambda < 0$ and $1 < p < 1+4/n$, then the problem* (4.57) *or equivalently* (4.58) *has a unique global solution $u(t)$ satisfying* (4.80).

Outline of proof of Theorems 4.2.2 *and* 4.2.3. The local well-posedness in L^2 and H^1 has been established in Theorem 4.2.1. Thus, we only consider the global well-posedness. By Remark 4.2.1(ii), it follows that

$$M(u(t)) = M(\varphi) = \int_{\mathbb{R}^n} |\varphi|^2 dx < \infty, \tag{4.81}$$

$$E(u(t)) \triangleq \int_{\mathbb{R}^n} \left(\frac{1}{2}|\nabla u|^2 + \frac{2\lambda}{p+1}|u|^{p+1}\right) dx = E(\varphi) < \infty. \tag{4.82}$$

We first prove Theorem 4.2.2. Since $1 < p < 1+4/n$, it is known that $T = T(\|\varphi\|_2)$, that is, T depends only on $\|\varphi\|_2$. Thus, the existence of global solutions to the problem (4.57) or (4.58) follows from (4.81).

We now prove Theorem 4.2.3. If $\lambda > 0$, we derive by (4.81) and (4.82) that

$$\|u(t)\|_{H^1} \le C(\|\varphi\|_{H^1}). \tag{4.83}$$

Since $1 < p < 1+4/(n-2)$ corresponds to the subcritical case, then $T = T(\|\varphi\|_{H^1})$ depends only on $\|\varphi\|_{H^1}$. Thus, from (4.83), the statement (i) of Theorem 4.2.3 follows.

If $\lambda < 0$, by the Gagaliardo–Nirenberg inequality, it is seen that

$$\int_{\mathbb{R}^n} |u|^{p+1} dx \leq C\|u\|_2^{(1-\theta)(p+1)} \|\nabla u\|_2^{\theta(p+1)}, \tag{4.84}$$

where $\theta = n/2 - n/(p+1)$. From (4.82) and (4.84), it is found that only if $(p+1)\theta < 2$ or $1 < p < 1 + 4/n$, one can use the Hölder inequality with ε to deduce (4.83), from which the statement (ii) of Theorem 4.2.3 follows. □

Remark 4.2.2.

(i) It follows from the outline of the proof above that the conservation (4.81) and (4.82) play a key part in establishing the L^2 and H^1 global well-posedness of solutions to the nonlinear Schrödinger equations. When $\varphi \in H^\theta$ with $0 < \theta < 1$, can (4.57) or (4.58) determine a global continuous flow $u(t) \in C(\mathbb{R}; H^\theta)$ in the case for $\lambda > 0$? The similar question also arises for other dispersive wave equations, classical wave equations and the Klein–Gordon equation. In [19], Bourgain proposed a general method in showing that, when θ is suitably close to 1; (4.57) or (4.58) determines a global continuous flow $u(t) \in C(\mathbb{R}, H^\theta)$. For example, in the case when $n = 3, p = 3$, if $\theta > 11/13$, then (4.57) or (4.58) determines a global flow $u(t) \in C(\mathbb{R}, H^\theta)$. The key idea is to decompose the initial data into the high and low frequency parts and then with the help of the Bourgain space and bilinear estimates to complete the proof. This will be discussed later. Interested readers are also referred to [19, 137, 197].

(ii) For the L^2 critical growth case ($p = 1 + 4/n$) or the H^1 critical growth case ($p = 1 + 4/(n-2)$), the maximal existence time $T^* = T(\varphi)$ of local solutions depends on φ itself other than only on $\|\varphi\|_2$ or $\|\varphi\|_{H^1}$. Thus, how to establish the L^2 and H^1 global well-posedness in the critical growth case is an hard question. For the critical wave equation,

$$\begin{cases} u_{tt} - \Delta u = -|u|^{2^*-2}u, \quad 2^* = 2n/(n-2), \quad n \geq 3 \\ u(0) = \varphi, \quad u_t(0) = \psi(x), \end{cases} \tag{4.85}$$

Grillakis proved in [111] that, in the case when $n = 3$, the smooth solution of the problem (4.85) is globally well posed. The method is to eliminate the infinite times concentration effect of energy by the Morawetz inequality [204]. For the general-dimensional case (e. g., $3 \leq n \leq 7$), the corresponding results can be seen in [236]. Shatah and Struwe proved the global well-posedness of solutions for the critical wave equation in the energy space H^1 [237]. However, for the critical Schrödinger equations the corresponding Morawetz estimate (e. g., $n = 3$)

$$\int_{\mathbb{R}} \int_{\mathbb{R}^3} \frac{|u|^6}{|x|} dx dt < \infty \tag{4.86}$$

is not enough to eliminate the energy concentration effect. But, in the case when initial data is radially symmetric, Bourgain established the H^1 global well-posedness for the critical Schrödinger equation by employing the localized form of the Morawetz inequality, weak L^2 dispersive estimates and the Littlewood–Paley dyadic decomposition. See Section 4.5 for details.

(iii) For the problem,

$$\begin{cases} iu_t + \frac{1}{2}\Delta u = \lambda |u|^{p-1}u, \quad \lambda > 0, \quad p > 1 + 4/(n-2) \\ u(0) = \varphi, \end{cases} \tag{4.87}$$

it is easy to show by the compact method (see Lions' book [176]) that there exists a global weak solution $u(t) \in (L^\infty \cap C_w)(\mathbb{R}; H^1 \cap L^{p+1})$ satisfying the energy inequality

$$E(u(t)) = \int_{\mathbb{R}^n} \left[\frac{1}{2}|\nabla u|^2 + \frac{2\lambda}{p+1}|u|^{p+1}\right] dx \le E(\varphi), \quad t \in \mathbb{R}.$$

However, we do not know whether the energy is conservative and whether the solution is unique. It is clear that, if the solution is unique, then

$$E(\varphi) \le E(u(t)),$$

which implies that the energy is conservative. From this, we know that the energy conservation is a necessary condition for the Cauchy problem of the supercritical Schrödinger equations to be well posed.

We now recall the scattering theory for the nonlinear Schrödinger equation. For other dispersive equations, the classical wave equation and the Klein–Gordon equation, the scattering theory is similar. Here, we consider the problem (4.57) as an example.

Formulation of the problem

Assume that the problem (4.57) has a global solution $u(t,x) \in C(\mathbb{R}; H^1(\mathbb{R}^n)) \cap X^1_{\text{loc}}(\mathbb{R})$, where $X^1_{\text{loc}}(\mathbb{R}) = L^q_{\text{loc}}(\mathbb{R}; W^{1,r}(\mathbb{R}^n))$, $(q,r) \in \Lambda$. What is the asymptotic behavior of $u(t)$ as $t \to \pm\infty$? Does the solution converge to the solution $u_0(t) = S(t)\varphi$ of the corresponding free Schrödinger equation in some sense, that is,

$$\lim_{t\to\pm\infty} \|u(t) - S(t)\varphi\|_{H^1} = 0?$$

In general, the answer to the second question is no. In fact, when $1 < p \le 1 + 2/n$, there exists $\varphi \in S(\mathbb{R}^n)$ satisfying that, no matter how small $\|\varphi\|_{H^1}$ is, we have

$$\lim_{t\to\pm\infty} \|u(t) - S(t)\varphi\|_{H^1} \ne 0,$$

where $u(t)$ is the global solution of (4.57) (see [9, 109]). Thus, it is necessary to seek for appropriate conditions under which the asymptotic behavior at $t \to \pm\infty$ of the global solution $u(t)$ can be established. Note that, though the global solution $u(t)$ of (4.57), in general, does not approach $S(t)\varphi$ as $t \to \pm\infty$, $u(t)$ may tend to $v_\pm(t) = S(t)\varphi_\pm$ for some functions $\varphi_\pm$ as $t \to \pm\infty$, where $v_\pm(t)$ is the solution to the following problem:

$$\begin{cases} iv_t + \frac{1}{2}\Delta v = 0, \\ v(0) = \varphi_\pm(x). \end{cases} \tag{4.88}$$

We now present the scattering theory for the problem (4.57), which is mainly concerned with the following two aspects.

(a) Wave operators $\Omega_\pm$

Let $v_\pm(t) = S(t)\varphi_\pm$ with $\varphi_\pm \in Y$ (or a dense subset of Y), where $Y \subset L^2(\mathbb{R}^n)$ is an appropriate Banach space. If there exists a solution $u(t)$ to the problem (4.57) such that

$$\|u(t) - v_\pm(t); Y\| = \|u(t) - S(t)\varphi_\pm(x); Y\| \to 0, \quad t \to \pm\infty \tag{4.89}$$

or

$$\|S(-t)u(t) - \varphi_\pm(x); Y\| \to 0, \tag{4.90}$$

then the wave operators can be defined as

$$\Omega_\pm : \varphi_\pm \longmapsto u(0) \stackrel{\triangle}{\equiv} \varphi \in H^1(\mathbb{R}^n), \tag{4.91}$$

which map from Y (or a dense subset of Y) into $H^1(\mathbb{R}^n)$. In particular, Ω_+ is called the positive wave operator and Ω_- is called as the negative wave operator. In general, $\varphi_\pm$ is called as the asymptotic states at $t = \pm\infty$ of $u(t)$. Note that if $S(t)$ does not generate a bounded operator group on Y, then (4.89) is usually replaced by (4.90).

(b) Asymptotic completeness

Let $u(t)$ be the global solution to the problem (4.57). Then the following question arises naturally: Are there any asymptotic states $\varphi_\pm \in Y$ such that (4.89) or (4.90) holds? If for any $\varphi \in H^1(\mathbb{R}^n)$, there always exist $\varphi_\pm \in Y$ (or a dense subset of Y) such that (4.89) or (4.90) holds, then (4.57) is said to be asymptotically complete.

If both (a) and (b) are satisfied, then the wave operators $\Omega_\pm$ satisfy

$$\Omega_+(\varphi_+) = \Omega_-(\varphi_-) = \varphi.$$

On the other hand, if $\varphi \in H^1$ and the Cauchy problem (4.57) has a unique global solution $u(t)$, then by the asymptotic completeness, there exist asymptotic states $\varphi_\pm \in Y$ such that (4.89) or (4.90) holds. Thus, the scattering operator S can be defined as

$$S = \Omega_+^{-1} \circ \Omega_- : Y \longmapsto Y.$$

This can be illustrated in the following diagram (see Figure 4.1).

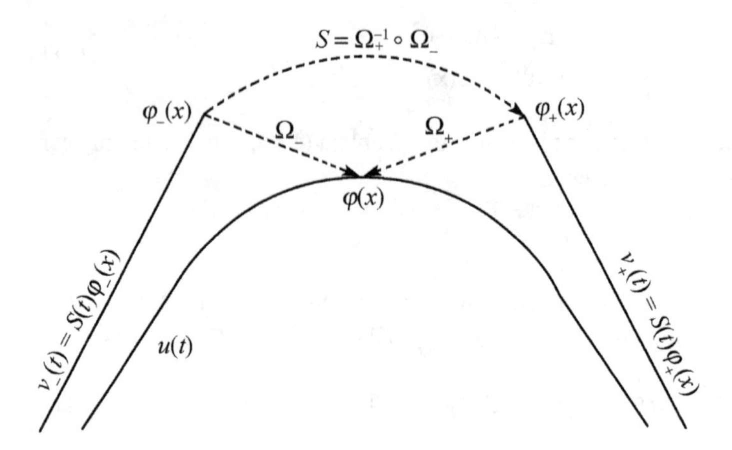

Figure 4.1: Scattering operator S.

In the scattering theory, the study on continuity, analyticity and homeomorphism of the scattering operator is a challenging task.

The existence of wave operators is essentially equivalent to the well-posedness of the problem

$$u(t) = S(t)\varphi_\pm + i \int_t^{\pm\infty} S(t-\tau)f(u(\tau))d\tau, \tag{4.92}$$

where $f(u) = \lambda|u|^{p-1}u$. On the other hand, in view of (4.58), the problem (4.92) is equivalent to

$$u(t) = S(t)\left[\varphi - i\int_0^{\pm\infty} S(-\tau)f(u(\tau))d\tau\right] + i\int_t^{\pm\infty} S(t-\tau)f(u(\tau))d\tau. \tag{4.93}$$

Therefore, the asymptotic completeness is equivalent to the following results for the global solution $u(t)$ of (4.57):

$$\begin{cases} \varphi - i\int_0^{\pm\infty} S(-\tau)f(u(\tau))d\tau \in Y, \\ \| \int_t^{\pm\infty} S(t-\tau)f(u(\tau))d\tau\|_Y \to 0, \quad t \to \pm\infty. \end{cases} \tag{4.94}$$

This gives the composition of the scattering operator S as follows:

$$S : \varphi_- \xrightarrow{\Omega_-} \Omega_-(\varphi_-) = \varphi_- + i \int_0^{-\infty} S(-\tau) f(u(\tau)) d\tau \equiv \varphi \tag{4.95}$$

$$\xrightarrow{\Omega_+^{-1}} \varphi - i \int_0^{+\infty} S(-\tau) f(u(\tau)) d\tau = \varphi_+.$$

Remark 4.2.3.

(i) For small solutions, the existence of wave operators implies the asymptotic completeness of wave operators. In general, however, the asymptotic completeness is a more difficult question than the existence of wave operators and requires not only the nonlinear term $f(u)$ to be of repulsion property but also a priori estimates of solutions to the nonlinear evolution equation. There is usually a unified method to deal with the existence of wave operators.

(ii) If, in the definition of wave operators, (4.89) is replaced by

$$\|u(t) - e^{i\theta(x,t)} S(t)\varphi_\pm; Y\| \to 0, \quad t \to \pm\infty,$$

where $\theta(x,t) \neq 0$ is a modified phase function, then modified wave operators can be defined and the corresponding modified scattering theory can be established similar to the definition of wave operators and scattering theory. Interested reader is referred to [100, 184, 212].

The study of scattering theory originates from Segal's conjecture in 1963, which has experienced mainly two stages. The first stage was concerned with studying scattering theory for small solutions in the classical sense, and the representative figures in this aspect are Segal, Strauss, Reed and Simon (see [225, 226, 231, 232]). In the second stage, the scattering theory was studied in the energy space, employing the Strichartz estimates established by Segal and Strichartz and the quasiconformal conservation integrals established by Ginibre and Velo (see [102, 233, 250] for details). We now briefly review the main work (including the scattering theory for wave equations) and progress in the studies of scattering theory. For simplicity, we only consider the defocusing case ($\lambda > 0$).

(i) Strauss and Glassey proved in [249] that, when

$$\begin{cases} 1 < p \leq 1 + 2/n, & n \geq 2, \\ 1 < p \leq 2, & n = 1, \end{cases} \tag{4.96}$$

the scattering theory does not hold. Precisely, there is a $\varphi \in S(\mathbb{R}^n)$ such that for the global solution $u(t)$ of (4.57) there is no asymptotic states $\varphi_\pm$ satisfying

$$\lim_{t\to\pm\infty} \|u(t) - S(t)\varphi_\pm\|_2 = 0.$$

This result was extended to the case with $1 < p \leq 3$ and $n = 1$ by Barab [9] using the pseudoconformal conservation identity (4.62) or (4.63) of Ginibre and Velo. This together with (4.96) means that the asymptotic completeness does not hold if $1 < p \leq 1 + 2/n$, $n \geq 1$.

(ii) With the help of Strichartz estimates, Strauss established in [247] the scattering theory for small energy solutions to the nonlinear Schrödinger equation under the condition that

$$1 + \frac{4}{n} < p < 1 + \frac{4}{n-2}. \tag{4.97}$$

The scattering theory was also established in [247] for small energy solutions to the Klein–Gordon equation for the case when

$$1 + \frac{4}{n-1} < p < 1 + \frac{4}{n-2}, \tag{4.98}$$

which was subsequently improved by Pecher [216] and M. Tsutsumi to the case when, instead of (4.98), (4.97) is satisfied.

(iii) In 1984, Y. Tsutsumi and Yajima proved in [276] that if $\varphi \in \Sigma = H^1 \cap \mathcal{F}H^1$, then (4.57) or (4.58) has a unique global solution

$$u(t) \in C(\mathbb{R}; \Sigma), \quad u,\ \nabla u,\ J(t)u \in \bigcap_{(q,r)\in\Lambda} L^q_{\text{loc}}(\mathbb{R}; L^r), \tag{4.99}$$

where $\lambda > 0$ and $1 < p < 1 + 4/(n-2)$. Further, under the optimal condition,

$$1 + \frac{2}{n} < p < 1 + \frac{4}{n-2},$$

they proved that there exist asymptotic states $\varphi_\pm \in L^2(\mathbb{R}^n)$ such that

$$\lim_{t\to\pm\infty} \|u(t) - S(t)\varphi_\pm\|_2 = 0. \tag{4.100}$$

This is the best result so far for asymptotic completeness in some sense. Certainly, it may be more natural if (4.100) holds in the norm of Σ.

(iv) In 1985, under the condition (4.97), Ginibre and Velo [103] and Brenner [25] established the scattering theory in the energy space of solutions to the problem (4.57) and the following Cauchy problem for the Klein–Gordon equation, respectively:

$$\begin{cases} u_{tt} - \Delta u + m^2 u = \lambda |u|^{p-1} u, & m \neq 0, \lambda > 0, \\ u(0) = \varphi, \quad u_t(0) = \psi(x). \end{cases} \tag{4.101}$$

The method they used is the Strichartz estimates and the nonlinear estimate in fractional Besov spaces of the nonlinear term.

(v) For nonlinear Schrödinger equations, the asymptotic completeness has been established by Tsutsumi and Yajima in some sense. Thus, the remaining question is the existence of wave operators. In 1987, Tsutsumi[275] established the scattering theory in the sense of the Σ norm for the problem (4.57) under the condition

$$\gamma(n) < p < 1 + \frac{4}{n-2}$$

and proved that $S : \Sigma \longmapsto \Sigma$ is a homeomorphic mapping, where

$$\gamma(n) = \frac{1}{2} + \frac{1}{n} + \sqrt{\left(\frac{1}{2} + \frac{1}{n}\right)^2 + \frac{2}{n}}.$$

Clearly, $1+2/n < \gamma(n) < 1+4/n$. In [37], Cazenave and Weissler proved the existence of wave operators in the Σ norm for the problem (4.57) under the condition

$$1 + \max\left\{\frac{2}{n}, \frac{4}{n+2}\right\} < p < 1 + \frac{4}{n-2}. \tag{4.102}$$

It is clear that the condition (4.102) is reduced to the optimal one $1 + 2/n < p < 1 + 4/(n-2)$ in the case with $n = 1, 2$. In [101], by introducing generalized Besov spaces, Ginibre, Ozawa and Velo established the existence of wave operators in the norm of $\Sigma^{\rho,\rho} = H^\rho \cap \mathcal{F}H^\rho$ for the problem (4.57) under the optimal condition $1 + 2/n < p < 1 + 4/(n-2)$, $n = 3$.

Remark 4.2.4. Interpretation of $\gamma(n)$. For $(q, p+1) \in \Lambda$, we should have

$$\left\| \int_t^\infty S(t-\tau) f(u(\tau)) d\tau; L^q((t,\infty); L^{p+1}(\mathbb{R}^n)) \right\| \to 0, \quad t \to \infty.$$

In particular, the above result should also be true if $u(t) = S(t)\varphi$. By Strichatz estimates, it is enough to show

$$\left\| \int_t^\infty |t-\tau|^{-\delta(p+1)} \tau^{-p\delta(p+1)} d\tau \right\|_{L^q([t,\infty))} \|\varphi\|^p_{(p+1)'} \to 0, \quad t \to \infty, \tag{4.103}$$

where $\delta(p+1) = n(1/2 - 1/(p+1)) = 2/q$. Noting that $\|\varphi\|_{(p+1)'} \leq C\|\varphi\|_\Sigma$, it is seen that if $p\delta(p+1) > 1$, then (4.103) can be guaranteed to hold. Solving $p\delta(p+1) = 1$ for p gives

$$\gamma(n) = \frac{n + 2 + \sqrt{n^2 + 12n + 4}}{2n}.$$

In what follows, we collect and briefly discuss certain classical results on the scattering theory for nonlinear Schrödinger equations. We are restricted to the defocusing case, that is,

$$\begin{cases} iu_t + \frac{1}{2}\Delta u = \lambda|u|^{p-1}u, \quad \lambda > 0, \\ u(0) = \varphi. \end{cases} \tag{4.104}$$

Theorem 4.2.4. *Let $1 + 2/n < p < 1 + 4/(n-2)$. Then for any $\varphi \in \Sigma$, there exist unique asymptotic states $\varphi_\pm \in L^2(\mathbb{R}^n)$ such that*

$$\lim_{t\to\pm\infty} \|u(t) - S(t)\varphi_\pm\|_2 = 0,$$

where $u(t)$ is the global solution to the problem (4.104) *and satisfies* (4.99).

Sketch of proof. From the expression $S(t) = M(t)D(t)\mathcal{F}M$, it is seen that the free solution $S(t)f$ is of the asymptotic surface

$$\left(\frac{1}{it}\right)^{\frac{n}{2}} \exp\left(\frac{i|x|^2}{2t}\right)\hat{f}\left(\frac{x}{t}\right);$$

see [226]. Thus, we can define the conformal transform

$$u(t) = \mathcal{C}v(x,t) = \left(\frac{1}{it}\right)^{\frac{n}{2}} \exp\left(\frac{i|x|^2}{2t}\right)\overline{v\left(\frac{1}{t}, \frac{x}{t}\right)}. \tag{4.105}$$

Under this transform, the nonlinear Schrödinger equation in the problem (4.104) can be changed into

$$iv_t + \frac{1}{2}\Delta v = \lambda|t|^{\frac{n(p-1)}{2}-2}|v|^{p-1}v. \tag{4.106}$$

So, the asymptotic completeness is equivalent to

$$\lim_{t\to\pm 0} \|v(t) - v_\pm(0)\|_2 = 0. \tag{4.107}$$

In fact, under the conformal transform (4.105), if $u(t) \in C(\mathbb{R}, \Sigma)$ satisfies (4.104), then it can be deduced that $v(t) = \mathcal{C}^{-1}u(t) \in C(\mathbb{R}^\pm; \Sigma)$ satisfies (4.106). Therefore, one only needs to prove

$$\lim_{t,s\to\pm 0} \|v(t) - v(s)\|_2 = 0, \tag{4.108}$$

which implies that there exist $v_\pm(0) \in L^2(\mathbb{R}^n)$ such that

$$\lim_{t\to 0^\pm} \|v(t) - v_\pm(0)\|_2 = 0.$$

Hence, letting $\varphi_\pm = \check{\overline{V}}_\pm(0)$, we complete the proof of Theorem 4.2.4. Now with the help of the weak limit

$$\lim_{t\to 0^\pm} v(t) \overset{w}{=} v_\pm(0),$$

we can prove (4.108), which proves the theorem; see [276] for details. □

Remark 4.2.5.

(i) From the results of Strauss and Barab, it is seen that the assumption on the nonlinear growth index p in Theorem 4.2.4 is optimal.

(ii) From (4.107), we see that when $p \le 1 + 2/n$, $t^{n(p-1)/2-2}$ has a singularity at $t = 0$, which cannot be removed. This is the reason why the scattering theory does not hold for the case $p \le 1 + 2/n$, even for the scattering of small solutions.

(iii) From another point of view, the scattering theory (asymptotic completeness) requires that (4.94) holds. If we take $Y = L^2$, then the limit in (4.94) must also be true for $v(t) = S(t)\varphi$, that is, $\|f(S(t)\varphi)\|_2$ must be integrable with respect to t. Note that

$$\|f(S(t)\varphi)\|_2 \le \|S(t)\varphi\|_{2p}^p \le \|S(t)\varphi\|_\infty^{p-1}\|S(t)\varphi\|_2 \le C|t|^{-\frac{n}{2}(p-1)}\|\varphi\|_\Sigma^{p-1}\|\varphi\|_2.$$

Therefore, when $(p-1)\frac{n}{2} > 1$, it is guaranteed that

$$\int_t^\infty \|f(S(t)\varphi)\|_2 d\tau \to 0, \quad t \to \infty.$$

However, the condition $(p-1)\frac{n}{2} > 1$ is exactly the condition $p > 1 + 2/n$.

(iv) Similar to the nonlinear Schrödinger equation, we can study the scattering theory for the Hartree-type equation

$$iu_t + \frac{1}{2}\Delta u = \lambda(|x|^{-\rho} * |u|^2)u, \quad \lambda > 0. \tag{4.109}$$

Similarly as before, $\rho = 1$ is the critical index for scattering. If $\rho \le 1$, then the asymptotic completeness does not hold [115]. Roughly speaking, since

$$\begin{aligned}
&\|(|x|^{-\rho} * |S(t)\varphi|^2)S(t)\varphi\|_2 \\
&\quad \le \||x|^{-\rho} * |S(t)\varphi|^2\|_{\frac{2n}{\rho}}\|S(t)\varphi\|_{\frac{2n}{n-\rho}} \\
&\quad \le C\|S(t)\varphi\|_{\frac{4n}{2n-\rho}}^2\|S(t)\varphi\|_{\frac{2n}{n-\rho}} \\
&\quad \le C\|S(t)\varphi\|_\infty^{\frac{\rho}{n}}\|S(t)\varphi\|_2^{2(1-\frac{\rho}{2n})}\|S(t)\varphi\|_\infty^{\frac{\rho}{n}}\|S(t)\varphi\|_2^{1-\frac{\rho}{n}} \\
&\quad \le C|t|^{-\rho}\|\varphi\|_1^{\frac{2\rho}{n}}\|\varphi\|_2^{3-\frac{2\rho}{n}} \le C|t|^{-\rho}\|\varphi\|_\Sigma^{\frac{2\rho}{n}}\|\varphi\|_2^{3-\frac{2\rho}{n}},
\end{aligned}$$

then in order that

$$\int_t^{\pm\infty} \|(|x|^{-\rho} * |u(t)|^2)u(t)\|_2 dt \longrightarrow 0, \quad t \to \pm\infty,$$

it is necessary that $\rho > 1$ so the asymptotic completeness can be guaranteed to hold.

Theorem 4.2.5. *Let $\gamma(n) < p < 1+4/(n-2)$. Then we have the following scattering results:*
(i) *For any $\varphi_\pm \in \Sigma$, there exists a unique $\varphi \in \Sigma$ such that*

$$\lim_{t\to\pm\infty} \|\varphi_\pm - S(-t)u(t)\|_\Sigma = 0, \tag{4.110}$$

where $u(t) \in C(\mathbb{R}, \Sigma)$ is a solution to the equation (4.104) with the initial data φ.
(ii) *For any $\varphi \in \Sigma$, let $u(t)$ be the solution of (4.104) with the initial data φ. Then there exist unique $\varphi_\pm \in \Sigma$ such that (4.110) holds.*
(iii) *$\Omega_+^{-1} \circ \Omega_-$ is a homeomorphic mapping from Σ to Σ.*

Remark 4.2.6.
(i) When $\gamma(n) < p < 1 + 4/(n-2)$, the statement (ii) of Theorem 4.2.5 implies Theorem 4.2.4.
(ii) The statement (i) of Theorem 4.2.5 implies that $\Omega_\pm : \Sigma \longmapsto \Sigma$ is well-defined, while the statement (ii) of Theorem 4.2.5 implies

$$\mathrm{Range}(\Omega_+) = \mathrm{Range}(\Omega_-) = \Sigma,$$

which determines the scattering operator $S = \Omega_+^{-1} \circ \Omega_- : \Sigma \longmapsto \Sigma$.

Sketch of the proof of Theorem 4.2.5. From Ginibre–Velo's quasiconformal conservation integral

$$\|xS(-t)u\|_2^2 + \frac{2\lambda}{p+1} t^2 \|u(t)\|_{p+1}^{p+1} + \frac{(np-n-4)}{p+1}\lambda \int_0^t \tau \|u\|_{p+1}^{p+1} d\tau = \|x\varphi\|_2^2$$

and the energy conservation identity

$$\|\nabla u\|_2^2 + \frac{2\lambda}{p+1}\|u(t)\|_{p+1}^{p+1} = \|\nabla\varphi\|_2^2 + \frac{2\lambda}{p+1}\|\varphi\|_{p+1}^{p+1},$$

we can show that

$$\|u(t)\|_{p+1} \le C(1+|t|)^{-\theta}, \quad \theta = \frac{n(p-1)}{2(p+1)},$$
$$\|xS(-t)u\|_2 \le C(1+|t|)^{a(p)}, \quad a(p) = \max\left(0, 1 - \frac{n}{4}(p-1)\right),$$
$$\|J_j u; L^q(\mathbb{R}; L^{p+1}(\mathbb{R}^n)\|, \ \|\nabla u; L^q(\mathbb{R}; L^{p+1}(\mathbb{R}^n))\| \le C.$$

With these three inequalities, Theorem 4.2.5 can be proved; see [184] for details. □

The existence of wave operators can be dealt with using unified methods for both the focusing and defocusing cases. Moreover, the following general condition ($\tilde{\mathrm{H}}1$) can be used to replace (H1) to study the question of existence of wave operators:
($\tilde{\mathrm{H}}1$) $f \in C^1(\mathbb{C},\mathbb{C})$ satisfies that $f(0) = 0$ and

$$|f'(z)| \le \max\left(\left|\frac{\partial f}{\partial z}\right|, \left|\frac{\partial f}{\partial \bar{z}}\right|\right) \le C(|z|^{p_1-1} + |z|^{p_2-1}), \tag{4.111}$$

where

$$1 < p_1 \le p_2 < 1 + \frac{4}{n-2}.$$

As we know, the existence of wave operators is equivalent to the existence of solutions to the integral equation

$$u(t) = S(t)\varphi_\pm(x) + i\int_t^{\pm\infty} S(t-\tau)f(u(\tau))d\tau, \tag{4.112}$$

which is essentially equivalent to the final value problem

$$\begin{cases} iu_t + \frac{1}{2}\Delta u = f(u), \\ u|_{t=\pm\infty} = \varphi_\pm. \end{cases} \tag{4.113}$$

We now take $t = +\infty$ as an example to demonstrate that the solution $u(t)$ defined by (4.112) satisfies the final value condition. In fact, this is equivalent to the result

$$\lim_{t\to+\infty} S(t)\varphi_+ = \varphi_+.$$

Noting that

$$S(t)\varphi_+(x) = (2\pi i t)^{-\frac{n}{2}} \exp\left(i\frac{|x|^2}{2t}\right) * \varphi_+(x),$$

$$\int \frac{1}{(2\pi i)^{n/2}} \exp\left(i\frac{|x|^2}{2}\right)dx = 1,$$

we have by the unity approximation method in L^2 that

$$\lim_{t\to+\infty} \psi_{\sqrt{\frac{1}{t}}} * \varphi_+(x) = \lim_{t\to\infty} S(t)\varphi_+(x) \overset{L^2}{=} \varphi_+(x),$$

where $\psi(x) = (2\pi i)^{-\frac{n}{2}} \exp(i\frac{|x|^2}{2})$.

It should be pointed out that the main problem in scattering theory remains unsolved and is also an active research topic at present. We shall discuss Bourgain's method and the relevant new results in the later sections and chapters.

Remark 4.2.7.

(i) For the focusing case such as

$$\begin{cases} iu_t + \frac{1}{2}\Delta u = \lambda|u|^{p-1}u, \quad \lambda < 0, \\ u(0) = \varphi(x), \end{cases} \tag{4.114}$$

if $p \geq 1 + 4/n$, then there is a $\varphi \in S(\mathbb{R}^n)$ (φ satisfies certain appropriate conditions) such that the solution of the problem (4.114) blows up. In particular, if $p = 1 + 4/n$, F. Merle constructed explicitly blow up solutions by studying quasiconformal transforms and standing wave solutions of the form $u(t) = Q(x)e^{it}$ to the problem (4.114), where $Q(x)$ satisfies

$$\frac{1}{2}\Delta Q = \lambda Q^{1+\frac{4}{n}} + Q$$

with $Q > 0$ satisfying the radial symmetric condition. We refer to [109, 181, 182].

(ii) For the following Cauchy problem of general nonlinear Schrödinger equations with derivatives,

$$\begin{cases} iu_t + \frac{1}{2}\Delta u = F(u, \bar{u}, \nabla_x u, \nabla_x \bar{u}), \\ u(0) = \varphi(x), \end{cases} \tag{4.115}$$

Kenig, Ponce and Vega in [136] established the global well-posedness of small solutions and the general local well-posedness for any initial data by using Kato's local smoothness effect, which can compensate the first-order derivative loss of the nonlinear term.

4.3 Well-posedness and scattering theory below energy norm

Bourgain develops a general method, which can be used to deal with the global well-posedness and scattering theory of solutions to nonlinear dispersive equations in low regularity space [19]. In this section, we consider the cubic nonlinear Schrödinger equation as an example, and establish the global well-posedness and scattering of solution in H^s ($s > \frac{11}{13}$),

$$\begin{cases} iu_t - \Delta u = -|u|^2 u, & t \in \mathbb{R},\ x \in \mathbb{R}^3, \\ u(0) = \varphi(x) \in H^s(\mathbb{R}^3), & s < 1. \end{cases} \tag{4.116}$$

Similarly, it can be used to deal with the general case $f(u) = |u|^{p-1}u$ and $x \in \mathbb{R}^n$ for $s > s_0 = s_0(p, n)$. The global well-posedness for wave equation in low energy space will be discussed in Chapter 5.

For this purpose, we need to introduce a new kind of function spaces and establish the associated Strichartz estimates.

Definition 4.3.1 (Bourgain space). We define

$$X_{s,b} = \Big\{ f(x,t) \in S'(\mathbb{R}^{n+1});\ \|f\|_{X_{s,b}} = \Big(\int_{\mathbb{R}} \int_{\mathbb{R}^n} (1 + |\tau - |\xi|^2|^2)^b (1 + |\xi|^2)^s |\hat{f}(\xi,\tau)|^2 d\xi d\tau \Big)^{1/2} < \infty \Big\},$$

where ^ denotes the Fourier transform with respect to (x,t).

Remark 4.3.1.

(a) If we denote $\langle \xi \rangle = (1 + |\xi|^2)^{\frac{1}{2}}$, $\langle \tau \rangle = (1 + |\tau|^2)^{\frac{1}{2}}$, then we have

$$\|f\|_{X_{s,b}(\phi)} = \|\langle \tau - \phi(\xi) \rangle^b \langle \xi \rangle^s \hat{f}\|_{L^2_\tau L^2_\xi}.$$

(b) We can also define Bourgain space $X_{s,b}$ for a more general dispersive equation. For example, for the dispersive equation

$$iu_t + \phi(D)u = N(u), \quad D_x = \frac{1}{i}\partial_x, \tag{4.117}$$

the corresponding Bourgain space can be defined:

$$X_{s,b} = \Big\{ f \in S'(\mathbb{R}^{n+1});\quad \|f\|_{X_{s,b}} = \Big(\int_{\mathbb{R}} \int_{\mathbb{R}^n} (1 + |\tau - \phi(\xi)|^2)^b \times (1 + |\xi|^2)^s |\hat{f}(\xi,\tau)|^2 d\xi d\tau \Big)^{\frac{1}{2}} < \infty \Big\}.$$

Clearly, (4.116) corresponds to $\phi(\xi) = |\xi|^2$ of (4.117) in the definition of $X_{s,b}$.

Proposition 4.3.1.

(1) *Let $W(t) = e^{it\phi(D)}$ be the unitary group generated by*

$$iu_t + \phi(D)u = 0.$$

Then

$$\|f\|_{X_{s,b}} = \|W(-t)f\|_{H^b_t(\mathbb{R};H^s)}. \tag{4.118}$$

In particular, for the Schrödinger equation (4.116), we have

$$\|f\|_{X_{s,b}} = \|W(-t)f\|_{H^b_t(\mathbb{R};H^s)}, \quad W(t) = e^{-it\triangle}. \tag{4.119}$$

(2) *For $s, b, \sigma, \rho \in \mathbb{R}$, let $J^\sigma_x f = \mathcal{F}^{-1}_x \langle \xi \rangle^\sigma \mathcal{F}_x f$, $J^\rho_t f = \mathcal{F}^{-1}_t \langle \tau \rangle^\rho \mathcal{F}_t f$, $\mathcal{J}^\rho f = W_\phi(t) J^\rho_t W_\phi(-\cdot) f$. Then*

$$J_x^\sigma : X_{s,b}(\phi) \longrightarrow X_{s-\sigma,b}(\phi)$$

and

$$\mathcal{J}^\rho : X_{s,b}(\phi) \longrightarrow X_{s,b-\rho}(\phi)$$

are the isometric isomorphisms.

Proof. Denoting by $\hat{}$ the Fourier transform with respect to (x,t), $^{\wedge(x)}$ to $x = (x_1, \dots, x_n)$, $^{\wedge(t)}$ to t, respectively. By the property of Fourier transform, it is easy to show that

$$\begin{aligned}
\|f\|_{X_{s,b}} &= \|W(-t)f\|_{H_t^b(\mathbb{R};H^s)} = \|e^{-it\phi(D)}f\|_{H_t^b(\mathbb{R};H^s)} \\
&= \|(1+|\xi|^2)^{\frac{s}{2}}(1+|\tau|^2)^{\frac{b}{2}}(e^{-it\phi(\xi)}\hat{f}^{(x)}(\xi,t))^{\wedge(t)}\|_{L_\tau^2 L_\xi^2} \\
&= \|(1+|\xi|^2)^{\frac{s}{2}}(1+|\tau|^2)^{\frac{b}{2}}\hat{f}(\xi,\tau+\phi(\xi))\|_{L_\tau^2 L_\xi^2} \\
&= \|(1+|\xi|^2)^{\frac{s}{2}}(1+|\tau-\phi(\xi)|^2)^{\frac{b}{2}}\hat{f}(\xi,\tau)\|_{L_\tau^2 L_\xi^2},
\end{aligned}$$

which implies (1).

By the definition of Bourgain space, we know that the Bessel operator $J_x^\sigma : X_{s,b} \longrightarrow X_{s-\sigma,b}$ is an isometric isomorphism. It suffices to show that $\mathcal{J}^\rho : X_{s,b} \longrightarrow X_{s,b-\rho}$. In fact,

$$\begin{aligned}
\|\mathcal{J}^\rho f\|_{X_{s,b-\rho}(\phi)} &= \|W_\phi J_t^\rho W_\phi(-\cdot)f\|_{X_{s,b-\rho}(\phi)} = \|J_t^\rho W_\phi(-\cdot)f\|_{H_t^{b-\rho}H_x^s} \\
&= \|W_\phi(-\cdot)f\|_{H_t^b H_x^s} = \|f\|_{X_{s,b}(\phi)}, \\
\|\mathcal{J}^{-\rho} f\|_{X_{s,b}(\phi)} &= \|W_\phi J_t^{-\rho} W_\phi(-\cdot)f\|_{X_{s,b}(\phi)} = \|J_t^{-\rho} W_\phi(-\cdot)f\|_{H_t^b H_x^s} \\
&= \|W_\phi(-\cdot)f\|_{H_t^{b-\rho}H_x^s} = \|f\|_{X_{s,b-\rho}(\phi)},
\end{aligned}$$

which concludes the proof. □

Remark 4.3.2.

(i) Let $\phi_i(\xi) : \mathbb{R}^n \to \mathbb{R}$, $i = 1,2$ be measurable functions with $\phi_1(\xi) - \phi_2(x) \in L^\infty(\mathbb{R}^n)$, then the corresponding $X_{s,b}(\phi_i)$-norms are equivalent, that is, there exists a $c \in \mathbb{R}$ such that

$$\frac{1}{c}\|f\|_{X_{s,b}(\phi_2)} \le \|f\|_{X_{s,b}(\phi_1)} \le c\|f\|_{X_{s,b}(\phi_2)}.$$

(ii) Let $\langle \cdot,\cdot \rangle$ denote the inner product on L_{xt}^2 and let $\Phi : X_{-s,-b}(\phi) \to (X_{s,b}(\phi))'$ be defined by

$$\Phi(g)[f] \triangleq \langle J_x^s \mathcal{J}^b f, J_x^{-s}\mathcal{J}^{-b} g \rangle. \tag{4.120}$$

Then Φ is an isometric isomorphism.

(iii) Let $\phi_i : \mathbb{R}^n \to \mathbb{R}, i = 1, 2$ be continuous functions with $\|\phi_1(\xi) - \phi_2(\xi)\|_{L^\infty} = \infty$. Then, for all $c \in \mathbb{R}$, $b \neq 0$, the estimate

$$\frac{1}{c}\|f\|_{X_{s,b}(\phi_2)} \leq \|f\|_{X_{s,b}(\phi_1)} \leq c\|f\|_{X_{s,b}(\phi_2)}$$

fails. This means that the space $X_{s,b}(\phi)$ are in general not invariant under time reversion and complex conjugation.

(iv) For all $s \in \mathbb{R}$ and any measurable function $\phi(\xi)$, the following properties hold true:

$$\begin{aligned} X_{s,b}(\phi) &\subset C_t(\mathbb{R}; H^s(\mathbb{R}^n)), \quad \forall\, b > \frac{1}{2}, \\ X_{s,b}(\phi) &\subset L_t^q(\mathbb{R}; H^s(\mathbb{R}^n)), \quad \forall\, 2 \leq q < \infty,\ b \geq \frac{1}{2} - \frac{1}{q}, \\ \|f\|_{X_{s,b}(\phi)} &\leq C\|f\|_{L_t^1(\mathbb{R};H^s(\mathbb{R}^n))}, \quad \forall\, b < -\frac{1}{2}, \\ \|f\|_{X_{s,b}(\phi)} &\leq C\|f\|_{L_t^q(\mathbb{R};H^s(\mathbb{R}^n))}, \quad \forall\, 1 < q \leq 2,\ b \leq \frac{1}{2} - \frac{1}{q}. \end{aligned}$$

In the applications, we always choose the parameters $b' < -\frac{1}{2}$ to ensure that the embedding $L_t^1(\mathbb{R}; H^s(\mathbb{R}^n)) \hookrightarrow X_{s,b'}(\phi)$ holds. However, when $b' = -\frac{1}{2}$, the above embedding does not holds. Therefore, the auxiliary space $Y_s(\phi)$ turns out to be useful and can be used to deal with this case, which are defined as completion of

$$\{f(x,t) \in \mathcal{S}(\mathbb{R}^n \times \mathbb{R}) : \|\cdot\|_{Y_s(\phi)} = \left\|\langle\xi\rangle^s \langle\tau - \phi(\xi)\rangle^{-1} \mathcal{F}\cdot\right\|_{L_\xi^2(L_\tau^1)} < \infty\} \tag{4.121}$$

with respect to the norm $\|\cdot\|_{Y_s(\phi)}$. By Hölder's inequality, we have the embedding $X_{s,b'}(\phi) \hookrightarrow Y_s(\phi)$ for $b' > -\frac{1}{2}$.

Proposition 4.3.2. *Let $0 < \delta \leq 1$, $\psi(t) \in C_c^\infty(\mathbb{R})$ be a bump function satisfying $0 \leq \psi(t) \leq 1$ and*

$$\psi = \begin{cases} 1, & |t| \leq 1, \\ 0, & |t| \geq 2. \end{cases}$$

Let $W_\phi(t) = e^{it\phi(D)}$ and

$$W_{\phi\star R}F = \int_0^t W_\phi(t-\tau)F(\tau)d\tau. \tag{4.122}$$

Then we have the following estimates:

(1) *(Estimates for the homogeneous part)*

$$\|\psi(t/\delta)W_\phi(t)\varphi\|_{X_{s,b}(\phi)} \leq C\delta^{\frac{1-2b}{2}}\|\varphi\|_{H^s(\mathbb{R}^n)}, \quad b \geq 0, \tag{4.123}$$

$$\|\psi(t/\delta)F\|_{X_{s,b}(\phi)} \leq C\delta^{\frac{1-2b}{2}}\|F\|_{X_{s,b}(\phi)}, \quad b > \frac{1}{2}. \tag{4.124}$$

(2) *(Estimates for the inhomogeneous part) Let $b' + 1 \geq b \geq 0 \geq b'$. Then*

$$\|\psi(t/\delta)W_{\phi\star R}F\|_{X_{s,b}(\phi)} \leq C\delta^{1+b'-b}\|F\|_{X_{s,b'}(\phi)} + C_1\delta^{\frac{1-2b}{2}}\|F\|_{Y_s(\phi)}. \tag{4.125}$$

If in addition $b' > -\frac{1}{2}$, (4.125) holds with $C_1 = 0$.

(3) *(Continuity of the inhomogeneous part in H^s) Let $F \in Y_s(\phi)$, then $W_{\phi\star R}F \in C_t([-T,T]; H^s(\mathbb{R}^n))$ for all $0 < T < \infty$, and the estimate*

$$\sup_{|t|\leq T}\|W_{\phi\star R}F\|_{H^s} \leq C\langle T\rangle\|F\|_{Y_s(\phi)}, \quad s \in \mathbb{R} \tag{4.126}$$

holds true.

Proof. (i) On one hand, by

$$\|\psi(\delta^{-1}t)\|_{H_t^b(\mathbb{R})} \lesssim \delta^{\frac{1}{2}}\|\psi\|_2 + \delta^{\frac{1-2b}{2}}\|\psi\|_{\dot{H}_t^b(\mathbb{R})} \lesssim \delta^{\frac{1-2b}{2}}\|\psi\|_{H_t^b(\mathbb{R})},$$

and the definition of the Bourgain space, we obtain

$$\begin{aligned}\|\psi(\delta^{-1}t)W_\phi\varphi\|_{X_{s,b}(\phi)} &= \|W_\phi(-\cdot)\psi(\delta^{-1}t)W_\phi\varphi\|_{H_t^b(H_x^s)} = \|\psi(\delta^{-1}t)\varphi\|_{H_t^b(H_x^s)}\\ &\leq \|\psi(\delta^{-1}t)\|_{H_t^b}\|\varphi\|_{H_x^s} \leq C\delta^{\frac{1-2b}{2}}\|\varphi\|_{H_t^b(\mathbb{R})}.\end{aligned}$$

On the other hand, noticing that

$$F(t,x) = \int_{\mathbb{R}} e^{it\tau}W_\phi(t)(\mathcal{F}_tW_\phi(-\cdot)F)(\tau)d\tau = W_\phi(t)\mathcal{F}_t^{-1}\{\mathcal{F}_tW_\phi(-\cdot)F\}(\tau),$$

and by the Sobolev embedding inequality, we obtain for $b > \frac{1}{2}$,

$$\begin{aligned}\|\psi(\delta^{-1}t)F\|_{X_{s,b}(\phi)} &= \|\psi(\delta^{-1}t)W_\phi(-\cdot)F\|_{H_t^b(H_x^s)}\\ &\leq \|\psi(\delta^{-1}t)\|_{H_t^b(\mathbb{R})}\|W_\phi(-\cdot)F\|_{H_t^b(H_x^s)}\\ &\leq C\delta^{\frac{1-2b}{2}}\|F\|_{X_{s,b}(\phi)}.\end{aligned}$$

(ii) Without loss of generality, we may assume $F \in \mathcal{S}(\mathbb{R}^n \times \mathbb{R})$, since the general case then follows from an approximation argument.

Let $Kg \triangleq \psi(\delta^{-1}t)\int_0^t g(\tau)d\tau$, we first show that

$$\|Kg\|_{H_t^b} \leq C\delta^{1+b'-b}\|g\|_{H_t^{b'}} + C_0\delta^{\frac{1-2b}{2}}\|\langle\tau\rangle^{-1}\mathcal{F}_tg\|_{L_\tau^1}, \tag{4.127}$$

where we may choose $C_0 = 0$ if $b' > -\frac{1}{2}$.

Noticing that

$$\int_0^t g(\tau)d\tau = g * \chi_{[0,t]}(t),$$

we have

$$\mathcal{F}_t\{g * \chi_{[0,t]}(t)\}(\tau) = C\mathcal{F}_t g(\tau)\mathcal{F}_t\chi_{[0,t]}(\tau) = C\frac{1-e^{-it\tau}}{i\tau}\mathcal{F}_t g(\tau),$$

and thus

$$\int_0^t g(\tau)d\tau = C\int_{\mathbb{R}} \frac{e^{it\tau}-1}{i\tau}\mathcal{F}_t g(\tau)d\tau. \tag{4.128}$$

By Taylor's expansion, we have

$$\begin{aligned} Kg &= \psi(\delta^{-1}t)\sum_{k\geq 1}\frac{t^k}{k!}\int_{|\tau|\delta\leq 1}(i\tau)^{k-1}\mathcal{F}_t g(\tau)d\tau - \psi(\delta^{-1}t)\int_{|\tau|\delta\geq 1}(i\tau)^{-1}\mathcal{F}_t g(\tau)d\tau \\ &\quad + \psi(\delta^{-1}t)\int_{|\tau|\delta\geq 1}(i\tau)^{-1}\exp(it\tau)\mathcal{F}_t g(\tau)d\tau \triangleq \mathrm{I} + \mathrm{II} + \mathrm{III}. \end{aligned} \tag{4.129}$$

By the fact that $\operatorname{supp}\psi(t) \subset (-2,2)$, we have

$$\begin{aligned} \int_{|\tau|\delta\leq 1}|\tau|^{k-1}|\mathcal{F}_t g(\tau)|d\tau &\leq \delta^{1-k}\int_{|\tau|\delta\leq 1}\langle\tau\rangle^{-b'}\langle\tau\rangle^{b'}|\mathcal{F}_t g(\tau)|d\tau \\ &\leq \delta^{1-k}\Big(\int_{|\tau|\delta\leq 1}\langle\tau\rangle^{-2b'}d\tau\Big)^{\frac{1}{2}}\|g\|_{H_t^{b'}} \\ &\leq C\delta^{\frac{1}{2}+b'-k}\|g\|_{H_t^{b'}}, \end{aligned}$$

and

$$\begin{aligned} \|t^k\psi(\delta^{-1}t)\|_{H_t^b}^2 &\leq \int_{\mathbb{R}}\langle\tau\rangle^{2b}|(\partial_\tau^k(\mathcal{F}_t\psi(\delta^{-1}\cdot)))(\tau)|^2d\tau \\ &= \delta^{2k+2}\int_{\mathbb{R}}\langle\tau\rangle^{2b}|(\mathcal{F}_t\psi)^{(k)}(\delta\tau)|^2d\tau \\ &\leq C\delta^{2k-2b+1}\int_{\mathbb{R}}\langle\tau\rangle^{2b}|(\mathcal{F}_t\psi)^{(k)}(\tau)|^2d\tau = C\delta^{2k-2b+1}\|t^k\psi\|_{H_t^b}^2 \\ &\leq C\delta^{2k-2b+1}\|t^k\psi\|_{H_t^1}^2 \leq C\delta^{2k-2b+1}2^k(k+1)\|\psi\|_{H_t^1}^2. \end{aligned}$$

Thus, we obtain

$$\|\mathrm{I}\|_{H_t^b} \leq \sum_{k\geq 1} \frac{1}{k!} \|t^k \psi(\delta^{-1}t)\|_{H_t^b} \int_{|\tau|\delta\leq 1} |\tau|^{k-1} |\mathcal{F}_t g(\tau)| d\tau \leq C\delta^{1+b'-b} \|g\|_{H_t^{b'}}. \tag{4.130}$$

Next, we consider the second contribution II. For $b \geq 0$, we have

$$\|\mathrm{II}\|_{H_t^b} \leq C\|\psi(\delta^{-1}t)\|_{H_t^b} \int_{|\tau|\delta\geq 1} |\tau|^{-1} |\mathcal{F}_t g(\tau)| d\tau \leq C_0 \delta^{\frac{1-2b}{2}} \|\langle\tau\rangle^{-1} \mathcal{F}_t g\|_{L_\tau^1}. \tag{4.131}$$

For $b' > -\frac{1}{2}$, we use Cauchy–Schwarz's inequality to obtain

$$\begin{aligned} \|\mathrm{II}\|_{H_t^b} &\leq C\|\psi(\delta^{-1}t)\|_{H_t^b} \int_{|\tau|\delta\geq 1} |\tau|^{-1} |\mathcal{F}_t g(\tau)| d\tau \\ &\leq C\delta^{\frac{1-2b}{2}} \|g\|_{H_t^{b'}} \Big(\int_{|\tau|\delta\geq 1} |\tau|^{-2} \langle\tau\rangle^{-2b'} d\tau \Big)^{\frac{1}{2}} \\ &\leq C\delta^{1+b'-b} \|g\|_{H_t^{b'}}. \end{aligned} \tag{4.132}$$

Finally, we consider the third contribution III. It is easy to check that

$$\int_{|\tau|\delta\geq 1} (i\tau)^{-1} \exp(it\tau) \mathcal{F}_t g(\tau) d\tau = C\mathcal{F}_t^{-1} (i\tau)^{-1} \chi_{|\tau|\delta\geq 1} \mathcal{F}_t g \triangleq J.$$

Thus, we have

$$\|J\|_{H_t^b}^2 \leq C \int_{|\tau|\delta\geq 1} \langle\tau\rangle^{2b-2-2b'} \langle\tau\rangle^{2b'} |\mathcal{F}_t g(\tau)|^2 d\tau \leq C \sup_{|\tau|\delta\geq 1} |\tau|^{2b-2-2b'} \|g\|_{H_t^{b'}}.$$

For all $b, b' \in \mathbb{R}$ satisfying that $b - b' < 1$, this gives

$$\|J\|_{H_t^b} \leq C\delta^{1+b'-b} \|g\|_{H_t^{b'}}. \tag{4.133}$$

For the Fourier transform of the product $\psi(\delta^{-1}\cdot)J$, we have

$$\begin{aligned} \langle\tau\rangle^b \mathcal{F}_t(\psi(\delta^{-1}\cdot)J)(\tau) &= \langle\tau\rangle^b \int_{\mathbb{R}} \mathcal{F}_t\psi(\delta^{-1}\cdot)(\tau_1) \mathcal{F}_t J(\tau - \tau_1) d\tau_1 \\ &\leq C \int_{\mathbb{R}} |\tau_1|^b |\mathcal{F}_t\psi(\delta^{-1}\cdot)(\tau_1) \mathcal{F}_t J(\tau - \tau_1)| d\tau_1 \\ &\quad + C \int_{\mathbb{R}} |\mathcal{F}_t\psi(\delta^{-1}\cdot)(\tau_1)| \langle\tau - \tau_1\rangle^b |\mathcal{F}_t J(\tau - \tau_1)| d\tau_1, \end{aligned}$$

which gives

$$\begin{aligned}\|\mathrm{III}\|_{H_t^b} &\lesssim \big\|(|\tau|^b|\mathcal{F}_t\psi(\delta^{-1}\cdot)|) * |\mathcal{F}_tJ|\big\|_{L^2_\tau} + \big\||\mathcal{F}_t\psi(\delta^{-1}\cdot)| * (\langle\tau-\tau_1\rangle^b|\mathcal{F}_tJ|)\big\|_{L^2_\tau}\\ &\lesssim \big\||\tau|^b|\mathcal{F}_t\psi(\delta^{-1}\cdot)|\big\|_{L^1_\tau}\|\mathcal{F}_tJ\|_{L^2_\tau} + \|\mathcal{F}_t\psi(\delta^{-1}\cdot)\|_{L^1_\tau}\|J\|_{H^b_\tau}\\ &\le C(\delta^{-b}\|J\|_{L^2_\tau} + \|J\|_{H^b_\tau}) \le C\delta^{1+b'-b}\|g\|_{H_t^{b'}}.\end{aligned} \tag{4.134}$$

And so (4.127) follows. Squaring both sides of (4.127), multiplying by $\langle\xi\rangle^{2s}$ and integrating with respect to $\xi \in \mathbb{R}^n$, we obtain

$$\|Kg\|^2_{H_t^bH_x^s} \le C\delta^{2(1+b'-b)}\|g\|^2_{H_t^{b'}H_x^s} + 2C_0\delta^{1-2b}\|\langle\xi\rangle^s\langle\tau\rangle^{-1}\mathcal{F}g\|^2_{L^2_\xi L^1_t}.$$

This implies

$$\|Kg\|_{H_t^bH_x^s} \le C\delta^{1+b'-b}\|g\|_{H_t^{b'}H_x^s} + C_1\delta^{\frac{1-2b}{2}}\|\langle\xi\rangle^s\langle\tau\rangle^{-1}\mathcal{F}g\|_{L^2_\xi L^1_t}, \tag{4.135}$$

with $C_1 = \sqrt{2}C_0$. Applied to $g(t) = W_\phi(-t)F(t)$, we get (4.125).

(iii) As for $F(x,t) \in \mathcal{S}(\mathbb{R}^n \times \mathbb{R})$, it is easy to check that $W_{\phi\star R}F \in C_t([-T,T]; H^s(\mathbb{R}^n))$. The general case follows by (4.126) and an approximation argument. It suffices to show (4.126). In fact, by decomposing the frequency $|\tau| \le 1$ and $|\tau| \ge 1$, we have

$$\left|\frac{e^{it\tau}-1}{\tau}\right| \le |t|\chi_{|\tau|\le 1} + 2|\tau|^{-1}\chi_{|\tau|\ge 1} \le C(1+|t|)\langle\tau\rangle^{-1}.$$

Using (4.128) and Plancherel's theorem, we see that

$$\begin{aligned}\left\|\int_0^t g(\tau)d\tau\right\|^2_{L^2_x} &= \int_{\mathbb{R}^n}\int_0^t\int_0^t \frac{e^{it\tau}-1}{i\tau}\mathcal{F}_tg(\tau)\frac{e^{-it\tau'}-1}{-i\tau'}\overline{\mathcal{F}_tg(\tau')}d\tau'd\tau dx\\ &\le C(1+t^2)\int_{\mathbb{R}^n}\int_0^t\int_0^t \langle\tau\rangle^{-1}|\mathcal{F}_tg(\tau)|\langle\tau'\rangle^{-1}|\mathcal{F}_tg(\tau')|d\tau'd\tau dx\\ &\le C(1+t^2)\|\langle\tau\rangle^{-1}|\mathcal{F}_tg(\tau)|\|^2_{L^2_\xi L^1_\tau}.\end{aligned} \tag{4.136}$$

Applying to $g(t) = J_x^sW_\phi(-t)F(t)$, we obtain (4.126). □

Remark 4.3.3.

(i) Let $s \in \mathbb{R}$, $\delta \in (0,1]$, $b \in (1/2,1]$ and $b' = 1-b$. Then, we have

$$\|\psi(\delta^{-1}t)W_{\phi\star R}F\|_{X_{s,b}(\phi)} \le C\|F\|_{X_{s,b'}(\phi)} \le C\delta^{\frac{1-2b}{2}}\|F\|_{X_{s,b-1}(\phi)}. \tag{4.137}$$

Noticing that $W_\phi(t)$ is a unitary group on $H^s(\mathbb{R}^n)$, we have by the Sobolev inequality

$$\sup_t \|\psi(\delta^{-1}t)W_{\phi\star R}F(\tau,x)\|_{H^s} \le C\delta^{\frac{1-2b}{2}}\|F\|_{X_{s,b-1}}. \tag{4.138}$$

Regarding $\int_0^t$ as the inverse of the derivative abstractly, we obtain from (4.124)

$$\begin{aligned}\left\|\psi(\delta^{-1}t)\int_0^t e^{i(t-\tau)\phi(D)}F(\tau,x)d\tau\right\|_{X_{s,b}} &\le \delta^{\frac{1-2b}{2}}\left\|\int_0^t e^{i\tau\triangle}F(\tau,x)d\tau\right\|_{X_{s,b}}\\ &\le \delta^{\frac{1-2b}{2}}\|F\|_{X_{s,b-1}},\end{aligned}$$

which is (4.137).

(ii) The inhomogeneous term $W_{\phi\star R}F$ satisfies the following:

$$W_{\phi\star R}F(t+t_1) = W_\phi(t)W_{\phi\star R}F(t_1) + W_{\phi\star R}(\tau_{-t_1}F)(t), \tag{4.139}$$

where $\tau_{-t_1}F(t) = F(t+t_1)$.

(iii) The estimate (4.124) have the following generalization. Let $0 \le b' < b < \frac{1}{2}$ or $-\frac{1}{2} < b' < b \le 0$. Then

$$\|\psi(\delta^{-1}t)F\|_{X_{s,b'}(\phi)} \le C\delta^{b-b'}\|F\|_{X_{s,b}(\phi)}, \tag{4.140}$$

$$\|\psi(\delta^{-1}t)F\|_{X_{s,\frac{1}{2}}(\phi)} \le C(\varepsilon)\delta^{-\varepsilon}\|F\|_{X_{s,\frac{1}{2}}(\phi)}, \quad \varepsilon > 0. \tag{4.141}$$

It is easy to see that (4.141) is the conclusion of the multiplier estimate:

$$\|fg\|_{H_t^{\frac{1}{2}}} \le C\|f\|_{H_t^{\frac{1}{2}+\varepsilon}}\|g\|_{H_t^{\frac{1}{2}}}, \quad \forall\varepsilon > 0. \tag{4.142}$$

While the proof of (4.140) may turn to the multiplier estimate

$$\|fg\|_{H_y^s} \le C\|f\|_{H_y^{s_1}}\|g\|_{H_y^{s_2}}, \tag{4.143}$$

where

$$0 \le s \le \min(s_1, s_2), \quad s < s_1 + s_2 - \frac{n}{2} \tag{4.144}$$

or

$$0 \le s < \min(s_1, s_2), \quad s \le s_1 + s_2 - \frac{n}{2}. \tag{4.145}$$

In fact, if $b > b' \ge 0$, let $s = b'$, $s_1 = \frac{1}{2} - (b - b')$, $s_2 = b$, then we have

$$\|\psi(\delta^{-1}\cdot)g\|_{H_t^{b'}} \le C\|\psi(\delta^{-1}t)\|_{H_t^{s_1}}\|g\|_{H_t^b} \le C\delta^{b-b'}\|g\|_{H_t^b},$$

where we have used

$$\|\psi(\delta^{-1}t)\|_{H_t^{s_1}} = C\delta^{\frac{1}{2}-s_1}\|\psi\|_{H_t^{s_1}}.$$

Hence, for $F \in X_{s,b}(\phi)$, we have

$$\begin{aligned}\|\psi(\delta^{-1}\cdot)F\|_{X_{s,b'}(\phi)} &\le C\|W_\phi(-\cdot)\psi(\delta^{-1}\cdot)F\|_{H_t^{b'}(H_x^s)}\\ &= C\|\psi(\delta^{-1}t)W_\phi(-\cdot)F\|_{H_t^{b'}(H_x^s)}\\ &\le C\delta^{b-b'}\|W_\phi(-\cdot)F\|_{H_t^b(H_x^s)} = C\delta^{b-b'}\|F\|_{X_{s,b}(\phi)}.\end{aligned}$$

As for $-\frac{1}{2} < b' < b \le 0$, by duality, we get (4.140).

Before discussing Bourgain's bilinear Strichartz estimate, we first introduce some notation. Assume $\chi(\xi) \in C_c(\mathbb{R}^n)$ is a bump function satisfying

$$\chi(\xi) = 1, \quad |\xi| < 1; \quad \chi(\xi) = 0, \quad |\xi| \ge 2.$$

Let $\psi(\xi) = \chi(\frac{\xi}{2}) - \chi(\xi), \chi_j(\xi) = \chi(2^{-j}\xi), \psi_j(\chi) = \psi(2^{-1}\xi)$, then the operator $S_0 = \mathcal{F}^{-1}\chi(\xi)\mathcal{F}\cdot$, $\Delta_j = \mathcal{F}^{-1}\psi_j(\xi)\mathcal{F}\cdot$ can construct a standard Littlewood–Paley decomposition, that is,

$$\sum_{j\in\mathbb{Z}} \Delta_j = I \text{ (homog. decomp.)}, \quad S_0 + \sum_{j\in\mathbb{N}} \Delta_j = I \text{ (inhomog. decomp.)}. \tag{4.146}$$

Lemma 4.3.1. *Let $n = 3$. Then, for $\ell \ge m$, the estimate*

$$\|(e^{-it\Delta}\Delta_\ell\varphi_1)(e^{-it\Delta}\Delta_m\varphi_2)\|_{L^2(\mathbb{R}\times\mathbb{R}^3)} \lesssim 2^{m-\frac{\ell}{2}}\|\Delta_\ell\varphi_1\|_{L^2(\mathbb{R}^3)}\|\Delta_m\varphi_2\|_{L^2(\mathbb{R}^3)}, \tag{4.147}$$

holds.

Proof. By the standard Strichartz estimate, we have

$$\begin{aligned}&\|(e^{-it\Delta}\Delta_\ell\varphi_1)(e^{-it\Delta}\Delta_m\varphi_2)\|_2\\ &\quad\le \|e^{-it\Delta}\Delta_\ell\varphi_1\|_4 \times \|e^{-it\Delta}\Delta_m\varphi_2\|_4\\ &\quad\le C\|\Delta_\ell\varphi_1\|_{\dot H^{\frac14}}\|\Delta_m\varphi_2\|_{\dot H^{\frac14}} \le 2^{\frac{\ell+m}{4}}\|\Delta_\ell\varphi_1\|_2\|\varphi_2\|_2\\ &\quad\lesssim 2^{\frac34(\ell-m)}2^{m-\frac{\ell}{2}}\|\Delta_\ell\varphi_1\|_2\|\Delta_m\varphi_2\|_2,\end{aligned}$$

which implies (4.147) when $\ell \sim m$.

Now, we assume $m \ll \ell$. By the Fourier transform and the Cauchy–Schwarz inequality, we have

$$\begin{aligned}&\hat\varphi_1(\xi_1)\hat\varphi_2(\xi_2)\psi_\ell(\eta_1)\psi_m(\eta_2)\cdot\hat\varphi_1(\eta_1)\hat\varphi_2(\eta_2)\psi_\ell(\xi_1)\psi_m(\xi_2)\\ &\quad\le \frac12|\hat\varphi_1(\xi_1)\hat\varphi_2(\xi_2)|^2\psi_\ell(\eta_1)\psi_m(\eta_2) + \frac12|\hat\varphi_1(\eta_1)\hat\varphi_2(\eta_2)|^2\psi_\ell(\xi_1)\psi_m(\xi_2),\end{aligned}$$

which implies that

$$\begin{aligned}
&\|(e^{-it\Delta}\Delta_\ell\varphi_1)(e^{-it\Delta}\Delta_m\varphi_2)\|_2^2 \\
&= C\int_{\mathbb{R}^3}\int_{\xi_1+\xi_2=\xi}\int_{\eta_1+\eta_2=\xi}\delta\Big(\sum_{i=1}^2(|\xi_i|^2-|\eta_i|^2)\Big)\prod_{i=1}^2\hat{\varphi}_i(\xi_i)\hat{\varphi}_i(\eta_i) \\
&\quad\times\psi_\ell(\xi_1)\psi_m(\xi_2)\psi_\ell(\eta_1)\psi_m(\eta_2)d\xi_1 d\eta_1 d\xi \\
&\le \frac{1}{2}C(I_1+I_2) = CI_1 \quad \text{(symmetry of integral variables)},
\end{aligned} \tag{4.148}$$

where

$$\begin{aligned}
I_1 = &\int_{\mathbb{R}^3}\int_{\xi_1+\xi_2=\xi}\int_{\eta_1+\eta_2=\xi}\delta\Big(\sum_{i=1}^2(|\xi_i|^2-|\eta_i|^2)\Big)\psi_\ell(\eta_1)\psi_m(\eta_2)d\eta_1 \\
&\times|\hat{\varphi}_1(\xi_1)\hat{\varphi}_2(\xi_2)|^2 d\xi_1 d\xi.
\end{aligned}$$

Noticing that

$$\int_{\mathbb{R}^3}\delta(P(x))f(x)dx = \int_{\{x:P(x)=0\}}\frac{f(x)}{|\nabla P(x)|}d\sigma_x, \tag{4.149}$$

hence we have

$$\begin{aligned}
J(\xi,\xi_1) &\triangleq \int_{\eta_1+\eta_2=\xi}\delta(|\xi_1|^2+|\xi_2|^2-|\eta_1|^2-|\eta_2|^2)\psi_\ell(\eta_1)\psi_m(\eta_2)d\eta_1 \\
&= \int_{\{\eta_1:P(\eta_1)=0\}}\psi_\ell(\eta_1)\psi_m(\xi-\eta_1)\frac{d\sigma_{\eta_1}}{|\nabla_{\eta_1}P(\eta_1)|},
\end{aligned} \tag{4.150}$$

where $P(\eta_1) \triangleq |\eta_1|^2 + |\xi-\eta_1|^2 - |\xi_1|^2 - |\xi_2|^2$. Hence,

$$|\nabla P(\eta_1)| = |4\eta_1 - 2\xi| = 2|\eta_1-\eta_2| \gtrsim 2^\ell.$$

This gives

$$|J(\xi,\xi_1)| \le C2^{-\ell}\int_{\{\eta_1:P(\eta_1)=0\}}\psi_m(\xi-\eta_1)d\sigma_{\eta_1} \lesssim 2^{2m-\ell}, \tag{4.151}$$

since $\int_{\{\eta_1:P(\eta_1)=0\}}\psi_m(\xi-\eta_1)d\sigma_{\eta_1}$ is the area of the intersection of $\{\eta_1 : P(\eta_1) = 0\}$. Finally, we conclude that

$$I_1 \lesssim 2^{2m-\ell}\|\varphi_1\|_2^2\|\varphi_2\|_2^2. \qquad \square$$

By Lemma 4.3.1, we can obtain the following generalized Strichartz inequality.

Theorem 4.3.1 (Bilinear Strichartz estimate). *Let $0 \le \rho < 1/2$ and let us assume that $a_+ = a + \varepsilon$, $0 < \varepsilon \ll 1$. Then*

$$\|D_x^\rho((e^{-it\Delta}\varphi_1)\cdot(e^{-it\Delta}\varphi_2))\|_{L^2_{t,x}} \le C\|\varphi_1\|_{H^{\frac{1}{2}+\rho}}\|\varphi_2\|_{L^2_x}, \tag{4.152}$$

$$\|D_x^\rho(u_1u_2)\|_{L^2_{t,x}(I\times\mathbb{R}^3)} \le C\|u_1\|_{X_{\frac{1}{2}+\rho,\frac{1}{2}+}(I)}\|u_2\|_{X_{0,\frac{1}{2}+}(I)}. \tag{4.153}$$

Moreover, for $\frac{1}{2} \le \rho < \sigma + \frac{1}{2}$ and $\sigma \le 1$ we have

$$\|D_x^\rho(e^{-it\Delta}\varphi_1\cdot e^{-it\Delta}\varphi_2)\|_{L^2_{t,x}} \le C\|\varphi_1\|_{H^\sigma}\|\varphi_2\|_{H^{\rho+\frac{1}{2}-\sigma}}, \tag{4.154}$$

$$\|D_x^\rho(u_1u_2)\|_{L^2_{t,x}(I\times\mathbb{R}^3)} \le C\|u_1\|_{X_{\sigma,\frac{1}{2}+}(I)}\|u_2\|_{X_{\rho+\frac{1}{2}-\sigma,\frac{1}{2}+}(I)}, \tag{4.155}$$

where $X_{s,b}(I)$ is the restriction space of $X_{s,b}$ on I, the corresponding norm is defined by

$$\|u\|_{X_{s,b}(I)} = \inf_{\tilde{u}|_I=u}\|\tilde{u}\|_{X_{s,b}}. \tag{4.156}$$

Proof. We first show (4.152) and (4.154). By the Littlewood–Paley decomposition, we have

$$\varphi_1 = \sum_{j\in\mathbb{Z}}\Delta_j\varphi_1, \quad \varphi_2 = \sum_{k\in\mathbb{Z}}\Delta_k\varphi_2,$$

where $\Delta_j f(x) = \mathcal{F}^{-1}\hat{\psi}_j\mathcal{F}f$, $\operatorname{supp}\hat{\psi}(\xi) = \{\xi, |\xi| \sim 2^j\}$. Therefore,

$$\begin{aligned}
&\|D_x^\rho(e^{-it\Delta}\varphi_1\cdot e^{-it\Delta}\varphi_2)\|_2\\
&\quad\le \sum_j\sum_k\|D_x^\rho(e^{-it\Delta}\Delta_j\varphi_1\cdot e^{-it\Delta}\Delta_k\varphi_2)\|_2\\
&\quad\lesssim \sum_j\sum_{k\le j}2^{j\rho}\|e^{-it\Delta}\Delta_j\varphi_1\cdot e^{-it\Delta}\Delta_k\varphi_2\|_2\\
&\qquad+\sum_j\sum_{k\ge j}2^{k\rho}\|e^{-it\Delta}\Delta_j\varphi_1\cdot e^{-it\Delta}\Delta_k\varphi_2\|_2.
\end{aligned} \tag{4.157}$$

Note that

$$\begin{aligned}
&2^{j\rho}\cdot\frac{2^j}{2^{\frac{k}{2}}},\quad 2^{k\rho}\cdot\frac{2^j}{2^{\frac{k}{2}}} \le 2^{j(\rho+\frac{1}{2})},\quad j\le k,\ 0\le\rho<\frac{1}{2},\\
&2^{j\rho}\cdot\frac{2^k}{2^{\frac{j}{2}}},\quad 2^{k\rho}\cdot\frac{2^k}{2^{\frac{j}{2}}} \le 2^{j(\rho+\frac{1}{2})},\quad j\ge k,\ 0\le\rho<\frac{1}{2},
\end{aligned}$$

which, together with (4.148), (4.157) and the Littlewood–Paley characterization of the Sobolev space, implies that

$$\|D_x^\rho(e^{-it\Delta}\varphi_1 \cdot e^{-it\Delta}\varphi_2)\|_2 \le \sum_j \sum_k 2^{j(\rho+\frac{1}{2})}\|\Delta_j\varphi_1\|_2 \cdot \|\Delta_k\varphi_2\|_2$$
$$\le C\|\varphi_1\|_{H^{\frac{1}{2}+\rho}}\|\varphi_2\|_2, \quad 0 \le \rho < \frac{1}{2}. \tag{4.158}$$

On the other hand, using

$$2^{j\rho} \cdot \frac{2^j}{2^{\frac{k}{2}}}, \quad 2^{k\rho} \cdot \frac{2^j}{2^{\frac{k}{2}}} \le 2^{j\sigma}2^{k(\rho+\frac{1}{2}-\sigma)}, \quad j \le k, \ \frac{1}{2} \le \rho < \sigma + \frac{1}{2}, \sigma \le 1,$$
$$2^{j\rho} \cdot \frac{2^k}{2^{\frac{j}{2}}}, \quad 2^{k\rho} \cdot \frac{2^k}{2^{\frac{j}{2}}} \le 2^{j\sigma}2^{k(\rho+\frac{1}{2}-\sigma)}, \quad j \ge k, \ \frac{1}{2} \le \rho < \sigma + \frac{1}{2}, \sigma \le 1,$$

together with (4.148), (4.157) and the Littlewood–Paley characterization of the Sobolev space, we get

$$\|D_x^\rho(e^{-it\Delta}\varphi_1 \cdot e^{-it\Delta}\varphi_2)\|_2 \le \sum_j \sum_k 2^{j\sigma}\|\Delta_j\varphi_1\|_2 \cdot 2^{k(\rho+\frac{1}{2}-\sigma)}\|\Delta_k\varphi_2\|_2$$
$$\le C\|\varphi_1\|_{H^\sigma}\|\varphi_2\|_{H^{\rho+\frac{1}{2}-\sigma}}. \tag{4.159}$$

By (4.152), (4.154) and transfer principle, we get (4.153) and (4.155). □

Remark 4.3.4. For $s \ge \max(s_c, 0)$, by Theorem 2.2.1 in Section 2.2, we can obtain the local well-posedness of nonlinear Schrödinger in the following space:

$$X(I) = C([0,T^*); H^s) \cap \bigcap_{(q,r)\in\Lambda} L^q([0,T^*); B^s_{r,2}).$$

See details in Section 2.2. For $s_c < 0$, we hope to obtain the local well-posedness of nonlinear dispersive equation in H^s, $s \ge s_c$. In this case, we need to turn to Bourgain's restriction norm method (make full use of the algebraic structure, multilinear multiplier estimate). The general method to establish the local well-posedness of Cauchy problem of nonlinear dispersive equation is described shortly. Consider the Cauchy problem

$$iu_t + \phi(D)u = N(u), \quad u(0) = u_0(x). \tag{4.160}$$

The local well-posedness on $I = (-\delta, \delta)$ is equivalent to solving the integral equation

$$u(t) = \psi(\delta^{-1}t)W_\phi(t)u_0 + \psi(\delta^{-1}t)W_{\phi\star R}N(u)(t)(t-\tau)d\tau. \tag{4.161}$$

(i) For the case $b + b' = 1$, $b > \frac{1}{2}$, by Proposition 4.3.2 and the analogue nonlinear estimate as

$$\|W_{\phi\star R}(N(u) - N(v)\|_{X^I_{s,b}(\phi)} \le C\delta^{1-b-b'}C_0(\|u\|_{X^I_{s,b}(\phi)}$$
$$+ \|v\|_{X^I_{s,b}(\phi)})\|u - v\|_{X^I_{s,b}(\phi)}, \tag{4.162}$$

we can show that (4.160) is locally well posed in $X^I_{s,b}(\phi) \subset C(I; H^s(\mathbb{R}^n))$, where $X^I_{s,b}(\phi)$ is the restriction space of $X_{s,b}(\phi)\mathbb{R}\times\mathbb{R}^n$ on $I\times\mathbb{R}^n$. We remark that $\delta = \delta(\|u_0(x)\|_{H^s}) > 0$ is monotone decreasing.

(ii) For the case $b + b' = 1$, $b = \frac{1}{2}$. $X^I_{s,b}(\phi)$ is the critical space corresponding to the time variable. Now, we consider another equivalent form of (4.160)

$$u(t) = \psi(\delta^{-1}t)W_\phi(t)u_0 + \psi(\delta^{-1}t)W_{\phi\star R}N(\psi(2\delta)^{-1}t)u(t)(t-\tau)d\tau.$$

The reason is that we cannot obtain the contraction factor from the linear estimate, but obtain it from nonlinear estimate. In order to obtain the well-posednes of (4.162) in the critical space, we should use Proposition 4.3.2 and the analogue estimate as

$$\begin{aligned} &\|W_{\phi\star R}(N(\psi((2\delta)^{-1}t)u) - N(\psi((2\delta)^{-1}t)v)\|_{X^I_{s,b}(\phi)\cap Y^I_s(\phi)} \\ &\quad \le C\delta^\theta C_0(\|u\|_{X^I_{s,b}(\phi)} + \|v\|_{X^I_{s,b}(\phi)})\|u-v\|_{X^I_{s,b}(\phi)}, \end{aligned} \tag{4.163}$$

where $\theta > 0$; See the details in [111].

Based on the above analysis, we aim to establish the global well-posedness of nonlinear Schrödinger (4.116) in $H^s(\mathbb{R}^3)$ $(s < 1)$. By Bourgain's method, we discuss the following steps.

Step 1 (Decomposition of initial data). For any $\varphi(x) \in H^s(\mathbb{R}^3)$, we decompose φ into the high frequency part and low frequency part, that is,

$$\varphi = \varphi_0(x) + \psi_0(x), \quad \varphi_0(x) = (2\pi)^{-\frac{3}{2}} \int_{|\xi|\le N_0} \hat\varphi(\xi)e^{ix\cdot\xi}d\xi, \tag{4.164}$$

where N_0 is a cutoff (to be specified largely). It is easy to check that

$$\hat\varphi(\xi) = \hat\varphi_0(\xi) + \hat\psi_0(\xi) \equiv \hat\varphi(\xi)\chi_{|\xi|\le N_0} + \hat\varphi(\xi)(1-\chi_{|\xi|\le N_0}), \tag{4.165}$$

and we have

$$\begin{aligned} \|\varphi_0(x)\|_{H^1} &= \|\xi\hat\varphi_0\|_2 + \|\hat\varphi_0\|_2 \le N_0^{1-s}\|\xi^s\hat\varphi_0\|_2 + \|\hat\varphi_0\|_2 \\ &\lesssim N_0^{1-s}, \quad 0 < s < 1. \end{aligned} \tag{4.166}$$

Since $H^s \hookrightarrow L^4$ for $s \ge \frac{3}{4}$, we have

$$E(\varphi_0) = \frac{1}{4}\int_{\mathbb{R}^3} |\varphi_0|^4 dx + \frac{1}{2}\int_{\mathbb{R}^3} |\nabla\varphi_0|^2 dx \lesssim N_0^{2(1-s)}. \tag{4.167}$$

Step 2 (Estimate of low frequency evolution (NLS) on $I = [0, \triangle T]$). By $\varphi_0(x) \in C^\infty(\mathbb{R}^3) \cap H^k(\mathbb{R}^3)$, $\forall k \in \mathbb{Z}$ and the classical well-posedness theory, the equation

$$\begin{cases} iu_{0t}(t) - \Delta u_0 = -|u_0|^2 u_0, \\ u_0(0,x) = \varphi_0(x), \end{cases} \tag{4.168}$$

has a global smooth solution $u_0(t) \in C(\mathbb{R}; H^1(\mathbb{R}^3)) \cap C^1(\mathbb{R}; L^2(\mathbb{R}^3))$. Let $I = [0, \triangle T]$, $\triangle T \sim N_0^{-(1-s)\frac{9}{2}-}$. Then $u_0(t)$ satisfies

$$\|u_0(t)\|_{L^5_{t,x}(I\times\mathbb{R}^3)}, \quad \|D_x^{\frac{3}{10}} u_0(t)\|_{L^5_t(I;L_x^{\frac{10}{3}})} = o(1), \tag{4.169}$$

$$\|u_0(t)\|_{X_{0,\frac{1}{2}+}(I)} < C, \quad \|u_0(t)\|_{X_{1,\frac{1}{2}+}(I)} < N_0^{1-s}. \tag{4.170}$$

In fact, by supp $\varphi_0(x) \subset \{\xi||\xi| \leq N_0\}$ and Sobolev's inequality, we have

$$\begin{aligned} \|e^{-it\Delta}\varphi_0\|_{L^5_{t,x}(I\times\mathbb{R}^3)} &\lesssim \|D_x^{\frac{3}{10}}(e^{-it\Delta}\varphi_0)\|_{L^5_t(I;L_x^{\frac{10}{3}}(\mathbb{R}^3))} \\ &\lesssim |I|^{\frac{1}{5}} \sup_t \|D_x^{\frac{3}{10}}(e^{-it\Delta}\varphi_0)\|_{\frac{10}{3}} \lesssim |I|^{\frac{1}{5}} \|\varphi_0\|_{\dot{H}^{\frac{9}{10}}} \\ &\lesssim |I|^{\frac{1}{5}} N_0^{\frac{9}{10}-s} < |I|^{\frac{1}{5}} N_0^{\frac{9}{10}(1-s)}, \end{aligned}$$

where we have used $\dot{H}^{\frac{9}{10}} \hookrightarrow \dot{W}^{\frac{3}{10},\frac{10}{3}}$. Note that $|I| = \triangle T = N_0^{-(1-s)\frac{9}{2}-}$. This implies

$$\|D_x^{\frac{3}{10}}(e^{-it\Delta}\varphi_0)\|_{L^5_t(I;L_x^{\frac{10}{3}}(\mathbb{R}^3))} = o(1). \tag{4.171}$$

Hence, by Sobolev's inequality,

$$\|u_0(t)\|_{L^5_{t,x}(I\times\mathbb{R}^3)} \lesssim \|D_x^{\frac{3}{10}} u_0(t)\|_{L^5_t(I;L_x^{\frac{10}{3}})},$$

and by (4.171), Hardy–Littlewood–Sobolev's inequality, we obtain

$$\begin{aligned} \|D_x^{\frac{3}{10}} u_0(t)\|_{L^5_t(I;L_x^{\frac{10}{3}})} &\lesssim o(1) + \left\| \int_0^t |t-\tau|^{-\frac{3}{5}} \|D_x^{\frac{3}{10}} u_0(t)\|_{L_x^{\frac{10}{3}}} \|u_0(t)\|^2_{L^5_x} d\tau \right\|_{L^5_t} \\ &\lesssim o(1) + \|D_x^{\frac{3}{10}} u_0(t)\|_{L^5_t(I;L_x^{\frac{10}{3}})} \|u_0(t)\|^2_{L^5_t(I;L^5(\mathbb{R}^3))} \\ &\lesssim o(1) + \|D_x^{\frac{3}{10}} u_0(t)\|^3_{L^5_t(I;L_x^{\frac{10}{3}})}, \end{aligned}$$

which implies (4.169) if we choose $|I|$ sufficiently small (i. e., N_0 is sufficiently large).

Now we estimate (4.170). By Strichartz's estimate,

$$\|e^{\pm it\Delta}\varphi\|_{L^q(I;L^r(\mathbb{R}^3))} \lesssim \|\varphi\|_2, \quad (q,r) \in \Lambda$$

and Sobolev's inequality, we have

$$\|F(x,t)\|_{L^q(I;L^r(\mathbb{R}^3))} = \|e^{-it\Delta}e^{it\Delta}F(x,t)\|_{L^q(I;L^r(\mathbb{R}^3))} \lesssim \|e^{it\Delta}F(x,t)\|_{L^2(\mathbb{R}^3)}$$
$$\lesssim \|e^{it\Delta}F(x,t)\|_{H^{\frac{1}{2}+}(I;L^2(\mathbb{R}^3))} \lesssim \|F\|_{X_{0,\frac{1}{2}+}},$$

which implies that by duality

$$\|F\|_{X_{0,-\frac{1}{2}-}} \lesssim \|F(x,t)\|_{L^{q'}(I;L^{r'}(\mathbb{R}^3))}, \quad (q,r) \in \Lambda. \tag{4.172}$$

Interpolating (4.172) with $\|F\|_{X_{0,0}} = \|F(x,t)\|_{L^2(I;L^2(\mathbb{R}^3))}$, we obtain

$$\|F\|_{X_{0,-\frac{1}{2}+}} \lesssim \|F(x,t)\|_{L^{q'+}(I;L^{r'+}(\mathbb{R}^3))}, \quad (q,r) \in \Lambda. \tag{4.173}$$

Particularly, we have

$$\|F\|_{X_{0,-\frac{1}{2}+}} \lesssim \|F\|_{L^{\frac{10}{7}+}(I;L^{\frac{10}{7}+})}. \tag{4.174}$$

Let $D = (-\Delta)^{\frac{1}{2}}$ (or $(I-\Delta)^{\frac{1}{2}}$), $(q,r) \in \Lambda$. By the analogue estimate as in (4.172) and (4.173), we have

$$\|D^\rho F(x,t)\|_{L^q(I;L^r(\mathbb{R}^3))} = \|e^{-it\Delta}D^\rho e^{it\Delta}F\|_{L^q(I;L^r(\mathbb{R}^3))} \lesssim \|D^\rho e^{it\Delta}F\|_{L^2(\mathbb{R}^3)}$$
$$\lesssim \|e^{it\Delta}F\|_{H^{\frac{1}{2}+}(I;H^\rho(\mathbb{R}^3))} \lesssim \|F\|_{X_{\rho,\frac{1}{2}+}}, \quad 0 \le \rho \le 1. \tag{4.175}$$

By Sobolev's inequality, we have

$$\|F(x,t)\|_{L^q(I;L^{\tilde{r}}(\mathbb{R}^3))} \lesssim \|F\|_{X_{\rho,\frac{1}{2}+}}, \quad \frac{2}{q} = 3\left(\frac{1}{2} - \frac{\rho}{3} - \frac{1}{\tilde{r}}\right), \quad 0 \le \rho \le 1. \tag{4.176}$$

Particularly, we have

$$\|F\|_{L^5(I;L^5(\mathbb{R}^3))} \lesssim \|F\|_{X_{\frac{1}{2},\frac{1}{2}+}(I)}, \quad \|F\|_{L^{10}(I;L^{10}(\mathbb{R}^3))} \lesssim \|F\|_{X_{1,\frac{1}{2}+}(I)}. \tag{4.177}$$

Note that $\|\varphi_0\|_2 \le C$, $\|\varphi_0\|_{H^1} \le N_0^{1-s}$. Taking $\psi(t) \in C_c^\infty(\mathbb{R})$ in Proposition 4.3.2 (without loss of generality, assume that $|I| = \triangle T < 1/2$), by (4.174) and the interpolation

$$\|u_0(t)\|_{L^{5+}(I\times\mathbb{R}^3)} \le \|u_0(t)\|^{1-}_{L^5(I\times\mathbb{R}^3)}\|u_0(t)\|^{0+}_{L_t^{10}(I;L^{10}(\mathbb{R}^3))}, \tag{4.178}$$

we have

$$\|u_0(t)\|_{X_{1,\frac{1}{2}+}(I)}$$
$$\lesssim \|\varphi_0\|_{H^1} + \||u_0(t)|^2 Du_0(t)\|_{X_{0,-\frac{1}{2}+}(I)}$$
$$\lesssim \|\varphi_0\|_{H^1} + \||u_0|^2 Du_0\|_{L^{\frac{10}{7}+}(I\times\mathbb{R}^3)}$$

$$
\begin{aligned}
&\lesssim \|\varphi_0\|_{H^1} + \|u_0(t)\|_{L^5(I\times\mathbb{R}^3)}\|u_0(t)\|_{L^{5+}(I\times\mathbb{R}^3)}\|Du_0(t)\|_{L^{\frac{10}{3}}(I\times\mathbb{R}^3)} \\
&\lesssim N_0^{1-s} + \|u_0(t)\|^{2-}_{L^5(I\times\mathbb{R}^3)}\|u_0(t)\|^{0+}_{L_t^{10}(I;L^{10}(\mathbb{R}^3))}\|Du_0(t)\|_{L^{\frac{10}{3}}(I\times\mathbb{R}^3)} \\
&\lesssim N_0^{1-s} + o(1)\|u_0(t)\|^{1+}_{X_{1,\frac{1}{2}+}}.
\end{aligned}
$$

From the above estimate, we obtain the second estimate in (4.170):

$$
\begin{aligned}
&\|u_0(t)\|_{X_{0,\frac{1}{2}+}(I)} \\
&\quad\lesssim \|\psi(t)u_0(t)\|_{X_{0,\frac{1}{2}+}(I)} \lesssim \|\varphi_0\|_2 + \||u_0|^2u_0\|_{X_{0,-\frac{1}{2}+}(I)} \\
&\quad\lesssim \|\varphi_0\|_2 + \||u_0|^2u_0\|_{L_t^{\frac{10}{7}+}(I;L^{\frac{10}{7}+}(\mathbb{R}^3))} \\
&\quad\lesssim \|\varphi_0\|_2 + \|u_0(t)\|_{L^5(I\times\mathbb{R}^3)}\|u_0(t)\|_{L^{5+}(I\times\mathbb{R}^3)}\|u_0(t)\|_{L^{\frac{10}{3}}(I\times\mathbb{R}^3)} \\
&\quad\lesssim \|\varphi_0\|_2 + o(1)N_0^{-\frac{1}{5+}(1-s)\frac{9}{2}-}\|u_0(t)\|_2^{\frac{1}{10}-}\|u_0(t)\|_6^{\frac{9}{10}+}\|u_0(t)\|_{X_{0,\frac{1}{2}+}(I)} \\
&\quad\lesssim \|\varphi_0\|_2 + o(1)N_0^{-(1-s)\frac{9}{10}-}\|u_0(t)\|^{\frac{9}{10}+}_{X_{1,\frac{1}{2}+}}\|u_0(t)\|_{X_{0,\frac{1}{2}+}(I)} \\
&\quad\lesssim \|\varphi_0\|_2 + o(1)N_0^{-(1-s)\frac{9}{10}-}N_0^{\frac{9}{10}(1-s)+}\|u_0(t)\|_{X_{0,\frac{1}{2}+}(I)} \\
&\quad\lesssim \|\varphi_0\|_2 + o(1)\|u_0(t)\|_{X_{0,\frac{1}{2}+}(I)},
\end{aligned}
$$

which implies the first estimate in (4.170).

Step 3 (A priori estimate of the high frequency difference evolution). Let

$$u(t) = u_0(t) + v(t). \tag{4.179}$$

Then v satisfies (the Cauchy problem corresponding to the difference equation)

$$\begin{cases} i\partial_t v - \Delta v = -[2|u_0|^2 v + 2u_0^2\bar{v} + 2\bar{u}_0 v^2 + 2u_0|v|^2 + |v|^2 v], \\ v(0) = \psi_0(x). \end{cases} \tag{4.180}$$

The corresponding integral equation to (4.180) is

$$v(t) = e^{-it\Delta}\psi_0(x) + \omega(t), \tag{4.181}$$

where

$$\omega(t) = i\int_0^t e^{-i(t-\tau)\Delta}[2|u_0|^2 v + 2u_0^2\bar{v} + 2\bar{u}_0 v^2 + 2u_0|v|^2 + |v|^2 v]d\tau. \tag{4.182}$$

Since

$$\|\psi_0\|_{H^s} \le C, \quad \|\psi_0\|_2 = \||\xi|^{-s}|\xi|^s\psi_0\|_2 \lesssim N_0^{-s}, \tag{4.183}$$

combining this with (4.174), (4.177), (4.183) and Hölder's inequality, we have for $\forall s > 1/2$,

$$\begin{aligned} &\|v\|_{X_{0,\frac{1}{2}+}(I)} \\ &\quad \lesssim \|\psi_0\|_2 + \big\|2|u_0|^2 v + 2u_0^2\bar{v} + 2\bar{u}_0 v^2 + 2u_0|v|^2 + |v|^2 v\big\|_{X_{0,-\frac{1}{2}+}(I)} \\ &\quad \lesssim \|\psi_0\|_2 + \|v\|_{X_{0,\frac{1}{2}+}(I)}\big[o(1) + o(1)\|v\|_{X_{\frac{1}{2},\frac{1}{2}+}(I)} + \|v\|^2_{X_{\frac{1}{2},\frac{1}{2}+}(I)}\big], \end{aligned} \tag{4.184}$$

$$\begin{aligned} \|v\|_{X_{s,\frac{1}{2}+}(I)} &\lesssim \|\psi_0\|_{H^s} + \|v\|_{X_{s,\frac{1}{2}+}(I)}\big[o(1) + o(1)\|v\|_{X_{\frac{1}{2},\frac{1}{2}+}(I)} + \|v\|^2_{X_{\frac{1}{2},\frac{1}{2}+}(I)}\big] \\ &\quad + \|u_0\|_{X_{s,\frac{1}{2}+}(I)}\big[o(1)\|v\|_{X_{\frac{1}{2},\frac{1}{2}+}(I)} + \|v\|^2_{X_{\frac{1}{2},\frac{1}{2}+}(I)}\big]. \end{aligned} \tag{4.185}$$

By $X_{s,\frac{1}{2}+}(I) \hookrightarrow X_{\frac{1}{2},\frac{1}{2}+}(I)$ and the interpolation,

$$\|u_0\|_{X_{s,\frac{1}{2}+}(I)} \lesssim \|u_0(t)\|^{1-s}_{X_{0,\frac{1}{2}+}(I)}\|u_0(t)\|^{s}_{X_{1,\frac{1}{2}+}} \lesssim N_0^{(1-s)s}, \tag{4.186}$$

we obtain from (4.184) and (4.185) ($s > \frac{1}{\sqrt{2}}$) that

$$\|v\|_{X_{0,\frac{1}{2}+}(I)} \lesssim N_0^{-s}, \quad \|v\|_{X_{s,\frac{1}{2}+}(I)} \lesssim C, \tag{4.187}$$

which, together with the interpolation, gives

$$\|v\|_{X_{\frac{1}{2},\frac{1}{2}+(I)}} \lesssim \|v\|^{1-\frac{1}{2s}}_{0,\frac{1}{2}+}\|v\|^{\frac{1}{2s}}_{X_{s,\frac{1}{2}+}} \lesssim N_0^{\frac{1}{2}-s}. \tag{4.188}$$

Now, we estimate the increment of the energy of the difference equation. Consider

$$\begin{aligned} &\|D_x\omega(t)\|_{L^\infty(I;L^2(\mathbb{R}^3))} \\ &\quad \le \sup_{\|\psi(x)\|_2\le 1}\int_I (e^{i\tau\Delta}\psi, D_x[|2|u_0|^2 v + 2u_0^2\bar{v} + 2\bar{u}_0 v^2 + 2u_0|v|^2 + v|v|^2])d\tau \\ &\quad \le \sup_{\|W\|_{X_{0,\frac{1}{2}+}}\le 1}\int (W, D_x[2|u_0|^2 v + 2u_0^2\bar{v} + 2\bar{u}_0 v^2 + 2u_0|v|^2 + v|v|^2])dt. \end{aligned} \tag{4.189}$$

Fix $\rho = \frac{1}{2}-$. By (4.170) and

$$\begin{aligned} (f, D_x g) &= \int_{\mathbb{R}^n} f D_x\bar{g}dx = \int_{\mathbb{R}^n} \hat{f}\cdot|\xi|\bar{\hat{g}}d\xi \\ &= \int_{\mathbb{R}^n} |\xi|^\rho\hat{f}\cdot|\xi|^{|1-\rho|}\bar{\hat{g}}d\xi = (D_x^\rho f, D_x^{1-\rho}g), \end{aligned}$$

we have

$$\int_I |(W, D_x(|u_0|^2 v))| d\tau \tag{4.190}$$
$$= \int_I |(WD_x u_0, u_0 v) + (Wu_0, D_x(u_0 v))| d\tau$$
$$\leq \iint_{I\times\mathbb{R}^3} [|W||u_0||v||D_x u_0| + |D_x^\rho(Wu_0)| \cdot |D_x^{1-\rho}(vu_0)|] dx d\tau.$$

By Hölder's estimate and Strichartz estimate (4.176), we have

$$\iint_{I\times\mathbb{R}^3} |W| \cdot |u_0(t)| \cdot |v| \cdot |D_x u_0| dx d\tau$$
$$\lesssim \|W\|_{\frac{10}{3}} \cdot \|u_0(t)\|_{10} \cdot \|v\|_{\frac{10}{3}} \cdot \|D_x u_0(t)\|_{\frac{10}{3}}$$
$$\lesssim \|u_0(t)\|^2_{X_{1,\frac{1}{2}+}} \|v\|_{X_{0,\frac{1}{2}+}} \lesssim N_0^{2-3s}, \tag{4.191}$$

and

$$\iint_{I\times\mathbb{R}^3} |D_x^\rho(Wu_0)||D_x^{1-\rho}(vu_0)| dx d\tau \lesssim \|D_x^\rho(Wu_0)\|_2 \|D_x^{1-\rho}(vu_0)\|_2$$
$$\lesssim \|u_0\|_{X_{\frac{1}{2}+\rho,\frac{1}{2}+}} \|v\|_{X_{\frac{1}{2}-\rho,\frac{1}{2}+}} \|u_0\|_{X_{1,\frac{1}{2}+}} \lesssim N_0^{2-3s_+}, \tag{4.192}$$

where we used Theorem 4.3.1. Hence,

$$\int_I |(W, D_x(|u_0|^2 v))| dt \lesssim N_0^{2-3s_+}. \tag{4.193}$$

Similarly, we have

$$\int_I |(W, D_x(u_0^2 \bar{v}))| dt \lesssim N_0^{2-3s_+}. \tag{4.194}$$

Moreover, for the nonlinear term involved v^2, by Theorem 4.3.1, (4.174) and (4.176), we have

$$\int_I |(W, D_x(\bar{u}_0 v^2))| dt$$
$$\lesssim \iint_{I\times\mathbb{R}^3} |W||D_x \bar{u}_0||v|^2 dx dt + \|D_x^\rho(W\bar{u}_0)\|_2 \|D_x^{1-\rho}(v^2)\|_2$$
$$\lesssim \|u_0\|_{X_{1,\frac{1}{2}+}} \|v\|^2_{X_{\frac{1}{2},\frac{1}{2}+}} + \|u_0\|_{X_{\frac{1}{2}+\rho,\frac{1}{2}+}} \|v\|_{X_{s,\frac{1}{2}+}} \|v\|_{X_{\frac{3}{2}-\rho-s,\frac{1}{2}+}}$$
$$\lesssim N_0^{2-3s_+} \tag{4.195}$$

and

$$\int_I |(W, D_x(u_0|v|^2))|dt \lesssim N_0^{2-3s_+}. \tag{4.196}$$

For the nonlinear term involved v^3, we have

$$\begin{aligned}
\int_I |(W, D_x(|v|^2 v)\Big| d\tau &\lesssim \int_I \Big|(W, D_x v \cdot v\bar{v})\Big| dt + \int_I \Big|(W, v D_x |v|^2)|dt \\
&\lesssim \|D_x^{s-\frac{1}{2}}(Wv)\|_2 \|D_x^{\frac{3}{2}-s}(v^2)\|_2 + \|D_x^{s-\frac{1}{2}}(W\bar{v})\|_2 \|D_x^{\frac{3}{2}-s}(|v|^2)\|_2 \\
&\lesssim \|v\|_{X_{s,\frac{1}{2}+}} \|v\|_{X_{s,\frac{1}{2}+}} \|v\|_{X_{2(1-s),\frac{1}{2}+}} \\
&\lesssim N_0^{2-3s_+}.
\end{aligned} \tag{4.197}$$

Since $s > \frac{1}{\sqrt{2}} > \frac{2}{3}$ and (4.184), we obtain

$$\sup_{t\in I} \|\omega(t)\|_{H^1} \lesssim N_0^{2-3s_+}. \tag{4.198}$$

Step 4 (Reiteration and the increment estimate of the Hamiltonian energy). Consider $t_1 = \triangle T$, we have

$$u(t_1) = u_0(t_1) + e^{-it_1\Delta}\psi_0 + \omega(t_1) \overset{\triangle}{\equiv} \varphi_1(t_1) + \psi_1(t_1), \tag{4.199}$$

where

$$\varphi_1(t_1) = u_0(t_1) + \omega(t_1), \quad \psi_1 = e^{-it_1\triangle}\psi_0. \tag{4.200}$$

Thus, the pair (φ_0, ψ_0) can be replaced by $(\varphi_1(t_1), \psi_1(t_1))$ to reiterate above steps. $\psi_1(t_1)$ is similar to ψ_0. The Hamiltonian increment when replacing φ_0 by $\varphi_1(t_1)$ is bounded by

$$\begin{aligned}
|E(\varphi_1(t_1)) - E(\varphi_0)| &= |E(u_0(t_1) + \omega(t_1)) - E(u_0(t_1))| \\
&\lesssim (\|u_0(t_1)\|_{H^1} + \|\omega(t_1)\|_{H^1})\|\omega(t_1)\|_{H^1} \qquad (4.201) \\
&\quad + (\|u_0(t_1)\|_6 + \|\omega(t_1)\|_6)^3 \|\omega(t_1)\|_2 \\
&\lesssim N_0^{1-s} N_0^{2-3s_+} + N_0^{3(1-s)-s} \\
&\lesssim N_0^{3-4s_+},
\end{aligned} \tag{4.202}$$

where we have used (4.170), (4.188) and (4.198).

By (4.202), in order to extend the solution to any time $T > 0$, we need iterate the times

$$\frac{T}{\Delta T} \sim TN_0^{\frac{9(1-s)}{2}-}, \tag{4.203}$$

and iteration of the procedure leads to the condition

$$TN_0^{\frac{9(1-s)}{2}-}N_0^{3-4s_+} < E_1(\varphi_0) \sim N_0^{2(1-s)}, \tag{4.204}$$

which is equivalent to the condition

$$TN_0^{\frac{11-13s}{2}+} < 1. \tag{4.205}$$

Hence, if

$$s > \frac{11}{13}, \tag{4.206}$$

we can continue the iteration on $[0, T]$ and maintain the increment

$$N_0 = N_0(T) = T^{\frac{2}{13s-11}+}. \tag{4.207}$$

Thus, we can obtain the following global well-posedness.

Theorem 4.3.2. *Let $s > \frac{11}{13}$, $\varphi(x) \in H^s$, then* (4.116) *is global well-posed in $H^s(\mathbb{R}^3)$, that is, the solution $u(t) \in C(\mathbb{R}; H^s(\mathbb{R}^3))$ and has the form*

$$u(t) = e^{-it\Delta}\varphi + v(t), \tag{4.208}$$

$$\|v(t)\|_{H^1} \lesssim (1 + |t|)^{\frac{2(1-s)}{13s-11}+}. \tag{4.209}$$

Proof. From Step 1 to Step 4, we can obtain the main result. It suffices to show (4.209). Noticing that

$$\|v(t)\|_{H^1} \leq E_1(v(t))^{\frac{1}{2}} + \|v(t)\|_2 \lesssim N_0^{1-s} \lesssim (1 + |t|)^{\frac{2(1-s)}{13s-11}+}, \tag{4.210}$$

we obtain the desired result. □

Remark 4.3.5.

(i) In Theorem 4.3.2, we did not obtain the space-time integrability of solution, as a result we cannot prove the scattering result directly. But we can establish the scattering of (4.116) in H^s for ($s > \frac{5}{7}$) by means of the Morawetz inequality

$$\begin{aligned}\int_0^T |u(0,t)|^2 dt + \int_0^T \int_{\mathbb{R}^3} \frac{|\nabla u|^2 - u_r^2}{r} dxdt + \int_0^T \int_{\mathbb{R}^3} \frac{|u|^4}{r} dxdt \\ \lesssim \sup_{0 \leq t \leq T} \|u(t)\|_{H^{\frac{1}{2}}}^2,\end{aligned} \tag{4.211}$$

and

$$\sup_{r\geq 0} r|\varphi(r)| \leq C\|\varphi\|_{H^{\frac{1}{2}+}}, \quad \varphi(x) = \varphi(|x|) \in H^{\frac{1}{2}+}, \tag{4.212}$$

if the initial data $\varphi(x) = \varphi(|x|) \in H^s$ are radial function (see details in [21]).

(ii) Morawetz's inequality (4.211) will be shown in next section. The pointwise estimate (4.212) is the conclusion of Hardy' inequality

$$\left\|\frac{\varphi}{|x|}\right\|_2 \lesssim \|\nabla\varphi\|_2$$

and the interpolation. In fact, let $v = r\varphi$, then

$$\begin{aligned}\|v\|_{L^2(dr)} &= \left(\int_0^\infty |\varphi|^2 r^2 dr\right)^{1/2} \leq \left(\frac{1}{|\Sigma^2|}\int_{\Sigma^2}\int_0^\infty |\varphi|^2 r^2 dr d\sigma\right)^{\frac{1}{2}} \\ &\leq \|\varphi\|_{L^2(\mathbb{R}^3)},\end{aligned}$$

where Σ^2 is the unit sphere in $\mathbb{R}^3$, and

$$\|v'\|_{L^2(dr)} \leq \|r\varphi'\|_{L^2(dr)} + \|\varphi\|_{L^2(dr)} \sim \|\nabla\varphi\|_2 + \left\|\frac{\varphi}{|x|}\right\|_2 \lesssim \|\varphi\|_{\dot{H}^1}.$$

Interpolating the above two inequalities, we can obtain

$$\|v\|_{\dot{H}^{\frac{1}{2}+}} \lesssim \|\varphi\|_{\dot{H}^{\frac{1}{2}+}}.$$

By Sobolev's inequality, it follows that

$$\sup_r r|\varphi(r)| \leq \|v\|_\infty \lesssim \|v\|_{H^{\frac{1}{2}+}} \lesssim \|\varphi\|_{H^{\frac{1}{2}+}}.$$

(iii) Hardy's inequality has the generalization

$$\left\|\frac{\varphi}{|x|}\right\|_p \leq C(p,n)\|\nabla\varphi\|_p, \quad C(p,n) = (p/(n-p))^p, \quad 1 < p < n, \tag{4.213}$$

where $C(p,n)$ is the sharp constant. For convenience, we give an elementary proof. In fact, it suffices to show (4.213) for all $\varphi(x) \in C_c^\infty(\mathbb{R}^n)$. It follows that

$$|\varphi(x)|^p = -\int_1^\infty \frac{d}{d\lambda}|\varphi(\lambda x)|^p d\lambda = -p\int_1^\infty \varphi^{p-1}(\lambda x)\langle x, \nabla\varphi(\lambda x)\rangle d\lambda.$$

By Hölder's inequality, we have

$$\int_{\mathbb{R}^n} \frac{|\varphi|^p}{|x|^p} dx = -p \int_1^\infty \int_{\mathbb{R}^n} \frac{\varphi^{p-1}(\lambda x)}{|x|^{p-1}} \left\langle \frac{x}{|x|}, \nabla\varphi(\lambda x) \right\rangle dx d\lambda$$
$$= -p \int_1^\infty \frac{d\lambda}{\lambda^{n+1-p}} \int_{\mathbb{R}^n} \frac{\varphi(y)^{p-1}}{|y|^{p-1}} \frac{\partial\varphi(y)}{\partial r} dy$$
$$= -\frac{p}{n-p} \int_{\mathbb{R}^n} \frac{\varphi(y)^{p-1}}{|y|^{p-1}} \frac{\partial\varphi(y)}{\partial r} dy$$
$$= \frac{p}{n-p} \left(\int_{\mathbb{R}^n} \frac{|\varphi|^p}{|y|^p} dy \right)^{\frac{p-1}{p}} \left(\int_{\mathbb{R}^n} \left| \frac{\partial\varphi(y)}{\partial r} \right|^p dy \right)^{\frac{1}{p}},$$

which implies (4.213).

Remark 4.3.6.

(i) For $n = 2$, the analogue estimate as Lemma 4.3.1 is that for $\ell \geq m$

$$\|(e^{-it\Delta}\Delta_\ell\varphi_1)(e^{-it\Delta}\Delta_m\varphi_2)\|_{L^2(\mathbb{R}\times\mathbb{R}^2)} \lesssim 2^{m-\ell}\|\Delta_\ell\varphi_1\|_{L^2(\mathbb{R}^2)}\|\Delta_m\varphi_2\|_{L^2(\mathbb{R}^2)}. \tag{4.214}$$

By the Littlewood–Paley decomposition, the following bilinear Strichartz estimates hold true:

$$\|D_x^s(e^{it\Delta}\psi_1)\cdot(e^{it\Delta}\psi_2)\|_2 \lesssim \|\psi_1\|_{H^s}\|\psi_2\|_2, \tag{4.215}$$
$$\|D_x^s(u_1u_2)\|_{L^2(I\times\mathbb{R}^2)} \lesssim \|u_1\|_{X_{s,\frac{1}{2}+}}\|u_2\|_{X_{0,\frac{1}{2}+}}, \tag{4.216}$$

when $0 \leq s < 1/2$. and

$$\|D_x^s(e^{it\Delta}\psi_1\cdot e^{it\Delta}\psi_2)\|_2 \lesssim \|\psi_1\|_{H^s}\|\psi_2\|_2 + \|\psi_1\|_{H^{\frac{1}{2}+}}\|\psi_2\|_{H^{s-\frac{1}{2}}}, \tag{4.217}$$
$$\|D_x^s(u_1u_2)\|_{L^2(I\times\mathbb{R}^2)} \lesssim \|u_1\|_{X_{s,\frac{1}{2}+}}\|u_2\|_{X_{0,\frac{1}{2}+}} + \|u_1\|_{X_{\frac{1}{2},\frac{1}{2}+}}\|u_2\|_{X_{s-\frac{1}{2},\frac{1}{2}+}}, \tag{4.218}$$

when $s \geq \frac{1}{2}$ and

$$\|D_x^s(u_1u_2)\|_{L^2(I\times\mathbb{R}^2)} \leq C\|u_1\|_{X_{s,\frac{1}{2}+}}\|u_2\|_{X_{0,\frac{1}{2}-}}, \quad 0 \leq s < \frac{1}{2}, \tag{4.219}$$

where $\frac{1}{2}-$ depends on $s+$ in (4.219).

(ii) As a similar result as in $\mathbb{R}^3$, we can show that the Cauchy problem

$$\begin{cases} iu_t - \Delta u = -|u|^2u, & (x,t) \in \mathbb{R}^2 \times \mathbb{R}, \\ u(0) = \varphi(x), \end{cases} \tag{4.220}$$

is globally well posed in $H^s(s > 2/3)$, that is, for any $\varphi(x) \in H^s(\mathbb{R}^2)$, there exists a unique global solution $u(t) \in C(\mathbb{R}, H^s)$ of (4.220) satisfying

$$u(t) - e^{-it\Delta}\varphi \in H^1, \quad \forall t \in \mathbb{R} \tag{4.221}$$

and

$$\|u(t) - e^{-it\Delta}\varphi\|_{H^1} \le C(1 + |t|)^{\frac{1-s}{3s-2}+}. \tag{4.222}$$

See the details in [19, 21].

(iii) In [19], Bourgain established the well-known theorem: Assume that $u(t) \in C((-T^*, T^*); L^2(\mathbb{R}^2)) \cap L^4((-T^*, T^*) \times \mathbb{R}^2)$ is the solution of the Cauchy problem

$$\begin{cases} iu_t + \Delta u = \pm|u|^2 u, & (t,x) \in \mathbb{R} \times \mathbb{R}^2, \\ u(0) = \varphi(x), & x \in \mathbb{R}^2, \end{cases} \tag{4.223}$$

where $(-T^*, T^*)$ is the maximal lifespan. Then

$$\lim_{t \to T^*} \sup_{\substack{A \subset \mathbb{R}^2 \\ |A| < (T^*-t)^{1/2}}} \int_A |u(t)|^2 dx > C > 0, \tag{4.224}$$

where C is a constant.

(iv) For $s > 2/3$, $\varphi(x) \in L^2(\mathbb{R}^2)$ and $|x|^s\varphi \in L^2(\mathbb{R}^2)$, then there exists a global solution of (4.220) in $L^2(\mathbb{R}^2)$, such that $u(t,x) \in L^4(\mathbb{R} \times \mathbb{R}^2)$. Consequently, the scattering result can be obtained. See the details in [21]. For the initial data $\varphi(x) \in L^2(\mathbb{R}^2)$, Dodson [62] obtains the scattering result.

4.4 I-method

We know that Bourgain's Fourier truncation method (decomposition technique of high and low frequency) can deal with the problem of low regularity effectively. However, it is difficult to deal with the scattering problem. In this section, we focus on I-team's I-method, which is another effective method to deal with regularity and can be used to study the scattering result directly. But it is only restricted to some specific nonlinear term.

In the following, we take the cubic, defocusing nonlinear Schrödinger equation in $\mathbb{R}^3$,

$$\begin{cases} iu_t + \Delta u = |u|^2 u, & x \in \mathbb{R}^3, t \in \mathbb{R}, \\ u(0) = \varphi(x) \in H^s(\mathbb{R}^3), & x \in \mathbb{R}^3 \end{cases} \tag{4.225}$$

as an example to explain the I-method.

From the local well-posedness theory and the scaling analysis in Chapter 1 [37], if $s \ge 1/2$, the Cauchy problem (4.225) is local well-posed in H^s, that is, there exists $u(t) \in C(I; H^s) \cap \bigcap_{(q,r)\in\Lambda} L^q(I; B^s_{r,2})$ satisfying

$$u(t) = S(t)\varphi - i\int_0^t S(t-\tau)[|u|^2u](\tau)d\tau, \tag{4.226}$$

where

$$\begin{cases} T = T(\|\varphi\|_{H^s}), & s > \frac{1}{2}, \\ T = T(\varphi), & s = s_c = \frac{1}{2} \text{ (the critical value)} \end{cases}$$

and $(q, r) \in \Lambda$ denotes the space-time admissible pair.

Remark 4.4.1.

(i) For the smooth solution of (4.225), one has the following conservation laws:

$$M(u(t)) = \int_{\mathbb{R}^3} |u(t)|^2 dx = M(\varphi),$$

$$E(u(t)) = \int_{\mathbb{R}^3} \left[\frac{1}{2}|\nabla u|^2 + \frac{1}{4}|u|^4\right] dx = E(\varphi). \tag{4.227}$$

(ii) Using local well-posedness and conservation laws, one can deduce that: If $\varphi \in H^1(\mathbb{R}^3)$, then there exists a global solution $u(t) \in C(\mathbb{R}; H^1(\mathbb{R}^3))$ of (4.225) satisfying

$$u(t) \in C(\mathbb{R}; H^1(\mathbb{R}^3)) \cap \bigcap_{(q,r)\in\Lambda} L^q_{\mathrm{loc}}(\mathbb{R}; W^{1,r}(\mathbb{R}^3)).$$

(iii) The Cauchy problem (4.225) has the scattering result in H^1. In 1985, Ginibre–Velo obtained the scattering results of defocusing semilinear Schrödinger equation with $f(u) = |u|^{p-1}u$, $1 + \frac{4}{n} < p < 1 + \frac{4}{n-2}$.

(iv) Noticing that $f(u) = |u|^2u$ is a C^∞ function, therefore, for any $s > 1$ and $\forall\varphi(x) \in H^s$, (4.225) or (4.226) has a global solution $u(t) \in C(\mathbb{R}; H^s(\mathbb{R}^3))$ in H^s and the scattering result in H^s holds.

Now let us recall the concept of scattering. Let $S(t) = e^{it\Delta}$ be a free Schrödinger group, $S^{NL}(t)$ denote the nonlinear flow determined by (4.225). Then

$$u(t) = S^{NL}(t)\varphi = S(t)\varphi - i\int_0^t S(t-\tau)|u(\tau)|^2u(\tau)d\tau$$

is the solution of (4.225) or (4.226).

H^s-scattering

For any $\varphi_\pm(x) \in H^s$, there exists a global solution $u(t) \in C(\mathbb{R}, H^s(\mathbb{R}^3))$ of (4.225) or (4.226) such that

$$\lim_{t\to\pm\infty} S(-t)u(t) = \lim_{t\to\pm\infty} S(-t)S^{NL}(t)\varphi \overset{H^s}{=} \varphi_\pm(x). \tag{4.228}$$

We define wave operator $\Omega_\pm\colon H^s(\mathbb{R}^3) \to H^s(\mathbb{R}^3)$ as in the following:

$$\Omega^\pm \varphi_\pm(x) = u(0) = \varphi(x). \tag{4.229}$$

If the wave operator $\Omega^\pm$ is a surjection, then (4.225) or (4.226) is asymptotic completed in $H^s(\mathbb{R}^3)$.

Problems

(a) For any $\varphi(x) \in H^s(\mathbb{R}^3)$ and $s \geq \frac{1}{2}$, whether does Cauchy problem (4.225) or (4.226) determine a unique strong continuous flow $u(t) \in C(\mathbb{R}; H^s(\mathbb{R})) \cap \cdots$?

(b) Is it scattered in the topology of $H^s(\mathbb{R}^3)$?

If $s \geq 1$, the global well-posedness of (4.225) or (4.226) can be deduced from local well-posedness and energy conservation law. However, if $\frac{1}{2} \leq s < 1$, there is no energy conservation law. Consequently, it is very difficult to establish the global well-posedness in H^s (lower regularity problem). Of course, one also needs the global H^s space-time estimate in order to establish the scattering result in $H^s(\mathbb{R}^3)$, that is,

$$\|u(t); L^q(\mathbb{R}; B^s_{q,2}(\mathbb{R}^3))\| < \infty.$$

Bourgain used the Fourier truncation method to obtain the following result:

Bourgain theorem

(a) Let $s > \frac{11}{13}$ and $\varphi(x) \in H^s(\mathbb{R}^3)$. Then there exists a unique global solution $u(t) \in C(\mathbb{R}; H^s(\mathbb{R}^3))$ of (4.225), which can be decomposed as

$$u(t) = e^{it\Delta}\varphi + v(t), \quad \|v(t)\|_{H^1} \lesssim (1+|t|)^{\frac{2(1-s)+}{13s-11}}. \tag{4.230}$$

(b) Let $\varphi(x) = \varphi(|x|) \in H^s(\mathbb{R}^3)$ and $s > \frac{5}{7}$. Then there exists a unique global solution $u(t) \in C(\mathbb{R}; H^s(\mathbb{R}^3))$ of (4.225) satisfying

$$\|u(t) - e^{it\Delta}\varphi\|_{H^1} \leq C(\|\varphi\|_{H^s})$$

and it is scattered in H^s.

Remark 4.4.2.

(i) Bourgain first gave an efficient method to deal with the lower regularity problem. Its procedure is first to decompose the initial function into high and low frequency,

$$\varphi(x) = \varphi_0(x) + \psi_0(x), \quad \varphi_0(x) = \mathcal{F}^{-1}(\hat{\varphi}|_{|\xi|\leq N}) \in C^\infty(\mathbb{R}^3) \cap H^1(\mathbb{R}^3), \tag{4.231}$$

and then to operate it as the following steps on $I = [0, \triangle T]$:

(a) The evolution of high frequency by free Schrödinger group, that is, $e^{it\Delta}\psi_0(x)$.
(b) The evolution of low frequency by nonlinear equation as follows:

$$u_0(t) = e^{it\Delta}\varphi_0(x) - i\int_0^t e^{i(t-\tau)\Delta}[|u_0(\tau)|^2 u_0(\tau)]d\tau.$$

(c) Showing that the remainder $\omega(t) = u(t) - u_0(t) - e^{it\Delta}\psi_0(x) \in H^1$, which is very difficult since we need an equation that $\omega(t)$ satisfies to give the estimate.

Next, repeat the previous steps on the interval $[\triangle T, 2\triangle T]$ with

$$\varphi_1(x) = u_0(t_1) + \omega(t_1) \in H^1(\mathbb{R}^3), \quad \psi_1(x) = e^{it_1\Delta}\psi_0(x).$$

That is, adding the remainder into the low frequency part, and we still have $\varphi_1(x) \in H^1(\mathbb{R}^3)$, which can guarantee that the nonlinear evolution of $\varphi_1(x)$ is the global solution in H^1. The corresponding high frequency $\psi_1(x)$ has similar property as $\psi_0(x)$, which can evolve by the free operator group. Repeatedly, for any $T > 0$, one can obtain the H^s solution of (4.225) or (4.226) on $[0, T]$.

(ii) The key point that Bourgain's methods can apply lies in the following estimate:

$$\int_0^t S(t-\tau)|u(\tau)|^2 u(\tau)d\tau \in H^1(\mathbb{R}^3), \quad \forall u(\tau) \in H^s(\mathbb{R}^3),\ s > \frac{11}{13}. \tag{4.232}$$

Consequently, many dispersive and wave equations have such a smooth effect, such as the KdV equation, Benjamin–Ono equation, wave equation and Klein–Gordon equation, etc.; please refer to [137, 197] for details.

(iii) Not all of the evolution equations have the similar smoothing effect to (4.232), for example, the wave map equation

$$\begin{cases} u_{tt} - \Delta u = (|\nabla u|^2 - |u_t|^2)u, \\ u(0) = \varphi(x), \quad u_t(0) = \psi(x), \end{cases} \tag{4.233}$$

and the Maxwell–Klein–Gordon equation.

Problem

How can one study the low regular problem for the evolution equations that does not have the smoothing effect?

Motivated by Bourgain's Fourier truncation method, I-team developed the I-method [48].

Key point of I-method

Using a smooth operator I, one constructs the smooth part Iu of solution $u(t)$ satisfying the almost conservation integral

$$E(Iu) = E(I\varphi) + N^{-1+}C(\|\varphi\|_{H^s}), \tag{4.234}$$

where

$$Iu = \mathcal{F}^{-1}m_N(\xi) * u, \quad m_N(\xi) = m_N(|\xi|) \in C^\infty(\mathbb{R}^n),$$
$$m_N(\xi) = \begin{cases} 1, & |\xi| \le N, \\ (\frac{N}{|\xi|})^{1-s}, & |\xi| \ge 2N, \end{cases} \quad 0 \le m_N(\xi) \le 1, \tag{4.235}$$

and N is a sufficiently large constant determined latter. It is easy to check that

$$\|u\|_{H^s}^2 \sim E(Iu) + \|u\|_2^2.$$

For $s > s_0$, by the preconservation integral, interaction Morawetz estimate, scaling analysis and continuousness methods, one can obtain the boundness of $\|u(t)\|_{H^s}$ and the global space-time estimate of u, which implies the global well-posedness and scattering of (4.225) or (4.226) in H^s.

Remark 4.4.3. The key point of the I-method and its advantage lie in the following:
(i) Controlling the estimate of $\|u(t)\|_{H^s}$ (polynomial growth respect to t) by the almost energy conservation that the smooth part of solution satisfies and characterizing qualitatively the transferring of the energy of the solution from high frequency to low frequency by the estimate of smooth solution in a low, regular Sobolev space (polynomial growth respect to t).
(ii) I-team's I-energy method can be used to deal with the low, regular problem of the nonlinear (dispersive) wave, which may not have the similar smooth effect as (4.232). At the same time, the I-method can obtain the more profound result than Bourgain's Fourier truncation approach in some situations. For instance, one can obtain the global well-posedness and scattering of (4.225) in H^s by the I-team's method [51] for $s > \frac{4}{5}$, while one obtains the global well-posedness of the solution by Bourgain's method for $s > \frac{11}{13}$ only, and one cannot obtain the scattering. From another example, one can obtain the existence of a global regular solution of the wave map by the I-method.

Before stating the main result, we introduce some notation. Let $A, B > 0$.
(1) $A \lesssim B$ denotes that there exists $K > 2$, satisfying $A \le KB$.
(2) $A \sim B \Longleftrightarrow A \lesssim B$ and $B \lesssim A$.
(3) $A \ll B$ denotes that there exists $K > 2$, satisfying $B > KA$.
(4) $\langle A\rangle = (1 + A^2)^{\frac{1}{2}}$, $\langle\nabla\rangle = (1 + (-\Delta))^{\frac{1}{2}}$, the symbol of $\langle\nabla\rangle$ is $(1 + |\xi|^2)^{\frac{1}{2}}$.
(5) $\frac{1}{2}+ \equiv \frac{1}{2} + \varepsilon$, $\frac{1}{2}- \equiv \frac{1}{2} - \varepsilon$, where $0 < \varepsilon \ll 1$.

Theorem 4.4.1 ([51]). *Let $s > \frac{4}{5}$ and $\varphi(x) \in H^s(\mathbb{R}^n)$. Then*
(a) *there exists a unique global solution $u(t) \in C(\mathbb{R}; H^s(\mathbb{R}^3))$ of (4.225) or (4.226).*

(b) *It scatters in H^s. That is, for any $\varphi_\pm(x) \in H^s(\mathbb{R}^3)$, there exists a solution $u(t) = S^{NL}(t)\varphi$ of* (4.225) *or* (4.226) *satisfying*

$$\|u(t) - e^{it\Delta}\varphi_\pm\|_{H^s} \to 0, \quad t \to \pm\infty. \tag{4.236}$$

Consequently, we can define the wave operator $\Omega_\pm : \quad H^s(\mathbb{R}^3) \longrightarrow H^s(\mathbb{R}^3)$ as

$$\Omega_\pm : \varphi_\pm(x) \longmapsto u(0) = \varphi(x). \tag{4.237}$$

On the other hand, there exists unique $\varphi_\pm(x) \in H^s(\mathbb{R}^3)$ such that (4.236) *holds true for $\varphi(x) \in H^s(\mathbb{R}^3)$ (asymptotic completeness), which implies that $\Omega_\pm$ is a surjection.*

Remark 4.4.4.

(i) Using the I-method, we can obtain the global well-posedness and scattering for (4.225) or (4.226) in H^s (without the need of the radial assumption), which is better than Bourgain's Fourier truncation method. However, it is an interesting problem whether the I-method can deal with the general nonlinear term. We can refer to Visan–Zhang [281].

(ii) The results of Theorem 4.4.1 are not sharp, which should be that the Cauchy problem (4.225) or (4.226) is global well-posedness and scattering in H^s with $s > \frac{1}{2}$.

(iii) In order to prove Theorem 4.4.1, one need to establish interaction Morawetz estimate, I-energy conservation integral, etc., which are important for the study of the other partial differential equations.

Morawetz interaction potential and interaction Morawetz estimate

Based on Morawetz action, we construct Morawetz interaction potential, and deduce the $L^4_{t,x}(\mathbb{R} \times \mathbb{R}^3)$-Morawetz inequality. Moreover, we establish the global well-posedness and scattering in H^s.

Consider the Cauchy problem

$$\begin{cases} i\partial_t u + \alpha\Delta u = \mu f(|u|^2)u, & (x,t) \in \mathbb{R}^3 \times \mathbb{R}, \\ u(0) = \varphi(x), & x \in \mathbb{R}^3, \end{cases} \tag{4.238}$$

where $u : \mathbb{R} \times \mathbb{R}^3 \longmapsto \mathbb{C}, f : \mathbb{R}^+ \longmapsto \mathbb{R}^+$ is a smooth function, and $\alpha, \mu \in \mathbb{R}$ are real constants. We also define

$$F(z) = \int_0^z f(s)ds. \tag{4.239}$$

We will use polar coordinates

$$x = r\omega, \quad r > 0, \; \omega \in \Sigma^2,$$

where Σ^2 denotes the unit sphere, $\triangle_\omega$ Laplace–Beltrami operator on Σ^2, respectively. For ease of reference below, we record some alternate forms of the equation in (4.238):

$$u_t = ia\Delta u - i\mu f(|u|^2)u, \tag{4.240}$$

$$\bar{u}_t = -ia\Delta\bar{u} + i\mu f(|u|^2)\bar{u}, \tag{4.241}$$

$$u_t = iau_{rr} + i\frac{2a}{r}u_r + i\frac{a}{r^2}\Delta_\omega u - i\mu f(|u|^2)u, \tag{4.242}$$

$$ru_t = ia(ru)_{rr} + i\frac{a}{r}\Delta_\omega u - i\mu rf(|u|^2)u, \tag{4.243}$$

$$(r\bar{u}_t) = -ia(r\bar{u})_{rr} - i\frac{a}{r}\Delta_\omega\bar{u} + i\mu rf(|u|^2)\bar{u}. \tag{4.244}$$

Introduce the canonical Morawetz action

$$M_0[u](t) = \operatorname{Im}\int_{\mathbb{R}^3} \bar{u}(x,t)\partial_r u(x,t)dx = \int_{\mathbb{R}^3} \operatorname{Im}[\bar{u}(x,t)\nabla u(x,t)]\cdot\frac{x}{|x|}dx.$$

Physical interpretation of Morawetz action M_0

Computing (4.240) $\times\ \bar{u}$ + (4.241) $\times\ u$, we can deduce

$$\begin{aligned}\frac{\partial}{\partial t}|u|^2 &= ia(\Delta u\cdot\bar{u} - \Delta\bar{u}\cdot u) = -a\operatorname{Im}(\Delta u\cdot\bar{u} - \Delta\bar{u}\cdot u))\\ &= -a\operatorname{Im}\nabla\cdot(\bar{u}\nabla u - u\nabla\bar{u}) = -2a\nabla\cdot\operatorname{Im}[\bar{u}(x,t)\nabla u(x,t)],\end{aligned} \tag{4.245}$$

therefore we may interpret M_0 as the spatial average of the radial component of the L^2 mass current. We expect that M_0 will increase with time if the wave u scatters since such behavior involves a broadening redistribution of the L^2 mass.

Proposition 4.4.1. *If u is a smooth solution of* (4.238), *then Morawetz action $M_0[u]$ satisfies*

$$\begin{aligned}\frac{d}{dt}M_0[u](t) &= 4\pi a|u(t,0)|^2 + \int_{\mathbb{R}^3}\frac{2a}{|x|}|\nabla\!\!\!/_0 u(x,t)|^2dx\\ &\quad + \mu\int_{\mathbb{R}^3}\frac{2}{|x|}\{|u|^2f(|u|^2)(t) - F(|u|^2)\}dx,\end{aligned} \tag{4.246}$$

where $\nabla\!\!\!/_0$ denotes angular component of the derivative, that is,

$$\nabla\!\!\!/_0 u = \nabla u - \frac{x}{|x|}\left(\frac{x}{|x|}\cdot\nabla u\right). \tag{4.247}$$

In particular, if the nonlinear term satisfies the repulsion condition

$$\mu\{|u|^2f(|u|^2) - F(|u|^2)\} \geq 0, \tag{4.248}$$

then $M_0(t)$ is an increasing function of time.

Remark 4.4.5. If $f(|u|^2) = |u|^{p-1}$, then $F(|u|^2) = \frac{2}{p+1}|u|^{p+1}$, and the condition (4.248) holds

$$|u|^2 f(|u|^2) - F(|u|^2) = \frac{p-1}{p+1}|u|^{p+1} = \frac{p-1}{2}F(|u|^2) > 0.$$

Proof of Proposition 4.4.1. Clearly, we may write

$$\begin{aligned} M_0(t) &= \int_{\mathbb{R}^3} \operatorname{Im}(\bar{u}(x,t)\partial_r u(x,t))dx = \operatorname{Im} \int_{\mathbb{R}^3} \bar{u}(x,t)\left(\partial_r + \frac{1}{r}\right)u(x,t)dx \\ &= \operatorname{Im} \int_0^\infty \int_{\Sigma^2} \overline{ru}(ru)_r d\omega dr. \end{aligned} \tag{4.249}$$

Integrating by parts and using the equations (4.243) and (4.244) give

$$\begin{aligned} \frac{d}{dt}M_0(t) &= \operatorname{Im} \int_0^\infty \int_{\Sigma^2} \overline{ru}(ru_t)_r + \overline{ru}_t(ru)_r d\omega dr = -2\operatorname{Im} \int_0^\infty \int_{\Sigma^2} \overline{(ru)}_r (ru_t) d\omega dr \\ &= -2\operatorname{Im} \int_0^\infty \int_{\Sigma^2} \overline{(ru)}_r \left\{ i\alpha(ru)_{rr} + i\frac{\alpha}{r} \triangle_\omega u - i\mu r f(|u|^2)u \right\} d\omega dr \\ &= -2\alpha \operatorname{Re} \int_0^\infty \int_{\Sigma^2} (\overline{ru})_r (ru)_{rr} d\omega dr - 2\alpha \operatorname{Re} \int_0^\infty \int_{\Sigma^2} (\overline{ru})_r \frac{\triangle_\omega u}{r} d\omega dr \\ &\quad + 2\mu \operatorname{Re} \int_0^\infty \int_{\Sigma^2} (\overline{ru})_r r f(|u|^2) u d\omega dr = \mathrm{I} + \mathrm{II} + \mathrm{III}. \end{aligned} \tag{4.250}$$

Noticing that $\partial_r |(ru)_r|^2 = 2\operatorname{Re}\{(\overline{ru})_r (ru)_{rr}\}$, one has

$$\mathrm{I} = -\alpha \int_{\Sigma^2} \left|(ru)_r\right|^2 \Big|_0^\infty d\sigma = 4\pi\alpha \left|u(t,0)\right|^2.$$

Next, writing $\triangle_\omega = \nabla_\omega \cdot \nabla_\omega$ and integrating by parts to get

$$\begin{aligned} \mathrm{II} &= -2\alpha \operatorname{Re} \int_0^\infty \int_{\Sigma^2} (\bar{u} + r\bar{u}_r)\frac{1}{r}\nabla_\omega \nabla_\omega u d\omega dr \\ &= \alpha \operatorname{Re} \int_0^\infty \int_{\Sigma^2} \left[\frac{2}{r}|\nabla_\omega u|^2 + \partial_r |\nabla_\omega u|^2\right] d\omega dr. \end{aligned}$$

Since $\nabla_\omega u = r \not{\nabla}_0 u$, we know that

$$\mathrm{II} = \int_{\mathbb{R}^3} \frac{2a}{|x|} |\nabla_0 u(x,t)|^2 dx.$$

Finally, we estimate III, and using the Leibniz rule to find

$$(\overline{ru})_r r f(|u|^2)u = (\bar{u} + r\bar{u}_r) r f(|u|^2)u = r|u|^2 f(|u|^2) + r^2 f(|u|^2) u\bar{u}_r,$$
$$2\mu \operatorname{Re}\{(\overline{ru})_r r f(|u|^2)u\} = 2\mu r|u|^2 f(|u|^2) + \mu r^2 (F(|u|^2))_r.$$

Integrating by parts, we have

$$\mathrm{III} = \mu \int_{\mathbb{R}^3} \left[\frac{2}{|x|} |u|^2 f(|u|^2)(t) - \frac{2}{|x|} F(|u|^2) \right] dx.$$

(4.246) follows from the above deductions. □

Now, we define the Morawetz action centered at y to be

$$M_y[u](t) = \int_{\mathbb{R}^3} \operatorname{Im}\left[\bar{u}(x,t) \frac{x-y}{|x-y|} \cdot \nabla u(x,t) \right] dx. \tag{4.251}$$

Then we have the following.

Corollary 4.4.1. *If $u(x,t)$ solves (4.238), the Morawetz action at y satisfies the identity*

$$\begin{aligned} \frac{d}{dt} M_y[u](t) = 4\pi a |u(y,t)|^2 &+ \int_{\mathbb{R}^3} \frac{2a}{|x-y|} |\nabla_y u(x,t)|^2 dx \\ &+ \int_{\mathbb{R}^3} \frac{2\mu}{|x-y|} \{|u|^2 f(|u|^2) - F(|u|^2)\} dx, \end{aligned} \tag{4.252}$$

where

$$\nabla_y u(x,t) = \nabla u - \frac{x-y}{|x-y|} \left(\frac{x-y}{|x-y|} \cdot \nabla u \right). \tag{4.253}$$

In particular, $M_y(t)$ is an increasing function of time if the nonlinearity satisfies the repulsive condition (4.248).

Proposition 4.4.2. *Assume $u(x,t)$ is a solution of* (4.238)*. Then*

$$|M_y[u](t)| \triangleq |M_y(t)| \lesssim \|u(t)\|^2_{\dot{H}_x^{\frac{1}{2}}}. \tag{4.254}$$

Proof. Without loss of generality, we take $y = 0$. By duality, we have

$$\left| \operatorname{Im} \int_{\mathbb{R}^3} \overline{u(x,t)} \partial_r u(x,t) dx \right| \lesssim \|u\|_{\dot{H}^{\frac{1}{2}}(\mathbb{R}^3)} \|\partial_r u\|_{\dot{H}^{-\frac{1}{2}}(\mathbb{R}^3)}. \tag{4.255}$$

It suffices to show $\|\partial_r u\|_{\dot H^{-\frac12}(\mathbb{R}^3)} \le \|u\|_{\dot H^{\frac12}(\mathbb{R}^3)}$. Note that $\partial_r = \frac{x}{|x|}\cdot\nabla$, it remains to prove

$$\left\|\frac{x}{|x|}f\right\|_{\dot H^{\frac12}(\mathbb{R}^3)} \lesssim \|f\|_{\dot H^{\frac12}(\mathbb{R}^3)}. \tag{4.256}$$

(4.256) follows from interpolating between the following two bounds:

$$\left\|\frac{x}{|x|}f\right\|_2 \le \|f\|_2, \quad \left\|\frac{x}{|x|}f\right\|_{\dot H^1} \le \left\|\frac{x}{|x|}\nabla f\right\|_2 + \left\|\frac{f}{|x|}\right\|_2 \lesssim \|\nabla f\|_2. \qquad \square$$

Remark 4.4.6. Alternative proof of $\|\partial_r u\|_{\dot H^{-\frac12}} \le \|u\|_{\dot H^{\frac12}}$. In fact, $\forall \psi(x) \in S(\mathbb{R}^3)$, we have

$$\begin{aligned}(\psi, \partial_r u) &= \left(\psi, \frac{x}{|x|}\cdot\nabla u\right) = \left(\frac{x}{|x|}\psi, \nabla u\right)\\ &\le \left\|\frac{x}{|x|}\psi\right\|_{\dot H^{\frac12}}\|\nabla u\|_{\dot H^{-\frac12}} \lesssim \|\psi\|_{\dot H^{\frac12}}\|u\|_{\dot H^{\frac12}}.\end{aligned} \tag{4.257}$$

By duality, we have

$$\|\partial_r u\|_{\dot H^{-\frac12}} \lesssim \|u\|_{\dot H^{\frac12}}.$$

Corollary 4.4.2 (Morawetz inequalities). *Suppose $u(x,t)$ solve* (4.238). *Then* $\forall\, y \in \mathbb{R}^3$,

$$\begin{aligned}&4\pi a\int_0^T |u(y,t)|^2 dt + \int_0^T\int_{\mathbb{R}^3} \frac{2a}{|x-y|}|\nabla\!\!\!/_y u(x,t)|^2 dxdt\\ &+\int_0^T\int_{\mathbb{R}^3}\frac{2\mu\{|u|^2 f(|u|^2) - F(|u|^2)\}}{|x-y|}dxdt \lesssim 2\sup_{t\in[0,T]}\|u(t)\|^2_{\dot H^{\frac12}}.\end{aligned} \tag{4.258}$$

Remark 4.4.7.

(a) Morawetz estimate plays an important role in the study of scattering of (4.238) or local decay estimate. In particular, the terms in the LHS of (4.258) are positive and the classical Morawetz estimate is obtained, if the nonlinearity $f(u)$ satisfies the repulsive condition (4.248). For example, if we take $f(u) = |u|^{p-1}u$, then for any $T > 0$,

$$\int_0^T\int_{\mathbb{R}^3}\frac{|u(x,t)|^{p+1}}{|x-y|}dxdt \lesssim \sup_{t\in[0,T]}\|u(t)\|^2_{\dot H^{\frac12}}, \quad \forall\, y \in \mathbb{R}^3, \tag{4.259}$$

$$\int_0^T\int_{\mathbb{R}^3}\frac{|\nabla\!\!\!/_y u(x,t)|^2}{|x-y|}dxdt \lesssim \sup_{t\in[0,T]}\|u(t)\|^2_{\dot H^{\frac12}}, \quad \forall\, y \in \mathbb{R}^3, \tag{4.260}$$

$$\int_0^T |u(y,t)|^2 dt \lesssim \sup_{t\in[0,T]}\|u(t)\|^2_{\dot H^{\frac12}}, \quad \forall\, y \in \mathbb{R}^3. \tag{4.261}$$

If $\varphi \in H^1(\mathbb{R}^3)$, by the energy conservation and the repulsive condition on $f(u)$, we can deduce the classical Morawetz estimate

$$\int_{\mathbb{R}}\int_{\mathbb{R}^3} \frac{|u(x,t)|^{p+1}}{|x-y|}dxdt \lesssim \|\varphi\|_{H^1}^2, \quad \forall\, y \in \mathbb{R}^3. \tag{4.262}$$

(b) In 1985, Ginibre–Velo in [104] first established the scattering theory of nonlinear Schrödinger equation with

$$f(u) = |u|^{p+1}u, \quad 1 + \frac{4}{n} < p < 1 + \frac{4}{n-2}.$$

They obtained the global integrability of solution u in $L^q(\mathbb{R}, L^r(\mathbb{R}^3))$ by the localized Morawetz estimate, the dispersive estimate and Strichartz estimate, which implies the scattering result of (4.238).

Morawetz interaction potential

Given a solution $u(t)$ of (4.238), we define the Morawetz interaction potential to be

$$M(t) = \int_{\mathbb{R}^3} |u(y,t)|^2 M_y(t)dy. \tag{4.263}$$

The bound (4.254) immediately implies

$$|M(t)| \lesssim \|u(t)\|_{L^2}^2 \|u(t)\|_{\dot{H}^{\frac{1}{2}}}^2. \tag{4.264}$$

If u solves (4.238), then the identity (4.252) gives us the following identity:

$$\begin{aligned}\frac{d}{dt}M(t) &= 4\pi a \int_{\mathbb{R}^3} |u(y,t)|^4 dy + \int_{\mathbb{R}^3}\int_{\mathbb{R}^3} \frac{2a}{|x-y|}|u(y,t)|^2 |\nabla u(x,t)|^2 dxdy \\ &+ \int_{\mathbb{R}^3}\int_{\mathbb{R}^3} \frac{2\mu|u(y,t)|^2}{|x-y|}\{|u(x,t)|^2 f|u(x,t)|^2) - F(|u(x,t)|^2)\}dxdy \\ &+ \int_{\mathbb{R}^3} \partial_t(|u(t,y)|^2)M_y(t)dy =: \mathrm{I} + \mathrm{II} + \mathrm{III} + \mathrm{IV}.\end{aligned} \tag{4.265}$$

Proposition 4.4.3. *Assume $u(x,t)$ is a solution of (4.238), then*

$$\begin{aligned}&4\pi a \int_{\mathbb{R}^3} |u(y,t)|^4 dy + \int_{\mathbb{R}^3}\int_{\mathbb{R}^3} \frac{2\mu|u(y,t)|^2}{|x-y|}\{|u|^2 f(|u|^2) - F(|u|^2)\}dxdy \\ &\le \frac{dM(t)}{dt}.\end{aligned} \tag{4.266}$$

In particular, $M(t)$ is monotone increasing for equations with repulsive nonlinearities.

Corollary 4.4.3. *Assume that $u(x,t)$ is a smooth solution to (4.238), and $f(u)$ satisfies the repulsive condition (4.248). Then we have the following interaction Morawetz inequalities:*

$$\begin{aligned} &4\pi a\int_0^T\int_{\mathbb{R}^3}|u(y,t)|^4dydt \\ &\quad+\int_0^T\int_{\mathbb{R}^3\times\mathbb{R}^3}\frac{2\mu|u(y,t)|^2}{|x-y|}(|u|^2f(|u|^2)-F(|u|^2))(x,t)dxdydt \\ &\leq 2\|\varphi(x)\|_2^2\sup_{t\in[0,T]}\|u(t)\|^2_{\dot{H}^{\frac{1}{2}}},\quad\forall\, T>0. \end{aligned}\tag{4.267}$$

In particular, we obtain

$$\int_0^T\int_{\mathbb{R}^3}|u(y,t)|^4dydt\lesssim\|\varphi\|_2^2\sup_{t\in[0,T]}\|u(t)\|^2_{\dot{H}^{\frac{1}{2}}},\quad\forall\, T>0.\tag{4.268}$$

Remark 4.4.8.

(i) The classical Morawetz estimate

$$\int_{\mathbb{R}}\int_{\mathbb{R}^3}\frac{|u(x,t)|^{p+1}}{|x|}dxdt\leq C(\|\varphi\|^2_{H^1}),$$

comes from the nonlinear term. However, the interaction Morawetz estimate comes from the linear term.

(ii) Estimate (4.258) means the decay of $u(x,t)$ at the singular point y, which can eliminate the energy concentration of $u(x,t)$ around y; however, it cannot remove the energy concentration of $u(x,t)$ away from y. Consequently, one cannot obtain the scattering results in the space H^s (except for the radial solution). The interaction Morawetz can deal with this problem under some nonlinear hypotheses.

(iii) For the case $\varphi(x)\in H^1(\mathbb{R}^3)$, one can obtain the new uniform Morawetz estimate:

$$\int_{\mathbb{R}}\int_{\mathbb{R}^3}|u(x,t)|^4dxdt\leq C(\|\varphi\|^2_{H^1})\tag{4.269}$$

by (4.267), from which we can give a simple proof for Ginibre–Velo's scattering results in H^1. In fact, applying the Strichartz estimate and the interpolation to the integral equation (4.226), we have

$$\begin{aligned} &\|u(t)\|_{L^{10}_{x,t}(\mathbb{R}^3\times\mathbb{R})}+\|\nabla u(t)\|_{L^{\frac{10}{3}}_{x,t}(\mathbb{R}^3\times\mathbb{R})} \\ &\quad\lesssim\|\varphi\|_{\dot{H}^1}+\||u|^2\nabla u(t)\|_{L^{\frac{10}{7}}_{x,t}(\mathbb{R}^3\times\mathbb{R})} \end{aligned}$$

$$\lesssim \|\varphi\|_{\dot H^1} + \|u\|^2_{L^5_{x,t}(\mathbb{R}^3\times\mathbb{R})}\|\nabla u(t)\|_{L^{\frac{10}{3}}_{x,t}(\mathbb{R}^3\times\mathbb{R})}$$

$$\lesssim \|\varphi\|_{\dot H^1} + \|u\|^{\frac{4}{3}}_{L^4_{x,t}(\mathbb{R}^3\times\mathbb{R})}\|u\|^{\frac{2}{3}}_{L^{10}_{x,t}(\mathbb{R}^3\times\mathbb{R})}\|\nabla u(t)\|_{L^{\frac{10}{3}}_{x,t}(\mathbb{R}^3\times\mathbb{R})}$$

$$\lesssim \|\varphi\|_{\dot H^1} + \|u\|^{\frac{4}{3}}_{L^4_{x,t}(\mathbb{R}^3\times\mathbb{R})}\big[\|u\|_{L^{10}_{x,t}(\mathbb{R}^3\times\mathbb{R})} + \|\nabla u(t)\|_{L^{\frac{10}{3}}_{x,t}(\mathbb{R}^3\times\mathbb{R})}\big]^{\frac{5}{3}}. \tag{4.270}$$

For any $\epsilon > 0$, there exists a finite decomposition $\mathbb{R} = \bigcup_{j=1}^{J_0} I_j$, such that

$$\|u\|^{\frac{4}{3}}_{L^4_{x,t}(\mathbb{R}^3\times I_j)} < \epsilon, \quad j = 1, 2, \ldots, J_0. \tag{4.271}$$

(iv) It is heuristic that it is an important step to deduce the generalized Morawetz estimate

$$\int_{\mathbb{R}}\int_{\mathbb{R}^3} |u|^4 dxdt \lesssim C(\|\varphi\|_{H^s}) \tag{4.272}$$

in the scattering theory of H^s. To do so, we need the scaling analysis, almost conservation technique.

Proof of Proposition 4.4.3. By (4.268), it suffices to show

$$\mathrm{IV} \geq -\mathrm{II}. \tag{4.273}$$

Using (4.245) to write

$$\begin{aligned}\mathrm{IV} &= -\int_{\mathbb{R}^3_y} \nabla\cdot \operatorname{Im}[2a\bar u(y,t)\nabla u(y,t)]M_y(t)dy\\ &= \int_{\mathbb{R}^3_y}\int_{\mathbb{R}^3_x} \partial_{y_\ell}\operatorname{Im}[2a\bar u(y,t)\partial_{y_\ell}u(y)]\operatorname{Im}\Big[\bar u(x,t)\frac{x_m-y_m}{|x-y|}\partial_{x_m}u(x)\Big]dxdy,\end{aligned} \tag{4.274}$$

where the repeated indices are implicitly summed. Write $P(x) = \operatorname{Im}(\bar u(x)\nabla u(x))$ for the mass current vector at x and integrate by parts in y in (4.274). Noticing that

$$\partial_{y_\ell}\Big(\frac{x_m-y_m}{|x-y|}\Big) = \frac{-\delta_{m\ell}}{|x-y|} + \frac{(x_\ell-y_\ell)(x_m-y_m)}{|x-y|^3}, \tag{4.275}$$

we have

$$\mathrm{IV} = -2a\int_{\mathbb{R}^3_y}\int_{\mathbb{R}^3_x}\Big[P(y)P(x) - P(y)\frac{x-y}{|x-y|}\Big(P(x)\frac{x-y}{|x-y|}\Big)\Big]\frac{dxdy}{|x-y|}. \tag{4.276}$$

The preceding integrand has a natural geometric interpretation. We are removing the inner product of the components of $P(y)$ and $P(x)$ parallel to the vector $\frac{x-y}{|x-y|}$ from the full inner product of $P(y)$ and $P(x)$. By the decomposition,

$$\begin{cases} P(y) = (P(y) \cdot \frac{x-y}{|x-y|})\frac{x-y}{|x-y|} + \Pi_{(x-y)^\perp} P(y), \\ P(x) = (P(x) \cdot \frac{x-y}{|x-y|})\frac{x-y}{|x-y|} + \Pi_{(x-y)^\perp} P(x), \end{cases} \tag{4.277}$$

we have

$$P(x)P(y) = P(y)\frac{x-y}{|x-y|} \cdot P(x)\frac{x-y}{|x-y|} + \Pi_{(x-y)^\perp} P(x) \cdot \Pi_{(x-y)^\perp} P(y). \tag{4.278}$$

But

$$\begin{aligned} |\Pi_{(x-y)^\perp} P(y)| &= \left|P(y) - \frac{x-y}{|x-y|}\left(\frac{x-y}{|x-y|} \cdot P(y)\right)\right| \\ &= |\mathrm{Im}(\bar{u}(y)\ \nabla_x u(y))| \le |u(y)||\nabla_x u(y)|, \end{aligned} \tag{4.279}$$

$$\mathrm{IV} = -2a \int_{\mathbb{R}^3_y}\int_{\mathbb{R}^3_x} \Pi_{(x-y)^\perp} P(x) \cdot \Pi_{(x-y)^\perp} P(y) \frac{dxdy}{|x-y|}, \tag{4.280}$$

which implies that

$$\begin{aligned} \mathrm{IV} &\ge -2a \int_{\mathbb{R}^3_y}\int_{\mathbb{R}^3_x} |u(x)||u(y)||\nabla_y u(x)||\nabla_x u(y)| \frac{dxdy}{|x-y|} \\ &\ge -a \int_{\mathbb{R}^3}\int_{\mathbb{R}^3} |u(y)|^2 |\nabla_y (u(x))|^2 \frac{dxdy}{|x-y|} \\ &\quad - a \int_{\mathbb{R}^3}\int_{\mathbb{R}^3} |u(x)|^2 |\nabla_x (u(y))|^2 \frac{dxdy}{|x-y|} \\ &= -\int_{\mathbb{R}^3}\int_{\mathbb{R}^3} \frac{2a}{|x-y|} |u(y)|^2 |\nabla_y u(x)|^2 dxdy. \end{aligned} \tag{4.281}$$

□

Almost conservation law

For $s < 1$, the energy of our solution $u(t) \in H^s(\mathbb{R}^3)$ might be infinite, that is, $E(u(t)) = \infty$. But the smoothing version $Iu(t)$ of $u(t)$ can satisfy $E(Iu(t)) < \infty$. Our aim will be to prove the almost conservation law of $E(Iu(t))$, namely control the growth in time of $E(Iu(t))$ as

$$E(Iu) = E(I\varphi) + O(N^{-1+}),$$

where N is sufficiently large number, which can imply the uniform bound on $\|u(t)\|_{H^s(\mathbb{R}^3)}$ and, furthermore, from which we can obtain the global well-posedness and scattering of the solution in $H^s(\mathbb{R}^3)$.

Definition 4.4.1. Let $s < 1$ and $N \gg 1$. Define the Fourier multiplier $m_N(\xi) \in C^\infty(\mathbb{R})$ to be radially symmetric, and nonincreasing in $|\xi|$, and

$$m_N(\xi) = \begin{cases} 1, & |\xi| \le N, \\ (\frac{N}{|\xi|})^{1-s}, & |\xi| > 2N. \end{cases} \tag{4.282}$$

Write I as the smoothing operator with symbol $m_N(\xi)$, that is,

$$Iu(t) = \mathcal{F}^{-1}(m_N(\xi)\hat{u}(\xi, t)). \tag{4.283}$$

Remark 4.4.9.

(a) Since

$$\left|\frac{\partial^\alpha m_N(\xi)}{\partial \xi^\alpha}\right| \lesssim |\xi|^{-|\alpha|}, \quad |\alpha| \le \left[\frac{n}{2}\right] + 1,$$

$m_N(\xi)$ is a L^p ($1 < p < \infty$) multiplier. In addition, $m_N(\xi)$ satisfies

$$m_N(\xi)\xi \lesssim \begin{Bmatrix} |\xi|, & |\xi| \le N, \\ |\xi|^s N^{1-s}, & |\xi| \ge 2N \end{Bmatrix} \le |\xi|^s N^{1-s},$$

$$|\xi|^s \lesssim m_N(\xi)|\xi| + 1 \cong \begin{cases} |\xi| + 1, & |\xi| \le N, \\ N^{1-s}|\xi|^s + 1, & |\xi| \ge 2N, \end{cases}$$

which implies

$$\begin{cases} E(Iu(t)) \lesssim (N^{1-s}\|u\|_{H^s})^2 + \|Iu(x,t)\|_{L^4}^4, \\ \|u(t)\|_{H^s}^2 \lesssim E(Iu) + \|u\|_2^2. \end{cases} \tag{4.284}$$

(b) We will control the growth of $E(Iu(t))$ to bound $\|u\|_{H^s}$. Since (4.225) is nonlinear equation, Iu is not the solution of (4.225). It only satisfies

$$\begin{cases} i(Iu)_t + \Delta(Iu) = I(|u|^2 u), \\ Iu(0) = I\varphi(x), \end{cases} \tag{4.285}$$

which implies that there is no energy conservation law for $Iu(t)$. In this section, we will focus on the proof of the almost conservation law of $E(Iu(t))$, which implies that $E(Iu(t))$ satisfies

$$E(Iu(t)) \le C(\|\varphi\|_{H^s}). \tag{4.286}$$

Proposition 4.4.4 (Almost conservation law). *Assume $s > 1/2$, $N \gg 1$, $0 < \varepsilon \ll 1$, $\varphi(x) \in C_c^\infty(\mathbb{R}^3)$, $u(t)$ is the solution of* (4.225) *or* (4.226) *on* $[0, T]$ *and satisfies*

$$\|u(t)\|_{L^4_{x,t}([0,T]\times\mathbb{R}^3)} \lesssim \varepsilon. \tag{4.287}$$

Assume in addition that

$$E(I\varphi) \lesssim 1, \tag{4.288}$$

then for $\forall t \in [0,T]$,

$$E(Iu(t)) = E(I\varphi) + O(N^{-1+}). \tag{4.289}$$

Remark 4.4.10.

(a) If one could show that

$$E(Iu(t)) = E(I\varphi) + O(N^{-\alpha}), \tag{4.290}$$

one can prove the global well-posedness of (4.225) or (4.226) in H^s for $s > \frac{3+\alpha}{3+2\alpha}$. In particular, if $E_1(Iu)$ were conserved (i. e., $\alpha = \infty$), one could show that (4.225) or (4.226) is globally well-posed for $s > 1/2$.

(b) The space-time norm plays an important role in establishing almost conservation law. Precisely, we will use the following norm:

$$Z_I(t) = \sup_{(q,r)\in\Lambda} \|\nabla(Iu)\|_{L^q([0,t);L^r(\mathbb{R}^3))}. \tag{4.291}$$

Lemma 4.4.1. *Assume that* $\varphi(x) \in C_c^\infty(\mathbb{R}^3)$, $0 < \varepsilon \ll 1$, $u(x,t)$ *is the solution of* (4.225), (4.226) *on* $[0,T^*]$ *and satisfies*

$$\|u(x,t)\|_{L^4([0,T^*]\times\mathbb{R}^3)} \lesssim \varepsilon. \tag{4.292}$$

Then for $\forall s > 1/2$ *and sufficiently large* N,

$$Z_I(t) \lesssim C(\|\varphi\|_{H^s(\mathbb{R}^3)}). \tag{4.293}$$

Proof. Applying $I\nabla$ to both sides of (4.225), we have

$$\begin{cases} i(I\nabla u)_t + \Delta(I\nabla u) = I\nabla(|u|^2u), \\ I\nabla u(0) = I\nabla\varphi. \end{cases} \tag{4.294}$$

Choosing $(q_2,r_2) = (\frac{10}{3},\frac{10}{3}) \in \Lambda$, and a fractional Leibniz rule give us that

$$\begin{aligned} Z_I(t) &\lesssim \|\nabla I\varphi\|_2 + \|\nabla I(|u|^2u)\|_{L^{\frac{10}{7}}_{x,t}([0,t]\times\mathbb{R}^3)} \\ &\lesssim \|\varphi\|_{H^s} + \|\nabla Iu\|_{L^{\frac{10}{3}}_{x,t}([0,t]\times\mathbb{R}^3)} \|u\|^2_{L^5_{x,t}([0,t]\times\mathbb{R}^3)}, \quad \forall t \in [0,T^*]. \end{aligned} \tag{4.295}$$

By the Littlewood–Paley decomposition,

$$
\begin{aligned}
u(x,t) &= \psi_0(x,t) + \sum_{i=1}^{\infty} \psi_j(x,t), \\
\operatorname{supp} \hat{\psi}_0(\xi,t) &\subset \{\xi : \langle\xi\rangle = (1+|\xi|^2)^{\frac{1}{2}} \leq N_0 = N\}, \\
\operatorname{supp} \hat{\psi}_j(t,\xi) &\subset \{\xi : \langle\xi\rangle \sim N_j = 2^{k_j},\ k_j \gtrsim \log_2(N)\}, \quad j = 1,2,\ldots.
\end{aligned}
$$

The strategy is to use $(L^4_{x,t}, L^{10}_{x,t})_{\theta_1} \sim L^5_{x,t}$ to estimate the low frequency component, and use $(L^{\frac{10}{3}}_{x,t}, L^{10}_{x,t})_{\theta_2} \sim L^5_{x,t}$ to estimate the high frequency component.

Specifically, the definition of I gives

$$
\|I\psi_j\|_{L^{10}_{x,t}} \sim \begin{cases} \|\psi_j\|_{L^{10}_{x,t}}, & j = 0, \\ N^{1-s} N_j^{s-1} \|\psi_j\|_{L^{10}_{x,t}}, & j \geq 1. \end{cases}
$$

Using Sobolev's inequality, we have

$$
\|\psi_j\|_{L^{10}_{x,t}([0,T^*]\times\mathbb{R}^3)} \leq \begin{cases} Z_I(T^*), & j = 0, \\ N_j^{1-s} N^{s-1} Z_I(T^*), & j \geq 1. \end{cases} \tag{4.296}
$$

Similarly,

$$
\|\nabla I\psi_j\|_{L^{\frac{10}{3}}_{x,t}} \sim N_j^s N^{1-s} \|\psi_j\|_{L^{\frac{10}{3}}_{x,t}}, \quad j = 1,2,\ldots.
$$

Hence

$$
\|\psi_j\|_{L^{\frac{10}{3}}_{x,t}} \lesssim N^{s-1} N_j^{-s} Z_I(T^*), \quad j \geq 1. \tag{4.297}
$$

By interpolation, we have

$$
\|\psi_0\|_{L^5_{x,t}} \lesssim \|\psi_0\|^{\frac{2}{3}}_{L^4_{x,t}} \|\psi_0\|^{\frac{1}{3}}_{L^{10}_{x,t}} \lesssim \varepsilon^{\frac{2}{3}} Z_I(T^*)^{\frac{1}{3}}, \tag{4.298}
$$

$$
\|\psi_j\|_{L^5_{x,t}} \lesssim \|\psi_j\|^{\frac{1}{2}}_{L^{\frac{10}{3}}_{x,t}} \|\psi_j\|^{\frac{1}{2}}_{L^{10}_{x,t}} \lesssim N^{s-1} N_j^{1/2-s} Z_I(T^*). \tag{4.299}
$$

Hence,

$$
\begin{aligned}
\|u(x,t)\|_{L^5_{x,t}} &\leq \sum_{j=0}^{\infty} \|\psi_j\|_{L^5_{x,t}} \leq \varepsilon^{\frac{2}{3}} Z_I(T^*)^{\frac{1}{3}} + \sum_{j=1}^{\infty} \|\psi_j\|_{L^5_{x,t}} \\
&\leq \varepsilon^{\frac{2}{3}} Z_I(T^*)^{\frac{1}{3}} + N^{s-1} Z_I(T^*).
\end{aligned} \tag{4.300}
$$

Choosing N sufficiently large (depending on ε), we have

$$\begin{aligned} Z_I(t) &\lesssim \|\varphi\|_{H^s} + \varepsilon^{\frac{2}{3}} Z_I(T^*)^{\frac{4}{3}} + N^{s-1} Z_I(T^*)^2 \\ &\lesssim 1 + \varepsilon^{\delta_1} Z_I(T^*)^{1+\delta_2}. \end{aligned} \tag{4.301}$$

Choosing ε sufficiently small, we obtain

$$Z_I(t) \leq C(\|\varphi\|_{H^s}).$$

□

Proof of Proposition 4.4.4. It suffices to consider the sufficiently smooth solution of (4.225). Integrating by parts and using (4.225), we have

$$\begin{aligned} \partial_t E(u(t)) &= \frac{d}{dt} \int_{\mathbb{R}^3} \left[\frac{1}{2} |\nabla u|^2 + \frac{1}{4} |u|^4 \right] dx = \operatorname{Re} \int_{\mathbb{R}^3} (\nabla \bar{u}_t \cdot \nabla u + |u|^2 u \bar{u}_t) dx \\ &= \operatorname{Re} \int_{\mathbb{R}^3} \bar{u}_t (-\Delta u + |u|^2 u) dx = \operatorname{Re} \int_{\mathbb{R}^3} \bar{u}_t \cdot i u_t dx \\ &= 0. \end{aligned}$$

Similarly, using (4.285), we obtain

$$\begin{aligned} \frac{d}{dt} E(Iu(t)) &= \operatorname{Re} \int_{\mathbb{R}^3} (\overline{Iu})_t (|Iu|^2 Iu - \Delta(Iu) - i(Iu)_t) dx \\ &= \operatorname{Re} \int_{\mathbb{R}^3} (\overline{Iu})_t (|Iu|^2 Iu - I(|u|^2 u)) dx. \end{aligned} \tag{4.302}$$

Note that

$$\begin{aligned} \int_{\mathbb{R}^3} fghl dx &= \int_{\mathbb{R}^3} \check{f}(\xi) \widehat{ghl}(\xi) d\xi \\ &= \int_{\mathbb{R}^3} \hat{f}(-\xi) \int_{\mathbb{R}^3} \widehat{fh}(\xi - \xi_4) \hat{l}(\xi_4) d\xi_4 d\xi \\ &= \int_{\mathbb{R}^3} \hat{f}(-\xi) \int_{\mathbb{R}^3} \int_{\mathbb{R}^3} \hat{g}(\xi - \xi_4 - \xi_3) \hat{h}(\xi_3) \hat{l}(\xi_4) d\xi_4 d\xi_3 d\xi, \\ (\xi_1 = -\xi) = (-1)^3 &\int_{\mathbb{R}^3} \hat{f}(\xi_1) \int_{\mathbb{R}^3} \int_{\mathbb{R}^3} \hat{g}(-\xi_1 - \xi_4 - \xi_3) \hat{h}(\xi_3) \hat{l}(\xi_4) d\xi_4 d\xi_3 d\xi_1 \\ &= \int_{\mathbb{R}^3 \times \mathbb{R}^3 \times \mathbb{R}^3} \hat{f}(\xi_1) \hat{g}(-\xi_1 - \xi_4 - \xi_3) \hat{h}(\xi_3) \hat{l}(\xi_4) d\xi_3 d\xi_4 d\xi_1 \\ &= \int_{\xi_1 + \xi_2 + \xi_3 + \xi_4 = 0} \hat{f}(\xi_1) \hat{g}(\xi_2) \hat{h}(\xi_3) \hat{l}(\xi_4) d\sigma, \end{aligned} \tag{4.303}$$

where $d\sigma = \delta(\xi_1 + \cdots + \xi_4) d\xi_1 d\xi_2 d\xi_3 d\xi_4$.

$$E(Iu(t)) - E(I\varphi) = \int_0^t \int_{\xi_1+\xi_2+\xi_3+\xi_4=0} \left(1 - \frac{m(\xi_2+\xi_3+\xi_4)}{m(\xi_2)m(\xi_3)m(\xi_4)}\right) \times \widehat{\overline{Iu_t}}(\xi_1)\widehat{Iu}(\xi_2)\widehat{\overline{Iu}}(\xi_3)\widehat{Iu}(\xi_4)d\sigma dt. \tag{4.304}$$

It remains to bound

$$\text{Term 1} + \text{Term 2} \lesssim N^{-1+}(Z_I(T))^{\rho}, \quad \rho > 0, \tag{4.305}$$

where

$$\text{Term 1} := \left| \int_0^T \int_{\xi_1+\xi_2+\xi_3+\xi_4=0} \left(1 - \frac{m(\xi_2+\xi_3+\xi_4)}{m(\xi_2)m(\xi_3)m(\xi_4)}\right) \times \widehat{\overline{\Delta(Iu)}}(\xi_1)\widehat{Iu}(\xi_2)\widehat{\overline{Iu}}(\xi_3)\widehat{Iu}(\xi_4)d\sigma dt \right|, \tag{4.306}$$

$$\text{Term 2} := \left| \int_0^T \int_{\xi_1+\xi_2+\xi_3+\xi_4=0} \left(1 - \frac{m(\xi_2+\xi_3+\xi_4)}{m(\xi_2)m(\xi_3)m(\xi_4)}\right) \times \widehat{\overline{I(|u|^2u)}}(\xi_1)\widehat{Iu}(\xi_2)\widehat{\overline{Iu}}(\xi_3)\widehat{Iu}(\xi_4)d\sigma dt \right|. \tag{4.307}$$

In both cases, we use the Littlewood–Paley decomposition

$$u = S_0 u + \sum_{j=1}^{\infty} \Delta_j u = \sum_{j=0}^{\infty} u_j, \quad \operatorname{supp} \hat{u}_0(\xi) \sim \{\xi, \langle\xi\rangle \lesssim N_0\},$$
$$\operatorname{supp} \hat{u}_j(\xi) \sim \{\xi, \langle\xi\rangle \sim 2^{k_j} \equiv N_j,\ k_j \in \{1,2,\dots\}\}$$

and a multilinear multiplier of Coifman–Meyer.

Consider an infinitely differentiable symbol $\sigma : \mathbb{R}^{nk} \longmapsto \mathbb{C}$ with

$$|\partial_\xi^\alpha \sigma(\xi)| \le C(\alpha)(1+|\xi|)^{-|\alpha|}, \quad \alpha \in \mathbb{N}^{nk},\ \xi = (\xi_1,\dots,\xi_k) \in \mathbb{R}^{nk}. \tag{4.308}$$

Define the multilinear operator

$$T[f_1,f_2,\dots,f_k](x) = \int_{\mathbb{R}^{nk}} e^{ix(\xi_1+\xi_2+\dots+\xi_k)}\sigma(\xi_1,\dots,\xi_k)\hat{f}_1(\xi_1)\hat{f}_2(\xi_2)\cdots\hat{f}_k(\xi_k)d\xi_1\cdots d\xi_k. \tag{4.309}$$

Theorem 4.4.2. *Suppose $p_j \in (1,\infty)$, $j = 1,2,\dots,k$ are such that*

$$\frac{1}{p} = \frac{1}{p_1} + \frac{1}{p_2} + \dots + \frac{1}{p_k} \le 1.$$

Assume that $\sigma(\xi_1, \dots, \xi_k)$ is a smooth symbol as in (4.308). Then there is a constant $C = C(p_j, n, k, C(\alpha))$ so that for all Schwartz class function $f_1, \dots, f_k$,

$$\|\mathcal{T}(f_1, \dots, f_k)\|_{L^p(\mathbb{R}^n)} \leq C\|f_1\|_{p_1} \cdots \|f_k\|_{p_k}. \tag{4.310}$$

Remark 4.4.11.

(a) Estimate (4.310) is also available for operators whose symbols obey much weaker bounds than (4.308); see Coifman–Meyer's result.

(b) In order to estimate (4.306) and (4.307), one should first seek the pointwise bound on the symbol

$$\left|1 - \frac{m(\xi_2 + \xi_3 + \xi_4)}{m(\xi_2)m(\xi_3)m(\xi_4)}\right| \leq B(N_2, N_3, N_4). \tag{4.311}$$

We factor $B(N_2, N_3, N_4)$ out of the left side, leaving a symbol σ that satisfies estimate (4.308). We are left to estimate a quantity of the form

$$\left|\int_0^T \int_{\mathbb{R}^3} \mathcal{T}[f_1, f_2, f_3]^\wedge(\xi_4)\overline{\hat{f}_4}(\xi_4)d\xi_4 dt\right| \cdot B(N_2, N_3, N_4), \tag{4.312}$$

where

$$\mathcal{T}[f_1, f_2, f_3] = \int_{\mathbb{R}^9} e^{i(\xi_1+\xi_2+\xi_3)x}\sigma(\xi_1, \dots, \xi_4)\hat{f}_1(\xi_1)\hat{f}_2(\xi_2)\hat{f}_3(\xi_3)d\xi_1 d\xi_2 d\xi_3. \tag{4.313}$$

We estimate this by using the Plancherel formula, Hölder's inequality, the Coifman–Meyer estimate and the Strichartz estimate.

(c) We can estimate u by the estimate of u_j. The advantage is that we can obtain a better decay estimate in the frequencies $|\xi| \sim N_j$. It suffices to obtain $O(N^{-1+})$ decay estimate for every kind of frequency interaction. For simplicity, we drop the complex conjugates since they do not affect the analysis used here. For example, we will conclude Term $1 \leq N^{-1+}$ once we prove

$$\begin{aligned}&\left|\int_0^T \int_{\xi_1+\xi_2+\xi_3+\xi_4=0} \left(1 - \frac{m(\xi_2 + \xi_3 + \xi_4)}{m(\xi_2)m(\xi_3)m(\xi_4)}\right)\widehat{\triangle Iu_1}(\xi_1)\widehat{Iu_2}(\xi_2)\widehat{Iu_3}(\xi_3)\widehat{Iu_4}(\xi_4)\right| \\ &\quad \lesssim N^{-1+}C(N_1, N_2, N_3, N_4)Z_I(T)^4,\end{aligned} \tag{4.314}$$

where $C(N_1, N_2, N_3, N_4)$ is sufficiently small. By symmetry, we may assume $N_2 \geq N_3 \geq N_4$. The precise extent to which $C(N_1, N_2, N_3, N_4)$ decays in this arguments, and the fact that this decay allows us to sum over all dyadic shells, will be described below.

Consider Term 1. By the symmetry $N_2 \geq N_3 \geq N_4$, we divide into the following cases.

Case I. $N \gg N_2$. By the definition of (4.282), the symbol

$$1 - \frac{m(\xi_2 + \xi_3 + \xi_4)}{m(\xi_2)m(\xi_3)m(\xi_4)} = 0.$$

Hence, the bound (4.314) holds trivially.

Case II. $N_2 \gtrsim N \gg N_3 \geq N_4$. Since $\xi_1 + \xi_2 + \xi_3 + \xi_4 = 0$, we have $N_1 \sim N_2$. We claim for (4.314) with

$$C(N_1, N_2, N_3, N_4) \sim N_2^{0-}. \tag{4.315}$$

With this decay factor and the fact that we are considering here terms where $N_1 \sim N_2$, we may immediately sum over all N_i.

By the mean value theorem, we have

$$\begin{aligned}\left|1 - \frac{m(\xi_2 + \xi_3 + \xi_4)}{m(\xi_2)m(\xi_3)m(\xi_4)}\right| &= \left|\frac{m(\xi_2) - m(\xi_2 + \xi_3 + \xi_4)}{m(\xi_2)}\right| \\ &\lesssim \frac{|\nabla m(\xi_2)| \cdot |\xi_3 + \xi_4|}{|m(\xi_2)|} \lesssim \frac{N_3}{N_2}.\end{aligned}$$

We view the N_3 in the numerator as resulting from a derivative falling on the Iu_3 factor in the integrand. Hence,

$$\begin{aligned}\text{LHS of (4.314)} &\lesssim \frac{N_3}{N_2}\left|\int_0^T \int_{\mathbb{R}^3} \mathcal{T}[\Delta I u_1, I u_2, I u_3] I u_4 dx dt\right| \\ &\lesssim \frac{1}{N_2}\|\Delta I u_1\|_{L_{x,t}^{\frac{10}{3}}} \|I u_2\|_{L_{x,t}^{\frac{10}{3}}} \|\nabla I u_3\|_{L_{x,t}^{\frac{10}{3}}} \|I u_4\|_{L_{x,t}^{10}} \\ &\lesssim \frac{N_1}{N_2 \cdot N_2}(Z_I(T))^4 \lesssim \frac{1}{N_2} Z_I(T)^4 \\ &\lesssim N^{-1+} N_2^{0-} Z_I(T)^4.\end{aligned} \tag{4.316}$$

Case III. $N_2 \geq N_3 \gtrsim N$. In this case, the only pointwise bound available for the symbol

$$\left|1 - \frac{m(\xi_2 + \xi_3 + \xi_4)}{m(\xi_2)m(\xi_3)m(\xi_4)}\right| = \left|1 - \frac{m(\xi_1)}{m(\xi_2)m(\xi_3)m(\xi_4)}\right|$$

is the straightforward one: when $|\xi_1|$ and $|\xi_2|$ are not comparable, no cancelation can occur in the numerator. When $|\xi_1| \sim |\xi_2|$, we then also need $|\xi_3|, |\xi_4| \leq N$ in order to get cancelation. If any of these conditions fail, our pointwise estimate will be simply

$$\left|1 - \frac{m(\xi_2 + \xi_3 + \xi_4)}{m(\xi_2)m(\xi_3)m(\xi_4)}\right| \lesssim \left|\frac{m(\xi_1)}{m(\xi_2)m(\xi_3)m(\xi_4)}\right|. \tag{4.317}$$

The frequency interactions here fall into two subcategories, depending on which frequency is comparable to N_2.

We further divide this case into (a) $N_1 \sim N_2 \gtrsim N_3 \gtrsim N$ and (b) $N_2 \sim N_3 \gtrsim N$.

Case III(a). $N_1 \sim N_2 \gtrsim N_3 \gtrsim N$. By assumption, $s = \frac{1}{2} + \delta$, $\delta > 0$. In this case, we prove the decay factor

$$N^{-1+}C(N_1, N_2, N_3, N_4) = N^{-1+2\delta} N_3^{0-2\delta} \tag{4.318}$$

in (4.314). This allows us to directly sum in N_3, N_4, and sum in N_1, N_2 after applying Cauchy–Schwarz to those factors. Estimate the symbol using (4.317), and using the Coifman–Meyer theorem and Hölder's inequality to take the factors involving u_i, $i = 1, 2, 3$, in $L_{t,x}^{10/3}$ and the u_4 factor in $L_{t,x}^{10}$.

$$\|\nabla Iu_1\|_{L_{x,t}^{\frac{10}{3}}} \|\nabla Iu_2\|_{L_{x,t}^{\frac{10}{3}}} \|\nabla Iu_3\|_{L_{x,t}^{\frac{10}{3}}} \|Iu_4\|_{L_{x,t}^{10}}.$$

In order to obtain (4.314), (4.318), it remains to show

$$\frac{m(N_1)N_1 N^{1-2\delta} N_3^{2\delta}}{m(N_2)m(N_3)m(N_4)N_2 N_3} \lesssim 1. \tag{4.319}$$

Note that $m(x)\langle x\rangle$ is bounded from below and for any $p > \frac{1}{2} - \delta$ (δ is sufficiently small), $m(x)|x|^p$ is increasing. The bound (4.319) is now straightforward:

$$\begin{aligned}
\text{LHS of (4.319)} &\lesssim \frac{N^{1-2\delta} N_3^{2\delta}}{m(N_3)m(N_4)N_3} \lesssim \frac{N^{1-2\delta} N_3^{2\delta}}{m(N_3)^2 N_3} \\
&\lesssim \frac{N^{1-2\delta}}{m(N_3)N_3^{\frac{1}{2}-\delta} m(N_3) N_3^{\frac{1}{2}-\delta}} \lesssim \frac{N^{1-2\delta} N_3^{2\delta}}{N^{1-2\delta} N_3^{2\delta}} \lesssim 1,
\end{aligned}$$

which gives (4.314) and (4.318).

Case III(b). $N_2 \sim N_3 \gtrsim N$. We aim in this case for the decay factor

$$N^{-1+}C(N_1, N_2, N_3, N_4) = N^{-1+2\delta} N_2^{-2\delta}, \tag{4.320}$$

where δ is sufficiently small. This will allow us to sum directly in all the N_i. As in case III(a) above, it suffices to show

$$\frac{m(N_1)N_1 N^{1-2\delta} N_2^{2\delta}}{m(N_2)m(N_3)m(N_4)N_2 N_3} \lesssim 1. \tag{4.321}$$

In fact, we have

$$\begin{aligned}
\text{LHS of (4.321)} &\lesssim \frac{m(N_1)N_1 N^{1-2\delta} N_2^{2\delta}}{m(N_2)^3 N_2^2} \lesssim \frac{m(N_2)N_2 N^{1-2\delta} N_2^{2\delta}}{(m(N_2))^3 N_2^2} \\
&\lesssim \frac{N^{1-2\delta} N_2^{2\delta}}{m(N_2)^2 N_2} \lesssim \frac{N^{1-2\delta} N_2^{2\delta}}{N_2^{2\delta} N^{1-2\delta}} \quad (N_2 \gtrsim N) \\
&\lesssim 1.
\end{aligned}$$

It remains to bound Term 2. When decomposing the integrand of Term 2 in frequency space, write N_{123} for the dyadic frequency into which we project the nonlinear factor (Iu^3). Noticing that in the treatment of Term 1 above, we always take

$$\|P_{N_1}(I\Delta u)\|_{L^{\frac{10}{3}}_{x,t}} \lesssim N_1 Z_I(T).$$

The analysis above for Term 1 therefore applies unmodified to Term 2 once we prove the following:

$$\|P_{N_{123}} I(u^3)\|_{L^{\frac{10}{3}}_{x,t}} \lesssim N_{123}(Z_I(T))^3.$$

Lemma 4.4.2. *Assume $u(t)$ is the smooth solution of* (4.225) *or* (4.226), *then*

$$\|P_{N_{123}}(I(u^3))\|_{L^{\frac{10}{3}}_{x,t}([0,T]\times\mathbb{R}^3)} \lesssim N_{123} Z_I(T)^3, \tag{4.322}$$

where P_M denotes the Littlewood–Paley projection onto the M frequency shell.

Proof. We write $u = u_\ell + u_h$, where

$$\operatorname{supp}\hat{u}_\ell(\xi,t) \subseteq \{\xi \mid |\xi| < 2\}, \quad \operatorname{supp}\hat{u}_h(\xi,t) \subseteq \{\xi \mid |\xi| > 1\}.$$

By $N_{123} \geq 1$, we have

$$\begin{aligned}\|P_{N_{123}}(I(u_\ell^3))\|_{L^{\frac{10}{3}}_{x,t}} &\lesssim \|u_\ell\|^3_{L^{10}_{x,t}} = \|Iu_\ell\|^3_{L^{10}_{x,t}} \\ &\lesssim (Z_I(T))^3 \lesssim N_{123}(Z_I(T))^3.\end{aligned} \tag{4.323}$$

By the Littlewood–Paley theory, Sobolev embedding and the Leibniz rule, we have

$$\begin{aligned}\left\|\frac{1}{N_{123}} P_{N_{123}}(I(u_h^3))\right\|_{L^{\frac{10}{3}}_{x,t}} &\lesssim \|\nabla^{-1} P_{N_{123}}(Iu_h^3)\|_{L^{\frac{10}{3}}_{x,t}} \\ &\lesssim \|\nabla^{\frac{1}{2}} I(u_h^3)\|_{L_t^{\frac{10}{3}} L_x^{\frac{10}{8}}} \lesssim \|\nabla^{\frac{1}{2}} Iu_h\|^3_{L_t^{10} L_x^{\frac{30}{8}}} \\ &\lesssim \|\nabla Iu_h\|^3_{L_t^{10} L_x^{\frac{30}{13}}} \lesssim Z_I(T)^3,\end{aligned} \tag{4.324}$$

where we used

$$H^{\frac{3}{2},\frac{10}{8}}(\mathbb{R}^3) \hookrightarrow L^{\frac{10}{3}}(\mathbb{R}^3), \quad H^{\frac{1}{2},\frac{30}{13}}(\mathbb{R}^3) \hookrightarrow L^{\frac{30}{8}}(\mathbb{R}^3), \quad \left(10, \frac{30}{13}\right) \in \Lambda.$$

Similarly,

$$
\begin{aligned}
&\left\| \frac{1}{N_{123}} P_{N_{123}} I(u_h \cdot u_h \cdot u_\ell) \right\|_{L^{\frac{10}{3}}_{x,t}} \\
&\quad \lesssim \| \nabla^{\frac{1}{2}} I(u_h \cdot u_h \cdot u_\ell) \|_{L_t^{\frac{10}{3}} L_x^{\frac{10}{8}}} \\
&\quad \lesssim \| \nabla^{\frac{1}{2}} I u_h \|_{L_t^{10} L_x^{\frac{30}{8}}} \| u_h \|_{L_t^{10} L_x^{\frac{30}{13}}} \cdot \| u_\ell \|_{L^{10}_{x,t}} \\
&\qquad + \| u_h \|_{L_t^{10} L_x^{\frac{30}{8}}} \| u_h \|_{L_t^{10} L_x^{\frac{30}{8}}} \| \nabla^{\frac{1}{2}} I u_\ell \|_{L_t^{10} L_x^{\frac{30}{8}}} \\
&\quad \lesssim \| \nabla I u_h \|_{L_t^{10} L_x^{\frac{30}{13}}} \| \nabla I u_h \|_{L_t^{10} L_x^{\frac{30}{13}}} \| I u_\ell \|_{L_t^{10} L_x^{10}} \\
&\qquad + \| \nabla^{\frac{1}{2}} I u_h \|_{L_t^{10} L_x^{\frac{30}{8}}} \| \nabla^{\frac{1}{2}} I u_h \|_{L_t^{10} L_x^{\frac{30}{8}}} \| \nabla I u_\ell \|_{L_t^{10} L_x^{\frac{30}{13}}} \\
&\quad \lesssim Z_I(T)^3,
\end{aligned}
\tag{4.325}
$$

where we use $\frac{2}{10} = 3(\frac{1}{2} - \frac{1}{6} - \frac{8}{30})$, and in the same way, we have

$$
\begin{aligned}
&\left\| \frac{1}{N_{123}} P_{N_{123}} I(u_h \cdot u_\ell \cdot u_\ell) \right\|_{L^{\frac{10}{3}}_{x,t}} \\
&\quad \lesssim \| u_h \cdot u_\ell \cdot u_\ell \|_{L_t^{\frac{10}{3}} L_x^{\frac{30}{19}}} \\
&\quad \lesssim \| u_h \|_{L_t^{10} L_x^{\frac{30}{13}}} \| u_\ell \|^2_{L^{10}_{x,t}} \lesssim \| \nabla (I u_h) \|_{L_t^{10} L_x^{\frac{30}{13}}} \| \nabla I u_\ell \|^2_{L_t^{10} L_x^{\frac{30}{13}}} \\
&\quad \lesssim Z_I(T)^3.
\end{aligned}
\tag{4.326}
$$

This completes the proof of Lemma 4.4.2. □

Therefore, we obtain the bound on Term 2. □

Proof of Theorem 4.4.1. We combine the interaction Morawetz estimate (4.267) or (4.268), almost conservation law with a scaling argument to prove the following statement, which gives uniform bounds in terms of the rough norm of the initial data.

Proposition 4.4.5. *Suppose $\varphi(x) \in C_c^\infty(\mathbb{R}^3)$, $u(t)$ is a global solution of* (4.225) *or* (4.226). *Then, so long as $s > \frac{4}{5}$, we have*

$$
\| u(t) \|_{L^4([0,\infty)\times\mathbb{R}^3])} \lesssim C(\|\varphi\|_{H^s}),
\tag{4.327}
$$

$$
\sup_{0 \le t < \infty} \| u(t) \|_{H^s(\mathbb{R}^3)} \lesssim C(\|\varphi\|_{H^s}).
\tag{4.328}
$$

Remark 4.4.12. As we know, $\forall \varphi(x) \in H^s$ $(s > 1/2)$, (4.225) or (4.226) is locally well posed and $T = T(\|\varphi\|_{H^s})$. The global well-posedness of (4.225) or (4.226) in H^s follows from density, and (4.328) involved in Proposition 4.4.5.

Proof. The first step is to scale the solution: if u is a solution of (4.225), then so is

$$
u_\lambda(x,t) = \lambda^{-1} u(\lambda^{-1} x, \lambda^{-2} t)
\tag{4.329}
$$

with initial data $\varphi_\lambda(x) = \lambda^{-1}\varphi(\lambda^{-1}x)$. Note that

$$\begin{aligned}
E(I\varphi_\lambda) &= \frac{1}{2}\|\nabla I\varphi_\lambda\|_2^2 + \frac{1}{4}\|I\varphi_\lambda\|_4^4 \le \frac{1}{2}\|\nabla I\varphi_\lambda\|_2^2 + \frac{1}{4}\|I\varphi_\lambda\|_6^2\|I\varphi_\lambda\|_3^2 \\
&\le \frac{1}{2}\|\nabla I\varphi_\lambda\|_2^2\left(1 + \frac{1}{2}\|I\varphi_\lambda\|_3^2\right) \\
&\lesssim N^{2(1-s)}\|\varphi_\lambda\|_{H^s}^2(1 + \|\varphi\|_3^2) \\
&\lesssim N^{2(1-s)}\lambda^{1-2s}(1 + \|\varphi\|_{H^s}^2)\|\varphi\|_{H^s}^2.
\end{aligned}$$

We choose

$$\lambda \approx N^{\frac{1-s}{s-\frac{1}{2}}}, \tag{4.330}$$

so that

$$E(I\varphi_\lambda) \le \frac{1}{2}. \tag{4.331}$$

We claim that the set W of times for which (4.327) holds is all of $[0,\infty)$. In the process of proving this, we will also show (4.328) holds on W.

For some universal constant C_1 to be chosen shortly, define

$$W \equiv \{T : \|u_\lambda\|_{L^4_{x,t}([0,T]\times\mathbb{R}^3)} \le C_1\lambda^{\frac{3}{8}}\}. \tag{4.332}$$

The set W is clearly closed and nonempty. It suffices to show it is open. Suppose T_0 satisfies that

$$\|u_\lambda\|_{L^4_{x,t}([0,T_0]\times\mathbb{R}^3)} \le 2C_1\lambda^{\frac{3}{8}}. \tag{4.333}$$

We claim that $T_0 \in W$. In fact, by the interaction Morawetz estimate,

$$\begin{aligned}
\|u_\lambda\|_{L^4_{x,t}([0,T_0]\times\mathbb{R}^3)} &\lesssim \|\varphi_\lambda\|_2^{\frac{1}{2}} \sup_{0\le t\le T_0}\|u_\lambda\|_{\dot H^{\frac{1}{2}}}^{\frac{1}{2}} \\
&\le C(\|\varphi\|_2)\lambda^{\frac{1}{4}} \sup_{0\le t\le T_0}\|u_\lambda\|_{\dot H^{\frac{1}{2}}}^{\frac{1}{2}},
\end{aligned} \tag{4.334}$$

where we have used the L^2 conservation. To bound the second factor in (4.334), decompose

$$u_\lambda(t) = P_{\le N}u_\lambda(t) + P_{\ge N}u_\lambda(t), \tag{4.335}$$

that is, a sum of functions supported on frequencies $|\xi| \le N$ and $|\xi| \ge N$, respectively. Interpolation and the fact that I is the identity on low frequencies give us the bound

$$\begin{aligned}\|P_{\leq N}u_\lambda(t)\|_{\dot H_x^{\frac12}} &\leq \|P_{\leq N}u_\lambda(t)\|_2^{\frac12}\|P_{\leq N}u_\lambda(t)\|_{\dot H_x^1}^{\frac12}\\ &\lesssim \|\varphi_\lambda\|_2^{\frac12}\|P_{\leq N}Iu_\lambda\|_{\dot H^1}^{\frac12}\\ &\leq C(\|\varphi\|_2)\lambda^{\frac14}\|Iu_\lambda\|_{\dot H^1}^{\frac12}.\end{aligned}\tag{4.336}$$

We interpolate the high-frequency component between $\dot H^s$ and L^2 and use the definition of I to get

$$\begin{aligned}\|P_{\geq N}u_\lambda(t)\|_{\dot H_x^{\frac12}} &\lesssim \|P_{\geq N}u_\lambda(t)\|_2^{1-\frac{1}{2s}}\|P_{\geq N}u_\lambda(t)\|_{\dot H_x^s}^{\frac{1}{2s}}\\ &\lesssim \|P_{\geq N}u_\lambda(t)\|_2^{1-\frac{1}{2s}}N^{\frac{s-1}{2s}}\|IP_{\geq N}u_\lambda(t)\|_{\dot H_x^1}^{\frac{1}{2s}}\\ &\lesssim C(\|\varphi\|_2)\|Iu_\lambda\|_{\dot H_x^1}^{\frac{1}{2s}},\end{aligned}\tag{4.337}$$

where we used

$$\lambda^{\frac12-\frac{1}{4s}} = N^{\frac{1-s}{s-\frac12}\cdot\frac{2s-1}{4s}} = N^{\frac{1-s}{2s}}.$$

By (4.334)~(4.337), we have

$$\|u_\lambda\|_{L^4_{x,t}([0,T_0]\times\mathbb{R}^3)} \leq C(\|\varphi\|_2)\Big(\lambda^{\frac38}\sup_{0\leq t\leq T_0}\|Iu_\lambda\|_{\dot H_x^1}^{\frac14} + \lambda^{\frac14}\sup_{0\leq t\leq T_0}\|Iu_\lambda(t)\|_{\dot H_x^1}^{\frac{1}{4s}}\Big).\tag{4.338}$$

We conclude $T_0 \in W$ if we establish

$$\sup_{0\leq t\leq T_0}\|Iu_\lambda\|_{\dot H^1} \lesssim 1\tag{4.339}$$

where $\lesssim$ only depends on $\|\varphi\|_{H^s(\mathbb{R}^3)}$. By (4.333), we may divide the time interval $[0,T_0]$ into subintervals

$$[0,T_0] = \bigcup_{j=1}^{L} I_j,$$

so that

$$\|u_\lambda\|_{L^4_{x,t}(I_j\times\mathbb{R}^3)} \leq \varepsilon, \quad j = 1,2,\dots,L.\tag{4.340}$$

Apply the almost conservation law on each of the subinterval I_j to get

$$\sup_{0\leq t\leq T_0}\|\nabla Iu_\lambda(t)\|_{L^2(\mathbb{R}^3)} \leq E(I\varphi_\lambda) + LN^{-1+}.\tag{4.341}$$

We get (4.339) if we can show

$$LN^{-1+} \le \frac{1}{4}. \tag{4.342}$$

Now recall L was defined essentially by (4.340). Since

$$\|u_\lambda\|^4_{L^4_{x,t}([0,T_0]\times\mathbb{R}^3)} \lesssim \lambda^{\frac{3}{2}},$$

we can be certain that

$$L \approx \lambda^{\frac{3}{2}}. \tag{4.343}$$

If we put this together with (4.342), we see that

$$\lambda^{\frac{3}{2}} N^{-1+} = N^{\frac{3-3s}{2s-1}} N^{-1+} \le \frac{1}{4}. \tag{4.344}$$

This is possible since $s > \frac{4}{5}$ the exponent on the left of (4.344) is negative. Notice that (4.328) holds on the set W using (4.339), L^2 conservation and the definition of I. □

The uniform estimate (4.226) and the local well-posedness theory imply the global well-posedness in H^s. It remains only to prove scattering result. Asymptotic completeness will follow quickly once we establish a uniform bound of the form

$$Z(t) \equiv \sup_{(q,r)\in\Lambda} \|\langle\nabla\rangle^s u\|_{L^q_t([0,t);L^r(\mathbb{R}^3))} \lesssim C(\|\varphi\|_{H^s}). \tag{4.345}$$

By the uniform estimate (4.327), we can decompose the time interval $[0,\infty]$ into a finite number of disjoint interval,

$$\mathbb{R} = \bigcup_{j=1}^{K} J_j, \quad j = 1,2,\dots,K,$$

where we have

$$\|u\|_{L^4_{x,t}(J_j\times\mathbb{R}^3)} \le \varepsilon(\|\varphi\|_{H^s}). \tag{4.346}$$

Applying $\langle\nabla\rangle^s$ to both sides of (4.225), and by Strichartz estimate, we have

$$\begin{aligned}
Z(t) &\le \|\langle\nabla\rangle^s\varphi\|_{L^2(\mathbb{R}^3)} + \|\langle\nabla\rangle^s(u\bar{u}u)\|_{L^{\frac{10}{7}}_{x,t}([0,t]\times\mathbb{R}^3)} \\
&\le \|\varphi\|_{H^s} + \|\langle\nabla\rangle^s u\|_{L^{\frac{10}{3}}_{x,t}} \|u\|^2_{L^5_{x,t}([0,t]\times\mathbb{R}^3)} \\
&\le \|\varphi\|_{H^s} + \|\langle\nabla\rangle^s u\|_{L^{\frac{10}{3}}_{x,t}} \|u\|^{\frac{4}{5}}_{L^4_{x,t}} \|u\|^{\frac{6}{5}}_{L^{10}_{x,t}},
\end{aligned}$$

where we use the interpolation. Hence,

$$Z(t) \le \|\varphi\|_{H^s} + \varepsilon^{\frac{4}{5}} Z(t)^{1+\frac{6}{5}}, \quad t \in J_1. \tag{4.347}$$

We conclude

$$Z(t) \le C(\|\varphi\|_{H^s}), \quad t \in J_1,$$

and we may repeat this argument to handle the remaining intervals J_j to yield (4.345).

The asymptotic completeness follows quickly from (4.345). Given $\varphi \in H^s(\mathbb{R}^n)$, set

$$\varphi_\pm(x) = \varphi(x) - i \int_0^{\pm\infty} S(-\tau)[|u|^2 u](\tau) d\tau \tag{4.348}$$

(where $u(t)$ is the global solution of (4.225) or (4.226)). We claim that $\varphi_\pm \in H^s$ and

$$\lim_{t\to\pm\infty} \|u(t) - S(t)\varphi_\pm\|_{H^s} = 0. \tag{4.349}$$

Since $S(t)$ is unitary group on L^2, equivalently, we want

$$\lim_{t\to\infty} \left\| \int_t^\infty \langle\nabla\rangle^s S(-\tau)[|u|^2 u](\tau) d\tau \right\|_{L^2(\mathbb{R}^3)} = 0. \tag{4.350}$$

$\forall F(x) \in L^2(\mathbb{R}^3)$, using the fractional Leibniz rule, we have

$$\begin{aligned}
&\left\langle F(x), \int_t^\infty \langle\nabla\rangle^s S(-\tau)(|u|^2 u) d\tau \right\rangle_{L^2(\mathbb{R}^3)} \\
&\approx \langle S(\tau)F(x), \langle\nabla\rangle^s u \cdot u^2 \rangle_{L^2_{x,t}([t,\infty)\times\mathbb{R}^3)} \\
&\le \|S(\tau)F(x)\|_{L^{\frac{10}{3}}_{x,t}([t,\infty)\times\mathbb{R}^3)} \|\langle\nabla\rangle^s u\|_{L^{\frac{10}{3}}_{x,t}([t,\infty)\times\mathbb{R}^3)} \|u\|^2_{L^5([t,\infty)\times\mathbb{R}^3)} \\
&\longrightarrow 0, \quad t \to +\infty,
\end{aligned} \tag{4.351}$$

where we used (4.345), the interpolation inequality and the Strichartz estimate.

For completeness, we conclude an argument proving the existence of wave operators Ω_+ on H^s. Given $\varphi_+(x) \in H^s(\mathbb{R}^3)$, we are looking for a solution $u(t)$ of (4.225) or (4.226) with initial data $\varphi(x)$) satisfying

$$\begin{aligned}
u(x,t) &= S(t)\varphi(x) - i \int_0^t S(t-\tau)|u|^2 u d\tau \\
&= S(t)(S^{NL}(-\infty)S(\infty)\varphi_+) - i \int_0^t S(t-\tau)|u|^2 u d\tau
\end{aligned}$$

$$= S(t)\left(\varphi_+(x) - i\int_\infty^0 S(-\tau)|u|^2 u d\tau\right) - i\int_0^t S(t-\tau)|u|^2 u d\tau$$

$$= S(t)\varphi_+(x) - i\int_t^\infty S(t-\tau)|u|^2 u d\tau. \tag{4.352}$$

We now sketch how this last integral equation is solved for u using a fixed-point argument and prove that

$$\lim_{t\to+\infty}\left\|\int_t^\infty S(t-\tau)|u|^2 u d\tau\right\|_{H^s} = 0. \tag{4.353}$$

By Strichartz's estimate and $(\frac{8}{3}, 4), (8, \frac{12}{5}) \in \Lambda$, we have

$$S(t)\varphi_+ \in L_t^{\frac{8}{3}}([0,\infty); W^{s,4}(\mathbb{R}^3)) \cap L_t^8((t,\infty), W^{s,\frac{12}{5}}(\mathbb{R}^3)). \tag{4.354}$$

Set

$$K_{t_0} = \|S(t)\varphi_+\|_{L^{\frac{8}{3}}([t_0,\infty);W^{s,4}(\mathbb{R}^3))} + \|S(t)\varphi_+\|_{L^8((t_0,\infty),W^{s,\frac{12}{5}}(\mathbb{R}^3))}. \tag{4.355}$$

Clearly,

$$\lim_{t_0\to\infty} K_{t_0} = 0. \tag{4.356}$$

Define

$$X = \{u(t) \in L_t^{\frac{8}{3}}([t_0,\infty); W^{s,4}(\mathbb{R}^3)) \cap L_t^8([t_0,\infty); W^{s,\frac{12}{5}}(\mathbb{R}^3)),$$
$$\|u\|_X = \|u\|_{L_t^{\frac{8}{3}}([t_0,\infty);W^{s,4}(\mathbb{R}^3))} + \|u\|_{L_t^8([t_0,\infty);W^{s,\frac{12}{5}}(\mathbb{R}^3))} \le 4K_{t_0}\},$$

with norm

$$d(u,v) = \|u-v\|_{L_t^{\frac{8}{3}}([t_0,\infty);W^{s,4}(\mathbb{R}^3))}.$$

Consider the mapping $\mathcal{T}$ on X,

$$\mathcal{T}u = S(t)\varphi_+(x) + i\int_t^\infty S(t-\tau)|u|^2 u d\tau. \tag{4.357}$$

$\forall u(t) \in X$, we have

$$
\begin{aligned}
\|\mathcal{T}u\|_X &\leq 2K_{t_0} + 2\||u|^2u\|_{L_t^{\frac{8}{5}}([t_1,\infty);W^{s,\frac{4}{3}})} \\
&\lesssim 2K_{t_0} + 2\|\langle\nabla\rangle^s u\|_{L_t^{\frac{8}{3}}L_x^4}\|u\|_{L_t^8L_x^4}^2 \\
&\leq 2K_{t_0} + C(2K_{t_0})^3,
\end{aligned}
\tag{4.358}
$$

where we use $W^{s,\frac{12}{5}}(\mathbb{R}^3) \hookrightarrow L_x^4(\mathbb{R}^3)$. In the same way, $\forall u(t), v(t) \in X$, we have

$$
d(\mathcal{T}u, \mathcal{T}v) \leq 2C(2K_{t_0})^2 d(u, v). \tag{4.359}
$$

Hence, for t_0 large enough, the map $\mathcal{T}u \in X$ and $\mathcal{T}$ is a contraction on X. We conclude that there is a unique solution $u(t)$ satisfying

$$
u(t) \in C([t_0, \infty); H^s(\mathbb{R}^3)) \cap X. \tag{4.360}
$$

By our global existence result of (4.225) and time reversibility, we may extend this solution $u(t)$, starting from data at time t_0, to all of $[0, \infty)$. By the Strichartz estimate, we conclude that

$$
\lim_{t\to\infty} \|u(t) - S(t)\varphi_+\|_{H^s(\mathbb{R}^3)} = 0. \tag{4.361}
$$

□

4.5 Global well-posedness and scattering for the energy-critical nonlinear Schrödinger equation

In this section, we mainly introduce the global well-posedness and scattering for the Cauchy problem of nonlinear Schrödinger equation in three dimension (3D)

$$
\begin{cases}
iu_t + \Delta u = |u|^4 u, & (x,t) \in \mathbb{R}^3 \times \mathbb{R}, \\
u(0) = \varphi(x), & \varphi(x) = \varphi(|x|) \in \dot{H}^1(\mathbb{R}^3).
\end{cases}
\tag{4.362}
$$

Moreover, the same statement holds for radial data $\varphi(x) \in H^s(\mathbb{R}^3)$, $s \geq 1$. We can also get the global solution and scattering of (4.362) in $H^s(\mathbb{R}^3)$.

As is well known, Struwe [248] proved that there exists a unique smooth global solution for the $\dot{H}^1$-critical wave equation with smooth radial data in 1988. Later in 1990, M. Grillakis [111] removed the symmetry assumption and proved that there exists a unique smooth global solution for the general smooth initial data $(\varphi(x), \psi(x))$; see also [236]. Concerned with the global well-posedness of the critical wave equation in $\dot{H}^1$ readers can refer to [237]. In the case of the wave equation, the proof is based on the following two different facts:

(i) As a consequence of the analysis of the local IVP, if global well-posedness fails, there is necessarily a "concentration" effect of the solution on small balls (that may be centered at 0 in the radial case), that is, there exists space-time cube Q, $\mathrm{mes}(Q) \sim \delta^3 \times \delta^2$, such that

$$\inf_{\delta>0} \|u(x,t)\|_{L^{10}(Q)} > 0,$$

or, there exists time interval I, $|I| \sim \delta^2$, such that

$$\inf_{\delta>0} \|u(x,t)\|_{L^\infty(I;L^6(\mathbb{R}^3))} > 0, \quad \text{or} \quad \inf_{\delta>0} \|u(x,t)\|_{L^\infty(I;H^1(\mathbb{R}^3))} > 0.$$

(ii) The Morawetz inequality, which excludes an infinite repetition of the effect described in (i).

Can we use the analogue method of critical wave equation to deal with the critical Schrödinger equation (4.362)? We know that the solution of (4.362) satisfies the following Morawetz inequality:

$$\int_{\mathbb{R}} \int_{\mathbb{R}^3} \frac{|u|^6}{|x|} dxdt \le C(\|\varphi\|_{H^1}),$$

which is not sufficient to disprove the "concentration" effect of energy. Essentially, the left side of the above inequality can be controlled by $\sup_t \|u\|^2_{H^{\frac{1}{2}}}$, that is,

$$\iint \frac{|u|^6}{|x|} dxdt \le \sup_t \|u(t)\|^2_{H^{\frac{1}{2}}}. \tag{4.363}$$

However, we can establish the $\dot{H}^1$ theory for radial data by the localized Morawetz estimate, L^2-weak dispersive estimate, the Strichartz estimate and the Littlewood–Paley decomposition.

We know that the Cauchy problem (4.362) is locally well posed in $\dot{H}^1(\mathbb{R}^3)$, and there exists a maximal interval of existence $I = (-T^*, T^*)$ such that

$$E(u(t)) = \frac{1}{2} \int_{\mathbb{R}^3} |\nabla u|^2 dx + \frac{1}{6} \int_{\mathbb{R}^3} |u|^6 dx = E(\varphi). \tag{4.364}$$

From which, we can deduce

$$\|u(t)\|_{\dot{H}^1} \le C(\|\varphi\|_{\dot{H}^1}), \quad \forall \varphi \in \dot{H}^1(\mathbb{R}^3). \tag{4.365}$$

While for general $\varphi \in H^1(\mathbb{R}^3)$, $T^* = T(\varphi)$ depends on φ rather than only on $\|\varphi\|_{\dot{H}^1}$ (For the subcritical case, $T^* = T(\|\varphi\|_{H^1})$, one can obtain global well-posedness of the subcritical nonlinear Schrödinger equation in H^1 by (4.365) and the L^2 conservation). Consequently, the boundness of $\|\varphi\|_{\dot{H}^1}$ is not enough to remove the "concentration" effect of energy in $\dot{H}^1$.

For radial data, Bourgain first developed the induction on energy method [22] to show the global well-posedness. And then, Tao [260] generalized Bourgain's result to a higher-dimensional case. We remark that Tao obtained the bounds on various space-time norms of the solution that are of exponential type in the energy, which improves on the tower-type bounds of Bourgain [22]. In the following, we follow the expositions of Tao [260].

Theorem 4.5.1. *Let $[t_-, t_+]$ be a compact interval, and let $u \in C([t_-, t_+]; H^1(\mathbb{R}^3)) \cap L^{10}_{t,x}([t_-, t_+] \times \mathbb{R}^3)$ be a spherically solution to (4.362) with energy $E(u) \leq E$ for some $E > 0$. Then we have*

$$\|u(t)\|_{L^{10}_{t,x}([t_-,t_+]\times\mathbb{R}^3)} \leq Ce^{CE^C}, \tag{4.366}$$

for some absolute constants C depending only on n (and thus independent of E, $t_\pm$, u).

Remark 4.5.1. Because the bounds are independent of the length of the time interval $[t_-, t_+]$, it is a standard matter to use this theorem, combined with the local well-posedness theory in [37], to obtain global well-posedness and scattering conclusions for large energy spherically symmetric data; see [21] for details.

Local mass conservation and Morawetz inequality

It is well known that there is no finite propagation speed for the Schrödinger equation. In order to establish the scattering result, we need to localize the mass and Morawetz estimate, where the localized mass conservation law can control the mass flow in the spatial region, and the localized Morawetz estimate can disprove the "concentration" effect of energy.

For the sake of convenience, we introduce the notation. Given I, write

$$\|u\|_{X(I)} \triangleq \|u\|_{L^{10}_{t,x}(I\times\mathbb{R}^3)}; \quad \|u\|_{W(I)} \triangleq \|\nabla u\|_{L^{10}_t L^{\frac{30}{13}}_x(I\times\mathbb{R}^3)}.$$

For the space-time slab $I \times \mathbb{R}^n$, define the norm of $\dot{S}^0(I)$ as

$$\|u\|_{\dot{S}^0(I)} := \sup_{(q,r)\in\Lambda} \|u\|_{L^q_t L^r_x(I\times\mathbb{R}^3)},$$

and the norm of $\dot{S}^1(I)$ as

$$\|u\|_{\dot{S}^1(I)} := \|\nabla u\|_{\dot{S}^0(I)}.$$

Local mass conservation

We first give the local mass conservation law (i. e., almost finite propagation speed). Let χ be a bump function supported on the ball $B_1(0)$, which equals one on the ball $B_{\frac{1}{2}}(0)$

and is nonincreasing in the radial direction. Observe that if u is a finite energy solution of (4.362), then

$$\partial_t |u(t,x)|^2 = -2\nabla \cdot \operatorname{Im}(\overline{u}\nabla u(t,x)). \tag{4.367}$$

Define

$$\operatorname{Mass}(u(t), B(x_0,R)) := \int \left|\chi\left(\frac{x-x_0}{R}\right)u(t,x)\right|^2 dx. \tag{4.368}$$

By integration by parts, we have

$$\begin{aligned}
\partial_t \operatorname{Mass}(u(t), B(x_0,R)) &= \int \left|\chi\left(\frac{x-x_0}{R}\right)\right|^2 \partial_t |u(t,x)|^2 dx \\
&= -2\int \left|\chi\left(\frac{x-x_0}{R}\right)\right|^2 \nabla \cdot \operatorname{Im}(\overline{u}\nabla u) dx \\
&= -\frac{4}{R}\int \chi\left(\frac{x-x_0}{R}\right)\nabla\chi\left(\frac{x-x_0}{R}\right)\operatorname{Im}(\overline{u}\nabla u) dx \\
&\lesssim \frac{1}{R}\|\nabla u(t)\|_{L^2}(\operatorname{Mass}(u(t), B(x_0,R)))^{1/2}.
\end{aligned}$$

So,

$$|\operatorname{Mass}(u(t_1), B(x_0,R))^{1/2} - \operatorname{Mass}(u(t_2), B(x_0,R))^{1/2}| \lesssim \frac{|t_1-t_2|}{R}. \tag{4.369}$$

This implies that if the local mass $\operatorname{Mass}(u(t), B(x_0,R))$ is large for some time t, then it can also be shown to be similarly large for times nearby t, by increasing the radius R if necessary to reduce the rate of change of the mass.

On the other hand, from the Sobolev and Hölder inequalities, we have

$$\operatorname{Mass}(u(t), B(x_0,R)) \le \left\|\chi\left(\frac{x-x_0}{R}\right)\right\|_{L_x^3}^2 \|u\|_{L_x^6}^2 \lesssim R^2\|\nabla u\|_{L_x^2}^2, \tag{4.370}$$

which gives the control of the mass on the small spatial volume.

Morawetz inequality

We now give the proof of the localized Morawetz estimate.

Proposition 4.5.1. *Assume that u is the solution of* (4.362) *on the space-time slab $I \times \mathbb{R}^3$, then for any $A \ge 1$, we have*

$$\int_I \int_{|x|\le A|I|^{1/2}} \frac{|u|^6}{|x|} dx dt \lesssim A|I|^{1/2} E(u). \tag{4.371}$$

Using the scale invariance, we may rescale so that $A|I|^{1/2} = 1$. By the local momentum conservation identity, we have

$$\partial_t \operatorname{Im}(\partial_k u\bar{u}) = -2\partial_j \operatorname{Re}(\partial_k u\overline{\partial_j u}) + \frac{1}{2}\partial_k \Delta(|u|^2) - \frac{2}{3}\partial_k |u|^6, \tag{4.372}$$

where j, k range over spatial indices 1, 2, 3, with the usual summation conventions. Observe that when u has finite energy, both sides of this equality make sense in the sense of distributions, so this identity can be justified in the finite energy case by the local well-posedness theory.

If we multiply the above identity by the weight $\partial_k a$ for some smooth, compactly supported weight $a(x)$, and then integrate in space, we obtain

$$\begin{aligned}\partial_t \int_{\mathbb{R}^3} (\partial_k a) \operatorname{Im}(\partial_k u\bar{u})dx = 2 \int_{\mathbb{R}^3} (\partial_j \partial_k a) \operatorname{Re}(\partial_k u\overline{\partial_j u})dx + \frac{1}{2} \int_{\mathbb{R}^3} (-\Delta\Delta a)(|u|^2)dx \\ + \frac{2}{3} \int_{\mathbb{R}^3} \Delta a|u|^6 dx.\end{aligned} \tag{4.373}$$

We apply this in particular to the C_0^∞ weight $a(x) := (\epsilon^2 + |x|^2)^{1/2}\chi(x)$, where χ is a bump function supported on $B_2(0)$, which equals 1 on $B_1(0)$, and $0 < \epsilon < 1$ is a small parameter, which will eventually be sent to zero. In the region $|x| \le 1$, we have

$$\begin{aligned}&(\partial_j \partial_k a) \operatorname{Re}(\partial_k u\overline{\partial_j u}) \ge 0,\\ &\Delta a = \frac{2}{(\epsilon^2 + |x|^2)^{1/2}} + \frac{\epsilon^2}{(\epsilon^2 + |x|^2)^{3/2}},\\ &-\Delta\Delta a = \frac{15\epsilon^4}{(\epsilon^2 + |x|^2)^{7/2}}.\end{aligned}$$

In the region $1 \le |x| \le 2$, a and all of its derivatives are bounded uniformly in ϵ, and so the integrals on the RHS of (4.373) are bounded by $O(E(u))$. Combining these estimates, we obtain

$$\partial_t \int_{|x|\le 2} (\partial_k a) \operatorname{Im}(\partial_k u\bar{u}) \ge c \int_{|x|\le 1} \frac{|u(t,x)|^6}{(\epsilon^2 + |x|^2)^{1/2}} dx - CE(u).$$

Integrating this in time on I, and then using the fundamental theorem of calculus and the observation that a is Lipschitz, we obtain

$$\sup_{t\in I} \int_{|x|\le 2} |\nabla u(t,x)||u(t,x)|dx \ge c \int_I \int_{|x|\le 1} \frac{|u(t,x)|^6}{(\epsilon^2 + |x|^2)^{1/2}} dxdt - C(E(u))|I|.$$

By (4.370), Cauchy–Schwarz and $|I| = A^{-2} < 1$, we have

$$\int_I \int_{|x|\le 1} \frac{|u(t,x)|^6}{(\epsilon^2+|x|^2)^{1/2}} dxdt \le C(E(u)).$$

Taking $\epsilon \to 0$, and using monotone convergence, (4.371) follows.

Local well-posedness

As a preliminary, we develop a local well-posedness and blowup criterion for (4.362).

Proposition 4.5.2 (Local well-posedness). *Let $u(t_0) \in \dot{H}^1$, and I be a compact time interval that contains t_0 such that*

$$\|U(t-t_0)u(t_0)\|_{X(I)} \le \eta, \tag{4.374}$$

for s sufficiently small absolute constant $\eta > 0$. Then there exists a unique solution $u \in C_t^0 \dot{H}_x^1$ to (4.362) on $I \times \mathbb{R}^3$ such that

$$\|u\|_{X(I)} \le C(\|u(t_0)\|_{\dot{H}^1}). \tag{4.375}$$

Proof. The proof of this proposition is standard and based on the contraction mapping arguments. We define the solution map to be

$$\Phi(u)(t) := U(t-t_0)u(t_0) - i\int_{t_0}^{t} U(t-s)f(u(s))ds \tag{4.376}$$

on $\mathcal{B}$ defined by

$$\mathcal{B} = \{u : \|u\|_{X(I)} \le 2\eta, \|u\|_{W(I)} \le 2C\|u(t_0)\|_{\dot{H}^1}\}$$

with the metric

$$\|u\|_{\mathcal{B}} = \|u\|_{X(I)} + \|u\|_{W(I)}.$$

For sufficiently small $\eta > 0$, $u \in \mathcal{B}$, we have

$$\begin{aligned}
\|\Phi(u)\|_{X(I)} &\le \|U(t-t_0)u(t_0)\|_{X(I)} + C\|u\|_{X(I)}^4\|u\|_{W(I)} \\
&\le \eta + 32C\eta^4\|u(t_0)\|_{\dot{H}^1} \le 2\eta; \\
\|\Phi(u)\|_{W(I)} &\le C\|u(t_0)\|_{\dot{H}^1} + C\|u\|_{X(I)}^4\|u\|_{W(I)} \\
&\le C\|u(t_0)\|_{\dot{H}^1} + 32C\eta^4\|u(t_0)\|_{\dot{H}^1} \le 2C\|u(t_0)\|_{\dot{H}^1}.
\end{aligned}$$

Hence, Φ maps $\mathcal{B}$ onto itself.

It suffices to show Φ is contractive. Let $u, v \in \mathcal{B}$, by nonlinear estimate, for sufficiently small $\eta > 0$, we have

$$\begin{aligned}\|\Phi(u)-\Phi(v)\|_{W(I)} &\le \left\| \int_0^t U(t-s)(|u|^4u-|v|^4v)ds \right\|_{W(I)} \\ &\le C\|u-v\|_{X(I)}(\|u\|_{W(I)}\|u\|_{X(I)}^3+\|v\|_{W(I)}\|v\|_{X(I)}^3) \\ &\quad + C\|u-v\|_{W(I)}(\|u\|_{X(I)}^4+\|v\|_{X(I)}^4) \\ &\le 16C\eta^3\|u(t_0)\|_{\dot H^1}\|u-v\|_{X(I)}+16\eta^4\|u-v\|_{W(I)} \\ &\le \frac{1}{4}(\|u-v\|_{X(I)}+\|u-v\|_{W(I)}).\end{aligned}$$

Similarly, we have

$$\begin{aligned}\|\Phi(u)-\Phi(v)\|_{X(I)} &\le 16C\eta^3\|u(t_0)\|_{\dot H^1}\|u-v\|_{X(I)}+16\eta^4\|u-v\|_{W(I)} \\ &\le \frac{1}{4}(\|u-v\|_{X(I)}+\|u-v\|_{W(I)}).\end{aligned}$$

Then the contraction mapping theorem implies the existence of the unique solution to (4.362) on I. □

Next, we give the blowup criterion of the strong solution for (4.362).

Proposition 4.5.3 (Blowup criterion). *Let $\varphi \in \dot H^1$, and u be a strong solution to* (4.362) *on the slab* $[0,T)\times\mathbb{R}^3$ *such that*

$$\|u\|_{X([0,T))}<\infty, \quad T<\infty.$$

Then there exists $\delta>0$ such that the solution u extends to a strong solution to (4.362) *on the slab* $[0,T+\delta]\times\mathbb{R}^3$.

Proof. By the absolute continuity of integrals, there exists $t_0\in[0,T)$, such that

$$\|u\|_{X([t_0,T))}\le \eta/4.$$

Then by Strichartz estimate, we have

$$\|u\|_{W([t_0,T))}\lesssim \|u(t_0)\|_{\dot H^1}+\|u\|_{X([t_0,T))}^4\|u\|_{W([t_0,T))}.$$

Therefore,

$$\|u\|_{W([t_0,T))}\lesssim \|u(t_0)\|_{\dot H^1}.$$

Now we write

$$u(t)=U(t-t_0)u(t_0)-i\int_{t_0}^t U(t-s)|u|^4u(s,x)ds. \tag{4.377}$$

Then for sufficiently small $\eta > 0$, we have

$$\begin{aligned}\|U(t-t_0)u(t_0)\|_{X([t_0,T))} &\leq \|u\|_{X([t_0,T))} + C\|u\|^4_{X([t_0,T))}\|u\|_{W([t_0,T))} \\ &\leq \frac{\eta}{4} + C\eta^4\|u(t_0)\|_{\dot{H}^1} \leq \frac{\eta}{2}.\end{aligned}$$

By the absolute continuity of integrals again, there exists $\delta > 0$, such that

$$\|U(t-t_0)u(t_0)\|_{X([t_0,T+\delta))} \leq \eta.$$

Thus, we may apply Proposition 4.5.2 on the interval $[t_0, T+\delta]$ to complete the proof. □

In other words, this lemma asserts that if $[0, T^*)$ is the maximal interval of existence of (4.362) and $T^* < \infty$, then

$$\|u(t)\|_{X([0,T^*))} = \infty.$$

Perturbation result

Now, we show the perturbation result, which shows that the solution cannot be large if the linear part of the solution is not large.

Lemma 4.5.1 (Perturbation lemma). *Let u be a solution to* (4.362) *on $I = [t_1, t_2]$ such that*

$$\frac{1}{2}\eta \leq \|u\|_{X(I)} \leq \eta, \tag{4.378}$$

where η is a sufficiently small constant depending on the norm of the initial data. Then

$$\|u\|_{\dot{S}^1(I)} \lesssim 1, \quad \|u_k\|_{X(I)} \geq \frac{1}{4}\eta, \tag{4.379}$$

where $u_k(t) = U(t-t_k)u(t_k)$, $k = 1, 2$.

Proof. From the Strichartz estimate and nonlinear estimate, we obtain

$$\begin{aligned}\|u\|_{\dot{S}^1(I)} &\lesssim \|u(t_1)\|_{\dot{H}^1} + \|u\|^4_{X(I)}\|u\|_{W(I)} \\ &\lesssim \|u(t_1)\|_{\dot{H}^1} + \|u\|^4_{X(I)}\|u\|_{\dot{S}^1(I)} \\ &\lesssim \|u(t_1)\|_{\dot{H}^1} + \eta^4\|u\|_{\dot{S}^1(I)}.\end{aligned}$$

If η is sufficiently small, we have the first claim

$$\|u\|_{\dot{S}^1(I)} \lesssim 1.$$

As for the second claim, we give the proof for $k = 1$; the case $k = 2$ is similar. Using the Strichartz estimate, we have

$$\|u - u_1\|_{X(I)} \lesssim \eta^4 \|u\|_{\dot{S}^1(I)} \lesssim \eta^4.$$

Therefore, the second claim follows by the triangle inequality and choosing $\eta > 0$ sufficiently small. □

Global well-posedness

Now we give the proof of Theorem 4.5.1. We will use the radial assumption to ensure the only possibility of the "concentration" of energy around zero (See Corollary 4.5.1 below).

For readability, we take some constants

$$C_1 = 40; \quad C_2 = 2; \quad C_3 = 150, \tag{4.380}$$

which come from several constraints in the rest of this section. All implicit constants in this section are permitted to depend on the energy.

Fix E, $[t_-, t_+]$, u. By the small energy theory, we may assume that the energy is large $E > c > 0$. From $E(u) < \infty$ and Sobolev embedding, for all $t \in [t_-, t_+]$, we have

$$\|u(t)\|_{\dot{H}_x^1} + \|u(t)\|_{L_x^6} \lesssim 1. \tag{4.381}$$

Assume that the solution u already exists on $[t_-, t_+]$. By Proposition 4.5.3, it suffices to obtain an a prior estimate

$$\|u\|_{X([t_-,t_+])} \leq O(1), \tag{4.382}$$

where $O(1)$ is independent of t_-, t_+.

Let $\eta > 0$ be small constant dependent of energy, we may assume that

$$\|u\|_{X([t_-,t_+])} \geq 2\eta,$$

otherwise it is trivial. We divide $[t_-, t_+]$ into J subintervals $I_j = [t_j, t_{j+1}]$ for some $J \geq 2$, such that

$$\frac{\eta}{2} \leq \|u\|_{X(I_j)} \leq \eta. \tag{4.383}$$

It suffices to estimate the number J.

Let $u_\pm = U(t - t_\pm)u(t_\pm)$. By Sobolev embedding and Strichartz estimates, we have

$$\|u_\pm\|_{X([t_-,t_+])} \lesssim 1. \tag{4.384}$$

We use the following definition of Tao [258].

Definition 4.5.1. We call I_j exceptional if

$$\|u_\pm\|_{X(I_j)} > \eta^{C_3}, \tag{4.385}$$

for at least one sign $\pm$. Otherwise, we call I_j unexceptional.

From (4.384), we obtain the upper bound on the number of exceptional intervals, $O(\eta^{-10C_3})$. We may assume that there exists unexceptional intervals, otherwise the claim would follow from this bound and (4.383). Therefore, it suffices to compute the number of unexceptional intervals.

We first prove the existence of a bubble of mass concentration in each unexceptional interval; see [22, 258].

Proposition 4.5.4 (Existence of a bubble). *Let I_j be an unexceptional interval. Then there exists $x_j \in \mathbb{R}^3$ such that*

$$\mathrm{Mass}(u(t), B(x_j, \eta^{-C_1}|I_j|^{1/2})) \gtrsim \eta^{C_1}|I_j| \tag{4.386}$$

for all $t \in I_j$.

Proof. By time translation invariance and scale invariance, we may assume that $I_j = [0,1]$. We subdivide I_j further into $[0,\frac{1}{2}]$ and $[\frac{1}{2},1]$. By (4.383) and the pigeonhole principle and time reflection symmetry, if necessary, we may assume that

$$\|u\|_{X([\frac{1}{2},1])} \geq \frac{\eta}{4}.$$

Thus by Lemma 4.5.1, we have

$$\left\|U\left(t-\frac{1}{2}\right)u\left(\frac{1}{2}\right)\right\|_{X([\frac{1}{2},1])} \geq \frac{\eta}{8}. \tag{4.387}$$

By the Duhamel formula, we have

$$u\left(\frac{1}{2}\right) = U\left(\frac{1}{2}-t_-\right)u(t_-) - i\int_{t_-}^{\frac{1}{2}} U\left(\frac{1}{2}-s\right)f(u(s))ds.$$

Applying the unitary group $U(t-\frac{1}{2})$ to the above formula, we have

$$\begin{aligned} U\left(t-\frac{1}{2}\right)u\left(\frac{1}{2}\right) &= U(t-t_-)u(t_-) - i\int_0^{\frac{1}{2}} U(t-s)f(u(s))ds \\ &\quad - i\int_{t_-}^0 U(t-s)f(u(s))ds. \end{aligned} \tag{4.388}$$

Since $[0,1]$ is unexceptional interval, we have

$$\|U(t-t_-)u(t_-)\|_{X([\frac{1}{2},1])} \triangleq \|u_-(t)\|_{X([\frac{1}{2},1])} \leq \|u_-(t)\|_{X([0,1])} \leq \eta^{C_3}. \tag{4.389}$$

On the other hand, by the nonlinear estimate and Lemma 4.5.1, we have

$$\left\| \int_0^{\frac{1}{2}} U(t-s)f(u(s))ds \right\|_{X([\frac{1}{2},1])} \lesssim \|u\|_{X([\frac{1}{2},1])}^4 \|u\|_{W([\frac{1}{2},1])}$$
$$\lesssim \eta^4 \|u\|_{\dot{S}^1([\frac{1}{2},1])} \lesssim \eta^4.$$

Thus, the triangle inequality implies that

$$\left\| \int_{t_-}^{0} U(t-s)f(u(s))ds \right\|_{X([\frac{1}{2},1])} \geq \frac{1}{100}\eta,$$

provided η is chosen sufficiently small. Hence, if we define

$$v(t) := \int_{t_-}^{0} U(t-s)f(u(s))ds,$$

then we have

$$\|v\|_{X([\frac{1}{2},1])} \geq \frac{1}{100}\eta.$$

Next, we estimate the upper bound on $\|v\|_{\dot{S}^1([\frac{1}{2},1])}$. By (4.383), (4.388), the Strichartz estimate and Lemma 4.5.1, we have

$$\begin{aligned} \|v\|_{\dot{S}^1([\frac{1}{2},1])} &\leq \left\| U\left(t-\frac{1}{2}\right)u\left(\frac{1}{2}\right)\right\|_{\dot{S}^1([\frac{1}{2},1])} + \|U(t-t_-)u(t_-)\|_{\dot{S}^1([\frac{1}{2},1])} \\ &\quad + \left\| \int_0^{\frac{1}{2}} U(t-s)f(u(s))ds \right\|_{\dot{S}^1([\frac{1}{2},1])} \\ &\lesssim \|u\|_{\dot{H}^1} + \|u\|_{X([0,\frac{1}{2}])}^4 \|u\|_{W([0,\frac{1}{2}])} \\ &\lesssim \|u\|_{\dot{H}^1} + \|u\|_{X([0,\frac{1}{2}])}^4 \|u\|_{\dot{S}^1([0,\frac{1}{2}])} \\ &\lesssim 1. \end{aligned} \tag{4.390}$$

We shall need some additional regularity control on v. For any $h \in \mathbb{R}^3$, let $u^{(h)}$ denote the translation of u by h, that is, $u^{(h)}(t,x) = u(t,x-h)$.

Lemma 4.5.2. *Let $\chi(x)$ be a bump function supported on the ball $B_1(0)$ of total mass one and define*

$$v_{av}(x) = \int \chi(y)v(x+\eta^{5C_2}y)dy. \tag{4.391}$$

Then we have

$$\|v - v_{av}\|_{X([\frac{1}{2},1])} \lesssim \eta^{C_2}. \tag{4.392}$$

Proof. By Hölder's inequality, it suffices to show that

$$\|v - v_{av}\|_{L^\infty L^{10}([\frac{1}{2},1]\times\mathbb{R}^3)} \lesssim \eta^{C_2}. \tag{4.393}$$

By (4.365), we only need to show that

$$\|v - v^{(h)}\|_{L^\infty L^{10}([\frac{1}{2},1]\times\mathbb{R}^3)} \lesssim \|\nabla u\|^5_{L^\infty L^2([\frac{1}{2},1]\times\mathbb{R}^3)} |h|^{\frac{1}{5}}. \tag{4.394}$$

In fact, we have

$$\begin{aligned}
\|v - v_{av}\|_{L^\infty L^{10}([\frac{1}{2},1]\times\mathbb{R}^3)} &= \left\| \int \chi(y)[v(x) - v(x + \eta^{5C_2} y)]\,dy \right\|_{L^\infty L^{10}([\frac{1}{2},1]\times\mathbb{R}^3)} \\
&\leq \int \chi(y) \|[v(x) - v(x + \eta^{5C_2} y)]\|_{L^\infty L^{10}([\frac{1}{2},1]\times\mathbb{R}^3)}\,dy \\
&\leq \int \chi(y) \|\nabla u\|^5_{L^\infty L^2([\frac{1}{2},1]\times\mathbb{R}^3)} \eta^{C_2} |y|^{\frac{1}{5}}\,dy \\
&\lesssim \eta^{C_2}.
\end{aligned}$$

By the dispersive estimate, Hölder's inequality and Sobolev embedding, we have

$$\begin{aligned}
&\|v - v^{(h)}\|_{L^\infty L^{10}([\frac{1}{2},1]\times\mathbb{R}^3)} \\
&\lesssim \left\| \int_{t_-}^{0} |t-s|^{-\frac{6}{5}} \left\| |u^{(h)}(s)|^4 u^{(h)}(s) - |u(s)|^4 u(s) \right\|_{L^{\frac{10}{9}}} ds \right\|_{L_t^\infty([\frac{1}{2},1])} \\
&\lesssim \|u\|^4_{L^\infty L^6} \left\| u^{(h)} - u \right\|_{L^\infty L^{\frac{30}{7}}} \\
&\lesssim \|\nabla u\|^4_{L^\infty L^2} \left\| u^{(h)} - u \right\|^{\frac{4}{5}}_{L^\infty L^6} \left\| u^{(h)} - u \right\|^{\frac{1}{5}}_{L^\infty L^2} \\
&\lesssim \|\nabla u\|^5_{L^\infty L^2} |h|^{\frac{1}{5}}.
\end{aligned} \tag{4.395}$$

This completes the proof of Lemma 4.5.2. □

Now, we return to the proof of Proposition 4.5.4. By Lemma 4.5.2 and (4.389), we have

$$\|v_{av}\|_{X([\frac{1}{2},1])} \gtrsim \eta. \tag{4.396}$$

On the other hand, by Hölder inequality, Young inquality and (4.390), we have

$$
\begin{aligned}
\|v_{av}\|_{L^6_{t,x}([\frac{1}{2},1]\times\mathbb{R}^3)} &\lesssim \|v_{av}\|_{L^\infty_t L^6_x([\frac{1}{2},1]\times\mathbb{R}^3)}\\
&\lesssim \|v\|_{L^\infty_t L^6_x([\frac{1}{2},1]\times\mathbb{R}^3)}\\
&\lesssim 1.
\end{aligned}
$$

Interpolating with (4.396) gives

$$
\|v_{av}\|_{L^\infty_{t,x}([\frac{1}{2},1]\times\mathbb{R}^n)} \gtrsim \|v_{av}\|^{\frac{5}{2}}_{X([\frac{1}{2},1])}\|v_{av}\|^{-\frac{3}{2}}_{L^6_{t,x}([\frac{1}{2},1]\times\mathbb{R}^n)} \gtrsim \eta^{\frac{5}{2}}. \tag{4.397}
$$

Thus, there exists $(s_j, x_j) \in [\frac{1}{2}, 1] \times \mathbb{R}^3$ such that

$$
|v_{av}(s_j, x_j)| \gtrsim \eta^{\frac{5}{2}}. \tag{4.398}
$$

Hence, by the Cauchy–Schwarz inequality, we have

$$
\begin{aligned}
|v_{av}(s_j, x_j)| &= \left|\int \chi(y)v(s_j, x_j + \eta^{5C_2}y)dy\right|\\
&= \eta^{-15C_2}\left|\int \chi\left(\frac{x-x_j}{\eta^{5C_2}}\right)v(s_j, x)dx\right|\\
&\lesssim \eta^{-15C_2}\eta^{\frac{15}{2}C_2}\,\mathrm{Mass}(v(s_j), B(x_j, \eta^{C_2}))^{1/2},
\end{aligned}
$$

that is,

$$
\mathrm{Mass}(v(s_j), B(x_j, \eta^{C_2})) \gtrsim \eta^{15C_2+5} \gtrsim \eta^{C_1}. \tag{4.399}
$$

If we take $R = \eta^{-C_1}$, and choose η sufficiently small, we have by (4.369) and the simple geometric observation

$$
\begin{aligned}
\mathrm{Mass}(v(t), B(x_j, \eta^{-C_1})) &\gtrsim \left(\mathrm{Mass}(v(s_j), B(x_j, \eta^{-C_1}))^{1/2} - \frac{1}{\eta^{-C_1}}\right)^2\\
&\gtrsim (\mathrm{Mass}(v(s_j), B(x_j, \eta^{C_2}))^{1/2} - \eta^{C_1})^2\\
&\gtrsim \eta^{C_1}
\end{aligned} \tag{4.400}
$$

for $t \in [0, 1]$,

The last step is to show that this mass concentration holds for u. We first show mass concentration for u at time $t = 0$.

Since $[0, 1]$ is an unexceptional interval, by the pigeonhole principle, there is $\tau_j \in [0, 1]$ such that

$$
\|u_-(\tau_j)\|_{L^{10}_x} \lesssim \eta^{C_3}, \tag{4.401}
$$

and so by Hölder's inequality,

$$\begin{aligned}\operatorname{Mass}(u_-(\tau_j), B(x_j, \eta^{-C_1})) &\lesssim \left\| \chi\Big(\frac{x - x_j}{\eta^{-C_1}}\Big) \right\|^2_{L_x^{\frac{5}{2}}} \|u_-(\tau_j)\|^2_{L_x^{10}} \\ &\lesssim \eta^{-\frac{12}{5}C_1 + 2C_3} \lesssim \eta^{2C_1}. \end{aligned} \tag{4.402}$$

From (4.369), we have

$$\operatorname{Mass}(u_-(0), B(x_j, \eta^{-C_1})) \lesssim \eta^{2C_1}. \tag{4.403}$$

Recall that $u(0) = u_-(0) - iv(0)$. Combining (4.400), (4.401) with the triangle inequality, we obtain

$$\operatorname{Mass}(u(0), B(x_j, \eta^{-C_1})) \gtrsim \eta^{C_1}. \tag{4.404}$$

Using (4.369) again, we obtain the result. □

Next, we use the radial assumption to show that the bubble of mass concentration must occur at the spatial origin.

Corollary 4.5.1 (Bubble at the origin). *Let I_j be an unexceptional interval, and u be a radial solution to* (4.362), *then*

$$\operatorname{Mass}(u(t), B(0, \eta^{-6C_1}|I_j|^{1/2})) \gtrsim \eta^{C_1}|I_j| \tag{4.405}$$

for all $t \in I_j$.

Proof. If x_j in Proposition 4.5.4 is within $\frac{1}{2}\eta^{-6C_1}|I_j|^{1/2}$ of the origin, then the result follows immediately. Otherwise by the radial assumption, there would be at least

$$O\Big(\frac{(\eta^{-6C_1}|I_j|^{1/2})^2}{(\eta^{-C_1}|I_j|^{1/2})^2}\Big) \approx O(\eta^{-10C_1}) \tag{4.406}$$

disjoint balls each containing at least $\eta^{C_1}|I_j|$ amount of mass. By Hölder's inequality, we get

$$\begin{aligned}\eta^{-10C_1} \times \eta^{C_1}|I_j| &\lesssim \int_{(\eta^{-6C_1} - \eta^{-C_1})|I_j|^{1/2} \le |x| \le (\eta^{-6C_1} + \eta^{-C_1})|I_j|^{1/2}} |u(t,x)|^2 dx \\ &\lesssim \|u\|^2_{L_x^6} \times \Big(\int_{(\eta^{-6C_1} - \eta^{-C_1})|I_j|^{1/2} \le |x| \le (\eta^{-6C_1} + \eta^{-C_1})|I_j|^{1/2}} dx \Big)^{2/3} \\ &\approx \|u\|^2_{L_x^6} \times ((\eta^{-6C_1}|I_j|^{1/2})^2 \times \eta^{-C_1}|I_j|^{1/2})^{\frac{2}{3}}, \end{aligned}$$

that is,

$$\|u\|^2_{L_x^6} \gtrsim \eta^{-\frac{1}{3}C_1}. \tag{4.407}$$

This contradicts with the boundedness on the energy of (4.381). This completes the proof. □

Next, we use Morawetz estimate (Proposition 4.5.1) to show that if there are many unexceptional intervals, they must form a cascade and must concentrate at some time t_*.

Corollary 4.5.2. *Assume that the solution u is spherically symmetric. For any interval $I \subseteq [t_-, t_+]$, and I be a union of consecutive unexceptional intervals I_j. Then*

$$\sum_{I_j \subseteq I} |I_j|^{1/2} \lesssim \eta^{-51C_1} |I|^{1/2} \tag{4.408}$$

and, moreover, there exists a j such that

$$|I_j| \gtrsim \eta^{102C_1} |I|. \tag{4.409}$$

Proof. For any unexceptional interval I_j, from Hölder inequality and Corollary 4.5.1, we have

$$\begin{aligned}
(\eta^{C_1}|I_j|)^3 &\leq \Big(\int_{|x|\leq 2\eta^{-6C_1}|I_j|^{\frac{1}{2}}} |u|^2 dx \Big)^3 \\
&\leq \Big(\int_{|x|\leq 2\eta^{-6C_1}|I_j|^{\frac{1}{2}}} \frac{|u|^6}{|x|} dx \Big) \cdot \Big(\int_{|x|\leq 2\eta^{-6C_1}|I_j|^{\frac{1}{2}}} |x|^{\frac{1}{2}} dx \Big)^2.
\end{aligned}$$

Therefore,

$$\begin{aligned}
\int_{|x|\leq 2\eta^{-6C_1}|I|^{1/2}} \frac{|u(t,x)|^6}{|x|} dx &\gtrsim (2\eta^{-6C_1}|I_j|^{1/2})^{-7} (\eta^{C_1}|I_j|)^3 \\
&\gtrsim \eta^{45C_1} |I_j|^{-\frac{1}{2}}.
\end{aligned} \tag{4.410}$$

We integrate this over each unexceptional interval I_j and sum over j,

$$\begin{aligned}
\eta^{45C_1} \sum_{I_j \subseteq I} |I_j|^{\frac{1}{2}} &\lesssim \sum_{I_j \subseteq I} \int_{I_j} \int_{|x|\leq 2\eta^{-6C_1}|I_j|^{1/2}} \frac{|u(t,x)|^6}{|x|} dx \\
&\lesssim \sum_{I_j \subseteq I} \int_{I_j} \int_{|x|\leq 2\eta^{-6C_1}|I|^{1/2}} \frac{|u(t,x)|^6}{|x|} dx \\
&\lesssim \int_I \int_{|x|\leq 2\eta^{-6C_1}|I|^{1/2}} \frac{|u(t,x)|^6}{|x|} dx \\
&\lesssim \eta^{-6C_1} |I|^{1/2}.
\end{aligned}$$

Note that

$$|I_j|^{1/2} \geq |I_j| \Big(\sup_{I_k \subseteq I} |I_k| \Big)^{-1/2}.$$

The second claim follows from the first claim (4.408)

$$\Big(\sup_{I_k \subseteq I} |I_k| \Big)^{-1/2} |I| = \Big(\sup_{I_k \subseteq I} |I_k| \Big)^{-1/2} \sum_{I_j \subset I} |I_j| \leq \sum_{I_j \subset I} |I_j|^{\frac{1}{2}} \lesssim \eta^{-51C_1} |I|^{\frac{1}{2}},$$

from which we obtain (4.409). □

Proposition 4.5.5 (Interval cascade). *Let I be an interval tiled by finitely many intervals $I_1, \ldots, I_N$. Suppose that for any contiguous family $\{I_j : j \in \mathcal{J}\}$ of the unexceptional intervals, there exists $j_* \in \mathcal{J}$ such that*

$$|I_{j_*}| \geq a \Big| \bigcup_{j \in \mathcal{J}} I_j \Big| \tag{4.411}$$

for some small $0 < a < 1$. Then there exist $K \geq \log(N)/\log(2a^{-1})$ distinct indices $j_1, \ldots, j_K$ such that

$$|I_{j_1}| \geq 2|I_{j_2}| \geq \cdots \geq 2^{K-1} |I_{j_K}|, \tag{4.412}$$

and for any $t_* \in I_{j_K}$,

$$\mathrm{dist}(I_{j_k}, t_*) \lesssim \frac{1}{a} |I_{j_k}| \tag{4.413}$$

holds for $1 \leq k \leq K$.

Proof. Here, we use an algorithm in [22, 258] to assign a generation to each I_j.

By hypothesis, I contains at least one interval of length $a|I|$. All intervals with a length larger than $a|I|/2$ belong to the first generation. By the total measure, we see that there are at most $2a^{-1} - 1$ intervals in the first generation. Removing these intervals from I leaves at most $2a^{-1}$ gaps, which are tiled by intervals I_j.

By (4.411) and the contradiction argument, we know that there is no gap with length larger than $|I|/2$, otherwise this will contract with

$$|I_j| \geq \frac{|I|}{2} > \frac{a}{2} |I|.$$

We now apply this argument recursively to all gaps generated by the previous iteration until every I_j has been labeled with a generation number.

Each iteration of the algorithm removes at most $2a^{-1} - 1$ intervals and produces at most $2a^{-1}$ gaps. Suppose that there are N consecutive unexceptional intervals initially, and we perform at most K times iteration. Then the number K obeys

$$N \le (2a^{-1} - 1) + (2a^{-1} - 1)2a^{-1} + \cdots + (2a^{-1} - 1)(2a^{-1})^{K-1}$$
$$\le (2a^{-1})^K, \tag{4.414}$$

which leads to the claim $K \ge \log(N)/\log(2a^{-1})$.

Let $I^{(K)}$ be a collection of the intervals obtained after $K-1$ iterations and I_{j_K} be any interval in $I^{(K)}$. For $1 \le i \le K-1$, let $I^{(i)}$ be the $(i-1)$-generation gap, which contains the I_{j_K}, and assign the I_{j_i} be any i-th generation interval which is contained in $I^{(i)}$. By the construction, for any $t_* \in I_{j_K}$, we have

$$\operatorname{dist}(t_*, I_{j_k}) \le |I^{(k)}| \le 2a^{-1}|I_{j_k}| \tag{4.415}$$

for all $1 \le k \le K$. □

Proposition 4.5.6 (Energy nonevacuation). *Let $I_{j_1}, \ldots, I_{j_K}$ be a disjoint family of unexceptional intervals obeying*

$$|I_{j_1}| \ge 2|I_{j_2}| \ge \cdots \ge 2^{K-1}|I_{j_K}|. \tag{4.416}$$

And for any $t_* \in I_{j_K}$

$$\operatorname{dist}(I_{j_k}, t_*) \lesssim \eta^{-102C_1}|I_{j_k}| \tag{4.417}$$

holds for $1 \le k \le K$. Then

$$K \le \eta^{-10^3 C_1}. \tag{4.418}$$

Proof. By Corollary 4.5.1, we have

$$\operatorname{Mass}(u(t), B(0, \eta^{-6C_1}|I_{j_k}|^{1/2})) \gtrsim \eta^{C_1}|I_{j_k}|$$

for all $t \in I_{j_k}$. By almost finite propagation speed (4.369), we have

$$\operatorname{Mass}(u(t), B(0, \eta^{-103C_1}|I_{j_k}|^{1/2})) \gtrsim \left((\eta^{C_1}|I_{j_k}|)^{1/2} - \frac{\operatorname{dist}(t_*, I_{j_k})}{\eta^{-103C_1}|I_{j_k}|^{1/2}} \right)^2 \tag{4.419}$$
$$\gtrsim \eta^{C_1}|I_{j_k}|. \tag{4.420}$$

On the other hand, from (4.370), we have

$$\operatorname{Mass}(u(t), B(0, 2\eta^{C_1}|I_{j_k}|^{1/2})) \lesssim \eta^{2C_1}|I_{j_k}|. \tag{4.421}$$

Define

$$A(k) = \{x : \eta^{C_1}|I_{j_k}|^{1/2} \le |x| \le \eta^{-103C_1}|I_{j_k}|^{1/2}\}, \tag{4.422}$$

and then we have

$$\int_{A(k)} |u(t_*,x)|^2 dx \gtrsim \mathrm{Mass}(u(t), B(0, \eta^{-103C_1}|I_{j_k}|^{1/2})) \\ - \mathrm{Mass}(u(t), B(0, 2\eta^{C_1}|I_{j_k}|^{1/2})) \\ \gtrsim \eta^{C_1}|I_{j_k}|. \tag{4.423}$$

By Hölder's inequality, we have

$$\int_{A(k)} |u(t_*,x)|^6 dx \gtrsim (\eta^{C_1}|I_{j_k}|)^3 (\eta^{-103C_1}|I_{j_k}|^{1/2})^{-6} \\ \gtrsim \eta^{700C_1}. \tag{4.424}$$

Choosing $M = -208C_1 \log \eta$, then we obtain by (4.416),

$$\eta^{-103C_1}|I_{j_{M+1}}|^{1/2} \leq \eta^{C_1}|I_{j_1}|^{1/2}; \\ \eta^{-103C_1}|I_{j_{2M+1}}|^{1/2} \leq \eta^{C_1}|I_{j_{M+1}}|^{1/2}; \\ \dots$$

Hence, the annuli $A(k)$ associated to $k = 1, M+1, 2M+1, \dots$ are disjoint. The number of such annuli is $O(K/M)$.

Therefore, from (4.382), we obtain

$$\frac{K}{M}\eta^{700C_1} \lesssim \int_{\mathbb{R}^n} |u(t_*,x)|^6 dx \lesssim 1.$$

That is,

$$K \lesssim M\eta^{-700C_1} \lesssim \eta^{-10^3 C_1}. \tag{4.425}$$

□

Now, we return back to the proof of Theorem 4.5.1. As explained at the beginning of this section, it suffices to bound the number of the unexceptional intervals.

Note that the number of exceptional interval is at most $O(\eta^{-10C_3})$. We first bound the number N of unexceptional intervals that can occur consecutively.

Let us denote the union of these consecutive unexceptional intervals by I. By Corollary 4.5.2, the hypotheses of Proposition 4.5.5 are satisfied with $a = \eta^{102C_1}$ and so we can find a cascade of K intervals and they satisfied the hypotheses of Proposition 4.5.6. The bound on K implies the bound on N, namely

$$N \lesssim (2\eta^{-102C_1})^K \approx (2\eta^{-102C_1})^{\eta^{-10^3 C_1}}. \tag{4.426}$$

At last, since there are at most $O(\eta^{-10C_3})$ exceptional intervals, the total number of intervals is

$$J \lesssim \eta^{-10C_3} + \eta^{-10C_3} N \lesssim e^{\eta^{-10^4 C_1}}.$$

This completes the proof of Theorem 4.5.1. □

Remark 4.5.2. For the general data, we must use the interaction Morawetz estimate, the localized Morawetz estimate in spatial and frequency space, the frequency localized L^2 almost conservation technique to remove the radial assumption and disprove the "concentration" of energy on any spatial point; see [53].

4.6 Notes on dynamics of NLS

As we pointed out before, most of the material in Section 4.3 stems from Bourgain [19]. We have followed the expositions of Colliander, Keel, Staffilani, Takaoka and Tao(I-team) [51] for Theorem 4.4.1 in Section 4.4, and Tao [260] for Theorem 4.5.1 in Section 4.5.

In the following of this note, we devote to summary the history of the dynamics of the nonlinear Schrödinger equation (NLS) of the form

$$\begin{cases} i\partial_t u + \Delta u = \mu |u|^{p-1} u, & (t,x) \in \mathbb{R} \times \mathbb{R}^n, \\ u(0,x) = u_0(x), \end{cases} \tag{4.427}$$

where $u(t,x) : \mathbb{R}\times\mathbb{R}^n \to \mathbb{C}, p > 1, \mu = 1$ corresponding to the defocusing case, and $\mu = -1$ corresponding to the focusing case. There are three important conserved quantities:

$$\textit{Mass:}\quad M(u) = \int_{\mathbb{R}^n} |u(t,x)|^2\, dx = M(u_0); \tag{4.428}$$

$$\textit{Energy:}\quad E(u) = \frac{1}{2}\int_{\mathbb{R}^n} |\nabla u(t,x)|^2\, dx + \frac{\mu}{p+1}\int_{\mathbb{R}^n} |u(t,x)|^{p+1}\, dx = E(u_0); \tag{4.429}$$

$$\textit{Momentum:}\quad P(u) = \operatorname{Im}\int_{\mathbb{R}^n} \nabla u \bar{u}\, dx = P(u_0). \tag{4.430}$$

Equation (4.427) admits a number of symmetries in energy space H^1, explicitly:

- *Space-time translation invariance*: if $u(t,x)$ solves (4.427), then so does $u(t+t_0, x+x_0)$, $(t_0, x_0) \in \mathbb{R}\times\mathbb{R}^n$;
- *Phase invariance*: if $u(t,x)$ solves (4.427), then so does $e^{i\gamma}u(t,x)$, $\gamma \in \mathbb{R}$;
- *Galilean invariance*: if $u(t,x)$ solves (4.427), then for $\beta \in \mathbb{R}^n$, so does $e^{i\frac{\beta}{2}\cdot(x-\frac{\beta}{2}t)}u(t,x-\beta t)$;

- *Scaling invariance*: if $u(t,x)$ solves (4.427), then so does $u_\lambda(t,x)$ defined by

$$u_\lambda(t,x) = \lambda^{\frac{2}{p-1}} u(\lambda^2 t, \lambda x), \quad \lambda > 0. \tag{4.431}$$

This scaling defines a notion of criticality for (4.427). In particular, one can check that the only homogeneous L_x^2-based Sobolev space that is left invariant under (4.431) is $\dot{H}_x^{s_c}(\mathbb{R}^n)$, where the critical regularity s_c is given by $s_c := \frac{n}{2} - \frac{2}{p-1}$,

$$\|u_\lambda(t,\cdot)\|_{\dot{H}_x^{s_c}(\mathbb{R}^n)} = \|u(\lambda^2 t,\cdot)\|_{\dot{H}_x^{s_c}(\mathbb{R}^n)}.$$

We introduce the notions as in the following:
- For $s_c < 0$, we call the problem (4.427) *mass subcritical* ($p < 1 + \frac{4}{n}$);
- For $s_c = 0$, we call the problem (4.427) *mass critical* ($p = 1 + \frac{4}{n}$);
- For $s_c \in (0,1)$, *interpolate between mass and energy critical* ($1 + \frac{4}{n} < p < 1 + \frac{4}{n-2}$);
- For $s_c = 1$, we call the problem (4.427) *energy critical* ($p = 1 + \frac{4}{n-2}$);
- For $s_c > 1$, we call the problem (4.427) *energy supercritical* ($p > 1 + \frac{4}{n-2}$).

For the general dispersive equations, there is a famous conjecture, called soliton resolution conjecture. Roughly speaking, that any reasonable (e. g., bounded energy) solution to semilinear dispersive equations eventually resolves into a superposition of a radiation component, which behaves like a solution to the free dispersive equation, plus a finite number of solitons. In the following, we will mainly state the history of the study for the solutions to Cauchy problem (4.427), including the well-posedness, scattering theory, blowup dynamics and stability of the special solution.

4.6.1 Energy-critical NLS

We first list the study of the Cauchy problem for energy-critical nonlinear Schrödinger equations

$$\begin{cases} (i\partial_t + \Delta)u = \mu|u|^{\frac{4}{n-2}}u, & (t,x) \in \mathbb{R} \times \mathbb{R}^n, \\ u(0,x) = u_0(x) \in \dot{H}^1(\mathbb{R}^n), \end{cases} \tag{4.432}$$

where $u : \mathbb{R}_t \times \mathbb{R}_x^n \to \mathbb{C}$, $n \geq 3$, $\mu = \pm 1$.

In the defocusing case, that is, $\mu = 1$, for any $u_0 \in \dot{H}^1(\mathbb{R}^n)$, the solution u to (4.432) with $\mu = 1$ is global and scatters in the sense that there exists $u_\pm \in \dot{H}^1(\mathbb{R}^n)$ such that

$$\lim_{t\to\pm\infty} \|u(t) - e^{it\Delta}u_\pm\|_{\dot{H}^1(\mathbb{R}^n)} = 0. \tag{4.433}$$

We list the history of the scattering result for the defocusing energy-critical NLS as in the following table (see Table 4.1).

Table 4.1: Defocusing energy-critical NLS.

	$n=3$	$n=4$	$n\geq 5$
radial	Bourgain [22]	Tao [260]	Tao [260]
nonradial	I-team [48], Killip–Visan [147]	Ryckman–Visan [228], Visan [280]	Visan [278, 279]

Next, we consider the focusing case, that is, $\mu=-1$. In this case, the energy becomes

$$E(u)=\frac{1}{2}\int_{\mathbb{R}^n}|\nabla u|^2\,dx-\frac{n-2}{2n}\int_{\mathbb{R}^n}|u|^{\frac{2n}{n-2}}\,dx.$$

Let W be the ground state of the elliptic equation

$$\Delta W+|W|^{\frac{4}{n-2}}W=0. \tag{4.434}$$

An explicit solution of (4.434) is the stationary solution in $\dot{H}^1(\mathbb{R}^n)$ (but in $L^2(\mathbb{R}^n)$ only if $n\geq 5$)

$$W:=\frac{1}{(1+\frac{|x|^2}{n(n-2)})^{\frac{n-2}{2}}}. \tag{4.435}$$

The works of Aubin and Talenti [5, 256] give the following characterization of W:

$$\|u\|_{L^{\frac{2n}{n-2}}(\mathbb{R}^n)}=C_n\|u\|_{\dot{H}^1(\mathbb{R}^n)}\Longrightarrow\exists\,\lambda_0>0,x_0,z_0,\quad u(x)=z_0W\Big(\frac{x+x_0}{\lambda_0}\Big), \tag{4.436}$$

where C_n is the sharp constant of the Sobolev embedding

$$\|u\|_{L^{\frac{2n}{n-2}}(\mathbb{R}^n)}\leq C\|u\|_{\dot{H}^1(\mathbb{R}^n)}.$$

Below threshold of ground state

It is easy to check that W is a global solution to (4.432) with $\mu=-1$ but does not scatter. It is natural to view the ground state W as the threshold of the scattering. Based on the energy induction of Bourgain [22], Kenig–Merle [132] developed concentration-compactness/rigidity method to obtain that for $E(u_0)<E(W)$ and u_0 is radial, if $\|\nabla u_0\|_{L_x^2}<\|\nabla W\|_{L_x^2}$, then the solution is global and scatters in the sense (4.433); if $\|\nabla u_0\|_{L_x^2}>\|\nabla W\|_{L_x^2}$, then the solution blows up in finite time. We list the history of the scattering result for the focusing energy-critical NLS as in the following table (see Table 4.2).

Table 4.2: Focusing energy-critical.

	$n = 3$	$n = 4$	$n \geq 5$
radial	Kenig–Merle [132]	Kenig–Merle [132]	
nonradial	Open!	Dodson [64]	Killip–Visan [145]

On the other hand, using the variational analysis, we have the following.

Lemma 4.6.1. *Let f be in $\dot{H}^1$. Then*

$$\|\nabla f\|_{L^2} < \|\nabla W\|_{L^2} \Longrightarrow \int_{\mathbb{R}^n} |f|^{\frac{2n}{n-2}}\, dx < \int_{\mathbb{R}^n} |W|^{\frac{2n}{n-2}}\, dx. \tag{4.437}$$

Assume furthermore that

$$E[f] \leq E[W], \tag{4.438}$$

then the reverse implication to (4.437) *holds, and we obtain*

$$\|\nabla f\|_{L^2} < \|\nabla W\|_{L^2} \iff \int_{\mathbb{R}^n} |f|^{\frac{2n}{n-2}}\, dx < \int_{\mathbb{R}^n} |W|^{\frac{2n}{n-2}}\, dx. \tag{4.439}$$

Moreover, (4.439) *also holds with nonstrict inequalities (in the case of equality, f is equal to W up to space translation, scaling and phase).*

Theorem 4.6.1 ([81]). *Let u be a solution of* (4.432) *with $\mu = -1$ and maximal time of existence $T_+(u)$, and assume*

$$\limsup_{t \to T_+(u)} \int_{\mathbb{R}^n} |u(x,t)|^{\frac{2n}{n-2}}\, dx < \int_{\mathbb{R}^n} |W(x)|^{\frac{2n}{n-2}}\, dx. \tag{4.440}$$

Assume furthermore that u is radial if $n = 3, 4$. Then $T_+(u) = +\infty$ and u scatters forward in time.

Remark 4.6.1. The assumption (4.440) is weaker than $\|\nabla u_0\|_{L^2} < \|\nabla W\|_{L^2}$, due to (4.437). On the other hand, in Killip–Visan [145], the assumption

$$\sup_{t \in I_{\max}} \|\nabla u(t,x)\|_{L^2_x} < \|\nabla W\|_{L^2}$$

is also weaker than the assumption

$$E(u_0) < E(W), \quad \|\nabla u_0\|_{L^2_x} < \|\nabla W\|_{L^2_x}.$$

Threshold solutions

In the threshold case: $E(u_0) = E(W)$, Duyckaerts–Merle gave the characterization of the dynamics of the threshold solution in [80] as follows.

Theorem 4.6.2 (Existence of special solutions (besides W) at threshold, [80]). *Let $n \in \{3, 4, 5\}$. There exist radial solutions W^- and W^+ of (4.432) with $\mu = -1$ such that*

$$E(W) = E(W^+) = E(W^-), \tag{4.441}$$

$$T_+(W^-) = T_+(W^+) = +\infty \quad \text{and} \quad \lim_{t\to+\infty} W^{\pm}(t) = W \quad \text{in} \quad \dot{H}^1, \tag{4.442}$$

$$\|W^-\|_{\dot{H}^1} < \|W\|_{\dot{H}^1}, \quad T_-(W^-) = +\infty, \quad \|W^-\|_{S((-\infty,0])} < \infty, \tag{4.443}$$

$$\|W^+\|_{\dot{H}^1} > \|W\|_{\dot{H}^1}, \quad \text{and, if} \quad n = 5, \quad T_-(W^+) < +\infty. \tag{4.444}$$

Remark 4.6.2. As for W, $W^+(t)$ and $W^-(t)$ belong to L^2 if and only if $n = 5$. One still expects $T_-(W^+) < +\infty$ for $n = 3, 4$.

Theorem 4.6.3 (Classification of solutions [80]). *Let $n \in \{3, 4, 5\}$. Let $u_0 \in \dot{H}^1$ radial, such that*

$$E(u_0) = E(W) = \frac{1}{nC_n^n}. \tag{4.445}$$

Let u be the solution of (4.432) with $\mu = -1$ and initial condition u_0 and I its maximal interval of definition. Then the following holds:

(i) *If $\|\nabla u_0\|_{L_x^2} < \|\nabla W\|_{L_x^2}$, then $I = \mathbb{R}$. Furthermore, either $u = W^-$ up to the symmetry of the equation, or $\|u\|_{S(\mathbb{R})} < \infty$.*

(ii) *If $\|\nabla u_0\|_{L_x^2} = \|\nabla W\|_{L_x^2}$, then $u = W$ up to the symmetry of the equation.*

(iii) *If $\|\nabla u_0\|_{L_x^2} > \|\nabla W\|_{L_x^2}$, and $u_0 \in L^2$, then either $u = W^+$ up to the symmetry of the equation, or I is finite.*

In the above theorem, when we say $u = v$ up to symmetries of the equation, we mean there exist $\theta_0, t_0 \in \mathbb{R}$, $\lambda_0 > 0$ such that

$$u(t,x) = e^{i\theta_0}\lambda_0^{-\frac{n-2}{2}} v\Big(\frac{t-t_0}{\lambda_0^2}, \frac{x}{\lambda_0}\Big) \quad \text{or} \quad u(t,x) = e^{i\theta_0}\lambda_0^{-\frac{n-2}{2}} \bar{v}\Big(\frac{t_0-t}{\lambda_0^2}, \frac{x}{\lambda_0}\Big).$$

Remark 4.6.3. In [166], Li and Zhang extend the results of Theorem 4.6.3 to all dimensions $n \geq 6$. On the other hand, for $n \geq 5$, Su and Zhao [254] remove the radial assumption for the subcritical threshold solutions as in Theorem 4.6.3(i).

Beyond threshold of ground state

In this part, we consider the beyond threshold case: $E(u_0) > E(W)$. Define the variance as

$$V(t) = \int_{\mathbb{R}^n} |x|^2 |u(x,t)|^2 \, dx. \tag{4.446}$$

Assuming $V(0) < \infty$ (referred to as finite variance), and letting $u(t,x)$ solves (4.432) with $\mu = -1$, then a simple computation shows the following virial identities:

$$V_t(t) = 4 \operatorname{Im} \int x \cdot \nabla u(x,t) \, \overline{u}(x,t) \, dx, \tag{4.447}$$

$$V_{tt}(t) = 8 \int |\nabla u(t)|^2 - 8 \int |u(t)|^{\frac{2n}{n-2}} \tag{4.448}$$

$$\equiv \frac{16n}{n-2} E[u] - \frac{16}{n-2} \|\nabla u(t)\|^2_{L^2(\mathbb{R}^n)}. \tag{4.449}$$

Theorem 4.6.4 (Classification of solutions [83]). *Let u be a solution of (4.432) with $\mu = -1$. Assume $V(0) < \infty$, $u_0 \in H^1(\mathbb{R}^n)$ and*

$$\mathcal{ME}[u] \left(1 - \frac{(V_t(0))^2}{32 E[u] \, V(0)} \right) \le 1, \quad \mathcal{ME}[u] := \frac{\|\nabla u\|_{L^2}}{\|\nabla W\|_{L^2}}. \tag{4.450}$$

(i) *(Blow up) If*

$$\int_{\mathbb{R}^n} |u_0|^{\frac{2n}{n-2}} \, dx > \int_{\mathbb{R}^n} |W|^{\frac{2n}{n-2}} \, dx \tag{4.451}$$

and

$$V_t(0) \le 0, \tag{4.452}$$

then $u(t)$ blows up in finite positive time, $T_+(u) < \infty$.

(ii) *(Boundedness and scattering) If*

$$\int_{\mathbb{R}^n} |u_0|^{\frac{2n}{n-2}} \, dx < \int_{\mathbb{R}^n} |W|^{\frac{2n}{n-2}} \, dx \tag{4.453}$$

and

$$V_t(0) \ge 0, \tag{4.454}$$

then

$$\limsup_{t \to T_+(u)} \int_{\mathbb{R}^n} |u(t)|^{\frac{2n}{n-2}} \, dx < \int_{\mathbb{R}^n} |W|^{\frac{2n}{n-2}} \, dx. \tag{4.455}$$

Furthermore, u scatters forward in time in $\dot{H}^1$ provided $n \ge 5$ or u is radial.

4.6.2 Mass-critical NLS

In this subsection, we review the study of Cauchy problem for mass-critical NLS

$$\begin{cases} (i\partial_t + \Delta)u = \mu|u|^{\frac{4}{n}}u, & (t,x) \in \mathbb{R} \times \mathbb{R}^n, \\ u(0,x) = u_0(x), \end{cases} \tag{4.456}$$

where $u : \mathbb{R}_t \times \mathbb{R}_x^n \to \mathbb{C}$, $\mu = \pm 1$. (4.456) admits a symmetry, which is not in energy space H^1, the so-called pseudoconformal transformation:

- Pseudoconformal transformation: if $u(t,x)$ solves (4.456), then so does

$$v(t,x) = \frac{1}{|t|^{n/2}}\bar{u}\left(\frac{1}{t}, \frac{x}{t}\right)e^{i\frac{|x|^2}{4t}}.$$

This additional symmetry yields the conservation of the pseudoconformal energy for initial data $u_0 \in \Sigma := H^1 \cap \{xu_0 \in L^2\}$, which is most frequently expressed as

$$\frac{d^2}{dt^2}\int_{\mathbb{R}^n} |x|^2|u(t,x)|^2\,dx = 4\frac{d}{dt}\operatorname{Im}\int_{\mathbb{R}^n} x \cdot \nabla u\bar{u}(t,x)\,dx = 16E(u_0). \tag{4.457}$$

In the defocusing case, that is, $\mu = 1$, for any $u_0 \in L^2(\mathbb{R}^n)$, the solution u to (4.456) exists globally and scatters in the sense that there exists $u_\pm \in L^2(\mathbb{R}^n)$ such that

$$\lim_{t\to\pm\infty}\|u(t,x) - e^{it\Delta}u_\pm\|_{L^2(\mathbb{R}^n)} = 0. \tag{4.458}$$

We list the history of the scattering result for the defocusing mass-critical NLS as in the following table (see Table 4.3).

Table 4.3: Defocusing mass-critical.

	$n = 1$	$n = 2$	$n \geq 3$
radial		Killip–Tao–Visan [144]	Tao–Visan–Zhang [267, 269]
nonradial	Dodson [63]	Dodson [62]	Dodson [59]

The proof follows the concentration-compactness approach [14, 132, 223] to induction on energy [22]; see Monica–Visan [148] and Planchon [217]. The argument is by contradiction. The failure of scattering result would imply the existence of very special class of solutions. On the other hand, these solutions have so many good properties that they do not exist. Thus, one can get a contradiction.

Next, we focus on the focusing mass critical NLS. In this case, special solutions play an important role. They are the so-called solitary waves of the form $u(t,x) = e^{i\omega t}Q_\omega(x)$, $\omega > 0$, where Q_ω solves the elliptic equation

$$\Delta Q_\omega + Q_\omega |Q_\omega|^{\frac{4}{n}} = \omega Q_\omega. \tag{4.459}$$

(i) In dimension $n = 1$, there exists a unique solution in H^1 up to translation to (4.459);
(ii) For $n \geq 2$, there are infinitely many solutions with growing L^2-norm for $n \geq 2$, but there is a unique positive solution Q_ω to (4.459) up to scaling translation, [14, 96, 156];
(iii) Q_ω is in addition radially symmetric, letting $Q_\omega(x) = \omega^{n/4} Q(\omega^{1/2} x)$, from scaling property, we know that $Q(x)$ is a unique positive solution to (4.459) with $\omega = 1$. Therefore, $\|Q_\omega\|_{L^2} = \|Q\|_{L^2}$.
(iv) Moreover, multiplying (4.459) by $\frac{n}{2} Q_\omega + x \cdot \nabla Q_\omega$ and integrating by parts yields the so-called Pohozaev identity $E(Q_\omega) = \omega E(Q) = 0$. In particular, none of the three conservation laws (mass, energy, momentum) in H^1 sees the variation of size of the stationary solutions Q_ω. These two facts are deeply related to the criticality of the problem, that is, the value $p = 1 + \frac{4}{n}$. Note that in dimension $n = 1$, Q can be written explicitly

$$Q(x) = \left(\frac{3}{\cosh^2(2x)}\right)^{\frac{1}{4}}.$$

Proposition 4.6.1 (Variational characterization of the ground state). *Let $v(x) \in H^1(\mathbb{R}^n)$ such that*

$$\int_{\mathbb{R}^n} |v(x)|^2 \, dx = \int_{\mathbb{R}^n} Q(x)^2 \, dx \quad \text{and} \quad E(v) = 0.$$

Then

$$v(x) = \lambda_0^{\frac{n}{2}} Q(\lambda_0 x + x_0) e^{i\gamma_0},$$

for some parameters $\lambda_0 \in \mathbb{R}_+^*$, $x_0 \in \mathbb{R}^n$, $\gamma_0 \in \mathbb{R}$.

Below threshold of ground state: $\|u_0\|_{L_x^2} < \|Q\|_{L_x^2}$

- From classical variational arguments, we obtain the global well-posedness in $H^1(\mathbb{R}^n)$. Indeed, this follows from the sharp Gagliardo–Nirenberg inequality:

$$\forall\, u \in H^1 : \; E(u) \geq \frac{1}{2}\left(\int_{\mathbb{R}^n} |\nabla u|^2 dx\right)\left[1 - \frac{\int_{\mathbb{R}^n} |u|^2 dx}{\int_{\mathbb{R}^n} Q^2 dx}\right]. \tag{4.460}$$

- If $u_0 \in L^2(\mathbb{R}^n)$ and $M(u_0) < M(Q)$, then the solution u to (4.456) with $\mu = -1$ exists globally and scatters in the sense (4.458). We list the history of the scattering result for the focusing mass-critical NLS as in the following table (see Table 4.4).

Table 4.4: Focusing mass-critical.

	$n = 1$	$n = 2$	$n \geq 3$
radial		Killip–Tao–Visan [144]	Killip–Visan–Zhang [149]
nonradial	Dodson [61]	Dodson [61]	Dodson [61]

We remark that the above assumption $M(u_0) < M(Q)$ is sharp, since blowup may occur when $M(u_0) \geq M(Q)$. Indeed, since $E(Q) = 0$ and $\nabla E(Q) = -Q$, there exists $u_{0\varepsilon} \in \Sigma$ with $\|u_{0\varepsilon}\|_{L^2} = \|Q\|_{L^2} + \varepsilon$ and $E(u_{0\varepsilon}) < 0$, and the corresponding solution must blow up from virial identity (4.457).

Threshold solution: $\|u_0\|_{L^2_x} = \|Q\|_{L^2_x}$

The pseudoconformal transformation applied to the stationary solution $e^{it}Q$ yields an explicit solution

$$S(t,x) = \frac{1}{|t|^{n/2}} Q\left(\frac{x}{t}\right) e^{-i\frac{|x|^2}{4t} + \frac{i}{t}}, \quad \|S(t)\|_{L^2} = \|Q\|_{L^2}, \tag{4.461}$$

which scatters as $t \to -\infty$, and blows up at $T = 0$ at the speed

$$\|\nabla S(t)\|_{L^2} \sim \frac{1}{|t|}.$$

An essential feature of (4.461) is compact up to the symmetries of the flow, meaning that all the mass goes into the singularity formation

$$|S(t)|^2 \rightharpoonup \|Q\|_{L^2}^2 \delta(x) \quad \text{as} \quad t \to 0. \tag{4.462}$$

It turns out that $S(t)$ is the unique minimal mass blowup solution in H^1 in the following sense: let $u(-1) \in H^1$ with $\|u(-1)\|_{L^2} = \|Q\|_{L^2}$, and assume that $u(t)$ blows up at $T = 0$, then $u(t) = S(t)$ up to the symmetries of the equation; see Theorem 4.6.5. Note that from direct computation

$$E(S(t,x)) > 0, \quad \text{and} \quad \|\nabla S(t)\|_{L^2} = \frac{C}{|t|}.$$

The general intuition is that such a behavior is exceptional in the sense that such minimal elements can be classified. The first result of this type was proved by Merle [182] using the pseudoconformal symmetry. This has been further simplified by Banica [8] and Hmidi–Keraani [204].

Theorem 4.6.5 (Determination of minimal blowup solutions, Merle [182]). *Let* $u_0 \in H^1(\mathbb{R}^n)$ *and*

$$\|u_0\|_{L^2_x} = \|Q\|_{L^2_x},$$

and assume the corresponding solution $u(t)$ blows up in finite time $0 < T < +\infty$. Then there exists $\theta \in \mathbb{R}$, $w > 0$, $x_0 \in \mathbb{R}^n$, $x_1 \in \mathbb{R}^n$ such that for $t < T$,

$$u(t,x) = \left(\frac{w}{T-t}\right)^{\frac{n}{2}} e^{i\theta + i|x-x_1|^2/4(t-T) - iw^2/(t-T)} Q\left(\frac{w}{T-t}((x-x_1)-(T-t)x_0)\right).$$

For the initial data $u_0 \in L^2(\mathbb{R}^n)$, there is a known rigidity conjecture as follows.

Conjecture 4.6.1 (Rigidity conjecture at the ground state mass). *Let $n \geq 1$. For general initial data, $u_0 \in L^2_x(\mathbb{R}^n)$ with $\|u_0\|_{L^2} = \|Q\|_{L^2}$. Then, either the corresponding solution scatters, or the nonscattering solution must be the solitary wave $e^{it}Q$ up to symmetries of the equation* (4.456) *with $\mu = -1$.*

We list the study of the above conjecture in the following table (see Table 4.5).

Table 4.5: History of the study of Conjecture 4.6.1.

	H^1	L^2
radial	Killip–Li–Visan-Zhang [140], Li–Zhang [168] $n \geq 2$	Li–Zhang [167] $n \geq 4$
nonradial	Merle [182] $T < +\infty$	Dodson [66, 67]

4.6.3 Mass supcritical and energy subcritical NLS

In this part, we study the Cauchy problem for nonlinear Schrödinger equations:

$$\begin{cases} (i\partial_t + \Delta)u = \mu|u|^{p-1}u, & (t,x) \in \mathbb{R} \times \mathbb{R}^n, \\ u(0,x) = u_0(x), \end{cases} \tag{4.463}$$

where $u : \mathbb{R}_t \times \mathbb{R}^n_x \to \mathbb{C}$, $1 + \frac{4}{n} < p < +\infty$ for $n \in \{1,2\}$, and $1 + \frac{4}{n} < p < 1 + \frac{4}{n-2}$ for $n \geq 3$, $\mu = \pm 1$. The equation (4.463) is $\dot{H}^{s_c}$-critical with

$$s_c = \frac{n}{2} - \frac{2}{p-1}.$$

First, we consider the defocusing case ($\mu = 1$).

Energy solution: $u_0 \in H^1(\mathbb{R}^n)$

Assume that $u_0 \in H^1(\mathbb{R}^n)$, $n \geq 3$ and $0 < s_c < 1$, Ginibre–Velo [103] proved that the solution u to (4.463) with $\mu = 1$ is global and scatters in the sense that there exists $u_\pm \in H^1(\mathbb{R}^n)$ such that

$$\lim_{t\to\pm\infty}\|u(t)-e^{it\Delta}u_\pm\|_{H^1(\mathbb{R}^n)}=0, \tag{4.464}$$

where they utilize the almost finite propagation speed

$$\int_{|x|\geq a}|u(t,x)|^2\,dx\leq\int\min\Big(\frac{|x|}{a},1\Big)|u(t_0)|^2\,dx+\frac{C}{a}\cdot|t-t_0|$$

for large spatial scale and the classical Morawetz inequality in [173]

$$\iint_{\mathbb{R}\times\mathbb{R}^n}\frac{|u(t,x)|^{p+2}}{|x|}\,dtdx\lesssim\|u\|^2_{L_t^\infty\dot{H}^{\frac{1}{2}}}\leq C(M(u_0),E(u_0)) \tag{4.465}$$

for small spatial scale to show

$$\lim_{t\to\pm\infty}\|u(t,x)\|_{L_x^{p+1}(\mathbb{R}^n)}=0.$$

Laterly, Tao–Visan–Zhang [268] gave another simple proof by using the following interaction Morawetz estimate:

$$\big\||\nabla|^{-\frac{n-3}{4}}u\big\|^4_{L^4_{t,x}(I\times\mathbb{R}^n)}\leq C\|u_0\|^3_{L_x^2}\|u\|_{L_t^\infty(I,\dot{H}_x^1(\mathbb{R}^n))},\quad n\geq 3.$$

And we refer to Planchon and Vega [219], Colliander, Grillakis and Tzirakis [46] for the interaction Morawetz estimate in dimensions $n\in\{1,2\}$ and $n=2$, respectively.

Low regularity: $u_0\in H^s(\mathbb{R}^n)$ with $s_c\leq s<1$

Conjecture 4.6.2 (Low regularity conjecture). *Let $s_c\geq 0$, $\mu=1$ and $u_0\in\dot{H}^{s_c}$. Then u is global and scatters in the sense that there exist unique $u_\pm\in\dot{H}_x^{s_c}(\mathbb{R}^n)$ such that*

$$\lim_{t\to\pm\infty}\|u(t)-e^{it\Delta}u_\pm\|_{\dot{H}_x^{s_c}(\mathbb{R}^n)}=0.$$

The interaction Morawetz inequality also plays an important role in the study of a low regularity problem, where we ask what is the minimal s to ensure that problem (4.463) has either a local solution or a global solution for which the scattering hold? Such a problem was first considered by Cazenave and Weissler [37], who proved that problem (4.463) is locally well posed in $H^s(\mathbb{R}^n)$ with $s\geq\max\{0,s_c\}$ and globally well posed together with scattering for small data in $\dot{H}^{s_c}(\mathbb{R}^n)$ with $s_c\geq 0$. They used Strichartz estimates in the framework of Besov spaces. On the other hand, since the lifespan of local solutions depend only on the H^s-norm of the initial data for $s>\max\{0,s_c\}$, one can easily obtain the global well-posedness for (4.463) in two special cases: the mass subcritical case ($p<\frac{4}{n}$) for $L_x^2(\mathbb{R}^n)$-initial data and the energy-subcritical case (for $p<\frac{4}{n-2}$, if $n\geq 3$ or for $p<+\infty$ if $n\in\{1,2\}$) for $H_x^1(\mathbb{R}^n)$-initial data by using the conservation of mass and energy, respectively.

This leaves the open problem on global well-posedness in $H^s(\mathbb{R}^n)$ in the intermediate regime $0 \le s_c \le s < 1$. The first progress on this direction came from the Bourgain' Fourier truncation method [19] where refinements of Strichartz' inequality [20], high-low frequency decompositions and perturbation methods were used to show that problem (4.463) with $p = 3$ is globally well posed in $H^s(\mathbb{R}^3)$ with $s > \frac{11}{13}$ such that

$$u(t) - e^{it\Delta}u_0 \in H^1(\mathbb{R}^3). \tag{4.466}$$

This leads to the I-method, which was derived by Keel and Tao in the study of wave maps [130]. Subsequently, the I-team developed the I-method to treat many low regularity problems including the nonlinear Schrödinger equations with derivatives, the one-dimensional quintic NLS, and the cubic NLS in two and three dimensions [47–50, 52]. Compared with the result in [19], the I-team also obtained the *scattering* in $H^s(\mathbb{R}^3)$ with $s > \frac{4}{5}$ by using the I-method and the interaction Morawetz estimate in [51]. Dodson [60] extended those results to $s > \frac{5}{7}$ by means of a linear–nonlinear decomposition, and then Su [253] to $s > \frac{49}{74}$.

Critical norm conjecture: $u_0 \in \dot{H}^{s_c}(\mathbb{R}^n)$

Inspired by the global well-posedness results for the mass and energy-critical cases, one is led to the following conjecture.

Conjecture 4.6.3 (Critical norm conjecture). *Let $s_c \ge 0$ and $\mu = 1$. Suppose $u : I \times \mathbb{R}^d \to \mathbb{C}$ is a maximal-lifespan solution to* (4.463) *such that*

$$u \in L_t^\infty \dot{H}_x^{s_c}(I \times \mathbb{R}^n). \tag{4.467}$$

Then u is global and scatters in the sense that there exist unique $u_\pm \in \dot{H}_x^{s_c}(\mathbb{R}^n)$ such that

$$\lim_{t\to\pm\infty} \|u(t) - e^{it\Delta}u_\pm\|_{\dot{H}_x^{s_c}(\mathbb{R}^n)} = 0.$$

The first result in this direction was due to Kenig and Merle [133], who treated the cubic problem in three dimensions. We list the study of the critical norm conjecture in the following table (see Table 4.6).

Table 4.6: Critical norm conjecture.

	$s_c = \frac{1}{2}$	$0 < s_c < 1$	$1 < s_c \le \frac{3}{2}$	$s_c > \frac{3}{2}$
$n = 3$	Kenig–Merle [133]	Murphy [208]	Murphy [208] radial	
$n = 4$	Murphy [207]	Murphy [206]	MMZ [186], DMMZ [69]	Lu–Zheng [178] radial
$n \ge 5$	Murphy [207]	Murphy [206]	Killip–Visan [146]	Killip–Visan [146]
$n = 2$	Yu [294]	Open		
$n = 1$		Open		

Next, we consider the energy-subcritical focusing NLS equations:

$$\begin{cases}(i\partial_t + \Delta)u = -|u|^{p-1}u, & (t,x) \in \mathbb{R} \times \mathbb{R}^n,\\ u(0,x) = u_0(x).\end{cases} \tag{4.468}$$

Let Q be the ground state of the elliptic equation

$$\Delta Q - Q + |Q|^{p-1}Q = 0. \tag{4.469}$$

Then the soliton solution $e^{it}Q$ is a global solution to (4.468) but does not scatter.

There are various ways to construct solutions to (4.469), the simplest one being to look for radial solutions via a shooting method [13].

Proposition 4.6.2 (Existence of solitary waves, [13]).

(i) *For $n = 1$, all solutions to* (4.469) *are translates of*

$$Q(x) = \left(\frac{p+1}{2\cosh^2(\frac{(p-1)x}{2})}\right)^{p-1}. \tag{4.470}$$

(ii) *For $n \geq 2$, there exist a sequence of radial solutions $\{Q_k\}_{k\geq 0}$ with increasing L^2 norm such that Q_k vanishes n times on $\mathbb{R}^n$.*

The exact structure of the set of solutions to (4.469) is not known in dimension $n \geq 2$. An important rigidity property, however, which combines nonlinear elliptic techniques and ODE techniques, is the uniqueness of the nonnegative solution to (4.469).

Theorem 4.6.6 (Uniqueness of the ground state). *All solutions to*

$$\Delta Q - Q + |Q|^{p-1}Q = 0, \quad Q \in H^1(\mathbb{R}^n),\ Q(x) > 0 \tag{4.471}$$

are a translate of an exponentially decreasing C^2 radial profile $Q(r)$ (see [96]*), which is the unique nonnegative radially symmetric solution to* (4.469) *(see* [156]*). Q is the so-called ground state solution.*

Let us now observe that we may let the full group of symmetries of (4.468) act on the solitary wave $u(t,x) = e^{it}Q$ to get a $2n+2$ parameters family of solitary waves: for $(\lambda_0, x_0, \gamma_0, \beta) \in \mathbb{R}_+ \times \mathbb{R}^n \times \mathbb{R} \times \mathbb{R}^n$,

$$u(t,x) = \lambda_0^{\frac{2}{p-1}} Q(\lambda_0(x+x_0) - \lambda_0^2\beta t)e^{i\lambda_0^2 t}e^{i\frac{\beta}{2}\cdot(\lambda_0(x+x_0)-\lambda_0^2\beta t)}.$$

Below threshold of ground state

By using compactness concentration, Duyckaerts, Homler and Roudenko obtained the following scattering/blowup dichotomy.

Theorem 4.6.7 (Scattering/blowup dichotomy, [75, 118]). *Let $p = n = 3$. Let $u_0 \in H^1_x(\mathbb{R}^3)$ satisfy $M(u_0)E(u_0) < M(Q)E(Q)$.*
(i) *If $\|u_0\|_{L^2_x}\|u_0\|_{\dot{H}^1_x} < \|Q\|_{L^2}\|Q\|_{\dot{H}^1}$, then the solution to* (4.468) *with initial data u_0 is global and scatters.*
(ii) *If $\|u_0\|_{L^2_x}\|u_0\|_{\dot{H}^1_x} > \|Q\|_{L^2}\|Q\|_{\dot{H}^1}$ and u_0 is radial or $xu_0 \in L^2_x(\mathbb{R}^3)$, then the solution to* (4.468) *with initial data u_0 blows up in finite time in both time directions.*

Furthermore, if $\psi \in H^1_x(\mathbb{R}^3)$ satisfies $\frac{1}{2}\|\psi\|^2_{L^2_x}\|\psi\|^2_{\dot{H}^1_x} < M(Q)E(Q)$, then there exists a global solution to (4.468) *that scatters to ψ forward in time. The analogous statement holds backward in time.*

We remark that Dodson and Murphy [70, 71] gave another simple proof of scattering below the ground state that avoids the use of concentration compactness by using the radial Sobolev embedding and a virial/Morawetz estimate. Theorem 4.6.7 holds for general $s_0 \in (0,1)$ under the assumption $M(u_0)^{s_c}E(u_0)^{1-s_c} < M(Q)^{s_c}E(Q)^{1-s_c}$.

The condition $\|u_0\|_{L^2_x}\|u_0\|_{\dot{H}^1_x} < \|Q\|_{L^2}\|Q\|_{\dot{H}^1}$ is equivalent to $\|u_0\|_{L^2_x}\|u_0\|^2_{L^4} < \|Q\|_{L^2}\|Q\|^2_{L^4}$. Indeed, we have the following.

Lemma 4.6.2 ([83]). *Let f be in H^1, $0 < s_c < 1$. Then*

$$\Big(\int_{\mathbb{R}^n} |\nabla f|^2 dx\Big)^{s_c} M[f]^{1-s_c} < \Big(\int_{\mathbb{R}^n} |\nabla Q|^2 dx\Big)^{s_c} M[Q]^{1-s_c}$$
$$\Longrightarrow \Big(\int_{\mathbb{R}^n} |f|^{p+1} dx\Big)^{s_c} M[f]^{1-s_c} < \Big(\int_{\mathbb{R}^n} |Q|^{p+1} dx\Big)^{s_c} M[Q]^{1-s_c}. \tag{4.472}$$

Assume furthermore that

$$M[f]^{\frac{1-s_c}{s_c}} E[f] \le M[Q]^{\frac{1-s_c}{s_c}} E[Q]. \tag{4.473}$$

Then the reverse implication to (4.472) *holds, and we obtain*

$$\Big(\int_{\mathbb{R}^n} |\nabla f|^2 dx\Big)^{s_c} M[f]^{1-s_c} < \Big(\int_{\mathbb{R}^n} |\nabla Q|^2 dx\Big)^{s_c} M[Q]^{1-s_c}$$
$$\Longleftrightarrow \Big(\int_{\mathbb{R}^n} |f|^{p+1} dx\Big)^{s_c} M[f]^{1-s_c} < \Big(\int_{\mathbb{R}^n} |Q|^{p+1} dx\Big)^{s_c} M[Q]^{1-s_c}. \tag{4.474}$$

Moreover, (4.474) *also holds with nonstrict inequalities (in the case of equality, f is equal to Q up to space translation, scaling and phase).*

Threshold of ground state

For $p = n = 3$, Duyckaerts and Rodenko give the classification of the solution at the mass-energy threshold: $M(u_0)E(u_0) = M(Q)E(Q)$ as follows.

Theorem 4.6.8 (Existence of special solutions (besides $e^{it}Q$) at the critical mass-energy threshold, [82]). *Let $p = n = 3$. There exist two radial solutions Q^+ and Q^- of (4.468) with initial conditions $Q_0^\pm$ such that $Q_0^\pm \in \bigcap_{s\in\mathbb{R}} H^s(\mathbb{R}^3)$ and*

(i) $M[Q^+] = M[Q^-] = M[Q]$, $E[Q^+] = E[Q^-] = E[Q]$, $[0, +\infty)$ *is in the (time) domain of definition of $Q^\pm$ and there exists $e_0 > 0$ such that*

$$\forall t \geq 0, \quad \|Q^\pm(t) - e^{it}Q\|_{H^1} \leq Ce^{-e_0 t},$$

(ii) $\|\nabla Q_0^-\|_2 < \|\nabla Q\|_2$, *$Q^-$ is globally defined and scatters for negative time,*
(iii) $\|\nabla Q_0^+\|_2 > \|\nabla Q\|_2$, *and the negative time of existence of Q^+ is finite.*

Next, we characterize all solutions at the critical mass-energy level as follows.

Theorem 4.6.9 (Classification of solution, [82]). *Let u be a solution of (4.468) satisfying $M(u_0)E(u_0) = M(Q)E(Q)$.*

(i) *If $\|\nabla u_0\|_2\|u_0\|_2 < \|\nabla Q\|_2\|Q\|_2$, then either u scatters or $u = Q^-$ up to the symmetries.*
(ii) *If $\|\nabla u_0\|_2\|u_0\|_2 = \|\nabla Q\|_2\|Q\|_2$, then $u = e^{it}Q$ up to the symmetries.*
(iii) *If $\|\nabla u_0\|_2\|u_0\|_2 > \|\nabla Q\|_2\|Q\|_2$ and u_0 is radial or of finite variance, then either the interval of existence of u is of finite length or $u = Q^+$ up to the symmetries.*

Beyond threshold of ground state

In [211], Nakanishi and Schlag described the global dynamics of H^1 solutions slightly above the mass-energy threshold, $\|u_0\|_{L^2_x}\|u_0\|_{\dot{H}^1_x} < (1+\epsilon)\|Q\|_{L^2}\|Q\|_{\dot{H}^1}$. Note that, in [83], Duyckaerts and Roudenko can describe solutions, which are not necessarily ϵ-close to the threshold. Define the variance as

$$V(t) = \int_{\mathbb{R}^n} |x|^2 |u(x,t)|^2 \, dx. \tag{4.475}$$

Assuming $V(0) < \infty$ (referred to as finite variance), and letting u be a solution of (4.468), then the following virial identities hold:

$$V_t(t) = 4\,\mathrm{Im} \int_{\mathbb{R}^n} x \cdot \nabla u(x,t)\, \overline{u}(x,t)\, dx, \tag{4.476}$$

$$V_{tt}(t) = 8 \int_{\mathbb{R}^n} |\nabla u(t)|^2 dx - \frac{4n(p-1)}{p+1} \int_{\mathbb{R}^n} |u(t)|^{p+1} dx \tag{4.477}$$

$$\equiv 4n(p-1)\,E[u] - 4(p-1)s_c \,\|\nabla u(t)\|^2_{L^2(\mathbb{R}^n)}. \tag{4.478}$$

Theorem 4.6.10. *Let u be a solution of (4.468). Assume $V(0) < \infty$, $u_0 \in H^1(\mathbb{R}^n)$ and*

$$\mathcal{ME}[u]\left(1 - \frac{(V_t(0))^2}{32\,E[u]\,V(0)}\right) \leq 1, \quad \mathcal{ME}[u] := \frac{M[u]^{\frac{1-s_c}{s_c}} E[u]}{M[Q]^{\frac{1-s_c}{s_c}} E[Q]}. \tag{4.479}$$

(i) (*Blowup*) *If*

$$M[u_0]^{1-s_c}\Big(\int_{\mathbb{R}^n}|u_0|^{p+1}\,dx\Big)^{s_c} > M[Q]^{1-s_c}\Big(\int_{\mathbb{R}^n}|Q|^{p+1}\,dx\Big)^{s_c} \tag{4.480}$$

and

$$V_t(0) \le 0, \tag{4.481}$$

then $u(t)$ *blows up in finite positive time,* $T_+(u) < \infty$.

(ii) (*Boundedness and scattering*) *If*

$$M[u_0]^{1-s_c}\Big(\int_{\mathbb{R}^n}|u_0|^{p+1}\,dx\Big)^{s_c} < M[Q]^{1-s_c}\Big(\int_{\mathbb{R}^n}|Q|^{p+1}\,dx\Big)^{s_c} \tag{4.482}$$

and

$$V_t(0) \ge 0, \tag{4.483}$$

then $T_+ = +\infty$, u *scatters forward in time in* H^1 *and*

$$\limsup_{t\to+\infty} M[u_0]^{1-s_c}\Big(\int_{\mathbb{R}^n}|u(t)|^{p+1}\,dx\Big)^{s_c} < M[Q]^{1-s_c}\Big(\int_{\mathbb{R}^n}|Q|^{p+1}\,dx\Big)^{s_c}. \tag{4.484}$$

For the study of the energy/mass critical Hartree equation

$$i\partial_t u - \Delta u = \pm(|x|^{-\gamma} * |u|^2)u, \quad (t,x) \in \mathbb{R}\times\mathbb{R}^n$$

with $\gamma \in \{4,2\}$, we refer to [165, 166, 191–195].

For the dynamics of the nonlinear Schrödinger equation with inverse square potential,

$$\begin{cases} u_{tt} - \Delta u + \frac{a}{|x|^2}u = \mu|u|^{p-1}u, & (t,x)\in\mathbb{R}\times\mathbb{R}^n,\\ (u,u_t)(0,x) = (u_0,u_1)(x), \end{cases} \tag{4.485}$$

we refer to [141, 143, 179, 188, 199, 202, 291, 295], and Chapter 6 for the energy critical case.

5 Wave equations

5.1 Restriction estimates and classical Strichartz estimates

Restriction theory in Fourier analysis is one of the fundamental problems in modern harmonic analysis. In this section, we will consider the Cauchy problem of the wave equation

$$\square u = F(t,x), \quad (u, \partial_t u)|_{t=0} = (f(x), g(x)), \tag{5.1}$$

where $\square = \partial_t^2 - \Delta$. The space-time integrability of the solution to this problem is closely related to the (L^p, L^2) restriction estimate of the Fourier transform. It essentially implies the Strichartz estimate of the linear wave equation. Similarly, we can derive the Strichartz estimate of the Schrödinger equation from the restriction estimate of the Fourier transform on paraboloid as discussed in the previous chapter.

Fourier restriction problem

Assume that $\Sigma \subset \mathbb{R}^n$ is a hypersurface. Let $Rf = \hat{f}|_\Sigma$ denote the Fourier transform in $\mathbb{R}^n$ $f \mapsto \hat{f} = \mathcal{F}f$ restricted on Σ. The question is whether there exists $p_0 \in [1,2)$, such that when $1 \le p \le p_0$, the restriction operator R is bounded from $L^p(\mathbb{R}^n)$ to $L^2(\Sigma)$, that is,

$$\|Rf\|_{L^2(\Sigma)} \le C\|f\|_{L^p(\mathbb{R}^n)}. \tag{5.2}$$

Indeed, when $p = 1$, according to Riemann–Lebesgue's lemma, we have $\hat{f} \in C_b(\mathbb{R}^n)$. Thus, R is a bounded operator from $L^1(\mathbb{R}^n)$ to $L^2(\Sigma)$. When $p = 2$, for any L^2 function $f(x)$, we have $\hat{f} \in L^2(\mathbb{R}^n)$. Since $m(\Sigma) = 0$ on $\mathbb{R}^n$, we know $\hat{f}|_\Sigma$ is meaningless. When $p > 2, \hat{f} \in \mathcal{S}'(\mathbb{R}^n)$ is out of the integrable function class. Therefore, the reasonable range of Fourier restriction problem is $1 \le p < 2$.

On the other hand, the boundedness of restriction operator R relies on the curvature of the hypersurface Σ as well. When Σ is a plane, the restriction theorem does not hold for any $p > 1$. For the unit sphere $\Sigma = \mathbb{S}^{n-1}$ in $\mathbb{R}^n$, Stein–Tomas [243] has proved that when

$$1 \le p \le p_0 = \frac{2(n+1)}{n+3}$$

restriction operator R is a bounded operator from $L^p(\mathbb{R}^n)$ to $L^2(\mathbb{S}^{n-1})$. Furthermore, Knapp illustrated that when $p > p_0$, Stein–Tomas restriction estimate does not hold.

Strichartz [252] realized that through duality principle—the famous TT^*—and the restriction estimate of Fourier transform on a conical surface, one can get the space-time integrability of the linear wave equation. This kind of space-time integrability is called Strichartz estimate. Although it has been generalized to many different types, we

https://doi.org/10.1515/9783111384726-005

still call them Strichartz estimates. For example, we will discuss the end-point Strichartz estimate, which was proved by Keel–Tao in the next section.

Before our discussion, we first present some classical results.

Proposition 5.1.1. *Suppose $n > 1$, $\mathbb{S}^{n-1}$ is the unit sphere in $\mathbb{R}^n$, $d\sigma$ denotes the measure on the unit sphere induced by Lebesgue measure in $\mathbb{R}^n$, then the Fourier transform of $d\sigma$ defined by*

$$\widehat{d\sigma}(x) = (2\pi)^{-\frac{n}{2}} \int\limits_{\mathbb{S}^{n-1}} e^{-ix\cdot\xi} d\sigma(\xi) \tag{5.3}$$

satisfies the estimate

$$\left|\widehat{d\sigma}(x)\right| \lesssim |x|^{-\frac{n-1}{2}}. \tag{5.4}$$

Proof. By direct computation, we have

$$\left|\widehat{d\sigma}(x)\right| \tilde{=} |x|^{-\frac{n-2}{2}} J_{\frac{n-2}{2}}(|x|),$$

where the Bessel function is defined by

$$J_m(r) = \frac{(\frac{r}{2})^m}{\Gamma(m+\frac{1}{2})\Gamma(\frac{1}{2})} \int\limits_{-1}^{1} e^{irs}(1-s^2)^m \frac{ds}{\sqrt{1-s^2}} \sim O(r^{-\frac{1}{2}}), \quad r \longrightarrow +\infty.$$

Thus, the estimate (5.4) is true. Furthermore, we have

$$\left|\widehat{d\sigma}(x)\right| \leq (1+|x|)^{-\frac{n-1}{2}}. \tag{5.5}$$

□

Theorem 5.1.1. *Assume Σ is a smooth hypersurface in $\mathbb{R}^n$, and at each point in Σ, there are at least k nonzero principal curvatures. Let $d\sigma$ denote the measure on hypersurface Σ induced by Lebesgue measure in $\mathbb{R}^n$, $\psi \in C_c^\infty(\mathbb{R}^n)$. Then $d\mu = \psi d\sigma$ satisfies*

$$\left|\widehat{d\mu}(x)\right| = O(|x|^{-\frac{k}{2}}), \quad |x| \longrightarrow \infty. \tag{5.6}$$

Naturally, we have the estimate

$$\left|\widehat{d\mu}(x)\right| \leq C(1+|x|)^{-\frac{k}{2}}, \tag{5.7}$$

where

$$\widehat{d\mu}(x) \sim \int\limits_{\Sigma} e^{-ix\xi}\psi(\xi)d\sigma(\xi). \tag{5.8}$$

Particularly, if at each point in the smooth hypersurface Σ, the Gaussian curvature is nonzero, then

$$|\widehat{d\mu}(x)| \le C(1+|x|)^{-\frac{n-1}{2}}. \tag{5.9}$$

See the proof in [243]. Next, we introduce an abstract TT^* method, which is a bridge connecting the Fourier restriction estimate and the Strichartz estimate.

Theorem 5.1.2. *Assume H is a Hilbert space (let $H^* = H$ for simplicity). B is a Banach space, B^* denotes the dual space of B. Consider the linear operator*

$$T : H \longrightarrow B,$$

then $T^* : B^* \longrightarrow H$, the conjugate operator of T, can be defined by

$$(T^*\varphi, h) = \varphi(Th), \quad \forall\, h \in H,\ \varphi \in B^* \tag{5.10}$$

and we have the following equivalent characterizations:
(a) *$T : H \longrightarrow B$ is a bounded linear operator.*
(b) *$T^* : B^* \longrightarrow H$ is a bounded linear operator.*
(c) *$TT^* : B^* \longrightarrow B$ is a bounded linear operator.*
(d) *The bilinear form*

$$(\varphi, \psi) \longmapsto \langle T^*\varphi, T^*\psi\rangle$$

is bounded on $B^* \times B^*$.
Furthermore,

$$\|T\|^2 = \|T^*\|^2 = \|TT^*\|. \tag{5.11}$$

Proof. (a) $\Longleftrightarrow$ (b):

$$\begin{aligned}\|T^*\| &= \sup_{\|\varphi\|_{B^*}=1} \|T^*\varphi\|_H = \sup_{\|\varphi\|_{B^*}=1, \|h\|_H=1} |\langle T^*\varphi, h\rangle| \\ &= \sup_{\|\varphi\|_{B^*}=1, \|h\|_B=1} |\varphi(Th)| = \sup_{\|h\|_H=1} \|Th\|_B = \|T\|.\end{aligned} \tag{5.12}$$

(a), (b) $\Longrightarrow$ (c): We can conclude it from $\|TT^*\| \le \|T\|\|T^*\|$.
(c) $\Longrightarrow$ (d): Noting that

$$\begin{aligned}|\langle T^*\varphi, T^*\psi\rangle| &= |\varphi(TT^*\psi)| \le \|\varphi\|_{B^*}\|TT^*\psi\|_B \\ &\le \|\varphi\|_{B^*}\|TT^*\|\cdot\|\psi\|_{B^*} \\ &= \|TT^*\|\cdot\|\psi\|_{B^*}\|\varphi\|_{B^*},\end{aligned}$$

then we get (d).

(d) $\Longrightarrow$ (b):

$$\|T^*\|^2 = \sup_{\|\varphi\|_{B^*}=1} \|T^*\varphi\|_H^2 = \sup_{\|\varphi\|_{B^*}=1} \langle T^*\varphi, T^*\varphi\rangle < \infty.$$

Furthermore, we have the identity

$$\|T^*\|^2 = \sup_{\|\varphi\|_{B^*}=1} \varphi(TT^*\varphi) = \sup_{\|\varphi\|_{B^*}=1} \|TT^*\varphi\|_B = \|TT^*\|, \tag{5.13}$$

thus we derive (5.11) from (5.12) and (5.13). □

For the Strichartz estimate, the question is how to derive the inhomogeneous Strichartz estimate from the homogeneous Strichartz estimate, that is, from the boundedness of operator TT^*, we derive the boundedness of the corresponding retarded operator $(TT^*)_R$. In the classical proof, it is done by considering three special cases and using interpolation methods. The proof is complicated. Christ–Kiselev's lemma [45] shows that the estimate of the retarded operator can be derived from the linear estimate and its dual type. Therefore, we first prove Christ–Kiselev's lemma.

Lemma 5.1.1. *Assume X, Y are two Banach spaces. $B(X,Y)$ denotes the Banach space consisted of all bounded linear operators from X to Y. Assume also $-\infty \le a < b \le \infty$, $K(t,s)$ is a continuous function defined on $[a,b]^2$ with a range in $B(X,Y)$. Define*

$$Tf(t) = \int_a^b K(t,s)f(s)ds,$$

and assume

$$\|Tf\|_{L^q([a,b];Y)} \le C\|f\|_{L^p([a,b];X)}. \tag{5.14}$$

Then when $1 \le p < q \le \infty$, the retarded operator

$$Wf(t) = \int_a^t K(t,s)f(s)ds, \quad a \le t \le b$$

satisfies

$$\|Wf\|_{L^q([a,b];Y)} \le \frac{2^{-2(\frac{1}{p}-\frac{1}{q})}\cdot 2C}{1-2^{-(\frac{1}{p}-\frac{1}{q})}}\|f\|_{L^p([a,b];X)}. \tag{5.15}$$

Proof. We adopt Smith–Sogge's method [240]. We need only to prove the case where $q < \infty$, while the case $q = \infty$ can be proved similarly. Without loss of generality, we assume that $\|f\|_{L^p([a,b],X)} = 1$ and $f(s) \in C([a,b];X)$. Let

$$F(t) = \int_a^t \|f(s)\|_X^p ds.$$

Then $F: [a,b] \longrightarrow [0,1]$ is a bijection. Note that when $I \subset [0,1]$, we have

$$\|\chi_{F^{-1}(I)}(s)f(s)\|_{L^p([a,b];X)} = |I|^{\frac{1}{p}}. \tag{5.16}$$

Actually, let $F^{-1}(I) = [a_1, b_1]$. Then

$$\begin{aligned}\|\chi_{F^{-1}(I)}(s)f(s)\|_{L^p([a,b];X)}^p &= \int_a^b \chi_{[a_1,b_1]}(s)\|f(s)\|_X^p ds \\ &= \int_a^{b_1} \|f(s)\|_X^p ds - \int_a^{a_1} \|f(s)\|_X^p ds \\ &= |I|.\end{aligned}$$

Now, we consider the set of all dyadic subintervals Ξ in $[0,1]$. For $I, J \in \Xi$, we say $I \sim J$, if

(i) $|I| = |J|$, $\max_{x\in I} x \le \min_{x\in J} x$.

(ii) I, J are mutually disjoint but have intersecting generality intervals, that is, $\exists I_0, J_0 \in \Xi$, satisfying $I \subset I_0, J \subset J_0$, $|I_0| = 2|I|$, $|J_0| = 2|J|$ and $I_0 \cap J_0 \ne \emptyset$.

It is easy to verify, for a fixed $J \in \Xi$, there exist at most two intervals I satisfying $I \sim J$. Furthermore, for each $(x,y) \in [0,1]^2, x < y$, there exists a unique pair of $I, J \in \Xi$ satisfying $I \sim J, x \in I, y \in J$.

For $x = F(s), y = F(t)$, using the above observation, we derive that the following equality holds almost everywhere:

$$\begin{aligned}\chi_{\{(s,t)\in[a,b]^2:s<t\}}(s,t) = \chi_{\{(x,y)\in[0,1]^2:x<y\}}(x,y) &= \sum_{I,J:I\sim J} \chi_I(x)\chi_J(y) \\ &= \sum_{I,J:I\sim J} \chi_{F^{-1}(I)}(s)\chi_{F^{-1}(J)}(t).\end{aligned}$$

Thus, applying the above argument to the representation formula W, we have

$$Wf = \sum_{I,J:I\sim J} \chi_{F^{-1}(J)}(t)T(\chi_{F^{-1}(I)}f).$$

Then we get

$$\|Wf\|_{L^q([a,b],Y)} \le \sum_{j=2}^{\infty} \Big\| \sum_{I,J:I\sim J,|I|=2^{-j}} \chi_{F^{-1}(J)}(t)T(\chi_{F^{-1}(I)}f)\Big\|_{L^q([a,b],Y)}. \tag{5.17}$$

Since there are at most two intervals I satisfying $I \sim J$, and the intervals J satisfying $|J| = 2^{-j}$ are mutually disjoint, we have

$$\Big\| \sum_{I,J:I\sim J,|I|=2^{-j}} \chi_{F^{-1}(J)}(t) T(\chi_{F^{-1}(I)} f) \Big\|_{L^q([a,b],Y)} \leq 2\Big(\sum_{I:|I|=2^{-j}} \|T(\chi_{F^{-1}(I)} f)\|^q_{L^q([a,b],Y)} \Big)^{\frac{1}{q}}.$$

From (5.14) and (5.16), we can observe:

$$\text{The right side} \leq 2C\Big(\sum_{I:|I|=2^{-j}} \|\chi_{F^{-1}(I)} f\|^q_{L^p([a,b],X)} \Big)^{\frac{1}{q}} \leq 2C\Big(\sum_{I:|I|=2^{-j}} 2^{-\frac{jq}{p}} \Big)^{\frac{1}{q}} \leq 2^{-j(\frac{1}{p}-\frac{1}{q})} \cdot 2C.$$

Note that $p < q < \infty$, substitute it into (5.17) and take summation with respect to j. Then we complete the proof of (5.15). □

When $p = q$, Lemma 5.1.1 may not keep true. For example, let $K(s,t) = (t-s)^{-1}$, $1 < p = q < \infty$.

Definition 5.1.1. We call $(q,r) \in \tilde{\Lambda}$ a wave admissible pair, if it satisfies $2 \leq q, r \leq \infty$ and

$$\frac{2}{q} \leq \gamma(r) \triangleq (n-1)\Big(\frac{1}{2} - \frac{1}{r}\Big), \quad (q,r,n) \neq (2,\infty,3). \tag{5.18}$$

Particularly, when the equality in (5.18) holds, we say (q,r) is a sharp wave admissible pair, and denote it by $(q,r) \in \Lambda$.

Remark 5.1.1.

(1) In the $(\frac{1}{q}, \frac{1}{r})$ coordinate system, the corresponding range of wave admissible pairs is:

(a) When $n = 2$, $(\frac{1}{q}, \frac{1}{r})$ corresponds with a triangle $\triangle OPR$, where

$$O = (0,0), \quad P = \Big(\frac{1}{4}, 0\Big), \quad R = \Big(0, \frac{1}{2}\Big).$$

(b) When $n = 3$, $(\frac{1}{q}, \frac{1}{r})$ corresponds with a triangle $\triangle OPR$ (except vertex P), where

$$O = (0,0), \quad P = \Big(\frac{1}{2}, 0\Big), \quad R = \Big(0, \frac{1}{2}\Big).$$

(c) When $n \geq 4$, $(\frac{1}{q}, \frac{1}{r})$ corresponds with a trapezoid $OPQR$, where

$$O = (0,0), \quad P = \left(\frac{1}{2}, 0\right), \quad Q = \left(\frac{1}{2}, \frac{n-3}{2(n-1)}\right), \quad R = \left(0, \frac{1}{2}\right).$$

(d) $(q,r) \in \Lambda$ means that $(\frac{1}{q}, \frac{1}{r})$ corresponds with the hypotenuse of a triangle or a right-angled trapezoid.

(2) It is easy to observe that the range of r defined by $(q,r) \in \tilde{\Lambda}$ is

$$\begin{cases} 2 \leq r \leq \frac{2(n-1)}{n-3}, & n \geq 4, \\ 2 \leq r < \infty, & n = 3, \\ 2 \leq r \leq \infty, & n = 2. \end{cases} \tag{5.19}$$

Conventionally, in the range of r determined by $(q,r) \in \tilde{\Lambda}$, we introduce symbols:

$$\beta(r) = \frac{n+1}{2}\left(\frac{1}{2} - \frac{1}{r}\right), \quad \delta(r) = n\left(\frac{1}{2} - \frac{1}{r}\right), \quad \gamma(r) = (n-1)\left(\frac{1}{2} - \frac{1}{r}\right). \tag{5.20}$$

Theorem 5.1.3 (Classical Strichartz estimate). *Suppose $n \geq 2$. The solution of Cauchy problem of linear wave equation (5.1)*

$$\begin{aligned} u(x,t) &= \mathcal{F}^{-1} \cos|\xi|t\mathcal{F}f + \mathcal{F}^{-1}\frac{\sin|\xi|t}{|\xi|}\mathcal{F}g + \int_0^t \mathcal{F}^{-1}\frac{\sin|\xi|(t-\tau)}{|\xi|}\mathcal{F}Fd\tau \\ &\triangleq \cos(-\Delta)^{\frac{1}{2}}tf + \frac{\sin(-\Delta)^{\frac{1}{2}}t}{(-\Delta)^{\frac{1}{2}}}g + \int_0^t \frac{\sin(-\Delta)^{\frac{1}{2}}(t-\tau)}{(-\Delta)^{\frac{1}{2}}}F(\tau,x)d\tau \\ &\triangleq \dot{K}(t)f + K(t)g + \int_0^t K(t-\tau)F(x,\tau)d\tau, \quad K(t) = \frac{\sin(-\Delta)^{\frac{1}{2}}t}{(-\Delta)^{\frac{1}{2}}} \end{aligned} \tag{5.21}$$

satisfies the following classical Strichartz estimate:

$$\|u(t,x)\|_{L_t^q L_x^r} \lesssim \|f\|_{\dot{H}^s} + \|g\|_{\dot{H}^{s-1}} + \|F\|_{L_t^{\bar{q}'}(L_x^{\bar{r}'})} \tag{5.22}$$

if and only if: $(q,r) \in \tilde{\Lambda}$, $(\bar{q},\bar{r}) \in \tilde{\Lambda}$ and they satisfy scaling gap condition,

$$\frac{1}{q} + \frac{n}{r} = \frac{n}{2} - s = \frac{1}{\bar{q}'} + \frac{n}{\bar{r}'} - 2, \tag{5.23}$$

or $\delta(r) - \frac{1}{q} = s$, $\delta(r) + \delta(\bar{r}) - \frac{1}{q} - \frac{1}{\bar{q}} = 1$. Here,

$$L_t^q L_x^r = L_t^q(\mathbb{R}; L_x^r(\mathbb{R}^n)), \quad or \quad L_t^q L_x^r = L_t^q(I; L_x^r(\mathbb{R}^n)), \quad I \subset \mathbb{R},\ 0 \in \bar{I}.$$

Remark 5.1.2.

(1) We will present the Strichartz estimate in the framework of Besov space in Section 5.2. Meanwhile, we will also discuss carefully the proof of Strichartz estimate in Section 5.2. In the proof, we emphasize how Fourier restriction estimate implies classical Strichartz estimate. We do not take the case of end-point into consideration in the proofs.

(2) Classical Strichartz's space-time estimate does not contain the end-point Q $(n \geq 4)$ or $P = Q$ $(n = 3)$. When $n \geq 4$, Keel and Tao have proved $Q = (\frac{1}{2}, \frac{n-3}{2(n-1)})$ is admissible. When $n = 3$, we can illustrate that $P = Q$ is not admissible by the counterexample.

(3) $S = (\frac{2(n+1)}{n-1}, \frac{2(n+1)}{n-1})$ corresponds to Strichartz's space-time estimate of symmetry; $R = (0, \frac{1}{2})$ corresponds to energy estimate. Actually, when $n = 3$, the original type of the Strichartz estimate given by Strichartz [249] is

$$\begin{aligned}\|u(t,x)\|_{L^4(\mathbb{R}^{1+3}_+)} + \|u(t)\|_{\dot{H}^{\frac{1}{2}}(\mathbb{R}^3)} + \|\partial_t u\|_{H^{-\frac{1}{2}}(\mathbb{R}^3)} \\ \leq C\|f\|_{\dot{H}^{\frac{1}{2}}(\mathbb{R}^3)} + C\|g\|_{\dot{H}^{-\frac{1}{2}}(\mathbb{R}^3)} + C\|F\|_{L^{\frac{4}{3}}(\mathbb{R}^{1+3}_+)}.\end{aligned} \tag{5.24}$$

(4) Readers will find that we only need to prove the case that $(q,r) \in \Lambda$, $(\bar{q},\bar{r}) \in \Lambda$, that is, $(\frac{1}{q}, \frac{1}{r}) \in \overline{QR}$. Other cases can be proved by using the Sobolev embedding.

(5) When $n = 2$, the space-time estimate of end-point holds merely in the sense of scaling. That means at point P we have

$$\|K(t)f + \dot{K}(t)g\|_{L^4_t \dot{C}^\alpha_x(\mathbb{R}^2)} \leqslant C\|f\|_{\dot{H}^{\alpha+\frac{1}{4}}} + C\|g\|_{\dot{H}^{\alpha-\frac{1}{4}}}, \quad 0 < \alpha < 1, \tag{5.25}$$

where

$$\|h\|_{\dot{C}^\alpha_x} = \sup_{x \neq y} \frac{|h(x+y) - h(x)|}{|y|^\alpha}. \tag{5.26}$$

Outline and proof of Theorem 5.1.3. The solution $u(t,x)$ of linear wave equation (5.1) can be divided into

$$u = v_+ + v_- + \frac{\omega_+ - \omega_-}{2}, \tag{5.27}$$

where

$$v_\pm(t,x) \cong \int e^{ix\xi} e^{\pm it|\xi|} \hat{f}_\pm(\xi) d\xi, \quad \hat{f}_\pm(\xi) = \frac{1}{2}\Big(\hat{f}(\xi) \pm \frac{\hat{g}(\xi)}{i|\xi|}\Big), \tag{5.28}$$

$$\omega_\pm(t,x) \cong \int_0^t \int_{\mathbb{R}^n} e^{ix\xi} e^{\pm i(t-\tau)|\xi|} \frac{\hat{F}(\xi,\tau)}{i|\xi|} d\xi d\tau. \tag{5.29}$$

Obviously, (5.28) means the restriction on conical surface $\tau = \pm|\xi|$ of the space-time Fourier transform of

$$v_{\pm}(t,x) \cong \int_{\mathbb{R}^{n+1}} e^{ix\cdot\xi}e^{it\cdot\tau}\delta(\tau \mp |\xi|)\hat{f}_{\pm}(\xi)d\tau d\xi,$$

that is,

$$\tilde{v}_{\pm}(\tau,\xi) \cong \delta(\tau \mp |\xi|)\hat{f}(\xi),$$

where "+" denotes the restriction on $\tau = |\xi|$, and "−" denotes that on $\tau = -|\xi|$. It is easy to verify that $v_{\pm}$ is just the conjugate operator of Fourier transform restricted on the conical surface. Since the restriction estimate on $\tau = |\xi|$ and $\tau = -|\xi|$ are all the same, and with the help of the density of Schwartz space $\mathcal{S}$ in the integrable space, the estimate in Theorem 5.1.3 can be simplified as

$$\|U(t)f\|_{L^q(L^r)} \triangleq \|e^{it(-\Delta)^{\frac{1}{2}}}f\|_{L^q(L^r)} \leq C\|f\|_{\dot{H}^s}, \quad f(x) \in \mathcal{S}(\mathbb{R}^n), \tag{5.30}$$

and

$$\left\|\int_0^t e^{i(t-\tau)(-\Delta)^{\frac{1}{2}}}F(\tau,x)d\tau\right\|_{L^q(L^r)} \leq C\|F\|_{L^{\bar{q}'}(L^{\bar{r}'})}, \quad F(t,x) \in \mathcal{S}(\mathbb{R}^{1+n}), \tag{5.31}$$

where $(q,r) \in \tilde{\Lambda}$, $(\bar{q},\bar{r}) \in \tilde{\Lambda}$ satisfy the scaling gap condition (5.23).

Step 1. Our idea is to prove (5.30) for a frequency localized f, and then prove (5.30) using the Littlewood–Paley decomposition. We now continue to use symbols about the Littlewood–Paley decomposition in Chapter 1. Let $\beta(\xi) \in C_c^{\infty}(\mathbb{R}^n)$, and consider the truncation wave operator:

$$Tf(t,x) = \int_{\mathbb{R}^n} e^{i(t|\xi|+x\xi)}\beta(\xi)\hat{f}(\xi)d\xi, \quad f(x) \in \mathcal{S}(\mathbb{R}^n). \tag{5.32}$$

We can choose $\tilde{\beta}(\xi) \in C_c^{\infty}(\mathbb{R}^n)$ identical to 1 in a neighborhood of $\operatorname{supp}\beta(\xi)$. Therefore, the proof of estimate (5.30) reduces to

$$\|Tf\|_{L^q(L^r)} \leq C\|f\|_2. \tag{5.33}$$

Actually, using the Littlewood–Paley decomposition, we have

$$f(x) = \sum_{j\in\mathbb{Z}} \Delta_j f = \sum_{j\in\mathbb{Z}} \tilde{\Delta}_j\Delta_j f, \quad \tilde{\Delta}_j = \Delta_{j-1} + \Delta_j + \Delta_{j+1}.$$

Note that (5.33) means

$$\|U(t)\Delta_0 f\|_{L^q(L^r)} = \|U(t)\tilde{\Delta}_0(\Delta_0 f)\|_{L^q(L^r)} \leq C\|\Delta_0 f\|_{L^2}. \tag{5.34}$$

Using scaling, we find

$$U(t)\Delta_j f = \int_{\mathbb{R}^n} e^{ix\xi} e^{it|\xi|} \tilde{\psi}_0(2^{-j}\xi)(\widehat{\Delta_j f})(\xi)d\xi$$
$$= 2^{jn} \int_{\mathbb{R}^n} e^{i2^j x\xi} e^{i2^j t|\xi|} \tilde{\psi}_0(\xi)(\widehat{\Delta_j f})(2^j\xi)d\xi$$
$$= [U(2^j t)\tilde{\Delta}_0(\Delta_j f)(2^{-j}\cdot)](2^j x). \tag{5.35}$$

Thus,

$$\|U(t)\Delta_j f\|_{L^q(L^r)} \leq C2^{-(\frac{1}{q}+\frac{n}{r})j}\|\Delta_j f(2^{-j}\cdot)\|_{L^2} \leq C2^{sj}\|\Delta_j f\|_2, \tag{5.36}$$

where $s = \delta(r) - \frac{1}{q}$. Noting the Sobolev embedding relationship

$$\dot{B}^s_{p,\min\{p,q\}} \hookrightarrow \dot{F}^s_{p,q} \hookrightarrow \dot{B}^s_{p,\max\{p,q\}}.$$

Using Minkowski's inequality, we get

$$\|U(t)f\|_{L^q(L^r)} \leq \left\|\Big(\sum_{j\in\mathbb{Z}}\|U(t)\Delta_j f\|^2_{L^r}\Big)^{\frac{1}{2}}\right\|_{L^q} \leq \Big(\sum_{j\in\mathbb{Z}}\|U(t)\Delta_j f\|^2_{L^qL^r}\Big)^{\frac{1}{2}}$$
$$\leq C\Big(\sum_{j\in\mathbb{Z}} 2^{2sj}\|\Delta_j f\|_2^2\Big)^{\frac{1}{2}} = C\|f\|_{\dot{H}^s}.$$

Step 2. The dual form T^* of T and TT^* principle. Rewrite the truncation wave operator Tf as

$$Tf = \int_{\mathbb{R}^{n+1}} e^{ix\cdot\xi} e^{it\cdot\tau} \delta(\tau - |\xi|)\beta(\xi)\hat{f}(\xi)d\tau d\xi. \tag{5.37}$$

According to duality Theorem 5.1.2, the space-time estimate (5.33) corresponding to truncation wave operator Tf can be converted into the space-time integrability of TT^* or T^*, where T^* is the conjugate of T. In detail, that is, the following.

Corollary 5.1.1. *The following propositions are equivalent:*
(1) $T : L^2(\mathbb{R}^n) \longrightarrow L^q_t(L^r_x)$ *is a bounded operator.*
(2) $T^* : L^{q'}_t(L^{r'}_x) \longrightarrow L^2(\mathbb{R}^n)$ *is a bounded operator.*
(3) $TT^* : L^{q'}_t(L^{r'}_x) \longrightarrow L^q_t(L^r_x)$ *is a bounded operator.*

Next, we calculate the expression of T^* and TT^*, and make some necessary and meaningful analysis. The conjugate operator of T should be $F(t,x) \mapsto (T^*F)(x)$, which should be determined by the following inner product:

$$\langle Tf, F\rangle = \langle f, T^*F\rangle, \quad f \in \mathcal{S}(\mathbb{R}^n),\ F \in \mathcal{S}(\mathbb{R}^{n+1}). \tag{5.38}$$

Here, $\langle\cdot,\cdot\rangle$ denotes L^2 inner product. In other words, that is,

$$\iint Tf\cdot\overline{F}dtdx=\int f\cdot\overline{T^*F}dx.$$

By definition, we have

$$\begin{aligned}\iint Tf\cdot\overline{F}dtdx&=\iint\widehat{Tf}\cdot\overline{\widehat{F}(t,\xi)}dtd\xi=\iint e^{it|\xi|}\beta(|\xi|)\hat{f}(\xi)\overline{\widehat{F}(t,\xi)}dtd\xi\\&=\int f(x)\Big(\int e^{it|\xi|-ix\xi}\beta(|\xi|)\overline{\widehat{F}(t,\xi)}dtd\xi\Big)dx.\end{aligned}$$

Therefore,

$$(T^*F)(x)=\int e^{ix\xi-it|\xi|}\overline{\beta(|\xi|)\widehat{F}(t,\xi)}d\xi dt=\int e^{ix\xi}\overline{\beta(|\xi|)\widetilde{F}(|\xi|,\xi)}d\xi,\tag{5.39}$$

where $\widetilde{F}$ denotes space-time Fourier transform.

Remark 5.1.3.

(i) The representation of T^* shows the connection between space-time estimate and the restriction problem on forward light cone in $\mathbb{R}^{n+1}$, that is,

$$\wedge=\{(\tau,\xi):\tau=|\xi|>0\}.$$

Indeed, from (5.39), we can see

$$\widehat{T^*F}(\xi)\cong\overline{\beta(\xi)\widetilde{F}(|\xi|,\xi)}=\overline{\beta(\xi)\widetilde{F}(\tau,\xi)\delta(\tau-|\xi|)}\triangleq RF(\xi)\tag{5.40}$$

is the product of the space-time Fourier transform of $F(t,x)$ restricted on forward light cone $\wedge$ and a smooth cut-off function. Note that $\wedge$ is the image of mapping: $\xi\mapsto(|\xi|,\xi)$. Under this parameterized representation, measure $d\sigma(\xi)$ and $d\xi$ differ only in a constant. Thus, by Plancherel's theorem, we have

$$\|T^*F\|_{L^2}\cong\|RF(\xi)\|_{L^2(\wedge,d\sigma)}.\tag{5.41}$$

Therefore, by the TT^* method, we know that the space-time estimate of the truncation half-wave operator (5.33) is equivalent to the following restriction result.

Theorem 5.1.4. *For any $(q,r)\in\tilde{\Lambda}$, $R:L^{q'}(L^{r'})\longrightarrow L^2(\wedge;d\sigma)$ is bounded, that is,*

$$\|RF(\xi)\|_{L^2(\wedge,d\sigma)}\cong\|T^*F\|_{L^2}\le C\|F(t,x)\|_{L^{q'}(L^{r'})}.$$

Now, we consider the expression of TT^*F. Substituting (5.39) into (5.32), we get

$$TT^*F = \int_{\mathbb{R}^n} e^{it|\xi|+ix\cdot\xi}|\beta|^2 \int_{\mathbb{R}^{n+1}} e^{-is\tau-iy\cdot\xi}F(s,y)dyds\delta(\tau-|\xi|)d\xi d\tau$$
$$= \int_{\mathbb{R}^n}\int_{\mathbb{R}^n}\int_{\mathbb{R}} e^{i(t-s)|\xi|+i(x-y)\cdot\xi}|\beta(\xi)|^2 F(s,y)dyd\xi ds \tag{5.42}$$

or

$$\widehat{TT^*F}(t,\xi) \cong e^{it|\xi|}\beta(\xi)\widehat{T^*F}(\xi) \cong \int e^{i(t-s)|\xi|}|\beta(\xi)|^2\widehat{F}(s,\xi)ds. \tag{5.43}$$

Then we get

$$TT^*F(t,x) = K*F \triangleq \int_{\mathbb{R}} W(t-s)F(s,x)ds, \tag{5.44}$$

where

$$W(t)f(x) = \int_{\mathbb{R}^n} e^{i(t|\xi|+x\cdot\xi)}|\beta(\xi)|^2\hat{f}(\xi)d\xi \triangleq K_t * f(x). \tag{5.45}$$

Essentially, W and the truncation half-wave operator T are analogous. The difference lies in the substitution of $\beta(\xi)$ with $|\beta(\xi)|^2$.

Step 3. According to TT^* principle, the space-time estimate (5.33) of the truncation half-wave operator T is reduced to

$$\|TT^*F\|_{L^q(L^r)} = \|K*F\|_{L^q(L^r)} \le C\|F\|_{L^{\bar{q}'}(L^{\bar{r}'})}, \quad (q,r),(\bar{q},\bar{r}) \in \tilde{\Lambda}. \tag{5.46}$$

Assertion. The space-time estimate (5.46) of the truncation type operator TT^* means that of inhomogeneous estimate (5.31).

Actually, according to Christ–Kiselev's theorem, choose $\beta(\xi) = \tilde{\psi}_0(\xi) > 0$, then (5.46) implies the estimate of the following truncation retarded operator:

$$\left\|\int_0^t U(t-s)\tilde{\Delta}_0^2 F(s,x)ds\right\|_{L^q(L^r)} \le C\|F\|_{L^{\bar{q}'}(L^{\bar{r}'})}. \tag{5.47}$$

Using scaling, we get

$$\int_0^t U(t-s)\Delta_j F ds = \int_0^t\int_{\mathbb{R}^n} e^{ix\cdot\xi}e^{i(t-s)|\xi|}|\tilde{\psi}_0(2^{-j}\xi)|^2(\widehat{\Delta_j F})(\xi,s)d\xi ds$$
$$= 2^{jn}\int_0^t\int_{\mathbb{R}^n} e^{i2^jx\cdot\xi}e^{i2^j(t-s)|\xi|}|\tilde{\psi}_0(\xi)|^2(\widehat{\Delta_j F})(2^j\xi,s)d\xi ds$$
$$= 2^{jn-j}\int_0^{2^jt}\int_{\mathbb{R}^n} e^{i2^jx\cdot\xi}e^{i(2^jt-s)|\xi|}|\tilde{\psi}_0(\xi)|^2(\widehat{\Delta_j F})(2^j\xi,2^{-j}s)d\xi ds.$$

Therefore,

$$\begin{aligned}
&\left\| \int_0^t U(t-s)\Delta_j F(x,s)ds \right\|_{L^q(L^r)} \\
&\leq C2^{-(\frac{1}{q}+\frac{n}{r})j} 2^{-j} \left\| \int_0^t U(t-s)\widetilde{\Delta}_0^2(\Delta_j F(2^{-j}x, 2^{-j}s))ds \right\|_{L^q(L^r)} \\
&\leq C2^{-(\frac{1}{q}+\frac{n}{r})j} 2^{-j} \|\Delta_j F(2^{-j}\cdot, 2^{-j}\cdot)\|_{L^{\overline{q}'}(L^{\overline{r}'})} \\
&\leq C2^{(\delta(r)+\delta(\overline{r})-\frac{1}{q}-\frac{1}{\overline{q}})j} \|\Delta_j F\|_{L^{\overline{q}'}(L^{\overline{r}'})}.
\end{aligned} \tag{5.48}$$

When $(q,r), (\overline{q},\overline{r}) \in \widetilde{\Lambda}$ satisfy the scaling gap condition (5.23), noting the Sobolev embedding relationship

$$\dot{B}^0_{p,\min\{p,2\}} \hookrightarrow \dot{F}^0_{p,2} = L^p \hookrightarrow \dot{B}^0_{p,\max\{p,2\}},$$

we get

$$\begin{aligned}
&\left\| \int_0^t U(t-s)F(x,s)ds \right\|_{L^q(L^r)} \\
&= \left\| \left(\sum_{j\in\mathbb{Z}} \left\| \int_0^t U(t-s)\Delta_j F(x,s)ds \right\|_{L^r}^2 \right)^{\frac{1}{2}} \right\|_{L^q} \\
&\leq \left(\sum_{j\in\mathbb{Z}} \left\| \int_0^t U(t-s)\Delta_j F(x,s)ds \right\|_{L^q(L^r)}^2 \right)^{\frac{1}{2}} \\
&= C\left(\sum_{j\in\mathbb{Z}} 2^{2j} \|\Delta_j F\|^2_{L^{\overline{q}'}(L^{\overline{r}'})} \right)^{\frac{1}{2}} \leq C\|F\|_{L^{\overline{q}'}(\dot{H}^{1,\overline{r}'})}.
\end{aligned}$$

Thus, we conclude the proof of inhomogeneous Strichartz estimate.

Remark 5.1.4.

(i) If we do not use Christ–Kiselev's theorem, we can derive the estimate of retarded operator (5.47) from (5.46) by the classical interpolation methods. Actually, according to Sobolev embedding theorem, we only need prove the case with the sharp admissible pairs $(q_1,r_1), (q_2,r_2) \in \Lambda$. For this purpose, we only need prove the following three cases.

Case I. $(q_1,r_1) = (q_2,r_2) = (q,r)$. This case corresponds to the diagonal case. Similarly, with TT^*, applying Hardy–Littlewood–Sobolev's inequality directly, we get

$$\|(TT^*)_R F\|_{L^q(I;L^r)} \leq C\|F\|_{L^{q'}(I;L^{r'})}. \tag{5.49}$$

Case II. $(q_1, r_1) = (\infty, 2)$, $(q_2, r_2) = (q, r)$. Noting that

$$(TT^*)_R F(t) = T(T^*(\chi_{[0,t)}F))(t)$$

and

$$\|T^* G\|_{L^2(\mathbb{R}^n)} \lesssim \|G\|_{L^{q'}(I;L^{r'})},$$

we can easily get

$$\begin{aligned}\|(TT^*)_R F(t)\|_{L^2} &\lesssim \|T^* \chi_{[0,t]} F\|_{L^2} \lesssim \|\chi_{[0,t]} F\|_{L_t^{q'} L_x^{r'}} \\ &\lesssim \|F\|_{L_t^{q'}(L^{r'})}.\end{aligned} \tag{5.50}$$

Case III. $(q_1, r_1) = (q, r)$, $(q_2, r_2) = (\infty, 2)$. Note that the dual operator of $(TT^*)_R$ is of the same form as $(TT^*)_R$, but the restriction function $\chi_{[0,t]}$ should be replaced by operator of $\chi_{[t,\infty)}$, that is,

$$((TT^*)_R)^* G(t) = \int_t^\infty W(t-\tau) G(\tau) d\tau.$$

By duality, we get

$$\|(TT^*)_R F(t)\|_{L_t^q L_x^r} \leq C \|F\|_{L_t^1(L_x^2)}. \tag{5.51}$$

Then we acquire the proof of (5.47) by interpolation formula.

Step 4. Now, we will prove $(q, r) \in \tilde{\Lambda}$ is the necessary condition that truncation operator T satisfies the estimate (5.46) or (5.33). Consider an example similar as Knapp's one. For a fixed $\delta > 0$, let

$$D = \{\xi \in \mathbb{R}^n,\ |\xi_1 - 1| < 1/2,\ |\xi'| < \delta\}.$$

Considering $\hat{f} = \chi_D$, we have

$$Tf(t,x) = e^{i(t+x_1)} \int_D e^{i[t(|\xi|-\xi_1)+(t+x_1)(\xi_1-1)+x'\cdot\xi']} d\xi. \tag{5.52}$$

Noting that

$$|\xi| - \xi_1 = \frac{|\xi'|^2}{|\xi| + \xi_1} \lesssim \delta^2,$$

we can choose in physical space

$$R = \{(t,x) \mid |t| \lesssim \delta^{-2},\ |t + x_1| \lesssim 1,\ |x'| \lesssim \delta^{-1}\},$$

such that when $(t,x) \in R$, $\xi \in D$, the factor in integral (5.52) can be controlled by a constant. Therefore,

$$|Tf(t,x)| \gtrsim |D|, \quad (t,x) \in \mathbb{R}.$$

Furthermore,

$$\frac{\|Tf\|_{L_t^q L_x^r}}{\|f\|_{L^2}} \gtrsim \frac{|D| \cdot \|\chi_R\|_{L_t^q L_x^r}}{|D|^{1/2}} \sim \delta^{\frac{n-1}{2} - \frac{2}{q} - \frac{n-1}{r}}. \tag{5.53}$$

Thus, let $\delta \to 0$, from (5.53), we can see the necessary condition ensuring estimate (5.33) is

$$\frac{2}{q} \leq (n-1)\left(\frac{1}{2} - \frac{1}{r}\right).$$

On the other hand, noting that TT^* is a convolution-type operator, we know that TT^* is translation-invariant. According to Hörmander's translation-invariant operator theory, if TT^* is a translation-invariant operator from $L_t^{q'} L_x^{r'}$ to $L_t^q L_x^r$, then we demand at least $q \geq 2$, $r \geq 2$. Therefore, we know that $(q,r) \in \widetilde{\Lambda}$ is the necessary condition to make sure Strichartz estimate holds.

Step 5. Derive the dispersive estimate from the restriction estimate and the stationary-phase method. In order to prove (5.46), we first prove $TT^*f = K_t * f$ satisfies the energy estimate

$$\|K_t * f\|_{L^2} \leq C\|f\|_{L^2} \tag{5.54}$$

and dispersive estimate

$$\|K_t * f\|_{L^\infty} \leq C(1 + |t|)^{-\frac{(n-1)}{2}} \|f\|_{L^1}. \tag{5.55}$$

Applying the Riesz–Thorin interpolation theorem to (5.54) and (5.55), we get $L^{r'} - L^r$ estimate:

$$\|K_t * f\|_{L^r} \leq C(1 + |t|)^{-\gamma(r)} \|f\|_{L^{r'}}, \quad 2 \leq r \leq \infty. \tag{5.56}$$

Then we can prove (5.46) by Young's inequality, Hardy–Littlewood–Sobolev's inequality and the interpolation method. Note that

$$\widehat{K_t}(\xi) \cong e^{it|\xi|} |\beta(\xi)|^2,$$

so the energy estimate (5.54) is obvious. By Young's inequality,

$$\|K_t * f\|_{L^\infty} \le \|K_t\|_{L^\infty}\|f\|_{L^1}.$$

We can reduce the proof of (5.55) to proving the following pointwise estimate:

$$\|K(t,x)\|_{L^\infty(\mathbb{R}^n)} \le C(1+|t|)^{-\frac{n-1}{2}}. \tag{5.57}$$

Method 1. Note that at each point of forward light cone $\wedge$ in $\mathbb{R}^{n+1}$, there are $n-1$ nonzero principal curvatures, and

$$\begin{aligned} K(t,x) &= \int_{\mathbb{R}^n} e^{it|\xi|}e^{ix\cdot\xi}|\beta(|\xi|)|^2 d\xi \\ &= \int_{\mathbb{R}^{n+1}} e^{i(t,x)\cdot(\tau,\xi)}|\beta(|\xi|)|^2\delta(\tau-|\xi|)d\tau d\xi, \end{aligned} \tag{5.58}$$

where $\delta(\tau-|\xi|)d\tau d\xi$ is the measure of $\wedge$ induced by the Lebesgue measure, then according to Theorem 5.1.1, we have

$$\|K(t,x)\|_{L^\infty(\mathbb{R}^n)} \le C(1+|(t,x)|)^{-\frac{n-1}{2}} \le C(1+|t|)^{-\frac{n-1}{2}}.$$

Method 2. Let $\sigma(x)$ be the measure on the unit sphere $\mathbb{S}^{n-1}$ induced by Lebesgue measure in $\mathbb{R}^n$, according to Proposition 5.1.1, we have

$$|\hat{\sigma}(\xi)| \le C(1+|\xi|)^{-\frac{n-1}{2}}. \tag{5.59}$$

Next, we will derive the pointwise estimate (5.57) using the special restriction estimate above and the stationary-phase argument. Note that $\beta(\xi)=\tilde{\psi}_0$ is a radial function with compact support, and its support does not contain zero point. We introduce polar coordinate transform $\xi=\rho\omega$, then $K(t,x)$ can be expressed as following in the polar coordinate:

$$K(t,x) = \int_0^\infty \int_{\mathbb{S}^{n-1}} e^{i\rho(x\cdot\omega+t)}a(\rho)d\sigma(\omega)d\rho = \int_0^\infty \hat{\sigma}(\rho x)e^{it\rho}a(\rho)d\rho, \tag{5.60}$$

where $a(\rho)=|\beta(\rho)|^2$ is a radial function with compact support, and its support does not contain point 0.

Case I. $|t|\ge 2|x|$. Introduce invariant derivative of $e^{i\rho(x\cdot\omega+t)}$,

$$D=\frac{\partial_\rho}{i(x\cdot\omega+t)} \quad \Longrightarrow \quad D^N e^{i\rho(x\cdot\omega+t)} = e^{i\rho(x\cdot\omega+t)},$$

where N is an arbitrary natural number. Applying integration by parts to the first equality in (5.60), we get

$$|I| \le C_N |t + x\cdot\omega|^{-N} \le C_N 2^N |t|^{-N}. \tag{5.61}$$

Case II. $|t| < 2|x|$. Using the restriction estimate on sphere (5.59), we get

$$\begin{aligned}|K(t,x)| &\le \int_0^\infty |\hat{\sigma}(\rho x)|\cdot|a(\rho)|d\rho \le C\int_0^\infty |\rho x|^{-\frac{n-1}{2}}|\cdot|a(\rho)|d\rho \\ &\le C|x|^{-\frac{n-1}{2}} \le C|t|^{-\frac{n-1}{2}}.\end{aligned} \tag{5.62}$$

Noting $|K(t,x)| < \infty$, we get the pointwise estimate (5.57) by combining Case I and Case II.

Step 6. The proof of (5.46). We only need to prove (5.46) for $(q,r), (\bar{q},\bar{r}) \in \Lambda$. Actually, assume $\tilde{\beta}(\xi) \in C_c^\infty$ and is identical to 1 near $\operatorname{supp}\beta(\xi)$. Therefore, we can substitute F with $\mathcal{F}^{-1}(\tilde{\beta}(\xi)\hat{F})$ in TT^*F. In other words, without loss of generality, we assume that the Fourier transform of F has compact support. Then for any $(q_1,r_1),(q_2,r_2) \in \tilde{\Lambda}$, there always exist $(q_1,\tilde{r}_1),(q_2,\tilde{r}_2) \in \Lambda$, satisfying $\tilde{r}_1 \le r_1, \tilde{r}_2 \le r_2$. Thus, using Bernstein's estimate and (5.43), we get

$$\begin{aligned}\|K_t * F\|_{L^{q_1}(L^{r_1})} &\le \|K * F\|_{L^{q_1}(L^{\tilde{r}_1})} \le C\|F\|_{L^{q_2'}(L^{\tilde{r}_2'})} \\ &\le C\|F\|_{L^{q_2'}(L^{r_2'})}, \quad (q_1,r_1),(q_2,r_2) \in \tilde{\Lambda}.\end{aligned}$$

Next, we will prove (5.46) for $(q,r),(\bar{q},\bar{r}) \in \Lambda$, $(\frac{2}{q},\gamma(r)) \ne (1,1)$ and $(\frac{2}{\bar{q}},\gamma(\bar{r})) \ne (1,1)$, (i. e., Keel–Tao's space-time estimate at end-point is not included). Introduce the bilinear operator

$$\begin{aligned}\mathcal{B}(F,G) &= \int_{\mathbb{R}} \langle K * F(t,x),\ G(t,x)\rangle dt = \int_{\mathbb{R}}\Big\langle \int_{\mathbb{R}} W(t-s)F(s,x)ds, G(t,x)\Big\rangle dt \\ &= \int_{\mathbb{R}^2} \langle T(t)T^*(s)F(s,x),\ G(t,x)\rangle dsdt \\ &= \int_{\mathbb{R}^2} \langle T^*(s)F(s,x), T^*(t)G(t,x)\rangle dsdt.\end{aligned} \tag{5.63}$$

Hence, we can reduce (5.46) to proving

$$|\mathcal{B}(F,G)| \le C\|F\|_{L^{\bar{q}'}(L^{\bar{r}'})}\|G\|_{L^{q'}(L^{r'})}, \quad (q,r),\ (\bar{q},\bar{r}) \in \Lambda. \tag{5.64}$$

Case 1. $(q_1,r_1) = (\infty,2)$, $(q_2,r_2) = (q,r)$. Using Hölder's inequality, (5.63) and Hardy–Littlewood–Sobolev's inequality, we can easily verify

$$\begin{aligned}\Big\|\int_{\mathbb{R}} T^*(t)F(t)dt\Big\|_2^2 &= \int_{\mathbb{R}^2} \langle T^*(s)F(s,x), T^*(t)F(t,x)\rangle dsdt \\ &= \int_{\mathbb{R}^2} \langle T(t)T^*(s)F(s,x),\ F(t,x)\rangle dsdt\end{aligned}$$

$$
\begin{aligned}
&= \int_{\mathbb{R}} \left\langle \int_{\mathbb{R}} W(t-s)F(s,x)ds, F(x,t) \right\rangle dt \\
&\leq \left\| \int_{\mathbb{R}} |t-s|^{-\gamma(r)} \|F(s,x)\|_{L^{r'}} ds \right\|_{L^q} \|F(t,x)\|_{L^{q'}(L^{r'})} \\
&\leq C \|F(t,x)\|^2_{L^{q'}(L^{r'})}.
\end{aligned}
$$

Thus,

$$
\left\| \int_{\mathbb{R}} T^*(t)F(t)dt \right\|_{L^2} \leq C \|F(t,x)\|_{L^{q'}(L^{r'})}. \tag{5.65}
$$

Therefore,

$$
\begin{aligned}
|\mathcal{B}(F,G)| &= \left| \int_{\mathbb{R}} \left\langle \int_{\mathbb{R}} T^*(s)F(s,x)ds,\ T^*(t)G(t,x) \right\rangle dt \right| \\
&\leq \left\| \int_{\mathbb{R}} T^*(t)F(t)dt \right\|_{L^2} \|G\|_{L^1(L^2)} \\
&\leq C \|F(t,x)\|_{L^{q'}(L^{r'})} \|G\|_{L^1(L^2)}.
\end{aligned} \tag{5.66}
$$

Case 2. $(q_1, r_1) = (q_2, r_2) = (q, r)$. Applying the $L^p - L^{p'}$ estimate (5.56) and Hardy–Littlewood–Sobolev's inequality directly, we get

$$
\begin{aligned}
|\mathcal{B}(F,G)| &= \left| \int_{\mathbb{R}} \left\langle \int_{\mathbb{R}} T(t)T^*(s)F(s,x)ds,\ G(t,x) \right\rangle dt \right| \\
&\leq C \int_{\mathbb{R}} (1+|t-s|)^{-\gamma(r)} \|F(s,x)\|_{L^{r'}} \|G(s,x)\|_{L^{r'}} ds \\
&\leq C \|F(t,x)\|_{L^{q'}(L^{r'})} \|G(t,x)\|_{L^{q'}(L^{r'})}.
\end{aligned} \tag{5.67}
$$

Case 3. $(q_1, r_1) = (q, r)$, $(q_2, r_2) = (\infty, 2)$. Similar to the proof of (5.66), we use energy estimate (5.65), and get

$$
\begin{aligned}
|\mathcal{B}(F,G)| &= \left| \int_{\mathbb{R}} \left\langle \int_{\mathbb{R}} T^*(s)F(s,x)ds,\ T^*(t)G(t,x) \right\rangle dt \right| \\
&\leq C \|F(t,x)\|_{L^1(L^2)} \|G(t,x)\|_{L^{q'}(L^{r'})}.
\end{aligned} \tag{5.68}
$$

By applying standard interpolation method and Ginibre–Velo's interpolation theorem (see [106]) to the estimate derived from Case I∼Case III, we can finish the proof of (5.64). □

5.2 Bilinear method and endpoint Strichartz estimate

In the previous section, we established the equivalence between the Fourier restriction estimates on the cones and Strichartz estimate. In addition, we presented the proof of the Strichartz estimate for free wave equations except the end-point case. The reason is that Hardy–Littlewood–Sobolev's inequality is not available when $\gamma(r) = 1$. To overcome this obstruction, Keel and Tao used the atomic decomposition of the L^p space such that the analogous bilinear operator corresponding to Strichartz estimate has fast decay bounds on each atom. The abstract Strichartz estimate was then established for a certain operator $U(t)$, which satisfies energy inequality and dispersive or truncated dispersive estimates. In particular, it covers the end-point Strichartz estimates for linear wave equations in $n \geq 4$ and for free Schrödinger equations in $n \geq 3$. In the mean time, we take examples to show the importance of the end-point estimates in the study of the low regularity problems on the nonlinear wave and Schrödinger equations.

On the other hand, as the approach to classical Strichartz estimate, the TT^* argument is based on the energy inequality and the dispersive estimate for dispersive equations where the energy inequality follows from Plancherel's formula. For the latter estimate, one needs to study the representation formula for free solutions. For the free Schrödinger operator, one has

$$u(t,x) = \mathcal{F}^{-1} e^{-i|\xi|^2 t} \mathcal{F}\varphi = (4\pi i t)^{-n/2} \int_{\mathbb{R}^n} e^{\frac{i|x-y|^2}{4t}} \varphi(y)dy \triangleq S(t)\varphi, \tag{5.69}$$

from which the dispersive estimate follows:

$$\|u\|_{L^\infty(\mathbb{R}^n)} = \|S(t)\varphi\|_{L^\infty(\mathbb{R}^n)} \leq C|t|^{-\frac{n}{2}} \|\varphi\|_{L^1(\mathbb{R}^n)}. \tag{5.70}$$

However, for the free wave equation,

$$\Box u = 0, \quad (u, \partial_t u)|_{t=0} = (f(x), g(x)). \tag{5.71}$$

It is not straightforward to get the corresponding dispersive estimate directly from $u(t,x) = v_+ + v_-$, where

$$v_\pm(t,x) \cong \int e^{ix\xi} e^{\pm it|\xi|} \hat{f}_\pm(\xi)d\xi, \quad \hat{f}_\pm(\xi) = \frac{1}{2}\left(\hat{f}(\xi) \pm \frac{\hat{g}(\xi)}{i|\xi|}\right). \tag{5.72}$$

It is easy to see that the dispersive estimate corresponding to the solution of the free wave equation can be reduced to the dispersive estimate corresponding to the half-wave operator $U(t)$,

$$U(t) = e^{it(-\Delta)^{\frac{1}{2}}} = \mathcal{F}^{-1} e^{i|\xi|t} \mathcal{F}. \tag{5.73}$$

By Young's inequality, we only need to show the following decay estimate for the oscillatory integrals:

$$\|K(t,x)\|_{L^\infty(\mathbb{R}^n)} \lesssim (1+|t|)^{-\frac{n-1}{2}}, \quad K(t,x) = \int_{\mathbb{R}^n} e^{it|\xi|} e^{ix\cdot\xi} \beta(\xi) d\xi, \tag{5.74}$$

where $\beta(\xi) \in C_c^\infty(\mathbb{R}^n)$. On account of this observation, we introduce the stationary phase method to estimate oscillatory integrals.

Consider the following oscillatory integral:

$$I_\psi(\tau) = \int_{\mathbb{R}^n} e^{i\tau\Phi(\xi)} \psi(\xi) d\xi, \tag{5.75}$$

where $\psi(\xi) \in \mathcal{D}(\mathbb{R}^n) \subset C_c^\infty(\mathbb{R}^n)$, and $\Phi(\xi)$ is a smooth real valued function defined in a neighborhood of $\operatorname{supp}\psi(\xi)$. We denote by $\mathcal{D}_K$ the test functions in $\mathcal{D}(\mathbb{R}^n)$, which are supported in a compact set K. In the subsequent theorems, C represents a constant depending only on $\partial^\alpha\psi$ and $\partial^\beta\Phi$ $(0 \le |\alpha| < \infty, 2 \le |\beta| < \infty)$. We shall start with discussing two different cases corresponding respectively $|\nabla\Phi| \neq 0$ and in a neighborhood of the points where $\nabla\Phi = 0$.

Lemma 5.2.1. *Suppose $K \subset\subset \mathbb{R}^n$ is compact, and let $c_0 > 0$ such that*

$$|\nabla\Phi| \ge c_0, \quad \xi \in K. \tag{5.76}$$

Then for any positive integer N, and any smooth function $\psi \in \mathcal{D}_K$, there is a constant C such that

$$|I_\psi(\tau)| \le C\tau^{-N}. \tag{5.77}$$

Proof. We introduce the operator leaving $e^{i\Phi(\xi)}$ invariant, that is,

$$L \triangleq -i\sum_{j=1}^n \frac{\partial_j\Phi}{|\nabla\Phi|^2}\partial_j \quad \Longrightarrow \quad L^* = -i\sum_{j=1}^n \partial_j\left(\frac{\partial_j\Phi}{|\nabla\Phi|^2}\right).$$

Obviously, $Le^{i\Phi(\xi)} = e^{i\Phi(\xi)}$, hence

$$e^{i\tau\Phi(\xi)} = \frac{1}{\tau} L e^{i\tau\Phi(\xi)}.$$

Integration by parts yields

$$I_\psi(\tau) = \frac{1}{\tau^N} \int_{\mathbb{R}^n} e^{i\tau\Phi(\xi)} (L^*)^N \psi(\xi) d\xi. \tag{5.78}$$

Noting that the coefficients of the operator with its derivatives are bounded in K, and hence $(L^*)^N\psi$ is bounded in K, and thus we complete the proof. □

Next, we consider the case when the phase function Φ has critical points at which $\nabla\Phi$ vanishes.

Lemma 5.2.2. *Suppose $K \subset\subset \mathbb{R}^n$ is compact, and let $c_0 > 0$ such that*

$$|\nabla\Phi| \leq c_0, \quad \xi \in K. \tag{5.79}$$

Then for any positive N, any smooth function $\psi \in \mathcal{D}_K$, there is a constant C such that

$$|I_\psi(\tau)| \leq C \int_K \frac{1}{(1+\tau|\nabla\Phi|^2)^N} d\xi. \tag{5.80}$$

Proof. The idea is to deduce asymptotic decay estimate based on integration by parts. To do this, we introduce the inhomogeneous derivative leaving $e^{i\tau\Phi(\xi)}$ invariant

$$L \triangleq \frac{1}{1+\tau|\nabla\Phi|^2}(1 - i\nabla\Phi\cdot\partial), \quad \nabla\Phi\cdot\partial = \sum_{j=1}^n \partial_j\Phi\partial_j.$$

Obviously, that $Le^{i\tau\Phi(\xi)} = e^{i\tau\Phi(\xi)}$, and thus $e^{i\tau\Phi(\xi)} = Le^{i\tau\Phi(\xi)}$. Integration by parts gives

$$I_\psi(\tau) = \int_{\mathbb{R}^n} e^{i\tau\Phi(\xi)} (L^*)^N \psi(\xi) d\xi.$$

Hence, we only need to show: for any integer N, there is $C > 0$, so that

$$|(L^*)^N \psi(\xi)| \leq \frac{C}{(1+\tau|\nabla\Phi|^2)^N}. \tag{5.81}$$

□

To estimate $(L^*)^N$, we need more analysis.

Definition 5.2.1. Let $N \in \mathbb{Z}$ and denote by S^N as the set of the smooth functions in $K\times\mathbb{R}^n$ satisfying

$$|\partial_\xi^\alpha \partial_\theta^\beta f(\xi,\theta)| \lesssim (1+|\theta|^2)^{\frac{N-|\beta|}{2}}, \quad \forall(\xi,\theta) \in \operatorname{supp}\psi\times\mathbb{R}^n,\ \forall(\alpha,\beta)\in\mathbb{N}^n\times\mathbb{N}^n. \tag{5.82}$$

Obviously, the space S^N expands as N increases, and $S^{N_1}\times S^{N_2} \subset S^{N_1+N_2}$. Meanwhile, $\partial_\theta^\beta(S^N) \subset S^{N-|\beta|}$.

Lemma 5.2.3. *Let $N \in \mathbb{Z}$. Then there is $f_N \in S^{-2N}$ so that*

$$(L_\tau^*)^N \psi = f_N(\xi, \tau^{\frac{1}{2}}\nabla\Phi). \tag{5.83}$$

Proof. By $\mathcal{D}_K \subset S^0$ and induction, we only need to prove: if $f \in S^M$, then

$$L_\tau^* f(\xi, \tau^{\frac{1}{2}} \nabla\Phi) = g(\xi, \tau^{\frac{1}{2}} \nabla\Phi), \quad \text{s.t.} \quad g(\xi, \theta) \in S^{M-2}. \tag{5.84}$$

For any $a(x) \in \mathcal{D}_K$, consider

$$\begin{aligned}
(f, L_\tau^* a) &\triangleq (L_\tau f, a) = \int_{\mathbb{R}^n} \frac{(I - i\nabla\Phi \cdot \partial) f}{1 + \tau|\nabla\Phi|^2} \bar{a}(\xi) d\xi \\
&= \int_{\mathbb{R}^n} \frac{\bar{a}}{1 + \tau|\nabla\Phi|^2} f(\xi) d\xi + i \sum_{j=1}^n \int_{\mathbb{R}^n} \partial_j \left(\frac{\bar{a}\partial_j\Phi}{1 + \tau|\nabla\Phi|^2} \right) f(\xi) d\xi \\
&= \int_{\mathbb{R}^n} \frac{\bar{a}}{1 + \tau|\nabla\Phi|^2} f(\xi) d\xi + i \sum_{j=1}^n \int_{\mathbb{R}^n} \overline{\frac{a\partial_j^2\Phi + \partial_j\Phi\partial_j a}{1 + \tau|\nabla\Phi|^2}} f(\xi) d\xi \\
&\quad - i \sum_{j=1}^n \int_{\mathbb{R}^n} 2 \overline{\frac{a\partial_j\Phi \cdot \tau\partial_j\partial_i\Phi \cdot \partial_i\Phi}{(1 + \tau|\nabla\Phi|^2)^2}} f(\xi) d\xi,
\end{aligned}$$

from which, we have

$$\begin{aligned}
L_\tau^* a &= \frac{a}{1 + \tau|\nabla\Phi|^2} - i \frac{a\Delta\Phi}{1 + \tau|\nabla\Phi|^2} - i \frac{\nabla\Phi\nabla a}{1 + \tau|\nabla\Phi|^2} + i \frac{2\tau\partial_j\Phi D_{ij}^2\Phi \cdot \partial_i\Phi}{(1 + \tau|\nabla\Phi|^2)^2} a \\
&= i \frac{\nabla\Phi\nabla a}{1 + \tau|\nabla\Phi|^2} + \sigma(\xi, \tau^{\frac{1}{2}}\nabla\Phi) a(\xi),
\end{aligned} \tag{5.85}$$

where

$$\sigma(\xi, \theta) = \frac{1 - i\Delta\Phi}{1 + |\theta|^2} + \frac{D^2\Phi_\xi(\theta, \theta)}{(1 + |\theta|^2)^2}. \tag{5.86}$$

Replacing $a(\xi)$ by $f(\xi, \tau^{\frac{1}{2}}\nabla\Phi)$, and using the chain rule of taking derivatives, we have

$$\nabla\Phi \cdot \nabla f(\xi, \tau^{\frac{1}{2}}\nabla\Phi) = (\nabla\Phi \cdot \nabla_\xi f + D^2\Phi(\xi)(\theta, \nabla_\theta f))(\xi, \tau^{\frac{1}{2}}\nabla\Phi). \tag{5.87}$$

Using (5.84)~(5.87), we get

$$g(\xi, \theta) = \frac{-i}{1 + |\theta|^2} [\nabla\Phi \cdot \nabla_\xi f(\xi, \theta) + D_\xi^2\Phi(\xi)(\theta, \nabla_\theta f(\xi, \theta))] + (\sigma f)(\xi, \theta), \tag{5.88}$$

which implies $g(\xi, \theta) \in S^{M-2}$, and Lemma 5.2.3 is proved. □

Theorem 5.2.1. *Let $K \subset\subset \mathbb{R}^n$ be compact, then for any $c_0 > 0$ given and arbitrary integers N_1 and N_2, there is a constant C such that*

$$|I_\psi(\tau)| \leq \frac{C}{\tau^{N_1}} + C \int_K \frac{I_{\{\xi \in \mathbb{R}^n, |\nabla\Phi| \leq c_0\}}}{(1 + \tau|\nabla\Phi|^2)^{N_2}} d\xi. \tag{5.89}$$

Proof. Choose $\chi(x) \in \mathcal{D}(\mathbb{R}^n)$, so that

$$\operatorname{supp}\chi(x) \subset B_1(0), \quad \chi(x) = 1, \quad \text{when} \quad x \in B_{\frac{1}{2}}(0).$$

We write $I(\tau) = I_1(\tau) + I_2(\tau)$, where

$$I_1 = \int e^{i\tau\Phi(\xi)}\Big(1 - \chi\Big(\frac{\nabla\Phi}{c_0}\Big)\Big)\psi(\xi)d\xi,$$

$$I_2 = \int e^{i\tau\Phi(\xi)}\chi\Big(\frac{\nabla\Phi}{c_0}\Big)\psi(\xi)d\xi.$$

Applying Lemma 5.2.1 and Lemma 5.2.2 to I_1 and I_2 respectively, we get the estimate in the theorem. □

Now we use Theorem 5.2.1 to deduce the dispersive estimate for free wave equations.

Theorem 5.2.2. *Let $n \geq 2$ and denote $\mathcal{C} \triangleq \{\xi \in \mathbb{R}^n \mid r \leq |\xi| \leq R\}$. If $\operatorname{supp}\hat{f}(\xi) \subset \mathcal{C}$ and $\operatorname{supp}\hat{g}(\xi) \subset \mathcal{C}$, then there is a constant $C > 0$, so that the solution $u(t,x)$ to the wave equation* (5.71) *satisfies*

$$\|u(t,x)\|_{L^\infty} \leq \frac{C}{(1+|t|)^{\frac{n-1}{2}}}\|f_\pm\|_{L^1} \leq \frac{C}{(1+|t|)^{\frac{n-1}{2}}}\|(f,g)\|_{L^1\times L^1}. \tag{5.90}$$

where $f_\pm$ is as in (5.72).

Proof. We choose a smooth cut-off function $\varphi(\xi) \in \mathcal{D}(\mathbb{R}^n \setminus \{0\})$ and $\varphi(\xi)$ is equal to 1 in the neighborhood of $\mathcal{C}$ and we have

$$\hat{u} = \sum_\pm e^{\pm it|\xi|}\hat{f}_\pm(\xi) = \sum_\pm e^{\pm it|\xi|}\varphi(\xi)\varphi(\xi)\hat{f}_\pm(\xi).$$

Hence,

$$u(t,x) = \sum_\pm K^\pm(t,\cdot) * \tilde{f}_\pm, \tag{5.91}$$

where

$$K^\pm(t,x) \cong \int_{\mathbb{R}^n} e^{ix\cdot\xi}e^{\pm it|\xi|}\varphi(\xi)d\xi, \quad \tilde{f}_\pm = \mathcal{F}^{-1}(\varphi(\xi)\hat{f}_\pm(\xi)). \tag{5.92}$$

Note that $\varphi(\xi) \in \mathcal{D}(\mathbb{R}^n \setminus \{0\})$, $\mathcal{F}^{-1}(|\xi|^{-1}\varphi) \in \mathcal{S}(\mathbb{R}^n)$ and

$$\sum_\pm \|\tilde{f}_\pm\|_{L^1} \leq C(\|f_+\|_{L^1} + \|f_-\|_{L^1}) \leq C(\|f\|_{L^1} + \|g\|_{L^1}).$$

Thus, by Young's inequality, we have

$$\|u\|_{L^\infty} \le \|K^\pm\|_{L^\infty} \|\widetilde{f}_\pm\|_{L^1}.$$

Notice that $|K(t,x)| \le C < \infty$, and we reduce (5.90) to the following pointwise estimate:

$$\|K^\pm(t,x)\|_{L^\infty} \le C|t|^{-\frac{n-1}{2}}. \tag{5.93}$$

Now, we consider

$$K^\pm(t,tx) \cong \int_{\mathbb{R}^n} e^{itx\cdot\xi \pm it|\xi|} \varphi(\xi)d\xi = \int_{\mathbb{R}^n} e^{it\xi\cdot(x\pm\frac{\xi}{|\xi|})} \varphi(\xi)d\xi. \tag{5.94}$$

By Theorem 5.2.1, we have

$$|K^\pm(t,tx)| \le \frac{C}{|t|^{\frac{n-1}{2}}} + C\int_{\mathcal{C}_x} \frac{d\xi}{(1+|t||x\pm\frac{\xi}{|\xi|}|^2)^n}, \tag{5.95}$$

where

$$\mathcal{C}_x = \left\{\xi \in \mathcal{C} : \left|x \pm \frac{\xi}{|\xi|}\right| \le \frac{1}{2}\right\}.$$

By the definition of $\mathcal{C}_x$, we have $x \ne 0$. Changing variables (split ξ as the summation of the components along x and the orthogonal direction, respectively):

$$\zeta_1 = \left(\xi \cdot \frac{x}{|x|}\right)\frac{x}{|x|}, \quad \zeta' = \xi - \left(\xi \cdot \frac{x}{|x|}\right)\frac{x}{|x|}.$$

Recalling the definition of $\mathcal{C}$ and $|\zeta| = |\xi|$, one has

$$\begin{aligned} |K^\pm(t,tx)| &\le \frac{C}{|t|^{\frac{n-1}{2}}} + C\int_{\mathcal{C}} \frac{d\zeta}{(1+|t|\frac{|\zeta'|^2}{|\xi|^2})^n} \\ &\le C|t|^{-\frac{n-1}{2}} + C\int_{\mathcal{C}} \frac{d\zeta}{(1+|t||\zeta'|^2)^n} \\ &\le C|t|^{-\frac{n-1}{2}} + C|t|^{-\frac{n-1}{2}} \int_{\mathcal{C}} \frac{d\tilde{\zeta}}{(1+|\tilde{\zeta}'|^2)^n} \\ &\le C|t|^{-\frac{n-1}{2}}. \end{aligned} \tag{5.96}$$

Here, we have used the changing of variables $\tilde{\zeta}' = t^{\frac{1}{2}}\zeta'$, ζ_1 is along x, that is, $x = |x|(\xi \cdot \frac{x}{|x|})^{-1}\zeta_1$, and correspondingly:

$$x \pm \frac{\xi}{|\xi|} = x \pm \frac{\zeta_1}{|\xi|} \pm \frac{\zeta'}{|\xi|} = \left[|x|\left(\xi \cdot \frac{x}{|x|}\right)^{-1} \pm \frac{1}{|\xi|}\right]\zeta_1 \pm \frac{\zeta'}{|\xi|},$$

which gives all the estimate used in (5.96),

$$\left|x \pm \frac{\xi}{|\xi|}\right|^2 \geq \frac{|\zeta'|^2}{|\xi|^2}, \quad (\zeta_1 \perp \zeta').$$

□

Before discussing Keel–Tao's end-point Strichartz estimate, let us recall the atomic decomposition for L^p spaces and take Hardy–Littlewood–Sobolev's inequality as an example to look inside the efficacy of atomic decomposition from which the idea originated from Keel–Tao's work [129] could be well motivated.

Theorem 5.2.3. *Let $p \in [1, \infty)$, then for any $f \in L^p(\mathbb{R}^n)$, we have the following decomposition:*

$$f(x) = \sum_{k=-\infty}^{\infty} c_k f_k, \quad \textit{orthogonal decomposition.} \tag{5.97}$$

Here, $\operatorname{supp} f_k$ *are disjoint, and satisfy:*

$$\begin{cases} \mu(\operatorname{supp} f_k) \leq 2^{k+1}, & \textit{support condition,} \\ \|f_k\|_\infty \leq 2^{-\frac{k}{p}}, & \textit{point wise estimate,} \\ \sum_{k\in\mathbb{Z}} |c_k|^p \leq C\|f\|_p^p, & \textit{equivalence of the norms.} \end{cases} \tag{5.98}$$

Proof. We first define the distribution function

$$f_*(\alpha) = \mu\{x : |f(x)| > \alpha\} \tag{5.99}$$

and the corresponding nonincreasing rearrangement (See Lieb–Loss [171])

$$f^*(\lambda) = \inf_{f_*(\alpha)\leq\lambda} \{\alpha\}. \tag{5.100}$$

It is easy to see

$$f^*(\lambda) \leq \alpha \Longleftrightarrow f_*(\alpha) \leq \lambda. \tag{5.101}$$

Define

$$\begin{cases} \lambda_k = \inf\{\alpha \mid f_*(\alpha) \leq 2^k\} = f^*(2^k), \\ c_k = 2^{\frac{k}{p}} \lambda_k, \\ f_k = \frac{1}{c_k} \chi_{(\lambda_{k+1}, \lambda_k]}(|f|) f. \end{cases} \tag{5.102}$$

Obviously, $\operatorname{supp} f_k$ are disjoint and

$$\|f_k\|_\infty \leq 2^{-\frac{k}{p}} \lambda_k^{-1} \|\chi_{\{\lambda_{k+1} < |f| \leq \lambda_k\}} \cdot f\|_\infty \leq 2^{-\frac{k}{p}}.$$

By the definition of $\lambda_k, f_*(\lambda_k) = \mu\{x \mid |f| > \lambda_k\} < 2^k$. We have

$$\mu\{x \mid \operatorname{supp}\chi_{(\lambda_{k+1},\lambda_k]}\} \le f_*(\lambda_{k+1}) \le 2^{k+1}.$$

Using Fubini's theorem, on can (directly) check that

$$\begin{aligned}\sum_{k\in\mathbb{Z}} c_k^p = \sum_{k\in\mathbb{Z}} 2^k\lambda_k^p &= p\sum_{k\in\mathbb{Z}} 2^k \int_0^\infty \lambda^{p-1} I_{\lambda\le\lambda_k}\, d\lambda \\ &= p\int_0^\infty \lambda^{p-1}\Big(\sum_{k:\,\lambda_k\ge\lambda} 2^k\Big)\, d\lambda.\end{aligned}$$

Noting that (5.101) implies

$$\lambda_k = f^*(2^k) \ge \lambda \Longleftrightarrow \mu\{x \mid |f| > \lambda\} = f_*(\lambda) \ge 2^k.$$

Using Fubini's theorem once more, we have

$$\begin{aligned}\sum_{k\in\mathbb{Z}} c_k^p &= p\int_0^\infty \lambda^{p-1} \sum_{k:\,\mu\{x:\,|f|>\lambda\}\ge 2^k} 2^k\, d\lambda \\ &\le 2p\int_0^\infty \lambda^{p-1}\mu\{x:\ |f| > \lambda\}\,d\lambda \le 2\|f\|_p^p. \end{aligned} \tag{5.103}$$

□

Remark 5.2.1. In the above theorem, $\|f\|_p \cong \|\{c_k\}\|_{\ell^p}$. Indeed, when $\lambda_{k+1} = f^*(2^{k+1}) < \alpha < f^*(2^k) = \lambda_k$, we have $f_*(\alpha) \cong 2^k$. Using Chebyshev's inequality, we have

$$\begin{aligned}\int_{\mathbb{R}^n} |f(x)|^p dx &= p\int_0^\infty f_*(\alpha)\alpha^{p-1}d\alpha = \sum_{k\in\mathbb{Z}} p\int_{\lambda_{k+1}}^{\lambda_k} f_*(\alpha)\alpha^{p-1}d\alpha \\ &\cong \sum_{k\in\mathbb{Z}} 2^k p\int_{\lambda_{k+1}}^{\lambda_k} \alpha^{p-1}d\alpha = \sum_{k\in\mathbb{Z}} 2^k(\lambda_k^p - \lambda_{k+1}^p) \\ &\le \sum_{k\in\mathbb{Z}} 2^k\lambda_k^p = \sum_{k\in\mathbb{Z}} c_k^p. \end{aligned} \tag{5.104}$$

It is known that Young's inequality cannot be adapted in a straightforward way to prove the famous Hardy–Littlewood–Sobolev's inequality,

$$\||\cdot|^{-\alpha} * f\|_q \le C\|f\|_p, \quad \frac{1}{q} + 1 = \frac{1}{p} + \frac{\alpha}{n}, \quad 0 < \alpha < n,\ p, q \in [1, \infty). \tag{5.105}$$

The atomic decomposition for L^p space, however, could help to conquer the difficulty. We next do mathematical analysis on this process to reveal the enlightenment from the approach to the end-point space-time estimates.

Without loss of generality, consider two positive functions $\|f\|_p = \|g\|_{q'} = 1$, and the bilinear operator

$$I(f,g) \triangleq \int_{\mathbb{R}^{2n}} |x-y|^{-\alpha} f(y)g(x)dydx. \tag{5.106}$$

Then Hardy–Littlewood–Sobolev's inequality is reduced to

$$|I(f,g)| \leq C < \infty. \tag{5.107}$$

Step 1. We decompose $I(f,g)$ as

$$I(f,g) = \sum_{j\in\mathbb{Z}} I_j(f,g), \quad I_j(f,g) = \int_{C_j} f(y)|x-y|^{-\alpha} g(x)dydx. \tag{5.108}$$

Here,

$$C_j = \{(x,y) \in \mathbb{R}^{2n} \mid 2^j \leq |x-y| \leq 2^{j+1}\}.$$

For any $(a,b) \in [1,\infty]^2$, $b \leq a'$, $(f,g) \in L^a(\mathbb{R}^n) \times L^b(\mathbb{R}^n)$, using Young's inequality, we have

$$\begin{aligned}
&\left|\int_{C_j} f(y)|x-y|^{-\alpha} g(x)dxdy\right| \\
&\quad = \left|\int_{\mathbb{R}^n} f(y) \int_{\mathbb{R}^n} \chi_{C_j} |x-y|^{-\alpha} g(x)dxdy\right| \\
&\quad \leq \|f\|_a \left\| \int_{\mathbb{R}^n} \chi_{C_j}(x,y)|x-y|^{-\alpha} g(x)dx \right\|_{L^{a'}} \\
&\quad \leq \|f\|_a \|g\|_b \left(\int_{|x-y|\sim 2^j} |x-y|^{-\alpha\ell} dx \right)^{\frac{1}{\ell}} \\
&\quad \lesssim 2^{jn(2-\frac{\alpha}{n}-\frac{1}{a}-\frac{1}{b})} \|f\|_a \|g\|_b, \quad 1 + \frac{1}{a'} = \frac{1}{b} + \frac{1}{\ell}.
\end{aligned} \tag{5.109}$$

In particular, taking $b = q'$, $a = p$, we have

$$|I_j(f,g)| \leq C 2^{jn(1-\frac{\alpha}{n}-\frac{1}{p}+\frac{1}{q})} \|f\|_p \|g\|_{q'} \leq C \|f\|_p \|g\|_{q'}. \tag{5.110}$$

It is impossible from (5.110) to conclude the convergency of the series. However, one may expect to decompose f and g by means of L^p atomic decomposition in order to obtain good coefficients, which decay fast enough on each atom to guarantee the convergence of the former series.

Step 2. We use atomic decomposition $f(x) = \sum_{k\in\mathbb{Z}} c_k f_k$, $g(x) = \sum_{k'\in\mathbb{Z}} d_{k'} g_{k'}$. Then

$$I(f,g) = \sum_{j,k,k'\in\mathbb{Z}} c_k d_{k'} I_j(f_k, g_{k'}). \tag{5.111}$$

Here, $I_j(f_k, g_{k'}) = \int_{C_j} f_k(y)|x-y|^{-\alpha} g_{k'}(x)dydx$. Note that

$$\|f_k\|_a \le \|f_k\|_\infty \mu(\operatorname{supp} f_k)^{\frac{1}{a}} \cong 2^{k(\frac{1}{a}-\frac{1}{p})}, \quad \|g_{k'}\|_b \cong 2^{k'(\frac{1}{b}-\frac{1}{q'})}.$$

Using (5.110), we have

$$\begin{aligned} I_j(f_k, g_{k'}) &\le 2^{jn(2-\frac{\alpha}{n}-\frac{1}{a}-\frac{1}{b})}\|f_k\|_a\|g_{k'}\|_b \\ &= C2^{(jn-k)(\frac{1}{p}-\frac{1}{a})}2^{(jn-k')(\frac{1}{q'}-\frac{1}{b})}. \end{aligned} \tag{5.112}$$

Note that the condition in (5.105) implies that $p < q$ or $q' < p'$. Hence, we can find (a,b) such that

$$\begin{cases} b \le a', & a < p, & b < q', \\ b \le a', & a < p, & b > q', \\ b \le a', & a > p, & b < q', \\ b \le a', & a > p, & b > q', \end{cases}$$

from which we deduce that there is $\varepsilon > 0$ so that

$$|I_j(f_k, g_{k'})| \le C2^{-2\varepsilon|jn-k|-2\varepsilon|jn-k'|} \le C2^{-\varepsilon|jn-k|-\varepsilon|jn-k'|-\varepsilon|k-k'|}. \tag{5.113}$$

Noting that $q' \le p'$, by weighted Hölder's inequality and discrete Young's inequality, we have

$$\begin{aligned} I(f,g) &\le C\sum_{j,k,k'\in\mathbb{Z}} c_k d_{k'} 2^{-\varepsilon|jn-k|-\varepsilon|jn-k'|-\varepsilon|k-k'|} \le C_\varepsilon \sum_{k,k'\in\mathbb{Z}} c_k d_{k'} 2^{-\varepsilon|k-k'|} \\ &\le C_\varepsilon \|\{c_k\}\|_{\ell^p} \left\| \left\{ \sum_{k'\in\mathbb{Z}} d_{k'} 2^{-\varepsilon|k-k'|} \right\} \right\|_{\ell^{p'}} \\ &\le C_\varepsilon \|\{c_k\}\|_{\ell^p} \|\{d_{k'}\}\|_{\ell^{p'}} \le C_\varepsilon \|\{c_k\}\|_{\ell^p} \|\{d_{k'}\}\|_{\ell^{q'}}. \end{aligned} \tag{5.114}$$

Therefore, we proved Hardy–Littlewood–Sobolev's inequality. We see from the above argument that an efficient technique in dealing with the end-point case is the various decomposition skills in functional spaces.

Next, we follow Keel–Tao's work in [129] to present the proof to the abstract Strichartz estimate with the end-point included.

Let (X, dx) be a measure space and H be a Hilbert space. Denote $L^p(X)$ the usual Lebesgue space, $\forall t \in \mathbb{R}$. We put the following assumption to the operator $V(t) : H \longmapsto L^2(X)$:

(H1) (Energy estimate). For any $t \in \mathbb{R}, f \in H$, $V(t)$ satisfies

$$\|V(t)f\|_{L^2(X)} \lesssim \|f\|_H. \tag{5.115}$$

(H2) (Decay estimate). There is $\sigma > 0$, for any $t \neq s, g \in L^1(X)$, $V(t)$ satisfies the following untruncated dispersive estimate:

$$\|V(s)V^*(t)g\|_\infty \lesssim |t-s|^{-\sigma}\|g\|_1, \tag{5.116}$$

or the truncated dispersive estimate

$$\|V(s)V^*(t)g\|_\infty \lesssim (1+|t-s|)^{-\sigma}\|g\|_1. \tag{5.117}$$

We are aiming at deducing the estimates for $\|V(t)f\|_{L_t^qL_x^r}$ in both the homogeneous and inhomogeneous cases on basis of the energy inequalities and dispersive estimates that $V(t)f$ satisfies.

Definition 5.2.2. We say (q, r) is σ-admissible, if $q, r \geq 2$, $(q, r, \sigma) \neq (2, \infty, 1)$ and

$$\frac{1}{q} \leq \sigma\left(\frac{1}{2} - \frac{1}{r}\right). \tag{5.118}$$

Moreover, when the equality holds in (5.118), (q, r) is called the sharp σ-admissible. In particular, when $\sigma > 1$, the end-point

$$P = \left(2, \frac{2\sigma}{\sigma-1}\right)$$

is the sharp σ-admissible.

Theorem 5.2.4. *Assume that $V(t)$ satisfies the energy inequality* (5.115) *and nontruncated decay estimate* (5.116), *then we have the following Strichartz estimates:*

$$\|V(t)f\|_{L_t^qL_x^r} \lesssim \|f\|_H, \tag{5.119}$$

$$\left\|\int V^*(s)F(s)ds\right\|_H \lesssim \|F\|_{L_t^{q'}L_x^{r'}}, \tag{5.120}$$

$$\left\|\int V(t)V^*(s)F(s)ds\right\|_{L_t^qL_x^r} \lesssim \|F\|_{L_t^{\tilde{q}'}L_x^{\tilde{r}'}}, \tag{5.121}$$

$$\Big\| \int_{s<t} V(t)V^*(s)F(s)ds \Big\|_{L_t^q L_x^r} \lesssim \|F\|_{L_t^{\tilde{q}'} L_x^{\tilde{r}'}}, \tag{5.122}$$

where (q,r) and $(\tilde{q},\tilde{r})$ are sharp σ-admissible.

Moreover, if the dispersive estimate (5.115) *can be sharpened to the truncated estimate* (5.117), *then Strichartz estimates* (5.119)~(5.122) *hold for all σ-admissible pairs (q,r), $(\tilde{q},\tilde{r})$.*

Remark 5.2.2.

(i) According to TT^* argument, we can obtain the equivalence of (5.73) and (5.121), which essentially corresponds to the homogeneous estimate. The inhomogeneous estimate (5.122) is called the retarded space-time estimate for wave equations. Using Christ–Kiselev's lemma in the previous section, (5.121) implies (5.122). Theorem 5.2.4 is reduced to the proof of (5.121). It is also accessible by using the classical interpolation method to steer clear of Christ–Kiselev's lemma; see [129] for details.

(ii) As for free wave equations, one has $\sigma = \frac{n-1}{2}$ and the operator $V(t)$ satisfies the truncated dispersive estimate. We also say the $\frac{n-1}{2}$-admissible pairs (q,r) is wave admissible pairs. In particular, when $n > 3$, $\sigma = \frac{n-1}{2} > 1$, Theorem 5.2.4 implies the Strichartz estimate at end-point $P = (2, \frac{2(n-1)}{n-3})$.

(iii) For Schrödinger equation, $\sigma = \frac{n}{2}$ and the operator $V(t)$ satisfies the nontruncated estimate. As a result, we say (q,r) is Schrödinger admissible, when (q,r) is sharp $\frac{n}{2}$-admissible. In particular, when $n > 2$, $\sigma = \frac{n}{2} > 1$, Theorem 5.2.4 involves the end-point Strichartz estimate for Schrödinger equations at $P = (2, \frac{2n}{n-2})$.

As an immediate consequence of Theorem 5.2.4, we have the following results.

Corollary 5.2.1. *Assume $n \geq 2$, (q,r), $(\tilde{q},\tilde{r})$ is wave admissible and $r, \tilde{r} < \infty$, $u(t)$ solves*

$$\begin{cases} u_{tt} - \Delta u = F(t,x), & (t,x) \in [0,T] \times \mathbb{R}^n, \\ u(0) = f, \quad \partial_t u(0) = g, & x \in \mathbb{R}^n. \end{cases} \tag{5.123}$$

Then for any $0 < T < \infty$, u satisfies the following Strichartz estimate:

$$\begin{aligned} &\|u\|_{L^q([0,T];L^r(\mathbb{R}^n))} + \|u\|_{C([0,T];\dot{H}^\gamma)} + \|\partial_t u\|_{C([0,T];\dot{H}^{\gamma-1})} \\ &\quad \lesssim \|f\|_{\dot{H}^\gamma} + \|g\|_{\dot{H}^{\gamma-1}} + \|F\|_{L^{\tilde{q}'}([0,T];L^{\tilde{r}'}(\mathbb{R}^n))}. \end{aligned} \tag{5.124}$$

Here, (q,r) and $(\tilde{q},\tilde{r})$ satisfies the "gap" condition

$$\frac{1}{q} + \frac{n}{r} = \frac{n}{2} - \gamma = \frac{1}{\tilde{q}'} + \frac{n}{\tilde{r}'} - 2. \tag{5.125}$$

Conversely, if (5.124) *is valid, then (q,r), $(\tilde{q},\tilde{r})$ are wave admissible satisfying* (5.125).

Corollary 5.2.2. *Assume $n \geq 1$, (q,r) and $(\tilde{q},\tilde{r})$ are Schrödinger admissible. Let $u(t)$ solve*

$$\begin{cases} i\partial_t u + \Delta u = F(t,x), & (t,x) \in [0,T] \times \mathbb{R}^n, \\ u(0,x) = f(x), \end{cases} \tag{5.126}$$

then for any $T, 0 < T < \infty$, we have

$$\begin{aligned} &\|u\|_{L_t^q([0,T];L_x^r(\mathbb{R}^n))} + \|u\|_{C([0,T];L^2(\mathbb{R}^n))} \\ &\quad \lesssim \|f\|_{L^2(\mathbb{R}^n)} + \|F\|_{L^{\tilde{q}'}([0,T];L^{\tilde{r}'}(\mathbb{R}^n))}. \end{aligned} \tag{5.127}$$

Conversely, if (5.127) *holds for all $f(x)$, $F(t,x)$, T, then (q,r) and $(\tilde{q},\tilde{r})$ must be Schrödinger admissible.*

Remark 5.2.3.

(i) Corollary 5.2.1 is nothing but Theorem 5.1.3 in the previous section and Corollary 5.2.2 is the classical Strichartz estimate for Schrödinger equations. One can see Chapter 4 or [108, 242] to find the corresponding results in the framework of Besov spaces. We will give the discussion in Besov spaces for wave equations in the subsequent sections.

(ii) In Corollary 5.2.1, replacing L^r with $\dot{B}_{r,2^0}$, one has (5.124) for $r = \infty$. The same argument yields that (5.124) is true for $\tilde{r} = \infty$ with $\dot{B}^0_{\tilde{r}',2}$ in place of $L^{\tilde{r}'}$.

(iii) In Corollary 5.2.1, the gap condition (5.125) can be written as

$$\delta(r) - \frac{1}{q} = \gamma, \quad \delta(r) + \delta(\tilde{r}) - \frac{1}{q} - \frac{1}{\tilde{q}} = 1,$$

which is determined exactly by the scaling argument. Indeed, if $u(t,x)$ solves (5.124), one easily gets that $u_\lambda(t,x) = u(\lambda t, \lambda x)$ solves

$$\begin{cases} u_{tt} - \Delta u = \lambda^2 F(\lambda t, \lambda x), \\ u(0) = \varphi(\lambda x), \quad u_t(0) = \lambda \psi(\lambda x) \end{cases}$$

and satisfies

$$\begin{aligned} &\|u_\lambda(t,x)\|_{L^q(\mathbb{R},L^r(\mathbb{R}^n))} + \|u_\lambda\|_{C(\mathbb{R};\dot{H}^\gamma)} + \|\partial_t u_\lambda\|_{C(\mathbb{R};\dot{H}^{\gamma-1})} \\ &\quad \leq \|\varphi(\lambda x)\|_{\dot{H}^\gamma} + \lambda \|\psi(\lambda x)\|_{\dot{H}^{\gamma-1}} + \lambda^2 \|F(\lambda t, \lambda x)\|_{L^{\tilde{q}'}(\mathbb{R};L^{\tilde{r}'}(\mathbb{R}^n))}. \end{aligned}$$

Thus, one is able to perform rescaling to put it back to (5.124) if and only if (q,r), $(\tilde{q},\tilde{r})$ satisfies (5.125) or the equivalent form above. The analogous argument can be applied equally well to the Schrödinger equation (5.126).

(iv) The end-point Strichartz estimates for 3-dimensional wave and 2-dimensional Schrödinger equations are not true. For counterexamples, see the counterexamples by Montgomerey–Smith [203] or Tao [261].

Remark 5.2.4. The following are several advantages of the abstract Stricartz estimates in Theorem 5.2.4:

(i) (Unified) It describes Stricartz estimates for both wave and Schrödinger equation in a unified way.

(ii) (Succinct) It excludes unnecessary assumptions and is more specified.

(iii) Theorem 5.2.4 presents the sharp result invariant left by scaling

$$\begin{cases} V(\frac{t}{\lambda}) \longmapsto V(t), & V^*(\frac{s}{\lambda}) \longmapsto V^*(s), \\ \lambda^\sigma dx \longmapsto dx, & \lambda^\sigma \langle f, g\rangle \longmapsto \langle f, g\rangle. \end{cases} \tag{5.128}$$

Indeed, set $u(x,t) = U(t)f$ solving the free wave equation, then u_λ still solves the free wave equation

$$u_\lambda(t,x) = \left(U\left(\frac{t}{\lambda}\right)f\right)\left(\frac{x}{\lambda}\right).$$

Direct computation yields

$$\begin{aligned} u_\lambda(t,x) &= \int_{\mathbb{R}^n} e^{i\xi\cdot\frac{x}{\lambda}} e^{\pm\frac{t}{\lambda}\cdot|\xi|i}\hat{f}(\xi)d\xi = \int_{\mathbb{R}^n} e^{ix\cdot\xi}e^{\pm i|\xi|t}\lambda^n\hat{f}(\lambda\xi)d\xi \\ &= \int_{\mathbb{R}^n} e^{ix\cdot\xi}e^{\pm i|\xi|t}\widehat{f\left(\frac{\cdot}{\lambda}\right)}(\xi)d\xi = (U\tilde{f})(x,t), \quad \tilde{f}(x) = f\left(\frac{x}{\lambda}\right), \end{aligned}$$

and

$$\|u_\lambda(t,x)\|_{L^q(\mathbb{R};L^r(X))} \lesssim \left\|f\left(\frac{x}{\lambda}\right)\right\|_H. \tag{5.129}$$

By rescaling, one has

$$\lambda^{\frac{1}{q}+\frac{\sigma}{r}}\|u\|_{L^q(\mathbb{R};L^r(X))} \lesssim \lambda^{\frac{\sigma}{2}}\|f(x)\|_H. \tag{5.130}$$

The equivalency between (5.129) and (5.130) calls for $\frac{1}{q} = \sigma(\frac{1}{2} - \frac{1}{r})$, $\sigma = \frac{n-1}{2}$, which is also sufficient.

Let $u(t,x)$ solve the free Schrödinger equation, then

$$u_\lambda(t,x) = u(\lambda^{-2}t, \lambda^{-1}x)$$

still solves free Schrödinger equation. Direct computation gives

$$\begin{aligned} u_\lambda(t,x) &= \left(S\left(\frac{t}{\lambda^2}\right)f\right)\left(\frac{x}{\lambda}\right) = \int_{\mathbb{R}^n} e^{i\lambda^{-1}x\cdot\xi}e^{i|\xi|^2\lambda^{-2}t}\hat{f}(\xi)d\xi \\ &= \int_{\mathbb{R}^n} e^{ix\cdot\xi}e^{i|\xi|^2t}\widehat{f\left(\frac{\cdot}{\lambda}\right)}(\xi)d\xi = \left[S(t)f\left(\frac{\cdot}{\lambda}\right)\right](x,t). \end{aligned}$$

By the Strichartz estimate, one has

$$\|u_\lambda(t,x)\|_{L^q(\mathbb{R};L^r(X))} \lesssim \left\|f\left(\frac{x}{\lambda}\right)\right\|_H. \tag{5.131}$$

Notice that the λ, which appeared in (5.128), turns to λ^2 (the Schrödinger equation has different scaling from the wave equation with respect to the variable t). By rescaling, one has

$$\lambda^{\frac{2}{q}+\frac{2\sigma}{r}}\|u(t,x)\|_{L^q(\mathbb{R};L^r(X))} \lesssim \lambda^\sigma \|f\|_X. \tag{5.132}$$

Therefore, (5.131) and (5.132) hold if and only if $\frac{1}{q} = \sigma(\frac{1}{2} - \frac{1}{r})$.

We split the proof of Theorem 5.2.4 into the following steps:

Step I. *Nonend-point case.* When $(q,r) \neq P$, the homogeneous Strichartz estimates (5.119), (5.120) and (5.121) are equivalent to each other, while the in-homogeneous Strichartz estimate (5.122) follows by Christ–Kiselev's lemma and (5.121). As a result, by TT^* argument, one reduces the proof to

$$T(F,G) \triangleq \left|\iint \langle V(s)^*F(s), V(t)^*G(t)\rangle ds dt\right| \lesssim \|F\|_{L_t^{q'}L_x^{r'}}\|G\|_{L_t^{q'}L_x^{r'}}. \tag{5.133}$$

By the energy estimate (5.115) and the dispersive estimate (5.116), one has

$$\begin{aligned}|\langle V(s)^*F(s), V(t)^*G(t)\rangle| &\lesssim \|V^*(t)F(t)\|_H\|V^*(s)G(s)\|_H \\ &\lesssim \|F(t)\|_2\|G(s)\|_2,\end{aligned} \tag{5.134}$$

and

$$|\langle V(s)^*F(s), V(t)^*G(t)\rangle| \lesssim |t-s|^{-\sigma}\|F(s)\|_1\|G(t)\|_1. \tag{5.135}$$

By interpolation, we have

$$|\langle V(s)^*F(s), V(t)^*G(t)\rangle| \lesssim |t-s|^{-1-\beta(r,r)}\|F(s)\|_{r'}\|G(t)\|_{r'}. \tag{5.136}$$

Here,

$$\beta(r,\tilde{r}) \triangleq \sigma - 1 - \frac{\sigma}{r} - \frac{\sigma}{\tilde{r}}. \tag{5.137}$$

It is easy to see that for any admissible pairs, $\beta(r,r) \leq 0$. When (q,r) is σ sharp, we have

$$\frac{1}{q'} - \frac{1}{q} = -\beta(r,r).$$

Thus, when $(q,r) \neq P$, we have $q > q'$, $\beta(r,r) < 0$. Hence, we get (5.133) by Hardy–Littlewood–Sobolev's inequality.

If $V(t)$ satisfies the truncated dispersive estimate, then (5.135) can be improved to

$$|\langle V(s)^* F(s), V(t)^* G(t)\rangle| \lesssim (1 + |t - s|)^{-1-\beta(r,r)} \|F(s)\|_{r'} \|G(t)\|_{r'}. \tag{5.138}$$

When (q, r) is not sharp σ-admissible, we have

$$-\beta(r, r) + \frac{1}{q} < \frac{1}{q'}.$$

And (5.133) follows by the generalized Young inequality.

Remark 5.2.5.

(i) That $V(t)$ satisfies the truncated dispersive estimate is aimed at wave equations. However, using the untruncated estimate and Sobolev's inequality, we can derive the same conclusion for wave equations as can be seen in the previous section.

(ii) In Step I, the Hardy–Littlewood–Sobolev inequality takes the form

$$\left|\int_{\mathbb{R}}\int_{\mathbb{R}} f(x)g(y)|x-y|^{-\lambda}dxdy\right| \lesssim \|f(x)\|_{L^p(\mathbb{R})} \cdot \|g(x)\|_{L^q(\mathbb{R})}. \tag{5.139}$$

Here, $f \in L^p(\mathbb{R})$, $g \in L^q(\mathbb{R})$, $q, p > 1$ satisfy

$$\frac{1}{p} + \frac{1}{q} + \lambda = 2.$$

The corresponding Young's inequality is

$$\left|\int_{\mathbb{R}}\int_{\mathbb{R}} f(x)g(y)h(x-y)dxdy\right| \lesssim \|f\|_{L^p}\|g\|_{L^q}\|h\|_{L^l}, \tag{5.140}$$

where $f \in L^p$, $g \in L^q$, $h \in L^l$, $p, q, l > 1$ satisfies

$$\frac{1}{p} + \frac{1}{q} + \frac{1}{l} = 2.$$

Step II. End-point case. Since

$$P = (q, r) = \left(2, \frac{2\sigma}{\sigma - 1}\right), \quad \sigma > 1$$

is sharp σ-admissible, we only need to prove the Strichartz estimate when $V(t)$ satisfies untruncated dispersive estimate. By the TT^* argument, it is equivalent to proving (5.133). Applying the dyadic decomposition, we have

$$T(F, G) = \sum_{j\in\mathbb{Z}} T_j(F, G). \tag{5.141}$$

Here,

$$T_j(F,G) = \int\limits_{t-2^{j+1}<s\le t-2^j} \langle V(s)^*F(s), V(t)^*G(t)\rangle dsdt.$$

Thus, (5.133) is reduced to show

$$\sum_{j\in\mathbb{Z}} |T_j(F,G)| \lesssim \|F\|_{L_t^2 L_x^{r'}} \|G\|_{L_t^2 L_x^{r'}}. \tag{5.142}$$

Lemma 5.2.4. *The following estimate holds:*

$$|T_j(F,G)| \lesssim 2^{-j\beta(a,b)} \|F\|_{L_t^2 L_x^{a'}} \|G\|_{L_t^2 L_x^{b'}}, \quad j \in \mathbb{Z}, \tag{5.143}$$

where $(\frac{1}{a}, \frac{1}{b})$ *is in a neighborhood* $\blacksquare$ *of* $(\frac{1}{r}, \frac{1}{r})$*, that is,*

$$\blacksquare = \left\{\left(\frac{1}{a}, \frac{1}{b}\right) \mid \left(\frac{1}{a}, \frac{1}{b}\right) \in \overline{ABCD} \setminus \{B, D\},\ A = (0,0), \right.$$
$$\left. B = \left(\frac{1}{2}, \frac{1}{r}\right),\ C = \left(\frac{1}{2}, \frac{1}{2}\right),\ D = \left(\frac{1}{r}, \frac{1}{2}\right)\right\}.$$

Proof. We first show that (5.142) is invariant under scaling (5.128) (take the wave equation, for example)

$$\begin{aligned} T_j(F,G) &= \int\limits_{t-2^{j+1}<s\le t-2^j} \langle V(s)^*F(s), \quad V(t)^*G(t)\rangle dsdt \\ &= 2^{2j} \int\limits_{2^j\tilde{t}-2^{j+1}<2^j\tilde{s}\le 2^j\tilde{t}-2^j} \langle V(2^j\tilde{s})^*F(2^j\tilde{s}), V(2^j\tilde{t})^*G(2^j\tilde{t})\rangle d\tilde{s}d\tilde{t} \\ &= 2^{2j} \int\limits_{t-2<s\le t-1} \langle V(2^js)F(2^js), V(2^jt)G(2^jt)\rangle dsdt. \end{aligned}$$

Note that

$$\begin{aligned} V(2^js)F(2^js) &= \int\limits_{\mathbb{R}^n} e^{ix\xi} e^{\pm i2^j|\xi|s} \hat{F}(2^js,\xi)d\xi \\ &= \int\limits_{\mathbb{R}^n} e^{i2^{-j}x\cdot\xi} e^{\pm i|\xi|s} (F(2^js, 2^jx))^\wedge(\xi)d\xi. \end{aligned}$$

Hence, by scaling (5.128), we see that

$$\begin{aligned} T_j(F,G) = 2^{2j} \int\limits_{t-2<s\le t-1} &\langle (V(s)F(2^js, 2^jx))(s, 2^{-j}x) \\ &\cdot (V(t)F(2^jt, 2^jx))(t, 2^{-j}x)\rangle dsdt \end{aligned}$$

$$\begin{aligned}
&\lesssim 2^{2j+j\sigma} \int\limits_{t-2<s\le t-1} \langle (V(s)F(2^j s, 2^j x))(s,x) \\
&\quad \cdot (V(t)F(2^j t, 2^j x))(t,x)\rangle ds dt \\
&\lesssim 2^{2j+j\sigma} \|F(2^j s, 2^j x)\|_{L_t^2 L_x^{a'}} \|G(2^j t, 2^j x)\|_{L_t^2 L_x^{b'}} \\
&\lesssim 2^{2j+j\sigma-\frac{1}{2}j-\frac{\sigma}{a'}j-\frac{1}{2}j-\frac{\sigma}{b'}j} \|F\|_{L_t^2 L_x^{a'}} \|G\|_{L_t^2 L_x^{b'}} \\
&\lesssim 2^{j-\sigma j+\frac{\sigma}{a}j+\frac{\sigma}{b}j} \|F\|_{L_t^2 L_x^{a'}} \|G\|_{L_t^2 L_x^{b'}} \\
&= 2^{-\beta(a,b)j} \|F\|_{L_t^2 L_x^{a'}} \|G\|_{L_t^2 L_x^{b'}}.
\end{aligned} \tag{5.144}$$

Therefore, $T_j(F,G)$, $\forall j \in \mathbb{Z}$ can be reduced to the estimate of $T_0(F,G)$. By interpolation, we only need to show (5.142) in the following three cases:

(i) $a = b = \infty$,

(ii) $2 \le a < r, b = 2$,

(iii) $2 \le b < r, a = 2$.

First, we prove (i). Integrate (5.135) on both sides with respect to t and s, we have

$$\begin{aligned}
|T_0(F,G)| &\lesssim \int\limits_{1\le|t-s|\le 2} \|V(t)V^*(s)F(t)\|_{L_x^\infty} \|G(s)\|_{L_x^1} dt ds \\
&\le C \int\limits_{\mathbb{R}} \int\limits_{1\le|t-s|\le 2} \|F(t)\|_{L_x^1} dt \|G(s)\|_{L_x^1} ds \\
&\le C \left\| \int\limits_{1\le|t-s|\le 2} \|F(t)\|_1 dt \right\|_{L_s^2} \|G(s)\|_{L_t^2(L_x^1)} \\
&\le C \|F\|_{L_t^2 L_x^1} \|G\|_{L_s^2 L_x^1}.
\end{aligned} \tag{5.145}$$

Next, we prove (ii). For any $r > a$, let $(q(a), a)$ is sharp σ-admissible, $F_t(s) = I_{1\le|t-s|\le 2} F(s)$. By energy inequality and the nonend-point Strichartz estimate, we have

$$\begin{aligned}
|T_0(F,G)| &= \left| \int\limits_{1\le|t-s|\le 2} \langle V(s)^* F(s), V(t)^* G(t)\rangle ds dt \right| \\
&= \left| \int\limits_{\mathbb{R}} \int\limits_{\mathbb{R}} \langle V(s)^* F_t(s), V(t)^* G(t)\rangle ds dt \right| \\
&\le \int\limits_{\mathbb{R}} \left\| \int\limits_{\mathbb{R}} V(s)^* F_t(s) ds \right\|_H \|V(t)^* G(t)\|_H dt \\
&\le \int\limits_{\mathbb{R}} \left(\int\limits_{1\le|t-s|\le 2} \|F(s)\|_{L^{a'}}^{q(a)'} ds \right)^{\frac{1}{q(a)'}} \|G(t)\|_2 dt
\end{aligned}$$

$$\leq \left\| \left(\int_{\mathbb{R}} I_{1\leq|t-s|\leq 2}(t-s)\|F(s)\|_{L^{a'}}^{q(a)'}\, ds \right)^{\frac{1}{q(a)'}} \right\|_{L_t^2} \|G(t)\|_{L^2(L^2)}$$

$$\lesssim \|F\|_{L_t^2 L_x^{a'}} \|G\|_{L_t^2 L_x^2}. \tag{5.146}$$

By symmetry, we can prove (iii). □

Lemma 5.2.5. *Let $j = 0, 1$, $1 \leq p_j, q_j \leq \infty$. Assume that the bilinear operator $\mathcal{T}(\cdot,\cdot)$ is continuous on $L_t^2(L^{p_j}) \times L_t^2(L^{q_j})$, then for $\forall\theta \in [0,1]$, the bilinear operator $\mathcal{T}(\cdot,\cdot)$ is continuous on $L_t^2(L^{p_\theta}) \times L_t^2(L^{q_\theta})$. Here,*

$$\left(\frac{1}{p_\theta}, \frac{1}{q_\theta}\right) = (1-\theta)\left(\frac{1}{p_0}, \frac{1}{q_0}\right) + \theta\left(\frac{1}{p_1}, \frac{1}{q_1}\right). \tag{5.147}$$

On account of the above bilinear estimate, we complete the proof of Lemma 5.2.4.
Step III. Complete the proof of Theorem 5.2.4. Take

$$a = r, \quad b = r, \quad \beta(r,r) = \sigma - 1 - \frac{2\sigma}{r} = 0. \tag{5.148}$$

Lemma 5.2.4 implies

$$|T_j(F,G)| \lesssim \|F\|_{L_t^2 L_x^{r'}} \|G\|_{L_t^2 L_x^{r'}}, \tag{5.149}$$

which is insufficient to prove (5.142). However, for any other $(\frac{1}{a}, \frac{1}{b}) \in \blacksquare$, one has the estimate like (5.143) with a exponent decaying coefficients, which means that we can further improve (5.149).

Apply atomic decomposition to $F(t), G(s) \in L^{r'}(\mathbb{R}^n)$, by the atomic characterization for $L^{r'}$, one has

$$F(t) = \sum_{k\in\mathbb{Z}} c_k(t) F_k(t,x), \quad G(s) = \sum_{k'\in\mathbb{Z}} d_{k'}(s) G_{k'}(s). \tag{5.150}$$

For $(\frac{1}{a}, \frac{1}{b})$ in the neighborhood $\blacksquare$ of $(\frac{1}{r}, \frac{1}{r})$, noticing that $\sigma - 1 = \frac{2\sigma}{r}$ and using Theorem 5.2.3 and Lemma 5.2.4, we get

$$T_j(F_k, G_{k'}) \lesssim \|c_k\|_{L_t^2} \|d_{k'}\|_{L_t^2} 2^{-j\beta(a,b)} 2^{-k(\frac{1}{r'}-\frac{1}{a'})} 2^{-k'(\frac{1}{r'}-\frac{1}{b'})}$$

$$\lesssim \|c_k\|_{L_t^2} \|d_{k'}\|_{L_t^2} 2^{(-j\sigma+k)(\frac{1}{r}-\frac{1}{a})} 2^{(-j\sigma+k')(\frac{1}{r}-\frac{1}{b})}. \tag{5.151}$$

For a fixed $r < \infty$, there is $\varepsilon > 0$, we can always choose a, b so that

$$(-j\sigma + k)\left(\frac{1}{r} - \frac{1}{a}\right) \leq -2\varepsilon|j\sigma - k|, \quad (-j\sigma + k')\left(\frac{1}{r} - \frac{1}{b}\right) \leq -2\varepsilon|j\sigma - k'|. \tag{5.152}$$

Hence,

$$|T_j(F_k, G_{k'})| \lesssim \|c_k\|_{L_t^2} \|d_{k'}\|_{L_t^2} 2^{-\varepsilon|j\sigma-k|} 2^{-\varepsilon|j\sigma-k'|} 2^{-\varepsilon|k-k'|}. \tag{5.153}$$

Using Hölder's inequality and Young's inequality in its discrete form, we have

$$\begin{aligned}
|T(F,G)| &\le C \sum_{j,k,k'\in\mathbb{Z}} \|c_k\|_{L_t^2(\mathbb{R})} \|d_{k'}\|_{L_t^2(\mathbb{R})} 2^{-\varepsilon|j\sigma-k|} 2^{-\varepsilon|k-k'|} \\
&\le C \sum_{k,k'\in\mathbb{Z}} \|c_k\|_{L_t^2(\mathbb{R})} \|d_{k'}\|_{L_t^2(\mathbb{R})} 2^{-\varepsilon|k-k'|} \\
&\le C_\varepsilon \Big(\sum_{k\in\mathbb{Z}} \|c_k\|_{L_t^2(\mathbb{R})}^2 \Big)^{\frac{1}{2}} \Big(\sum_{k'\in\mathbb{Z}} \|d_{k'}\|_{L_t^2(\mathbb{R})}^2 \Big)^{\frac{1}{2}} \\
&\le C_\varepsilon \Big(\int_{\mathbb{R}} \|\{c_k\}\|_{\ell^2}^2 dt \Big)^{\frac{1}{2}} \Big(\int_{\mathbb{R}} \|\{d_{k'}\}\|_{\ell^2}^2 dt \Big)^{\frac{1}{2}} \\
&\le C_\varepsilon \Big(\int_{\mathbb{R}} \|\{c_k\}\|_{\ell^{r'}}^2 dt \Big)^{\frac{1}{2}} \Big(\int_{\mathbb{R}} \|\{d_{k'}\}\|_{\ell^{r'}}^2 dt \Big)^{\frac{1}{2}} \\
&\le C_\varepsilon \|F\|_{L_t^2(L_x^{r'})} \|G\|_{L_t^2(L_x^{r'})},
\end{aligned} \tag{5.154}$$

where we have used the embedding $\ell^{r'} \hookrightarrow \ell^2$, $r' \le 2$. Thus, we complete the proof of Theorem 5.2.4.

Remark 5.2.6.

(i) It is easy to see from (5.153) that (5.149) attains its sharp only if the supports of F and G are contained in a set of measure $2^{j\sigma}$, otherwise the estimates like (5.153) always hold.

(ii) Noting that $r' < 2$, one has room in $\ell^{r'} \hookrightarrow \ell^2$. In fact, one can replace in the Strichartz estimate (5.119) L_x^r by the Lorentz space $L_x^{r,2}$ by interpolation.

(iii) We have a straightforward proof for (5.142) by means of the bilinear interpolation technique in [14]; see also [129].

Next, we apply the abstract Strichartz estimate to prove the corresponding space-time estimates for the wave and Schrödinger equation.

Proof of Corollary 5.2.1. "Proof of the necessity." From Remark 5.2.4, the gap condition (5.125) follows by the scaling argument (dimensional analysis), and the admissible relation

$$\frac{2}{q} \le (n-1)\Big(\frac{1}{2} - \frac{1}{r}\Big), \quad \frac{2}{\tilde{q}} \le (n-1)\Big(\frac{1}{2} - \frac{1}{\tilde{r}}\Big)$$

is due to Knapp's example (see the previous section):

$$(q, r, n) \ne (2, \infty, 3), \quad (\tilde{q}, \tilde{r}, n) \ne (2, \infty, 3) \tag{5.155}$$

is due to the counterexample in Montgomery–Smith [203]. By the TT^* method, the Strichartz estimate is reduced essentially to that the operator is bounded $V_\pm V_\pm^* = K*$: $L_t^{q'}L_x^{r'} \to L_t^q L_x^r$. In particular, when $T = \infty$, $V_\pm V_\pm^*$ is translation invariant, which implies $q > q'$. By symmetry, one has $\tilde{q} > \tilde{q}'$. Here, $V_\pm$ is the truncated wave evolution operator defined by

$$V_\pm(t)f(x) = \mathcal{F}^{-1}(\chi_{[0,T]}\beta(\xi)e^{\pm it|\xi|}\hat{f}(\xi)). \tag{5.156}$$

Here, $\beta(\xi) = \beta(|\xi|) \in C_0^\infty(\mathbb{R}^n)$ is bump function satisfying

$$\beta(\xi) = \begin{cases} 1, & \frac{3}{4} \le |\xi| \le \frac{8}{3}, \\ 0, & |\xi| \ge \frac{3}{8} \text{ or } |\xi| > \frac{16}{3}, \end{cases} \qquad \beta_j = \beta\Big(\frac{\xi}{2^j}\Big).$$

Indeed, $\beta(\xi)$ is $\tilde{\psi}_0(\xi) = \psi_{-1}(\xi) + \psi_0(\xi) + \psi_1(\xi)$ in the Littlewood–Paley decomposition.

“Proof of sufficiency”: Let $q, r, \gamma, \tilde{q}, \tilde{r}$ satisfy the condition in Corollary 5.2.1. We know that

$$\begin{aligned} u(t) &= \cos(t\sqrt{-\Delta})f + \frac{\sin(t\sqrt{-\Delta})}{\sqrt{-\Delta}}g + GF(t) \\ &\triangleq \cos(t\sqrt{-\Delta})f + \frac{\sin(t\sqrt{-\Delta})}{\sqrt{-\Delta}}g + \int_0^t \frac{\sin((t-s)\sqrt{-\Delta})}{\sqrt{-\Delta}}F(s)ds \end{aligned} \tag{5.157}$$

solves (5.69). If we choose $H = L^2(\mathbb{R}^n)$, $X = \mathbb{R}^n$, $\sigma = \frac{n-1}{2}$, by Plancherel's theorem, we have $V_\pm(t)$ satisfies the energy inequality (5.115). The standard stationary phase method gives that (i. e., the estimate on the oscillatory integral corresponding to the kernel of $V_\pm(t)(V_\pm(s))^*$), $V_\pm(t)$ satisfies the truncated dispersive estimate (5.117). According to Theorem 5.2.4, we conclude that $V_\pm(t)$ satisfies the following Strichartz estimate:

$$\|V_\pm(t)f\|_{C(I;L_x^2)}, \ \|V_\pm^*(t)f\|_{L_t^q L_x^r} \lesssim \|f\|_2, \tag{5.158}$$

$$\Big\|\int_{t>s} V_\pm(t)V_\pm^*(s)F(s)ds\Big\|_{C(I;L_x^2)\cap L_t^q L_x^r} \lesssim \|F\|_{L_t^{\tilde{q}'}L_x^{\tilde{r}'}}. \tag{5.159}$$

Applying the same argument as in the previous section, by Littlewood–Paley's theory, (5.157), (5.158), (5.159), we have the estimate (5.124). By Plancherel's theorem, we know that $V_\pm f$ is continuous in L^2. For the continuity of the nonhomogeneous part $G_\pm F(t)$, one only need to use that

$$G_\pm F = \int_{t>s} V_\pm(t)V_\pm^*(s)F(s)ds = e^{i\varepsilon\sqrt{-\Delta}}G_\pm F(t) - G_\pm(\chi_{[t,t+\varepsilon]}F)(t)$$

and

$$\|\chi_{[t,t+\varepsilon]}F\|_{L_t^{\tilde{q}'}L_x^{\tilde{r}'}} \longrightarrow 0, \quad \varepsilon \to 0,$$

to get the desired conclusion. □

Proof to Corollary 5.2.2. For necessity, by scaling argument, one easily sees that (q,r) and $(\tilde{q},\tilde{r})$ have to satisfy the admissible condition, while Montgomery–Smith's counterexample shows that when $n = 2$ the end-point $(q,r) = (2,\infty)$ is not admissible. For sufficiency, we note that

$$u(t) = S(t)f + GF(t) = \chi_{[0,T]}e^{it\Delta}f + \int\limits_{s<t} S(t)S^*(s)F(s)ds$$

solves (5.126), and we only have to take $H = L^2(\mathbb{R}^n)$, $X = \mathbb{R}^n$, $\sigma = n/2$. By the energy inequality $\|e^{it\Delta}f\|_2 \lesssim \|f\|_2$ and the dispersive estimate,

$$\|e^{i(t-s)\Delta}f\|_\infty \lesssim |t-s|^{-\frac{n}{2}}\|f\|_1,$$

where

$$e^{it\Delta}f = \frac{1}{(4\pi it)^{n/2}}\int\limits_{\mathbb{R}^n} e^{-\frac{|x-y|^2}{4it}}f(y)dy.$$

We conclude Corollary 5.2.2. □

Next, we perform the Strichartz estimate for wave equations in the framework of Besov spaces. We start with the discussion on each dyadic region in the Fourier transform side.

Theorem 5.2.5. *Assume that $u(t,x)$ solves the linear equation* (5.1), *that is, it is the solution of*

$$\square u = F(t,x), \quad \partial u|_{t=0} = \gamma(x) \triangleq (\nabla f, g)$$

where

$$\partial u = (\nabla u, \partial_t u) = (\partial_1 u, \dots, \partial_n u, \partial_t u).$$

Then for any wave admissible pair $(q_j, r_j) \in \tilde{\Lambda}$, $j = 1,2$, we have the following Strichartz estimate:

$$\|\partial\Delta_j u\|_{L^{q_1}(L^{r_1}(\mathbb{R}^n))} \leq C2^{j\mu_1}\|\Delta_j\gamma\|_{L^2} + C2^{j\mu_{12}}\|\Delta_j F\|_{L^{q_2'}(L^{r_2'})}. \tag{5.160}$$

Here,

$$\mu_1 = \delta(r_1) - \frac{1}{q_1}, \quad \mu_{12} = \delta(r_1) + \delta(r_2) - \frac{1}{q_1} - \frac{1}{q_2}. \tag{5.161}$$

Remark 5.2.7. It is clear and convenient to use the notation above by Y. Chemin. Using our familiar notation, (5.160) is

$$\|\Delta_j u\|_{L^{q_1}(L^{r_1}(\mathbb{R}^n))} \lesssim 2^{j\mu_1}\|\Delta_j f\|_{L^2} + 2^{j(\mu_1-1)}\|\Delta_j g\|_{L^2} + 2^{(\mu_{12}-1)j}\|\Delta_j F\|_{L^{q_2'}(L^{r_2'})}. \tag{5.162}$$

Proof. Denote

$$U(t)h = e^{it(-\Delta)^{\frac{1}{2}}}h = \mathcal{F}^{-1}(e^{i|\xi|t}\hat{h}(\xi)), \quad U^*(t)h = U(-t)h.$$

Therefore, the solution to (5.1) can be represented by

$$\begin{aligned} u(x,t) &= U(t)f_+ + U(-t)f_- + \int_0^t \frac{U(t-\tau) - U^*(t-\tau)}{2(-\Delta)^{\frac{1}{2}}i}F(t,x)d\tau \\ &= \frac{U(t)+U^*(t)}{2}f + \frac{U(t)-U^*(t)}{2(-\Delta)^{\frac{1}{2}}i}g + \int_0^t \frac{U(t-\tau)-U^*(t-\tau)}{2(-\Delta)^{\frac{1}{2}}i}F(t,x)d\tau \\ &= \cos(-\Delta)^{\frac{1}{2}}tf + \frac{\sin(-\Delta)^{\frac{1}{2}}t}{(-\Delta)^{\frac{1}{2}}}g + \int_0^t \frac{\sin(-\Delta)^{\frac{1}{2}}(t-\tau)}{(-\Delta)^{\frac{1}{2}}}F(\tau,x)d\tau, \end{aligned} \tag{5.163}$$

where $\hat{f}_\pm(\xi) = \frac{1}{2}(\hat{f}(\xi) \pm \frac{\hat{g}(\xi)}{i|\xi|})$. As in (5.156), we introduce the truncated wave operator

$$V_\pm(t)h(x) = \mathcal{F}^{-1}(\chi_{[0,T]}\beta(\xi)e^{\pm it|\xi|}\hat{h}(\xi)).$$

Take $H = L^2(\mathbb{R}^n)$, $\sigma = \frac{n-1}{2}$, $X = \mathbb{R}^n$, by the oscillatory integral estimate and Theorem 5.2.2, we deduce that $V_\pm$ verifies the energy inequality (5.115) and the truncated dispersive estimate (5.117). Thus, Keel–Tao's abstract Strichartz estimate implies

$$\|V_\pm(t)h\|_{C(I;L^2_x(\mathbb{R}^n))}, \quad \|V_\pm^*(t)h\|_{C(I;L^2_x(\mathbb{R}^n))} \lesssim \|h\|_{L^2(\mathbb{R}^n)}, \tag{5.164}$$

$$\Big\|\int_{s<t} V_\pm(t)V_\pm^*(s)F(s)ds\Big\|_{C(I;L^2_x)\cap L^{q_1}_t(L^{r_1}_x)} \lesssim \|F\|_{L^{q_2'}_t L^{r_2'}_x}, \tag{5.165}$$

where $(q_j, r_j) \in \tilde{\Lambda}$, $j = 1, 2$. Note that

$$\beta(\xi)\psi_0(\xi) = \psi_0(\xi), \quad |\beta(\xi)|^2\psi_0(\xi) = \psi_0(\xi).$$

Hence, we have

$$U_\pm\Delta_0 h = \mathcal{F}^{-1}(e^{\pm it|\xi|}\psi_0\hat{h}(\xi)) = \mathcal{F}^{-1}(e^{\pm it|\xi|}\beta(\xi)\psi_0\hat{h}(\xi)) = V_\pm(t)\Delta_0 h.$$

Therefore,

$$\|U_\pm(t)\Delta_0 h\|_{C(I;L^2_x(\mathbb{R}^n))}, \quad \|U^*_\pm(t)\Delta_0 h\|_{C(I;L^2_x(\mathbb{R}^n))} \lesssim \|\Delta_0 h\|_{L^2(\mathbb{R}^n)}, \tag{5.166}$$

$$\Big\| \int_{s<t} U(t)U^*(s)\Delta_0 F(s)ds \Big\|_{C(I;L^2_x(\mathbb{R}^n))\cap L^{q_1}_t(L^{r_1}_x)} \lesssim \|\Delta_0 F\|_{L^{q_2'}_t L^{r_2'}_x}. \tag{5.167}$$

On account of this, for any wave admissible pair $(q_1, r_1), (q_2, r_2) \in \tilde{\Lambda}$, we have

$$\|\Delta_0 u\|_{L^{q_1}(L^{r_1}(\mathbb{R}^n))} \lesssim \|\Delta_0 f\|_{L^2} + \|\Delta_0 g\|_{L^2} + \|\Delta_0 F\|_{L^{q_2'}(L^{r_2'})}. \tag{5.168}$$

By the scaling argument, we see

$$\begin{aligned}
\|\Delta_j \partial u(t,x)\|_{L^{q_1}(L^{r_1})} &\le \Big\| \int_{\mathbb{R}^n} 2^{jn}\psi_0(2^j y)\partial u(t, x-y)dy \Big\|_{L^{q_1}(L^{r_1})} \\
&\le \Big\| \int_{\mathbb{R}^n} \psi_0(y)(\partial u)(2^{-j}(2^j t), 2^{-j}(2^j x - y))dy \Big\|_{L^{q_1}(L^{r_1})} \\
&\le 2^{-\frac{n}{r_1}j-\frac{j}{q_1}} \Big\| \int_{\mathbb{R}^n} \psi_0(y)(\partial u)(2^{-j}t, 2^{-j}(x-y))dy \Big\|_{L^{q_1}(L^{r_1})} \\
&\le C2^{-\frac{n}{r_1}j-\frac{j}{q_1}} [\|\Delta_0 \gamma(2^{-j}x)\|_{L^2(\mathbb{R}^n)} + \|\Delta_0 F(2^{-j}x, 2^{-j}t)\|_{L^{q_2'}(L^{r_2'})}] \\
&\le C2^{-\frac{n}{r_1}j-\frac{j}{q_1}} \Big[\Big\| \int \psi_0(y)\gamma(2^{-j}(x-y))dy \Big\|_{L^2(\mathbb{R}^n)} \\
&\quad + \Big\| \int \psi_0(y)F(2^{-j}(x-y), 2^{-j}t) \Big\|_{L^{q_2'}(L^{r_2'})} \Big] \\
&\le C2^{-\frac{n}{r_1}j-\frac{j}{q_1}} \Big[\Big\| \int \psi_j(y)\gamma(2^{-j}x-y)dy \Big\|_{L^2(\mathbb{R}^n)} \\
&\quad + \Big\| \int \psi_j(y)F(2^{-j}x-y, 2^{-j}t)dy \Big\|_{L^{q_2'}(L^{r_2'})} \Big] \\
&\le C2^{-\frac{n}{r_1}j-\frac{j}{q_1}} [\|\Delta_j \gamma\|_{L^2(\mathbb{R}^n)} 2^{\frac{jn}{2}} + \|\Delta_j F\|_{L^{q_2'}(L^{r_2'})} 2^{\frac{j}{q_2'}+\frac{j}{r_2'}}] \\
&\le C2^{j\mu_1}\|\Delta_j \gamma\|_{L^2} + C2^{j\mu_{12}}\|\Delta_j F\|_{L^{q_2'}(L^{r_2'})}.
\end{aligned} \tag{5.169}$$

And similarly, we have (5.162).

Multiplying by $2^{j\sigma}$ on both sides of (5.169) and taking ℓ^2 norm, we get the Strichartz estimate in the mixed space-time:

$$\|\partial u\|_{\mathcal{L}^{q_1}(\dot{B}^{\sigma}_{r_1,2}(\mathbb{R}^n))} \le C\|\gamma\|_{\dot{H}^{\mu_1+\sigma}} + C\|F\|_{\mathcal{L}^{q_2'}(\dot{B}^{\sigma+\mu_{12}}_{r_2',2})}. \tag{5.170}$$

Noting that $q_1 \ge 2$, $q_2' \le 2$, by Minkowski's inequality, we get the Strichartz estimate in the framework of Besov space. □

Theorem 5.2.6. *Assume $\sigma \in \mathbb{R}$, under the condition of Theorem 5.2.5, we have the following Strichartz estimate:*

$$\|\partial u\|_{L^{q_1}(\dot{B}^{\sigma}_{r_1,2})} \leq C\|\gamma\|_{\dot{H}^{\mu_1+\sigma}} + C\|F\|_{L^{q_2'}(\dot{B}^{\sigma+\mu_{12}}_{r_2',2})} \tag{5.171}$$

or

$$\|u\|_{L^{q_1}(\dot{B}^{\sigma}_{r_1,2})} \leq C\|f\|_{\dot{H}^{\mu_1+\sigma}} + C\|g\|_{\dot{H}^{\mu_1+\sigma-1}} + C\|F\|_{L^{q_2'}(\dot{B}^{\sigma+\mu_{12}-1}_{r_2',2})}. \tag{5.172}$$

Remark 5.2.8.

(i) Noticing that $\dot{B}^0_{r_1,2} \hookrightarrow L^{r_1}, L^{r_2'} \hookrightarrow \dot{B}^0_{r_2',2}$, we see that Theorem 5.2.6 implies

$$\begin{aligned}&\|u\|_{L^{q_1}([0,T);L^{r_1})} + \|u\|_{C([0,T)(\dot{H}^{\mu_1})} + \|u\|_{C([0,T)(\dot{H}^{\mu_1-1})}\\ &\quad \leq C\|f\|_{\dot{H}^{\mu_1}} + C\|g\|_{\dot{H}^{\mu_1-1}} + C\|F\|_{L^{q_2'}([0,T);L^{r_2'})},\end{aligned} \tag{5.173}$$

where $(q_j, r_j) \in \tilde{\Lambda}, j = 1, 2$, and

$$\mu_1 = \delta(r_1) - \frac{1}{q_1}, \quad \delta(r_1) + \delta(r_2) - \frac{1}{q_1} - \frac{1}{q_2} = 1.$$

(ii) Noting the energy estimate $\dot{B}^0_{r,2} \hookrightarrow \dot{F}^0_{r,2} = L^r$, we derive that Theorem 5.2.6 implies

$$\begin{aligned}&\|u\|_{L^{q}([0,T);L^{r})} + \|u\|_{C([0,T)(\dot{H}^{\mu})} + \|u_t\|_{C([0,T)(\dot{H}^{\mu-1})}\\ &\quad \leq C\|f\|_{\dot{H}^{\mu}} + C\|g\|_{\dot{H}^{\mu-1}} + C\|F\|_{L^{1}([0,T);\dot{H}^{\mu-1})}.\end{aligned} \tag{5.174}$$

Here, $(q, r) \in \tilde{\Lambda}$ and $\mu = \delta(r) - \frac{1}{q}$.

Remark 5.2.9. That (5.168) holds for any sharp wave-admissible pairs. In fact, for any two wave-admissible pairs $(q_1, r_1), (q_2, r_2) \in \tilde{\Lambda}$, we can always find $\tilde{r}_1 \leq r_1$, $\tilde{r}_2 \leq r_2$ such that $(q_1, \tilde{r}_1), (q_2, \tilde{r}_2) \in \Lambda$ is sharp wave admissible. Hence, (5.168) holds for any $(q_1, \tilde{r}_1)$ and $(q_2, \tilde{r}_2)$. By Bernstein's estimate, we get

$$\begin{cases}\|\Delta_0 u\|_{L^{q_1}_t L^{r_1}_x} \lesssim \|\Delta_0 u\|_{L^{q_1}_t L^{\tilde{r}_1}_x},\\ \|\partial\Delta_0 u\|_{L^{q_1}_t L^{r_1}_x} \lesssim \|\Delta_0 \partial u\|_{L^{q_1}_t L^{\tilde{r}_1}_x},\\ \|\Delta_0 f\|_{L^{q_2'}_t L^{r_2'}_x} \gtrsim \|\Delta_0 f\|_{L^{q_2'}_t L^{\tilde{r}_2'}_x}.\end{cases}$$

On account of this, for any wave-admissible pairs (q_1, r_1), (q_2, r_2), we have

$$\|\Delta_0 u\|_{L^{q_1}_t L^{r_1}_x} \leq C(\|\Delta_0 u_0\|_2 + \|\Delta_0 u_1\|_2 + \|\Delta_0 f\|_{L^{q_2'}_t L^{r_2'}_x}).$$

Finally, we give an application of the end-point Strichartz estimate. We consider the semilinear wave equations

$$\begin{cases} u_{tt} - \Delta u = F_k(u), \quad k > 1, \\ u(x,0) = f(x) \in \dot{H}^{\gamma}(\mathbb{R}^n), \\ \partial_t u(x,0) = g(x) \in \dot{H}^{\gamma-1}(\mathbb{R}^n), \end{cases} \tag{5.175}$$

where u may take vector value. The nonlinearity $F_k(u) \in C^1$ satisfies

$$\begin{cases} |F_k(u)| \lesssim |u|^k, \\ |uF_k(u)| \sim F_k(u). \end{cases} \tag{5.176}$$

We always try to find the smallest $\gamma = \gamma(k,n)$, such that the local existence holds for (5.175). In other words, to ensure that (5.175) determines a local continuous flow in $\dot{H}^{\gamma}$. Kapitanskii [125] first studied that case when $n = 3$, then Lindblad and Sogge [174] gave almost the best result. Denote

$$k_0 = \begin{cases} \frac{(n+1)^2}{(n-1)^2+4}, & n \geq 3, \\ 3, & n = 2. \end{cases} \tag{5.177}$$

If

$$\begin{cases} k_0 < k < \infty, & n = 2,3, \\ k_0 < k \leq \frac{n+1}{n-3}, & n \geq 4, \end{cases} \tag{5.178}$$

in particular, when $n \geq 4$, $F_k(u) \in C^{\infty}$, (5.178) can also be transferred into

$$k_0 \leq k < \infty, \quad n \geq 4. \tag{5.179}$$

Then for

$$\gamma > \gamma(n,k) = \begin{cases} \frac{n+1}{4} - \frac{1}{k-1}, & k_0 < k \leq \frac{n+3}{n-1}, \\ \frac{n}{2} - \frac{2}{k-1}, & k \geq \frac{n+3}{n-1}, \end{cases} \tag{5.180}$$

(5.175) is locally well posed. When $k < k_0$, there is no good result by now, however, Lindblad and Sogge [174] proved the local well-posedness for (5.175) under

$$\gamma = \frac{n+1}{4} - \frac{(n+1)(n+5)}{4} \cdot \frac{1}{2nk - (n+1)}, \quad 1 + \frac{3}{n} < k < k_0. \tag{5.181}$$

Next, we use the end-point Strichartz estimate to establish the local well-posedness for (5.159) when $k = k_0$.

Theorem 5.2.7. *Assume* $n \geq 4$,

$$\gamma = \gamma_0 = \frac{n-3}{2(n-1)}, \quad k = k_0 = \frac{(n+1)^2}{(n-1)^2+4}. \tag{5.182}$$

Then for any $f(x) \in \dot{H}^{\gamma}$, $g(x) \in \dot{H}^{\gamma-1}$, *there is* $T = T(\|f\|_{\dot{H}^{\gamma}}, \|g\|_{\dot{H}^{\gamma-1}}) > 0$ *and a unique weak solution to* (5.175), $u(t) \in C([0,T), \dot{H}^{\gamma}) \cap C^1([0,T); \dot{H}^{\gamma-1})$ *satisfying*

$$u \in L_t^{q_0}([0,T); L^{r_0}(\mathbb{R}^n)), \quad q_0 = \frac{2(n+1)}{n-3}, \quad r_0 = \frac{2(n^2-1)}{(n-1)^2+4}. \tag{5.183}$$

Proof. We take $\gamma = \gamma_0$, $(q,r) = (q_0, r_0)$ and

$$(\tilde{q}, \tilde{r}) = \left(2, \frac{2(n-1)}{n-3}\right). \tag{5.184}$$

Obviously, (q,r) and $(\tilde{q},\tilde{r})$ are wave-admissible pairs of (5.125) satisfying the gap condition. We use the end-point Strichartz estimate (5.124) to prove Theorem 5.2.7.

By standard contraction argument, we introduce the complete metric space

$$\begin{aligned} X = X(T,M) = \{&u \in L_t^q([0,T); L_x^r),\ u(t) \in C([0,T); \dot{H}^{\gamma}) \\ &\cap C^1([0,T), \dot{H}^{\gamma-1}),\ \|u\|_{L^q((0,T],L^r)} + \|u\|_{C([0,T);\dot{H}^{\gamma})} \leq M\}, \\ d(u,v) = \|&u - v\|_{L^q([0,T],L^r)}, \end{aligned}$$

where T, M are to be determined later. As is well known, the integral equation equivalent to (5.175) is

$$\begin{aligned} u(t) &= \mathcal{F}^{-1}\cos|\xi|t\mathcal{F}f + \mathcal{F}^{-1}\frac{\sin|\xi|t}{|\xi|}\mathcal{F}g \\ &\quad + \int_0^t \mathcal{F}^{-1}\frac{\sin|\xi|(t-\tau)}{|\xi|}\mathcal{F}F_{k_0}(u(\tau))d\tau \\ &\equiv S(t)(f,g) + GF_{k_0}(u). \end{aligned} \tag{5.185}$$

Therefore, we reduce the original problem to studying whether or not the following nonlinear map:

$$\mathcal{T} : \mathcal{T}u(t) = S(t)(f,g) + GF_{k_0}(u) \tag{5.186}$$

has a fixed point in the space X. We take $M = 6(\|f\|_{\dot{H}^{\gamma}} + \|g\|_{\dot{H}^{\gamma-1}})$, using the end-point Strichartz estimate, we see

$$\begin{aligned}\|\mathcal{T}u(t)\|_X &\le 3(\|f\|_{\dot H^\gamma} + \|g\|_{\dot H^{\gamma-1}}) + 3C\|F_{k_0}(u)\|_{L^{\tilde q'} L_x^{\tilde r'}} \\ &\le M/2 + 3CT^{\frac{1}{\tilde q'} - \frac{k}{q}} \|u\|^k_{L^q([0,T);L^r)} \\ &= \frac{M}{2} + 3CT^{\frac{1}{2} - \frac{(n+1)(n-3)}{2(n-1)^2+8}} M^k. \end{aligned} \tag{5.187}$$

Here, we have used

$$r = \tilde r' k, \quad q > \tilde q' k. \tag{5.188}$$

On the other hand, note that

$$\begin{aligned}|F_k(u) - F_k(v)| &= \left| \int_0^t \frac{d}{d\theta} F_k(\theta u + (1-\theta)v) d\theta \right| \\ &= \left| \int_0^1 (u - v) \cdot \nabla F_k(\theta u + (1-\theta)v) d\theta \right| \\ &\lesssim |u - v| \big(|u| + |v|\big)^{k-1}, \end{aligned} \tag{5.189}$$

and

$$\begin{aligned} d(\mathcal{T}u, \mathcal{T}v) &\lesssim \big\| |u - v| \big(|u| + |v|\big)^{k-1} \big\|_{L^{\tilde q'} L_x^{\tilde r'}} \\ &\lesssim T^{\frac{1}{\tilde q'} - \frac{k}{q}} \|u - v\|_{L_t^q L_x^r} \big\| \big(|u| + |v|\big)^{k-1} \big\|_{L_t^{\frac{q}{k-1}} L_x^{\frac{r}{k-1}}} \\ &\lesssim M^{k-1} T^{\frac{1}{2} - \frac{(n+1)(n-3)}{2(n-1)^2+8}} d(u, v), \end{aligned} \tag{5.190}$$

where we used (5.188). Hence, we only need to take

$$M^{k-1} T^{1/2 - \frac{(n+1)(n-3)}{2(n-1)^2+8}} \ll 1. \tag{5.191}$$

Then (5.187) and (5.190) ensure that $\mathcal{T}$ is a contraction map from $X(T, M)$ to itself and, therefore, we prove Theorem 5.2.7. □

5.3 Energy solution for Cauchy problems of nonlinear Klein–Gordon equations

In this section, we consider the global well-posedness of the Cauchy problem in energy space $H^1 \oplus L^2(\mathbb{R}^n)$ for the following nonlinear Klein–Gordon equation:

$$\Box u = u_{tt} - \Delta u = -f(u), \quad (t, x) \in \mathbb{R} \times \mathbb{R}^n, \tag{5.192}$$

$$u(t_0) = \varphi(x), \quad u_t(t_0) = \psi(x), \quad x \in \mathbb{R}^n. \tag{5.193}$$

Here, u is a complex/real-valued function defined in $\mathbb{R} \times \mathbb{R}^n$, and $f(u)$ is a nonlinear function. If $f(u)$ takes the following specific form:

$$f(u) = \lambda_0 u + \lambda u|u|^{p-1}, \quad \lambda_0 \geq 0,\ 1 \leq p < \infty, \tag{5.194}$$

then (5.192) corresponds respectively to the classical nonlinear wave equation in case of ($\lambda_0 = 0$) and nonlinear Klein–Gordon equation when ($\lambda_0 > 0$). We keep on using the notation for the wave equations, such as

$$2\beta(r)/(n+1) = \gamma(r)/(n-1) = \delta(r)/n = \alpha(r) = (1/2 - 1/r) \tag{5.195}$$

to determine $\beta(r), \gamma(r), \delta(r)$ and $\alpha(r)$ with the rest part of the notation omitted. There are two typical approaches to problems (5.192), (5.193). The one is the compactness method and the other is the fixed-point argument. Both of them are based on the conservation of energy to prove the existence of the solutions to (5.192), (5.193) in subspace of $X_e = H^1 \oplus L^2$. In order to have the conservation of energy, it is natural to require the following conformal conditions for the nonlinear interacting terms to satisfy $f(e^{i\omega}u) = e^{i\omega}f(u)$. For instance,

$$f(u) = \sum_{j=1,2} \lambda_j |u|^{p_j-1}u, \quad 1 \leq p_1 \leq p_2 < \infty,\ \lambda_j \geq 0. \tag{5.196}$$

However, for the uniqueness, one needs to put restrictions on p_2. For example, for the nonlinear interacting function in (5.196), one can get the global existence of weak solutions for all $1 \leq p_1 \leq p_2 < \infty$ using the compactness method while the knowledge of the uniqueness develops slowly. Using the contraction method, Segal [231] proved the existence and uniqueness in X_e under the restriction $p_2 < 1 + \frac{2}{n-2}$ for global solutions. Glassey and Tsutsumi in [110] established the uniqueness with $p_2 \leq 1 + \frac{4n}{(n+1)(n-2)}$ for the global weak solutions to (5.192), (5.193). It has been studied in X_e to achieve the global well-posedness for $p_2 \leq 1 + \frac{4}{n-2}$. However, we proceeded much less due to the lack of suitable tools. The results only restrict to certain particular cases in the subspace of X_e (and $n \leq 3$), as can be seen in [123] and [215]. From the work of Segal [232] and the space-time estimate for wave equations established by Strichartz [252], the situation has been changed dramatically. Various refined Strichartz's estimate combined with partial contraction techniques completely resolved the problems on global well-posedness for the Klein–Gordon equation in subcritical cases. As for the critical case, the global well-posedness for wave equations are basically resolved ($n \leq 7$); see [111, 237]. This section is devoted to achieve the global well-posedness for (5.192), (5.193) in energy space. We introduce the following assumptions on $f(z)$ for simplicity:
(H_1) $f(z) \in C^1(\mathbb{C}, \mathbb{C}), f(0) = 0$ and for $1 \leq p < \infty$, it satisfies

$$|f'(z)| = \max\left\{\left|\frac{\partial f}{\partial z}\right|, \left|\frac{\partial f}{\partial \bar{z}}\right|\right\} \leq C(1 + |z|^{p-1}), \quad z \in \mathbb{C}. \tag{5.197}$$

(H$_2$) There is the function $V \in C^1(\mathbb{C}, \mathbb{R})$ such that $V(0) = 0$, $V(z) = V(|z|)$, $z \in \mathbb{C}$ and $f(z) = \frac{\partial V}{\partial \bar{z}}$. For any $R > 0$, $V(z)$ satisfies the following estimate:

$$V(R) > -a^2R^2, \quad a \geq 0. \tag{5.198}$$

Before our discussion, let us recall the existence theorem for weak solutions. Set $X = H^1 \cap L^{p+1}$, then $X' = H^1 + L^{(p+1)/p}$. We use $\langle \cdot, \cdot \rangle$ to denote the L^2 inner product. For $(u, u_t) \in (H^1 \cap L^{p+1}) \oplus L^2$, we know that (H$_2$) implies formally the conservation of energy

$$E(u, u_t) = \|u_t\|_2^2 + \|\nabla u\|_2^2 + \int_\Omega V(u)dx = E(\varphi, \psi). \tag{5.199}$$

The existence of weak solutions can be stated as the following theorem.

Theorem 5.3.1. *Assume $f(u)$ satisfies* (H$_1$) *and* (H$_2$). *If $p+1 > 2^* = \frac{2n}{n-2}$, we need to assume further that*

$$V(\rho) \geq -a^2\rho^2 + C\rho^{p+1}, \quad C > 0, \ \rho > 0. \tag{5.200}$$

Let $t_0 \in \mathbb{R}$, $(\varphi, \psi) \in X \oplus L^2$, then (5.192), (5.193) *have a solution u satisfying*

$$u \in (L^\infty_{\text{loc}} \cap C_w)(\mathbb{R}; X) \cap \text{Lip}(\mathbb{R}; L^2) \cap \bigcap_{2 \leq r < \max(p+1, 2^*)} C^{\mu(r)}(\mathbb{R}; L^2),$$

$$u_t \in L^\infty_{\text{loc}}(\mathbb{R}; X) \cap C_w(\mathbb{R}; X) \cap \text{Lip}(\mathbb{R}; X')$$

and the estimate

$$\|u(t)\|_2 \leq e(E, t - t_0), \tag{5.201}$$

$$\|u_t\|_2^2 + \|\nabla u\|_2^2 + C\|u(t)\|_{p+1}^{p+1} \leq \dot{e}(E, t - t_0)^2. \tag{5.202}$$

Here,

$$e(E, \tau) = \|\varphi\|_2 ch(a\tau) + (E(\varphi, \psi) + a^2\|\varphi\|_2^2)^{\frac{1}{2}} a^{-1} sh(a|\tau|). \tag{5.203}$$

Moreover, if $u \to \int V(u)dx$ is weak lower semicontinuous from a bounded set of X to $\mathbb{R}$, then $u(t)$ satisfies the energy inequality

$$E(u, u_t) \leq E(\varphi, \psi), \tag{5.204}$$

where $\mu(r) = 1 - \delta(r)\min\{1, \delta(p+1)^{-1}\}$, $\dot{e}(t)$ is the derivative of $e(t)$ with respect to t.

Remark 5.3.1.

(i) If we replace (H$_1$) with a weaker condition (H$_1$)$'$: $f(z) \in C(\mathbb{C}, \mathbb{C})$, and for $1 \leq p < \infty$, we have

$$|f(z)| \leq C(|z| + |z|^p). \tag{5.205}$$

Then Theorem 5.3.1 still holds.

(ii) We refer to [176] for the proof of Theorem 5.3.1. By means of Galerkin's method, for any t, the approximating solutions converges weakly in $X \cap L^2(\mathbb{R}^n)$, which implies (5.203) for any t. Estimates (5.201) and (5.202) follows from basic computations and finite-dimensional approximation. Formally, we have from (5.200) and the conservation of energy (5.204),

$$\begin{aligned} E = E(\varphi, \psi) &\geq \|u_t\|_2^2 + \|\nabla u\|_2^2 - a^2\|u\|_2^2 + C\|u\|_{p+1}^{p+1} \\ &\geq \|u_t\|_2^2 - a^2\|u\|_2^2. \end{aligned} \tag{5.206}$$

Let $y = a\|u(t)\|_2$. Then y satisfies

$$|\dot{y}| \leq a(E + y^2)^{\frac{1}{2}}. \tag{5.207}$$

Noticing that $\int 1/\sqrt{1+x^2}dx = \mathrm{sh}^{-1}x$, we integrate the above inequality to get

$$\mathrm{sh}^{-1}\left(\frac{y}{\sqrt{E}}\right) \leq a(t - t_0) + \mathrm{sh}^{-1}\left(\frac{y_0}{\sqrt{E}}\right).$$

It is nothing but

$$\begin{aligned} |y| &\leq \sqrt{E}\mathrm{sh}(a(t - t_0))\mathrm{ch}(\mathrm{sh}^{-1}(|y_0|/\sqrt{E})) + |y_0|\mathrm{ch}(a(t - t_0)) \\ &\leq \sqrt{E + |y_0|^2}\mathrm{sh}(a(t - t_0)) + |y_0|\mathrm{ch}(a(t - t_0)). \end{aligned}$$

Hence, we get (5.201). Finally, we substitute it into the first inequality of (5.206) and we obtain (5.202).

Next, we derive the well-posedness for (5.192)–(5.193) directly from the method based on the space-time estimate. For convenience, we prove the case when $n \geq 2$, and adapt it to prove the case when $n = 1$. As in the standard process, we consider the following integral equations, which are equivalent to (5.192), (5.193),

$$\begin{aligned} u(t, x) &= \dot{K}(t - t_0)\varphi + K(t - t_0)\psi - \int_{t_0}^{t} K(t - \tau)f(u(\tau))d\tau \\ &\equiv u^{(0)}(t) - F(t_0, u)(t) \equiv A(t_0, u^{(0)}(t), u(t)). \end{aligned} \tag{5.208}$$

Denote

$$G(t_1, t_2, u(t)) = -\int_{t_1}^{t_2} K(t - \tau)f(u(\tau))d\tau, \tag{5.209}$$

then

$$F(t_0,u)=G(t_0,t,u)(t),\quad F(t_2,u)-F(t_1,u)=G(t_1,t_2,u(t)). \tag{5.210}$$

For any interval $I\subset\mathbb{R}$ and suitable l, q, r, q_1, we set

$$\mathcal{X}_0(I)=L^q(I;L^l),\quad \mathcal{X}_1(I)=L^{q_1}(I,L^r). \tag{5.211}$$

Lemma 5.3.1. *Assume $n\ge 2$, $f(u)$ satisfies (H_1), and l, r, q_1, q satisfy*

$$\begin{cases}1\le l,r,q,q_1\le\infty, \text{ and when } n=2, & \text{we require further } 1<r<\infty,\\ \text{when } n>3, & \text{we require } |\gamma(r)|\le 1,\end{cases} \tag{5.212}$$

$$\frac{(p-1)n}{l}\le\min\{1+\gamma(r),n(1-\gamma(r))\}, \tag{5.213}$$

$$(p-1)/q+1/q_1\le 1, \tag{5.214}$$

$$\eta_1=2-(p-1)\left(\frac{n}{l}+\frac{1}{q}\right)>0. \tag{5.215}$$

Let I be a subinterval bounded in $\mathbb{R}$, $u\in\mathcal{X}_0(I)\cap\mathcal{X}_1(I)$. Then we have:

(1) *For any $t_0\in I$, $F(t_0,u)$ belongs to $\mathcal{X}_1(I)$ where $F(t_0,u)$ varies as a continuous function of t_0 taking values in $\mathcal{X}_1(I)$. For $t_0\in I$ and any $u_1,u_2\in\mathcal{X}_0(I)\cap\mathcal{X}_1(I)$, we have the following estimate:*

$$\|F(t_0,u_1)-F(t_0,u_2)\|_{\mathcal{X}_1(\mathcal{I})}\le C_1\|u_1-u_2;\mathcal{X}_1(I)\|\times\left\{|I|^2+|I|^{\eta_1}\sum_{j=1}^{2}\|u_j;\mathcal{X}_0(I)\|^{p-1}\right\}. \tag{5.216}$$

(2) *For any $t_1,t_2\in I$, $G(t_1,t_2,u)\in\mathcal{X}_{1,\mathrm{loc}}(\mathbb{R})$. For any bounded closed interval $J\subset\mathbb{R}$, $G(t_1,t_2,u)$ varies as a continuous function with arguments t_1, t_2 and takes value in $\mathcal{X}_1(J)$. For any $t_1,t_2\in I$, $u_1,u_2\in\mathcal{X}_1(I)\cap\mathcal{X}_0(I)$ and any bounded closed interval $J\supset I$, we have the following inequality:*

$$\begin{aligned}&\|G(t_1,t_2,u_1)-G(t_1,t_2,u_2);\mathcal{X}_1(J)\|\\&\quad\le C_1\|u_1-u_2;\mathcal{X}_1([t_1,t_2])\|\cdot\left\{|J|^2+|J|^{\eta_1}\sum_{j=1}^{2}\|u_j;\mathcal{X}_0([t_1,t_2])\|^{p-1}\right\}.\end{aligned} \tag{5.217}$$

(3) *For any $t_0\in I$, $F(t_0,u)$ satisfies $\Box F(t_0,u)=f(u)$ in $\mathcal{S}'(I\times\mathbb{R}^n)$, and for arbitrary $t_1,t_2\in I$, $G(t_1,t_2,u)$ satisfies $\Box G(t_1,t_2,u)=0$ in the sense of $\mathcal{S}'(\mathbb{R}^{n+1})$.*

Proof. We first prove (1) and (2). By (H_1), we write f as $f=f_1+f_2$, satisfying

$$|f_1'(z)|\le C,\quad |f_2'(z)|\le C|z|^{p-1}.$$

Hence, we estimate the contributions from f_1 and f_2, respectively, to the nonlinear functions F and G. From the $L^r - L^s$ estimate of linear Klein–Gordon, one sees that

$$\|K(t)u\|_r \leq C|t|^{1-\delta(r)+\delta(s)}\|u\|_s. \tag{5.218}$$

Here, we require

$$\begin{cases} 0 \leq \delta(r) - \delta(s) \leq \min\{1+\gamma(r), n(1-\gamma(r))\}, \\ 1 < r, s < \infty, \ n = 2. \end{cases} \tag{5.219}$$

For function $f_1(z)$, taking $r = s$, we have

$$\|K(t-\tau)(f_1(u_1(\tau)) - f_1(u_2(\tau)))\|_r \leq C|t-\tau| \cdot \|u_1(\tau) - u_2(\tau)\|_r. \tag{5.220}$$

And for $f_2(z)$, we have the estimate

$$\begin{aligned} &\|K(t-\tau)(f_2(u_1(\tau)) - f_2(u_2(\tau)))\|_r \\ &\quad \leq C|t-\tau|^{1-\delta(r)+\delta(s)}\|f_2(u_1(\tau)) - f_2(u_2(\tau))\|_s \\ &\quad \leq C|t-\tau|^{1-\delta(r)+\delta(s)}\|u_1(\tau) - u_2(\tau)\|_r \\ &\qquad \times \{\|u_1(\tau)\|_l^{p-1} + \|u_2(\tau)\|_l^{p-1}\}, \end{aligned} \tag{5.221}$$

where

$$(p-1)n/l = \delta(r) - \delta(s). \tag{5.222}$$

By (5.213), we know the s determined above satisfies exactly (5.219). Note that (5.214), (5.215) and

$$\begin{cases} \frac{1}{\tilde{q}_1} + 1 = \frac{1}{q_1} + 1, \\ \frac{1}{\tilde{q}_1} + 1 = (\frac{p-1}{q} + \frac{1}{q_1}) + \frac{1}{y}, \quad y = (1 - \frac{p-1}{q})^{-1}. \end{cases} \tag{5.223}$$

We substitute (5.220), (5.221), respectively, into F and G, and use Young's inequality with respect to t to get (5.216), (5.217). The continuity of $G(t_1, t_2, u(t))$ with respect to t_1, t_2 can be deduced from (5.217). While the continuity of $F(t_0, u(t))$ with respect to t_0 can be deduced from (5.210) and the continuity of $G(t_1, t_2, u(t))$. □

As for (3), it can be done by combining the duality argument technique with definition of distribution solutions and direct computation. It points out essentially equivalency between the integral equation (5.208) and (5.192), (5.193) in the sense of weak solutions and reveals the relations of (5.208) at different times. As a consequence of Lemma 5.3.1, we have the following corollary.

Corollary 5.3.1. *Assume $n \geq 2$, $f(u)$ satisfies (H_1), l, r, q and q_1 satisfy* (5.212)~(5.215). *Let I be an open interval in $\mathbb{R}$, $u \in \mathcal{X}_1(I) \cap \mathcal{X}_0(I)$. Then we have:*

(1) *That $u(t,x)$ solves (5.192) in the sense of $\mathcal{S}'(I \times \mathbb{R}^n)$ holds if and only if for any $t_0 \in I$, $u^{(0)} \equiv u(t) - F(t_0, u(t,x))$ solves $\Box u^{(0)} = 0$ in $\mathcal{S}'(I \times \mathbb{R}^n)$.*
(2) *If $u(t)$ satisfies (5.208), then for any $t_1 \in I$, $u(t)$ solves $u(t) = A(t_1, u^{(1)}; u)$, where*

$$u^{(1)} = u^{(0)} + G(t_0, t_1; u(t,x)), \tag{5.224}$$

$u^{(1)}$ is a continuous function taking values in $\mathcal{X}_1(I)$ with variable t_1.

Remark 5.3.2. In Lemma 5.3.1 and Corollary 5.3.1, when the interval I becomes infinite, we only need to replace $\mathcal{X}_j (j = 0, 1)$ with $\mathcal{X}_{j,\mathrm{loc}}$ and the corresponding results are valid.

Theorem 5.3.2. *Assume $n \geq 2$, and l, r, q, q_1 satisfy* (5.212)~(5.215), *$f(u)$ satisfies (H_1). Let $I \subset \mathbb{R}$ be an open interval, $t_0 \in I$, $u^{(0)} \in \mathcal{X}_{1,\mathrm{loc}}(I)$. Then equation* (5.208) *has at most one solution in $\mathcal{X}_{1,\mathrm{loc}}(I) \cap \mathcal{X}_{0,\mathrm{loc}}(I)$.*

Proof. Suppose u_1, u_2 are the solutions to (5.208) with the same initial data, then $u_1 - u_2$ satisfies

$$u_1 - u_2 = F(t_0, u_1) - F(t_0, u_2). \tag{5.225}$$

Take J as an interval containing t_0 small enough, satisfying

$$C_1 \left\{ |J|^2 + |J|^{\eta_1} \sum_{j=1}^{2} \|u_j; \mathcal{X}_0(J)\|^{p-1} \right\} \leq \frac{1}{2}. \tag{5.226}$$

Then, in (5.216), we replace I by J, using (5.226), and we have

$$\|u_1 - u_2\|_{\mathcal{X}_1(I)} \leq \frac{1}{2} \|u_1 - u_2\|_{\mathcal{X}_1(I)}. \tag{5.227}$$

Hence, we get $u_1 \equiv u_2$, $t \in J$. Repeat this procedure, we conclude that in the whole interval I, $u_1 \equiv u_2$. □

Remark 5.3.3. Theorem 5.3.2 applies to large p as well. Indeed, fix $1 < p < \infty$, we choose

$$\gamma(r) = \frac{n-1}{n+1}, q_1 = \infty \quad \text{(corresponds to the case } 1 - \gamma(r) = n(1 - \gamma(r))\text{)}.$$

Simultaneously, (5.213), (5.214) become

$$(p-1)n/l \leq \frac{2n}{n+1}, \quad \frac{p-1}{q} \leq 1. \tag{5.228}$$

Hence, it is obvious that (5.215) and (5.228) hold if l, q are large enough. However, in this situation, the finite energy solution u lies in $L^\infty_{\mathrm{loc}}(\mathbb{R}, L^r)$ ($\gamma(r) = \frac{n-1}{n+1}$). For l or q large enough, it is not necessary that the finite energy solution $u(t)$ still belongs to $\mathcal{X}_{0,\mathrm{loc}}(\mathbb{R})$. In fact, only when $1 - \frac{n}{2} \geq -\frac{n}{l} - \frac{1}{q}$, the finite energy solution $u(t)$ can be in $\mathcal{X}_{0,\mathrm{loc}}(\mathbb{R})$, which, together with (5.215), is equivalent to $p < 1 + \frac{4}{n-2}$, the uniqueness condition to guarantee the finite energy solution.

Lemma 5.3.2. *Assume that $n \geq 2$, $0 \leq \gamma(r) \leq 1$, then $K(t)$ satisfies the following estimate:*

$$\|K(t)u; \dot{B}^{\rho}_{r,2}\| \leq C|t|^{-\mu}\|u; \dot{B}^{\bar{\rho}}_{\bar{r},2}\|. \tag{5.229}$$

Here, we require ρ, $\bar{\rho}$, $\bar{r}$, μ to satisfy

$$0 \leq 1 + \mu = \delta(r) + \rho - \delta(\bar{r}) - \bar{\rho} \leq \frac{1}{2}(\gamma(r) - \gamma(\bar{r}))(1 + 1/\gamma(r)) \leq 1 + \gamma(r). \tag{5.230}$$

In particular, the above relation implies $|\gamma(\bar{r})| \leq \gamma(r)$.

Proof. Recall the $L^{r'} - L^r$ estimate for linear wave equations

$$\|K(t)u; \dot{B}^{\rho}_{r,2}\| \leq C|t|^{-\gamma(r)}\|u; \dot{B}^{\rho+2\beta(r)-1}_{r',2}\|, \quad \frac{1}{r} + \frac{1}{r'} = 1. \tag{5.231}$$

By the multiplier estimate and interpolation formula, one easily sees that

$$\|K(t)u; \dot{B}^{\rho}_{r,2}\| \leq C|t| \cdot \|u; \dot{B}^{\rho}_{r,2}\|. \tag{5.232}$$

Noting that $|\sin y| \leq |y|$ and the estimate on the oscillatory integrals

$$\begin{aligned} \|K(t)\varphi_k\|_\infty &= (2\pi)^{-\frac{n}{2}}\left|\int \frac{\sin t|\xi|}{|\xi|}\hat{\varphi}_k(\xi)e^{ix\cdot\xi}d\xi\right| \\ &\leq (2\pi)^{-\frac{n}{2}}2^{nk}|t| \cdot \|\hat{\varphi}_0\|_1, \end{aligned} \tag{5.233}$$

where $\{\varphi_j\}$ is the partition of unity in the definition of homogeneous Besov spaces, we have

$$\|K(t)(u * \varphi_j)\|_2 \leq |t| \cdot \|u * \varphi_j\|_2, \tag{5.234}$$

and

$$\begin{aligned} \|K(t)(u * \varphi_j)\|_\infty &\leq \sum_{|k-j|\leq 1} \|K * \varphi_k\|_\infty \|u * \varphi_j\|_1 \\ &\leq 2(2\pi)^{-\frac{n}{2}}|t|2^{nk}\|\hat{\varphi}_0\|_1\|u * \varphi_j\|_1. \end{aligned} \tag{5.235}$$

Interpolating between (5.234) and (5.235), by the definition of homogeneous Besov spaces, we have

$$\|K(t)u; \dot{B}^{\rho}_{r,2}\| \leq C|t| \cdot \|u; \dot{B}^{\rho+2\delta(r)}_{r',2}\|. \tag{5.236}$$

Interpolating (5.231) and (5.236), we have

$$\|K(t)u; \dot{B}^{\rho}_{r,2}\| \le C|t|^{-\gamma(r)\theta+(1-\theta)}\|u; \dot{B}^{\rho+2\delta(r)-\gamma(r)\theta-\theta}_{r',2}\|, \tag{5.237}$$

where $0 \le \theta \le 1$ is to be determined. Interpolating (5.232) and (5.237), we get (5.229), and the relations are

$$\begin{cases} (1-\frac{1}{r})\tilde{\theta} + \frac{1-\tilde{\theta}}{r} = \frac{1}{\bar{r}}, \\ (\rho + 2\delta(r) - \gamma(r)\theta - \theta)\tilde{\theta} + \rho(1-\tilde{\theta}) = \bar{\rho}, \\ -\mu = [-\gamma(r)\theta + (1-\theta)]\tilde{\theta} + (1-\tilde{\theta}). \end{cases} \tag{5.238}$$

By direct computation, the first equality in (5.238) implies $2\delta(r)\tilde{\theta} = \delta(r) - \delta(\bar{r})$. Substituting the last two formulas in (5.238), we get

$$\begin{cases} \bar{\rho} = \delta(r) - \delta(\bar{r}) - (\gamma(r)+1)\theta\tilde{\theta} + \rho, \\ 1 + \mu = (\gamma(r)+1)\theta\tilde{\theta}. \end{cases} \tag{5.239}$$

Hence, taking the first formula in (5.239) into the second, we get the first inequality in (5.230). Moreover, noting that $0 \le \theta \le 1$ and

$$[\delta(r) - \delta(\bar{r})]/2\delta(r) = [\gamma(r) - \gamma(\bar{r})]/2\gamma(r) = \tilde{\theta},$$

we prove that the conditions in (5.230) are satisfied. □

For any interval $I \subset \mathbb{R}$ and suitable ρ, r, q, denote $\mathcal{X}_2(I) = L^q(I, \dot{B}^{\rho}_{r,2})$ and let $B_j(I,R)$ be a closed ball centered at the origin with radii R in $\mathcal{X}_j(I)$ $(j = 1,2)$. Then we have the following nonlinear estimate.

Lemma 5.3.3. *Assume that $n \ge 2$, $f(u)$ satisfies* (H_1). *Let ρ, r, q satisfy $0 \le \rho < 1$ and*

$$0 \le \gamma(r) \le \frac{n-1}{n+1}, \tag{5.240}$$

$$(p-1)\left(\frac{n}{r} - \rho\right) \le 1 + \gamma(r), \tag{5.241}$$

$$p \le q, \tag{5.242}$$

$$\eta_2 \equiv 2 - (p-1)\left(\frac{n}{r} - \rho + \frac{1}{q}\right) > 0. \tag{5.243}$$

Let I be an open bounded interval, $t_0 \in I$, $u(t) \in \mathcal{X}_2(I)$. Then we have

(1) *$F(t_0,u) \in \mathcal{X}_2(I)$ and the following estimate holds:*

$$\|F(t_0,u); \mathcal{X}_2(I)\| \le C_2\{|I|^2\|u; \mathcal{X}_2(I)\| + |I|^{\eta_2}\|u; \mathcal{X}_2(I)\|^p\}. \tag{5.244}$$

(2) *For any bounded interval $J \supset I$ and any $t_1, t_2 \in I$, $G(t_1, t_2, \varphi)$ is a continuous function taking values in $\mathcal{X}_2(J)$ with arguments t_1, t_2, and the following estimate holds:*

$$\|G(t_1, t_2, u); \mathcal{X}_2(J)\| \leq C_2\{|J|^2\|u; \mathcal{X}_2([t_1, t_2])\| + |J|^{\eta_2}\|u; \mathcal{X}_2([t_1, t_2])\|^p\}. \tag{5.245}$$

Proof. Similar to the proof of Lemma 5.3.1, we write f as $f = f_1 + f_2$ and deal with f_1, f_2 differently. We only need to consider the estimate for $f_2(u)$ (take $p = 1$ we have the corresponding estimate for $f_1(u)$). From Lemma 5.3.2 and the nonlinear estimate in the homogeneous Besov space, one easily has

$$\begin{aligned}\|K(t-\tau)f_2(u); \dot{B}^{\rho}_{r,2}\| &\leq C|t-\tau|^{-\mu}\|f_2(u); \dot{B}^{\bar{\rho}}_{\bar{r},2}\| \\ &\leq C|t-\tau|^{-\mu}\|u; \dot{B}^{\rho}_{r,2}\|^p,\end{aligned} \tag{5.246}$$

where $\bar{\rho} \leq \rho$ and $p(\frac{n}{r} - \rho) = \frac{n}{\bar{r}} - \bar{\rho}$, or equivalently,

$$(p-1)\left(\frac{n}{r} - \rho\right) = \rho + \delta(r) - \bar{\rho} - \delta(\bar{r}) = 1 + \mu. \tag{5.247}$$

Essentially, we take $\bar{\rho} = \rho$, then (5.246) is the nonlinear estimate given by Cazenave and Weissler in [37], where $\bar{r}$ is determined by

$$0 \leq \delta(r) - \delta(\bar{r}) \leq (p-1)\left(\frac{n}{r} - \rho\right) \leq \frac{1}{2}(\gamma(r) - \gamma(\bar{r}))\left(1 + \frac{1}{\gamma(r)}\right) \leq 1 + \gamma(r).$$

For r in (5.240), one can easily find $\bar{r}$ satisfying the above requirement. Now we substitute (5.246) into F and G, noting that (5.242), (5.243) and Young's inequality, and we get (5.244), (5.245). □

Theorem 5.3.3. *Assume that $n \geq 2$, $f(u)$ satisfies (H_1), and ρ, r, q, q_1 satisfy $0 \leq \rho < 1$, $1 \leq q \leq q_1 \leq \infty$ and (5.240)~(5.243). Then for any $R > 0$, there is $T(R) > 0$, so that for $t_0 \in \mathbb{R}$ and $u^{(0)} \in B_2(I, R) \cap \mathcal{X}_1(I)$, $I = [t_0 - T(R), t_0 + T(R)]$, the integral equation (5.208) has a solution $u(t)$ in $B_2(I, 2R) \cap \mathcal{X}_1(I)$, which enjoys the following inequality:*

$$\|u; \mathcal{X}_1(I)\| \leq 2\|u^{(0)}; \mathcal{X}_1(I)\|. \tag{5.248}$$

Moreover, the solution is unique in $\mathcal{X}_1(I) \cap \mathcal{X}_2(I)$.

Proof. Take $\frac{n}{l} = \frac{n}{r} - \rho$, conditions (5.240)~(5.243) imply that (5.213)~(5.215) are valid when $\eta_1 = \eta_2 = \eta$. Using Sobolev's imbedding $\dot{B}^{\rho}_{r,2} \hookrightarrow L^l$, we have $\mathcal{X}_2(\cdot) \hookrightarrow \mathcal{X}_0(\cdot)$ and

$$\|u\|_l \leq C_3\|u; \dot{B}^{\rho}_{r,2}\|, \quad \|u; \mathcal{X}_0(I)\| \leq C_3\|u; \mathcal{X}_2(I)\|. \tag{5.249}$$

Take $R = 2\|u^{(0)}\|_{\mathcal{X}_1(I)}$, $T = T(R)$ small enough, so that

$$C_1\{(2T)^2 + (2T)^\eta 2(2C_3R)^{p-1}\} \le \frac{1}{2}, \tag{5.250}$$

$$C_2\{(2T)^2 + (2T)^\eta (2R)^{p-1}\} \le \frac{1}{2}. \tag{5.251}$$

Then, by Lemma 5.3.1 and Lemma 5.3.3, we have

$$\|A(t_0, u^{(0)}(t), u(t))\|_{\mathcal{X}_1(I)\cap\mathcal{X}_2(I)} \le 2R, \tag{5.252}$$

$$\|A(t_0, u^{(0)}(t), u(t)) - A(t_0, u^{(0)}(t), v(t))\|_{\mathcal{X}_1(I)} \le \frac{1}{2}\|u - v; \mathcal{X}_1(I)\|. \tag{5.253}$$

Hence, the map $A(t_0, u^{(0)}(t), u(t))$ is contraction from $S = B_2(I, 2R) \cap \mathcal{X}_1(I)$ to itself (in the norm of $\mathcal{X}_1(I)$). For any $R_1 > 0$, since $B_1(I, R_1) \cap B_2(I, 2R)$ is w^* compact in $\mathcal{X}_1(I) \cap \mathcal{X}_2(I)$, which is compact in the w^* topology in $\mathcal{X}_1(I)$. Therefore, it is closed in $\mathcal{X}_1(I)$, which means that S is closed in $\mathcal{X}_1(I)$. By the fixed-point theorem, we get Theorem 5.3.3. □

Remark 5.3.4. One sees from Theorem 5.3.3 that (5.241)~(5.243) gives essentially the upper bounds for p. The sharp case is that $\rho \lesssim 1$, $\gamma(r) = \frac{n-1}{n+1}$, $q = q_1 = \infty$ ($\lesssim$ means less or equal to and close arbitrarily to). Under this condition, (5.241)~(5.243) turns to

$$(p-1)\left(\frac{n}{2} - \frac{n}{n+1} - 1\right) < \frac{2n}{n+1},$$

or equivalently,

$$(p-1)\left(\frac{n}{2} - \frac{3}{2} - \frac{1}{n}\right) < 2. \tag{5.254}$$

Obviously, it is weaker than $p < 1 + \frac{4}{n-2}$. We see from this, when $n \le 3$, that there are no restrictions on the upper bounds for p. However, similar to Theorem 5.3.2, only when ρ, r, q are not large, the finite energy solution can be in $\mathcal{X}_2(I)$.

Now we discuss the global well-posedness for the finite energy solutions of (5.192), (5.193) or (5.208) for which we introduce the energy space

$$X_e = \{(\varphi, \psi); \varphi \in H^1, \psi \in L^2\} = H^1 \oplus L^2. \tag{5.255}$$

For any $(\varphi, \psi) \in X_e$, the Cauchy problems for the corresponding free Klein–Gordon equation has its solution

$$u^{(0)}(t) = \dot{K}(t - t_0)\varphi + K(t - t_0)\psi \in C(\mathbb{R}, H^1). \tag{5.256}$$

As a particular case of the space-time estimate for wave equation, $u^{(0)}(t)$ satisfies the following lemma.

Lemma 5.3.4. *Assume that $n \geq 2$, ρ, r, q satisfy*

$$\begin{cases} 0 \leq \delta(r) \leq \frac{n}{2}, \\ -1 \leq \sigma \equiv \rho + \delta(r) - 1 < \frac{1}{2}, \\ \sigma \leq \gamma(r)/2. \end{cases} \tag{5.257}$$

and

$$\frac{1}{q} = \max(0, \sigma). \tag{5.258}$$

Then for any $(\varphi, \psi) \in X_e$, we have $u^{(0)}(t) \in \mathcal{X}_2(\mathbb{R})$ and it satisfies the following space-time estimate:

$$\|u^{(0)}(t), \mathcal{X}_2(\mathbb{R})\| \leq C(\|\psi\|_2 + \|\nabla\varphi\|_2). \tag{5.259}$$

On account of the above space-time estimate, we have the following theorem.

Theorem 5.3.4.
(i) *Let $n \geq 2$, $f(u)$ satisfies* (H_1) *and $p < 1 + \frac{4}{n-2}$. Then we deduce that q, r, ρ $0 \leq \rho < 1$ satisfying* (5.240)~(5.243), (5.257) *and* (5.258).
(ii) *Denote $\mathcal{X}_1$ and $\mathcal{X}_2$ as the function space given by ρ, r, q $q_1 \geq q$ in* (i), *then, for any $t_0 \in \mathbb{R}$ and any $(\varphi, \psi) \in X_e$, there is $T = T(\|(\varphi, \psi), X_e\|) > 0$ such that* (5.208) *has a unique solution in $\mathcal{X}_1(I) \cap \mathcal{X}_2(I)$, where $I = [t_0 - T, t_0 + T]$.*
(iii) *For any $(\varphi, \psi) \in X_e$ and any interval I, if $t_0 \in I$, the integral equation* (5.208) *has at most one solution in $\mathcal{X}_{1,\mathrm{loc}}(I) \cap \mathcal{X}_{2,\mathrm{loc}}(I)$.*

Proof. (i) According to (5.257), condition (5.241) turns out to be

$$(p - 1)(n/2 - 1 - \sigma) \leq 1 + \gamma(r). \tag{5.260}$$

By (5.258), we have (5.241), (5.242) in the form of

$$p\sigma \leq 1, \tag{5.261}$$

$$p < 1 + 4/(n - 2). \tag{5.262}$$

Next, we prove that for p satisfying (5.262), r and σ can always be chosen to meet the remaining conditions, that is, $0 \leq \rho < 1$ and (5.240), (5.257), (5.260), (5.261). Indeed, if $p - 1 \leq 4/(n - 1)$, we take $\rho = 0$, then if σ is negative, it is obvious that the rest conditions hold true. If $p - 1 \geq 4/(n - 1)$, we can take $\gamma(r) = (n - 1)/(n + 1)$, then (5.260) becomes

$$\sigma \geq n/2 - 1 - 2n/[(n + 1)(p - 1)],$$

which is an increasing function of p. The upper bound for p in (5.262) corresponds to the lower bounds for σ, which is

$$\sigma \geq n/2 - 1 - n(n-2)/[2(n+1)] = (n-2)/2(n+1). \tag{5.263}$$

It is exactly compatible with the upper bound condition in (5.257) $\sigma \leq (n-2)/2(n+1)$ and the upper bound condition in (5.261) $\sigma \leq (n-2)/(n+2)$, while the condition $0 \leq \rho < 1$ is met naturally. Hence, we have (i).

(ii) For all r, q_1 satisfying (5.240), we have $u^{(0)}(t) \in \mathcal{X}_{1,\mathrm{loc}}(R)$. Thus, by the space-time estimate (5.259) and Theorem 5.3.3, we have (ii).

(iii) Similar to the proof in Theorem 5.3.3, taking $\frac{n}{l} = \frac{n}{r} - \rho$, by (i), Lemma 5.3.4 and Theorem 5.3.2, we get (iii). □

We next discuss the global well-posedness for (5.192), (5.193) or (5.208), under the condition (H_2) of $f(u)$, then the solution obtained in Theorem 5.3.4 satisfies the following conservation of energy:

$$E(u, u_t) = \|u_t\|_2^2 + \|\nabla u\|_2^2 + \int_{\mathbb{R}^n} V(u(t))dx = E(\varphi, \psi). \tag{5.264}$$

Formally, direct computation from (H_2) yields (5.264). We aim at giving strict proof of (5.264) in the sense of energy solution. In order to do this, we need to regularize the equation to reach our target by constructing approximating solutions. Denote $h_0(x) \in C_c^\infty(\mathbb{R}^n)$ and $\|h_0\|_1 = 1$. For any integer j, define $h_j(x) = j^n h_0(jx)$ and

$$f_j(u) = h_j * f(h_j * u), \tag{5.265}$$

$$E_j(u, v) = \|v\|_2^2 + \|\nabla u\|_2^2 + \int V(h_j * u)dx. \tag{5.266}$$

Consider the regularized form of (5.208),

$$u = h_j * u^{(0)} + F_j(t_0, u) \equiv A_j(t, u^{(0)}; u(t)), \tag{5.267}$$

where F_j is the F in (5.210) with $f(u)$ replaced by $f_j(u)$. For the regularized equation (5.267), we have the following.

Lemma 5.3.5. *Assume that* $n \geq 2$, $f(u)$ *satisfies* (H_1) *and* $p < 1 + \frac{4}{n-2}$. *Let* ρ, r, q, q_1 *meet the conditions in Theorem* 5.3.4(i), *then we have:*

(i) *The conclusion* (ii) *in Theorem* 5.3.4 *holds for the regularized equation* (5.267), *with the same* T *(independent of* j*).*

(ii) *The solution* u_j *of* (5.267) *converges to the solution of* (5.208) *in the sense of* $\mathcal{X}_1(I)$.

Proof. Note that the operator determined by $h_j *$ is contracting in L^r or $\dot{B}^\rho_{r,2}$. Thus, replacing $f(u)$ by $f_j(u)$, we have the estimates and results established previously, and hence (i) holds.

Next, we prove (ii). Note that

$$h_j * f(h_j * u_j) - f(u) = h_j * f(u) - f(u) + h_j * \left\{ \int_0^1 f'(\theta h_j * u_j \right.$$
$$\left. + (1-\theta)u) \cdot (h_j * u - u + h_j * (u_j - u))d\theta \right\}. \tag{5.268}$$

By Lemma 5.3.1 and the estimate (5.216), we see that

$$\begin{aligned} \|u_j - u; \mathcal{X}_1(I)\| \le{}& \|h_j * u^{(0)}(t) - u^{(0)}(t); \mathcal{X}_1(I)\| \\ &+ C_1[\|h_j * u - u; \mathcal{X}_1(I)\| + \|h_j * (u_j - u); \mathcal{X}_1(I)\|] \\ &\times \left\{ |I|^2 + |I|^{\eta_1} \sum_{j=1}^{2} [\|h_j * u_j; \mathcal{X}_0(I)\|^{p-1} + \|u; \mathcal{X}_0(I)\|^{p-1}] \right\}. \end{aligned} \tag{5.269}$$

Because $\|u_j; \mathcal{X}_0(I)\|$, $\|u; \mathcal{X}_0(I)\|$ is bounded, $I = [t_0 - T, t_0 + T]$, therefore, in $\tilde{I} = [t_0 - \frac{T}{m}, t_0 + \frac{T}{m}]$, taking m large enough, we have

$$\|u_j - u; \mathcal{X}_1(\tilde{I})\| \to 0, \quad j \to \infty.$$

Repeat the above process, translating t_0 to $t_0 - \frac{T}{m}$, or $t_0 + \frac{T}{m}$, we have

$$\|u_j - u; \mathcal{X}_1(I)\| \to 0, \quad j \to \infty.$$

□

Lemma 5.3.6. *Assume that $u_j(t,x)$ solves the regularized equation* (5.267)*, then we have:*

(i) *For any integer k, $(u_j, \dot{u}_j) \in C^1(I, H^{k+1} \oplus H^k)$ and u_j solves*

$$\Box u_j + f_j(u_j) = 0. \tag{5.270}$$

(ii) *Let $f(u)$ satisfy* (H_2)*, $p < 1 + \frac{4}{n-2}$. Then u_j satisfies the conservation of energy*

$$E_j(u_j(t), \dot{u}_j(t)) = E_j(h_j * \varphi, h_j * \psi) \equiv E_j \tag{5.271}$$

and energy inequality

$$\|u_j(t)\|_2 \le e(E_j, t - t_0) \le e(\bar{E}, t - t_0), \tag{5.272}$$
$$\|\dot{u}_j\|_2^2 + \|\nabla u_j(t)\|_2^2 \le \dot{e}(E_j, t - t_0)^2 \le \dot{e}(\bar{E}, t - t_0)^2, \tag{5.273}$$

where $e(E,t)$ is the same as in (5.203)*, $\bar{E} = \sup_j E_j < \infty$. In particular, $(u_j(t), \dot{u}_j(t))$ is uniformly bounded with respect to j in $H^1 \oplus L^2$.*

Proof. It is trivial for (i). For (ii), we take $k > \frac{n}{2} + 2$, u_j is the classical solution, and (5.271) follows by direct calculation. Noting that (H_2) and Remark 5.3.1, one has the estimate (5.272), (5.273). □

Remark 5.3.5. Note that $u_j \in C(I, L^r)$, $\forall j \in \mathbb{Z}^+$. By (ii) in Lemma 5.3.5 (corresponding to $q_1 = \infty$), we see that

$$u_j \xrightarrow{C(I,L^r)} u, \quad j \longrightarrow \infty. \tag{5.274}$$

Theorem 5.3.5. *Assume that $n \geq 2$, $f(u)$ satisfies (H_1), (H_2) and $p < 1 + \frac{4}{n-2}$. Suppose that $(\varphi, \psi) \in X_e$, I is an open interval, $t_0 \in I$. Let ρ, r, q and $0 \leq \rho < 1$, (5.240)~(5.242), (5.257) and (5.258), $q_1 = \infty$. Let $u^{(0)}(t)$ solve the free wave equation determined by (5.256), and $u(t)$ solves the integral equation (5.208) in $\mathcal{X}_1(I) \cap \mathcal{X}_2(I)$, then $(u(t), \dot{u}(t)) \in C(I, H^1 \oplus L^2)$ and u satisfy the conservation of energy*

$$E(u(t), \dot{u}(t)) = E(\varphi, \psi) \equiv E, \quad t \in I \tag{5.275}$$

and the estimate

$$\|u(t)\|_2 \leq e(E, t_0 - t), \quad \forall t \in I, \tag{5.276}$$

$$\|\dot{u}(t)\|_2^2 + \|\nabla u(t)\|_2^2 \leq \dot{e}(E, t_0 - t)^2, \quad \forall t \in I. \tag{5.277}$$

Proof. We only need to prove the case that for any subinterval $I' \subset\subset I$ including t_0. Set

$$R = \sup_{s \in I'} \|u^{(0)} + G(t_0, s, u); \mathcal{X}_2(I')\|. \tag{5.278}$$

By (5.245) in Lemma 5.3.3, we have $R < \infty$. Set $T = T(R)$ determined by (5.250), (5.251) in Theorem 5.3.3. Using Theorem 5.3.3 and Corollary 5.3.1, for any $t \in I'$, we can get the solution $u(t)$ in $I' \cap [t - T, t + T]$ by solving (5.208) with the initial data at t. Obviously, I' can be covered by finite intervals I_k, which is centered at $t_k = t_0 + (1 - \varepsilon)kT$ and has length $2T$. Here, $\varepsilon > 0$. Therefore, the results for I' can be deduced from that for I_k $(k = 0, \pm 1, \pm 2, \ldots)$. Thus, we only need to perform the fixed-point argument in a small interval I (to make sure the contraction holds) containing t_0 to prove Theorem 5.3.5, which is the task of the following section.

Let u_j solve the integral equation (5.267) in I. By (5.272), (5.273), Remark 5.3.5 and the standard compactness argument, one easily has

$$u_j \xrightarrow{w^*} u, \quad \text{in} \quad L^\infty(I; H^1); \quad \dot{u}_j \xrightarrow{w^*} \dot{u}, \quad \text{in} \quad L^\infty(I; L^2). \tag{5.279}$$

Here, $\{u_j\}$ denotes itself rather than the subsequence, as can be seen in Lions [176]. Thus, for any $t \in I$, by the uniform boundedness of $\|u_j\|_{H^1}$ and $u_j \xrightarrow{L^r} u$ $(j \to \infty)$, we have

$$u_j(t) \xrightarrow{w} u(t), \quad \text{in} \quad H^1,\ j \to \infty. \tag{5.280}$$

Next, we only need to show that $\dot{u}_j(t)$ converges to $\dot{u}(t)$ weakly in $L^2(\mathbb{R}^n)$ for any $t \in I$. Indeed, from equation (5.270), one gets the Lipschitz continuity of $\dot{u}_j(t)$ with respect to j in H^{-1}, and that (5.273) implies the uniform boundedness of $\|\dot{u}_j\|_2$. Hence, $\{\dot{u}_j\}$ is weakly compact in $L^2(\mathbb{R}^n)$. A compactness argument yields that there is a subsequence of $\{\dot{u}_j\}$ (denoted still by $\{\dot{u}_j\}$) such that

$$\dot{u}_j(t) \xrightarrow{w} \chi(t), \quad \chi \in L^2(\mathbb{R}^n). \tag{5.281}$$

Now for any $v \in H^1$ and $\theta > 0$ consider

$$\begin{aligned}\langle v, \dot{u} - \chi\rangle = 2\theta^{-1} \int_{t-\theta}^{t+\theta} \langle v, (\dot{u}(t) - \dot{u}(\tau)) + (\dot{u}(\tau) - \dot{u}_j(\tau)) \\ + (\dot{u}_j(\tau) - \dot{u}_j(t)) + (\dot{u}_j(t) - \chi)\rangle d\tau.\end{aligned} \tag{5.282}$$

Noting that $\dot{u}$, $\dot{u}_j$ are continuous uniformly in H^{-1}, we derive that the first and the third term in (5.282) tends to 0 as $\theta \to 0$. For the second term, for a fixed θ, by the weak * convergence of $\dot{u}_j$ to $\dot{u}$ in $L^\infty(I, L^2)$, we have that it tends to 0 ($j \to \infty$). Using (5.281), we deduce that the last term tends to 0 as $j \to \infty$. Since θ and j are arbitrary, (5.282) implies $\chi(t) = \dot{u}(t)$. Here, $\dot{u}$ is the weak derivative of u in the sense of $\mathcal{D}'(I, L^2(\mathbb{R}^n))$. Taking $j \to \infty$ on both sides of (5.272) and (5.273), we get $(u, \dot{u})$ satisfy (5.276), (5.277), for almost every t. Moreover, by $u \in C(I, L^r) \cap L^\infty(I, H^1)$ and $u_t \in L^\infty(I, L^2)$, we see

$$u(t) \in C(I, L^s) \cap C_w(I, H^1), \quad 2 \le s < \frac{2n}{n-2}. \tag{5.283}$$

Hence, u satisfies (5.192) in the sense of $\mathcal{D}'(I \times \mathbb{R}^n)$. Thus, $\ddot{u} \in L^\infty(I, H^{-1})$, $\dot{u} \in C(I, H^{-1})$, which means that $\dot{u} \in C_w(I, L^2)$. According to the continuity of $(u, \dot{u})$, we have (5.276), (5.277) for any $t \in I$.

Now we take limits as $j \to \infty$ on both sides of (5.271). Noting that u_j converges to u in $C(I; L^s)$ $(2 \le s \le p+1)$, we see that $\int V(u_j)dx \to \int V(u)dx$. Hence, the RHS of (5.271) converges to $E(\varphi, \psi)$. As for the LHS, applying the weak convergence of $(u_j, \dot{u}_j)$ in $H^1 \oplus L^2$, we have

$$E(u, \dot{u}) \le E(\varphi, \psi), \quad \forall t \in I. \tag{5.284}$$

Since the equation (5.192) is reversible with respect to time t, Theorem 5.3.3 and Corollary 5.3.1 yield the conservation of energy (5.275). Moreover, by (5.275) and $u \in C(I; L^s)$ $(2 \le s \le p+1)$, we have $\|(u, \dot{u})\|_{H^1 \oplus L^2}$ is continuous with respect to t. On account of this and the weak continuity of $(u, \dot{u})$ in $H^1 \oplus L^2$, we have that $(u, \dot{u})$ is a weak continuous function taking values in $H^1 \oplus L^2$ with variable t. Consequently, in the topology of $H^1 \oplus L^2$, we have $(u_j(t), \dot{u}_j(t)) \xrightarrow{j\to\infty} (u(t), \dot{u}(t))$. □

Theorem 5.3.6 (Global existence). *Assume that $n \geq 2, f(u)$ satisfies* (H_1), (H_2) *and* $p < 1 + \frac{4}{n-2}$. *Let* $(\varphi, \psi) \in X_e$, $t_0 \in \mathbb{R}$. *Then* (5.192), (5.193) *or* (5.208) *have a unique solution u satisfying* $(u, \dot{u}) \in C(\mathbb{R}, X_e)$ *and the energy conservation* (5.275) *and inequality* (5.276), (5.277). *Hence, if we let* ρ, r, q, q_1 $0 \leq \rho < 1$, (5.240), (5.242), (5.257) *and* (5.258), $q_1 \geq q$, *then* (5.192), (5.193) *or* (5.208) *have a unique solution in* $u \in \mathcal{X}_{1,\mathrm{loc}}(\mathbb{R}) \cap \mathcal{X}_{2,\mathrm{loc}}(\mathbb{R})$.

Proof. Suppose ρ, r, q and q_1 are as in Theorem 5.3.6. Then, by (ii) in Theorem 5.3.4, for any $(\varphi, \psi) \in X_e$, (5.192), (5.193) (5.208) have a unique local solution in $\mathcal{X}_1(I) \cap \mathcal{X}_2(I)$, where $T = T(\|(\varphi, \psi)\|_{X_e}) > 0$. By Theorem 5.3.5, the local solution $u(t) \in C(I, X_e)$ enjoys the estimates (5.276), (5.277). By the standard iteration method, we conclude that (5.192), (5.193) or (5.208) have global solution $u(t)$ in $\mathcal{X}_{1,\mathrm{loc}}(\mathbb{R}) \cap \mathcal{X}_{2,\mathrm{loc}}(\mathbb{R}) \cap C(I, X_e)$. By $L^\infty_{\mathrm{loc}}(\mathbb{R}, X_e) \hookrightarrow \mathcal{X}_{1,\mathrm{loc}}(\mathbb{R}) \cap \mathcal{X}_{2,\mathrm{loc}}(\mathbb{R})$ and (iii) in Theorem 5.3.4, we have that the uniqueness for the solution in $\mathcal{X}_{1,\mathrm{loc}}(\mathbb{R}) \cap \mathcal{X}_{2,\mathrm{loc}}(\mathbb{R})$ implies the uniqueness in $C(I, X_e)$. □

Remark 5.3.6. For the critical case ($p = 1 + \frac{4}{n-2}$), Struwe first established the existence for the global smooth solutions to (5.192), (5.193) in $\mathbb{R}^3$ under the assumption of radial initial data [251]. Thereafter, Grillakis [111] removed the restriction on the radial data, and established the well-posedness for the smooth solution when $n = 3$. Then Shatah and Struwe [237] established the existence and uniqueness for the Cauchy problems of critical wave equations when $n \leq 7$. For the supercritical case $p > 1 + \frac{4}{n-2}$, it is completely open.

Remark 5.3.7. Based on the space-time estimates obtained in the first two sections and Besov space theory, we can get the local well-posedness and the global existence for small data for the nonlinear wave and Klein–Gordon equations in $Y^{s+1} = H^{s+1} \times H^s (s > -1)$. The readers who are interested in this can see [126, 174].

5.4 Smooth solution of semilinear wave equation

In this section, we will study the Cauchy problem of semilinear wave equation

$$\begin{cases} u_{tt} - \Delta u + f(u) = 0, & (x,t) \in \mathbb{R}^n \times \mathbb{R}, \\ u(0) = \varphi(x), \quad u_t(0) = \psi(x), & x \in \mathbb{R}^n. \end{cases} \tag{5.285}$$

It is originated from Jörgen, where

$$f(u) = u|u|^{p-1}, \quad 1 < p \leq \frac{n+2}{n-2}, \; n \geq 3. \tag{5.286}$$

$p_c = \frac{n+2}{n-2}$ corresponds to H^1-critical growth of (5.285). When $n \leq 2$, the case with $1 < p < \infty$ belongs to the range of subcritical growth (there is no critical growth order). It is easier compared with the case of higher dimensions, thus we do not discuss it. When $n = 3$,

$1 < p < p_c = 5$, Jörgen has proved in [123] the global well-posedness of the smooth solution of (5.285) and (5.286). For the case of higher dimensions ($3 \le n \le 9$), Brenner, Wahl, Pecher, etc. have proved the global well-posedness of the smooth solution in [26, 215]. The global well-posedness of the energy solution of (5.285) and (5.286) is solved by Ginibre and Velo in [103, 107]. However, not much is known about the critical wave equation for a long time. When $n = 3$, $p_c = 5$, under the assumption of small energy condition

$$E(\varphi, \psi)(x) = \int_{\mathbb{R}^n} \left(\frac{1}{2}|\nabla\varphi|^2 + \frac{1}{2}\psi^2 + \frac{1}{p+1}\varphi^{p+1} \right) dx \ll 1, \tag{5.287}$$

Rauch [224] proved in 1984 the global well-posedness of the smooth solution of (5.285) and (5.286). In 1988, Struwe [251] proved: if $\varphi(x) = \varphi(|x|) \in C^3(\mathbb{R}^3)$, $\psi(x) = \psi(|x|) \in C^2(\mathbb{R}^2)$, then (5.285), (5.286) has a unique global smooth solution $u(t) \in C^2(\mathbb{R} \times \mathbb{R}^3)$. After that, Grillakis [111] removed Struwe's radial assumption on initial data, and proved with the help of Morawetz estimate that (5.285) and (5.286) with smooth initial data $\varphi(x) \in C^3(\mathbb{R}^3)$, $\psi(x) \in C^3(\mathbb{R}^3)$ have a unique global smooth solution $u(t) \in C^2(\mathbb{R} \times \mathbb{R}^3)$. Kapitanskii used Strichartz estimates subtly proved the existence and uniqueness of the partial regularity solution of (5.285), (5.286) in a completely different way [125]. Furthermore, by combining Strichartz estimate with the Morawetz estimate, Grillakis proved that the global well-posedness of the smooth solution of (5.285), (5.286) in the critical case when $3 \le n \le 5$. Meanwhile, he also proved the global well-posedness of (5.285) and (5.286) for $n \le 7$, $\varphi(x) = \varphi(|x|)$, $\psi(x) = \psi(|x|)$ (the radial initial value). Later Shatah and Struwe [237] proved the global well-posedness of (5.285) and (5.286) in the critical case for $n \le 7$. For the energy solution case, Shatah and Struwe, etc. proved the global well-posedness of (5.285), (5.286) in the energy norm sense in the critical case.

Problems, results and idea of proof

Taking $\mathbb{R}^3$ as an example, we consider the Cauchy problem of the semilinear wave equation

$$\begin{cases} \Box u = -f_k(u), & (x,t) \in \mathbb{R}^3 \times \mathbb{R}, \\ u(0,x) = \varphi(x), \quad u_t(0,x) = \psi(x), & x \in \mathbb{R}^3. \end{cases} \tag{5.288}$$

To ensure the global well-posedness of the smooth solution of (5.288), we need the following basic assumption:

(H1) $f_k(u) \in C^2(\mathbb{R})$ satisfies $f_k(0) = 0$ and growth condition of power function type:

$$|f_k'(u)| \le C_0(1 + |u|^{k-1}). \tag{5.289}$$

(H2) The repulsive condition:

$$F_k(u) = \int_0^u f_k(\tau)d\tau \geq 0, \quad |u|^{k+1} \leq C_1(1 + F_k(u)). \tag{5.290}$$

(H3) For $k = p_c = 5$ (the critical nonlinear growth), we, in addition, assume that

$$uf_k(u) - 4F_k(u) \geq 0, \quad |u| \gg 1. \tag{5.291}$$

Remark 5.4.1.

(a) (H1) implies that

$$F_k(u), uf_k(u) \leq C_0\big(|u| + |u|^{k+1}\big). \tag{5.292}$$

(b) Particularly when $f_k(u) = u^5, f_k$ satisfies the inequality (5.291) in (H3). It is necessary to establish the Morawetz estimate.

(c) Condition (H3) is not necessary for the nonlinear term with subcritical growth ($1 < k < 5$). Obviously, for the special nonlinear function $f_k(u) = |u|^{k-1}u$, condition (5.291) implies k is the superconformal index $k > 1 + \frac{4}{n-1} = 3$.

(d) For simplicity, we usually let u' denote $(\partial_t u, \partial_{x_1}, \ldots, \partial_{x_n})$.

(e) Formally, we know that the solution of (5.288) satisfies the following conservation integral:

$$E(u(t), u_t(t)) = \int_{\mathbb{R}^3} \left(\frac{1}{2}u_t^2 + \frac{1}{2}|\nabla u|^2 + F_k(u)\right)dx = E(\varphi, \psi). \tag{5.293}$$

In order to prove the global well-posedness of the solution, we need to ensure the contribution of nonlinear term in energy (potential energy)

$$\int_{\mathbb{R}^3} F_k(u)dx$$

can be controlled by the kinetic energy. This requires $k \leq 5$. In the mathematical view, it is equivalent to

$$\dot{H}^1(\mathbb{R}^3) \hookrightarrow L^q_{\mathrm{loc}}(\mathbb{R}^3), \quad \forall q \leq 6.$$

Theorem 5.4.1. *Assume* $1 < k \leq 5$, *and* $f_k(u)$ *satisfies* (H1), (H2). *In particular, assume that* (H3) *is satisfied when* $k = 5$. *Let* $\varphi(x) \in C^3(\mathbb{R}^3)$, $\psi(x) \in C^2(\mathbb{R}^3)$. *Then* (5.288) *has a unique global solution in forward direction*

$$u(x,t) \in C^2(\mathbb{R}_+ \times \mathbb{R}^3)$$

(we can seek for solution in forward direction without lose of generality). Furthermore, if assuming additionally that $f_k \in C^\infty(\mathbb{R})$, $\varphi(x) \in C^\infty(\mathbb{R}^3)$, $\psi(x) \in C^\infty(\mathbb{R}^3)$, then we have

$$u(x,t) \in C^\infty(\mathbb{R}_+ \times \mathbb{R}^3).$$

Remark 5.4.2.

(a) Obviously, the case considered in Theorem 5.4.1 includes the critical wave equation

$$\Box u = -u^5.$$

However, the solution of focusing on the nonlinear wave equation

$$\Box u = u^5, \tag{5.294}$$

will blow up, even though with a C^∞ initial function (or even C_c^∞ initial function). For example,

$$u(t,x) = (3/4)^{\frac{1}{4}}(1-t)^{-\frac{1}{2}} \tag{5.295}$$

is a smooth solution of (5.294) in $[0,1) \times \mathbb{R}^3$, and blows up as $t \nearrow 1$.

(b) To prove Theorem 5.4.1, we only need to prove the case with the initial function $(\varphi(x), \psi(x))$ supported in compact sets. Indeed, let $\chi(x) \in C_c^\infty(\mathbb{R}^3)$ be a radial symmetric function and satisfy

$$\chi(x) = 1, \quad |x| \le 1. \tag{5.296}$$

Set

$$\varphi_r(x) = \chi\left(\frac{x}{r}\right)\varphi(x), \quad \psi_r(x) = \chi\left(\frac{x}{r}\right)\psi(x). \tag{5.297}$$

Assume that Theorem 5.4.1 is valid for the above smooth initial data with compact support. Let $u_r(t,x)$ be the smooth solution of (5.288) with initial data $(\varphi_r(x), \psi_r(x))$. We claim that, as $r \to \infty$, $u_r(t,x)$ converges in $C^2(\mathbb{R}_+^{1+3})$ to a solution $u(t,x)$ of (5.288). To see this, $\forall t_0 \in \mathbb{R}_+$, let

$$\Lambda_{t_0,0} = \{(x,t),\ 0 \le t \le t_0,\ |x| \le t_0 - t\}$$

denote the backward light cone through $(t_0, 0)$, see Figure 5.1. Observe that when r_1, $r_2 > t_0$,

$$u_{r_1}(0,x) = u_{r_2}(0,x) = \varphi(x), \quad \dot{u}_{r_1}(0,x) = \dot{u}_{r_2}(0,x) = \psi(x), \quad x \in \Lambda_{t_0,0} \cap \mathbb{R}^3.$$

Then, by uniqueness, we have

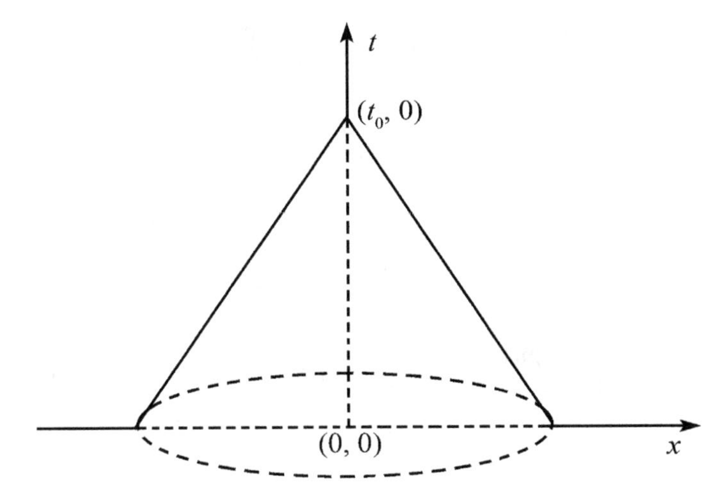

Figure 5.1: Backward light cone through $(t_0, 0)$.

$$u_{r_1}(t,x) = u_{r_2}(t,x), \quad (t,x) \in \Lambda_{t_0,0},$$
$$\lim_{r\to\infty} u_r(x,t) \equiv u(x,t), \quad (t,x) \in \Lambda_{t_0,0}.$$

Since

$$\mathbb{R}_+^{1+3} = \bigcup_{t_0>0} \Lambda_{t_0,0},$$

we have

$$\lim_{r\to\infty} u_r = u(x,t), \quad \text{in the sense of} \quad C^2(\mathbb{R}_+ \times \mathbb{R}^3).$$

Let us recall the local existence theorem of the semilinear wave equation before we discuss the proof of Theorem 5.4.1.

Theorem 5.4.2. *Consider the Cauchy problem of the semilinear wave equation*

$$\begin{cases} f(u) \in C^k(\mathbb{R}), \quad f(0) = 0, & k = 1, 2, \ldots, \\ \varphi(x) \in C_c^{k+1}(\mathbb{R}^3), \quad \psi(x) \in C_c^k(\mathbb{R}^3), & k = 1, 2, 3, \ldots. \end{cases} \tag{5.298}$$

Then there exists a $T^* > 0$ *and a unique solution of* (5.298),

$$u(t) =: u(x,t) \in C^k([0, T^*) \times \mathbb{R}^3) \tag{5.299}$$

satisfying the following alternative properties:
(i) $T^* = \infty$,
(ii) $T^* < \infty$ *and* $\lim_{t\to T^*} \sup_x |u(t,x)| = \infty$.

As we know, the solution of the inhomogeneous wave equation

$$\begin{cases} \Box v(x,t) = g(x,t), \\ v(0) = \varphi(x), \quad v_t(0) = \psi(x) \end{cases} \tag{5.300}$$

is

$$\begin{aligned} v(t,x) &= \mathcal{F}^{-1}\cos(|\xi|t)\mathcal{F}\varphi + \mathcal{F}^{-1}\frac{\sin|\xi|(t-\tau)}{|\xi|}\mathcal{F}\psi \\ &\quad + \int_0^t \mathcal{F}^{-1}\frac{\sin|\xi|(t-\tau)}{|\xi|}\mathcal{F}g(x,\tau)d\tau \end{aligned} \tag{5.301}$$

which satisfies the following Strichartz estimate:

$$\begin{aligned} \|v\|_{L_t^4(I,L_x^{12})} &\lesssim \|\varphi\|_{\dot H^1} + \|\psi\|_{L^2} + \|g\|_{L_t^1(I;L^2(\mathbb{R}^3))} \\ &\triangleq \|v'(0)\|_{L^2} + \|g\|_{L_t^1(I;L^2(\mathbb{R}^3))}, \end{aligned} \tag{5.302}$$

$$\|v\|_{L^\infty(I,L^6(\mathbb{R}^3))} \lesssim \|v'(0)\|_{L^2} + \|g\|_{L_t^1(I;L^2(\mathbb{R}^3))}. \tag{5.303}$$

Now, we analyze the idea of proving Theorem 5.4.1. According to Remark 5.4.2, we need only prove the case with initial data supported in compact sets. By the local existence theorem, if $T^* < \infty$, then $u(t,x) \notin L^\infty((0,T)\times\mathbb{R}^3)$. Our next step is to replace $\|\cdot\|_{L^\infty_{t,x}}$ with the mixed-norm $\|\cdot\|_{L_t^4 L_x^{12}}$, that is, to prove either $T^* = \infty$ or

$$T^* < \infty, \quad \|u(t,x)\|_{L_t^4([0,T^*);L_x^{12}(\mathbb{R}^3))} = \infty. \tag{5.304}$$

In other words, establishing the global well-posedness reduces to prove $u(t,x) \in L_t^4 L_x^{12}([0,T^*)\times\mathbb{R}^3)$.

Proposition 5.4.1. *Assume* $1 < k \le 5$ *and* $f_k(u) \in C^2$ *satisfies (H1),* $\varphi(x) \in C_c^3(\mathbb{R}^3)$, $\psi(x) \in C_c^2(\mathbb{R}^2)$. *Then there is a* $T^* > 0$ *and a unique solution of* $u(t) \in C^2([0,T^*)\times\mathbb{R}^3)$ *such that one of the following properties holds:*
(i) $T^* = +\infty$,
(ii) $T^* < \infty$, $u(t) \notin L_t^4([0,T^*);L_x^{12}(\mathbb{R}^3))$.

Proof. The existence part of Proposition 5.4.1 follows from local well-posedness, so we only need to prove the second part. We will prove it by contradiction. Suppose that $T^* < \infty$, and $u \in C^2([0,T)\times\mathbb{R}^3)$ is a solution of (5.288) satisfying

$$u \in L_t^4(I;L_x^{12}(\mathbb{R}^3)), \quad I = [0,T^*). \tag{5.305}$$

Then we will show that u can be extended to $C^2([0,T^*]\times\mathbb{R}^3)$, which implies

$$u(t) \in L^\infty([0,T^*) \times \mathbb{R}^3). \tag{5.306}$$

By the local existence theorem, we can deduce the contradiction.

Now, we will prove the theorem with this idea: let $0 < R < \infty$ be large enough that

$$\operatorname{supp} \varphi(x),\ \operatorname{supp} \psi(x) \subset B_R(0).$$

By the Huggens principle, we know that

$$\operatorname{supp} u(x,t) \subset B_{R+t}(0).$$

According to the assumption (H1), $\forall 0 \le t_0 < s < T^*$, we have

$$\begin{aligned}\sum_{|\alpha|\le 1} \|\partial_x^\alpha (f_k(u))\|_{L_t^1 L_x^2([t_0,s]\times\mathbb{R}^3)} &\lesssim C(T^* - t_0, R+T^*) \\ &\quad \times \sum_{|\alpha|\le 1} \|\partial^\alpha u\|_{L^\infty([t_0,s];L^6(\mathbb{R}^3))} + \sum_{|\alpha|\le 1} \|u^{k-1}\partial_x^\alpha u\|_{L_t^1 L_x^2},\end{aligned} \tag{5.307}$$

where

$$\lim_{t_0\to T^*} C(T^* - t_0, R+T^*) = 0. \tag{5.308}$$

By Strichartz estimate and Hölder's inequality, we obtain

$$\begin{aligned}&\sup_{t_0\le t\le s} \sum_{|\alpha|\le 1} \|\partial_x^\alpha u(t)\|_{L^6(\mathbb{R}^3)} \\ &\quad \lesssim \sum_{|\alpha|\le 1} \|(\partial_x^\alpha u)'(t_0)\|_{L^2(\mathbb{R}^3)} + C(T^* - t_0, R+T^*) \sum_{|\alpha|\le 1} \|\partial_x^\alpha u\|_{L^\infty([t_0,s],L^6(\mathbb{R}^3))} \\ &\qquad + \sum_{|\alpha|\le 1} \|\partial_x^\alpha u\|_{L_t^\infty([t_0,s],L^6(\mathbb{R}^3))} \|u\|^{k-1}_{L_t^{k-1}([t_0,s],L^{3(k-1)}(\mathbb{R}^3))}.\end{aligned} \tag{5.309}$$

Let us consider the case when $k = 5$. Observing (5.308) and

$$\lim_{s,t_0\to T^*} \|u\|_{L_t^4([t_0,s];L^{12}(\mathbb{R}^3))} = 0, \tag{5.310}$$

as t_0 gets closed to T^*, we obtain

$$\sup_{t_0\le t\le s} \sum_{|\alpha|\le 1} \|\partial_x^\alpha u(t,\cdot)\|_{L^6(\mathbb{R}^3)} \le 2 \sum_{|\alpha|\le 1} \|(\partial_x^\alpha u)'(t_0)\|_{L^2(\mathbb{R}^3)} = C(t_0) < \infty.$$

Let $s \to T^*$. Since $u(t,x) \in C^2([0,t_0] \times \mathbb{R}^3)$ and

$$\operatorname{supp} u(t,x) \subset \{x \mid |x| \le t_0 + R\}, \tag{5.311}$$

we conclude that

$$\sup_{0\le t\le T^*}\sum_{|\alpha|\le 1}\|\partial_x^\alpha u(\cdot,t)\|_{L^6(\mathbb{R}^3)}<\infty. \tag{5.312}$$

By the Sobolev embedding theorem, $W^{1,6}(\mathbb{R}^3)\hookrightarrow L^\infty(\mathbb{R}^3)$, we deduce that (5.306) holds for the critical case.

The subcritical case with $1<k<5$ can be handled similarly. Noting that u has compact support, we can apply Hölder's inequality to see that

$$\|u\|^{k-1}_{L^{k-1}([t_0,s],L^{3(k-1)}(\mathbb{R}^3))}\le\rho(T^*-t_0,T^*+R)\|u\|^{k-1}_{L^4([t_0,s];L^{12}(\mathbb{R}^3))}, \tag{5.313}$$

where

$$\lim_{t\to T^*}\rho(T^*-t_0,R+T^*)=0.$$

Noting that $k>1$, similarly as $k=5$, we can deduce that (5.306) holds for the subcritical case. □

On account of this proposition, to prove Theorem 5.4.1, we may assume that u is as above and then it suffices to show that (5.305) holds. We can observe that:

(i) When $1\le k<5$, we can prove (5.286) with energy conservation law and Strichartz estimate.

(ii) For $k=5$, besides the energy conservation law and Strichartz estimate, we also need tools such as the local energy estimate and Morawetz estimate.

Energy estimate and the subcritical case

Proposition 5.4.2. *Suppose that $f_k(u)$ is as in Theorem 5.4.1. $\varphi(x)\in C_c^3(\mathbb{R}^3)$, $\psi(x)\in C_c^2(\mathbb{R}^3)$. Suppose that $u\in C^2([0,T^*)\times\mathbb{R}^3)$ solves the Cauchy problem (5.288). Then*

$$\begin{aligned}E(u(t),u_t(t))&=\int_{\mathbb{R}^3}(\frac{1}{2}|u'(t,x)|^2+F_k(u(t,x))dx\\&=E(\varphi(x),\psi(x)),\quad 0<t<T^*,\quad u'=(u_t,\nabla u).\end{aligned} \tag{5.314}$$

Moreover, if

$$\operatorname{supp}\varphi,\ \operatorname{supp}\psi\subset\{x\mid |x|\le R\},$$

then

$$\int_{\mathbb{R}^3}(|u'(t,x)|^2+|u(t,x)|^{k+1})dx\le C_{R,T^*},\quad 0\le t<T^*. \tag{5.315}$$

Proof. Noting that $\operatorname{supp}u\subset\{x\mid |x|\le R+T^*\}$, we can deduce (5.315) from (5.290). To obtain energy estimate (5.314), we multiply the equation (5.288) by $\partial_t u$, then

$$0 = \partial_t u(\Box u + f_k(u)) = \operatorname{div}_{t,x} e(u), \tag{5.316}$$

where

$$e(u) = \left(\frac{1}{2}|u'|^2 + F_k(u), -\partial_t u \nabla_x u\right). \tag{5.317}$$

If we fix $0 < t < T^*$, then $u(t,x) \in C^2([0,T^*) \times \mathbb{R}^3)$ has compact support in $[0,t] \times \mathbb{R}^3$. Therefore, integrating (5.316) leads to

$$\begin{aligned} 0 &= \int_0^t \int_{\mathbb{R}^3} \operatorname{div}_{t,x} e(u) dx dt = \int_{\mathbb{R}^3} \int_0^t \frac{\partial}{\partial t}\left(\frac{1}{2}|u'|^2 + F_k(u)\right) d\tau dx \\ &= \int_{\mathbb{R}^3} \left(\frac{1}{2}|u'(t)|^2 + F_k(u(t))\right) dx - \int_{\mathbb{R}^3} \left(\frac{1}{2}|u'(0)|^2 + F_k(u(0))\right) dx. \end{aligned}$$

Therefore, (5.314) holds. □

Lemma 5.4.1. *Let $0 < C_0 < \infty$, and suppose that $0 \le y(s) \in C([a,b])$ satisfies*

$$y(a) = 0, \quad y(s) \le C_0 + \varepsilon y(s)^\sigma, \quad \sigma > 0. \tag{5.318}$$

Then, if $\varepsilon < 2^{-\sigma} C_0^{1-\sigma}$, it follows that

$$y(s) \le 2C_0, \quad s \in [a,b]. \tag{5.319}$$

Proof. Consider

$$h(x) = C_0 + \varepsilon x^\sigma - x.$$

If $x_1 = 2C_0$, $\varepsilon < 2^{-\sigma} C_0^{1-\sigma}$,

$$C_0 + \varepsilon x_1^\sigma - x_1 \equiv h(x_1) = h(2C_0) < C_0 + 2^{-\sigma} C_0^{1-\sigma} (2C_0)^\sigma - 2C_0 = 0.$$

Hence, to ensure that

$$C_0 + \varepsilon x^\sigma - x \ge 0, \quad \forall x \in [0, x_0],$$

it is necessary that $x_0 < 2C_0$. Notice that $y(s)$ must be smaller than the supremum of such x_0 that makes the inequality above hold. Thus, we obtain

$$y(s) \le 2C_0, \quad \forall s \in [a,b].$$ □

Proof of Theorem 5.4.1 (*Subcritical case*). As analyzed before, the problem reduces to: under the condition,

$$\operatorname{supp}\varphi(x),\ \operatorname{supp}\psi(x)\subset\{x\mid |x|\le R\},\tag{5.320}$$

and

$$u\in C^2([0,T^*)\times\mathbb{R}^3),\quad 0<T^*<\infty\tag{5.321}$$

prove

$$u(t)\in L_t^4L_x^{12}([0,T^*)\times\mathbb{R}^3).\tag{5.322}$$

This means $T^*=\infty$.

For $0\le t_0<s<T^*$, by the Strichartz estimate and Hölder's inequality, we have

$$\begin{aligned}\|u\|_{L_t^4L_x^{12}([t_0,s]\times\mathbb{R}^3)}&\lesssim\|u'(t_0)\|_{L^2(\mathbb{R}^3)}+\||u|+|u|^k\|_{L_t^1L_x^2([t_0,s]\times\mathbb{R}^3)}\\&\lesssim\|u'(t_0)\|_{L^2(\mathbb{R}^3)}+C(R,T^*)+\||u|^k\|_{L_t^1L_x^2[t_0,s]\times\mathbb{R}^3)}\\&\le C(R,T^*)+(2E(\varphi,\psi))^{\frac12}+\||u|^k\|_{L_t^1L_x^2[t_0,s]\times\mathbb{R}^3)}.\end{aligned}\tag{5.323}$$

By Hölder's inequality and

$$1=\frac{k-1}{4}+\frac{5-k}{4},\quad\frac12=\frac{7-k}{12}+\frac{k-1}{12},\tag{5.324}$$

we get

$$\||u|^k\|_{L_t^1L_x^2([t_0,s]\times\mathbb{R}^3)}\lesssim\|u\|_{L_t^{\frac{4}{5-k}}L_x^{\frac{12}{7-k}}([t_0,s]\times\mathbb{R}^3)}\|u\|^{k-1}_{L_t^4L_x^{12}([t_0,s]\times\mathbb{R}^3)}.\tag{5.325}$$

Noticing that

$$\frac{12}{7-k}<k+1,\quad 1<k<5\tag{5.326}$$

and

$$\operatorname{supp}u(t,x)\subset\{x\mid |x|\le t+R\},\tag{5.327}$$

we obtain

$$\begin{aligned}\|u\|_{L_t^{\frac{4}{5-k}}L_x^{\frac{12}{7-k}}([t_0,s]\times\mathbb{R}^3)}&\le(T^*-t_0)^{\frac{5-k}{4}}\sup_{t_0\le t\le s}\|u(t)\|_{L_x^{\frac{12}{7-k}}}\\&\lesssim(T^*-t_0)^{\frac{5-k}{4}}(T_0^*+R)^{3(\frac{7-k}{12}-\frac{1}{k+1})}\sup_{t_0\le t\le s}\|u\|_{L^{k+1}(\mathbb{R}^3)}\\&\le\rho(R,T^*)(T^*-t_0)^{\frac{5-k}{4}},\end{aligned}\tag{5.328}$$

where we have used in the last step the following equality:

$$\frac{7-k}{12} = \frac{1}{k+1} + \frac{1}{\chi}, \quad \frac{1}{\chi} = \left(\frac{7-k}{12} - \frac{1}{k+1}\right).$$

Let

$$\varepsilon(t_0) = \rho(R, T^*)(T^* - t_0)^{\frac{5-k}{4}}. \tag{5.329}$$

Then (5.323) becomes

$$\begin{aligned}\|u\|_{L_t^4 L_x^{12}([t_0,s]\times\mathbb{R}^3)} \le C(R, T^*) + C(2E(\varphi, \psi))^{\frac{1}{2}} \\ + \varepsilon(t_0)\|u(t)\|^{k-1}_{L_t^4 L_x^{12}([t_0,s]\times\mathbb{R}^3)}.\end{aligned} \tag{5.330}$$

Note that

$$\lim_{t_0 \nearrow T^*} \varepsilon(t_0) = 0, \quad k < 5. \tag{5.331}$$

Therefore, Lemma 5.4.1 implies that

$$\|u\|_{L_t^4 L_x^{12}([t_0,T^*)\times\mathbb{R}^3)} \le 2C(R, T^*) + 2C(2E(\varphi, \psi))^{\frac{1}{2}}. \tag{5.332}$$

This clearly gives (5.322), since u is bounded in $[0, t_0] \times \mathbb{R}^3$ and compactly supported. □

Remark 5.4.3.

(a) According to the proof above, if $E(u(0)) \ll 1$ and $f_k(u) = u^5$, we can obtain the global existence of the smooth solution, which is the result of Rauch. In this case, the constant with form $C(R, T^*)$ does not appear in the estimate above, thus we can prove the theorem with the initial data that have no compact supports.

(b) From the technique of proof above, we know that when $k = 5$, $\varepsilon(t_0)$ may be a constant large enough. Therefore, it is not available for problems with large initial data. In order to obtain the estimate $\|u\|_{L_t^4 L_x^{12}}$ of the solution u on the backward light cone, we need to modify the definition of $\varepsilon(t_0)$, so that $\varepsilon(t_0)$ contains the integral $\int_{|x-x_0|\le|T^*-t|} |u(t)|^6 dx$. If we can prove that the integral goes to 0 as $t \longmapsto T^*$, then we can acquire the norm $\|u\|_{L_t^4 L_x^{12}}$ of u on the backward light cone. Noticing the condition that $f_k(u)$ satisfies, we know that $\int_{|x-x_0|\le|T^*-t|} |u(t)|^6 dx$ is equivalent to $\int_{|x-x_0|\le|T^*-t|} F_k(u) dx$. To prove this, we need some tools like the local energy inequality.

Decay estimate and the critical case

We first introduce some symbols. Fix $x_0 \in \mathbb{R}^3$, $0 \le t_0 < s < T^*$ and $\delta > 0$. Let

$$\Lambda(\delta, t_0, s) = \{(t, x) \mid t_0 \le t \le s,\ |x - x_0| \le \delta + T^* - t\}$$

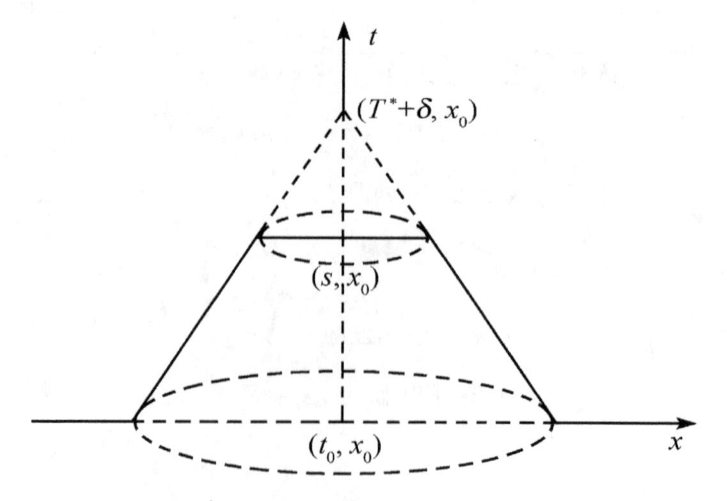

Figure 5.2: The region of frustum of the backward light cone.

be the frustum of the backward light cone interpreted by hyperplane $t = t_0$, $t = s$ through $(T^* + \delta, x_0)$,

$$\begin{aligned}
D_{t_0} &= \{(t,x) \in \Lambda(\delta, t_0, s),\ t = t_0\},\\
D_s &= \{(t,x) \in \Lambda(\delta, t_0, s),\ t = s\},\\
M_{t_0}^s &= \{(t,x) \in \Lambda(\delta, t_0, s),\ t_0 \le t \le s,\ |x - x_0| = \delta + T^* - t\},
\end{aligned}$$

see Figure 5.2.

Let $u(t,x)$ be the solution obtained in Proposition 5.4.2. Let

$$E(u, D_t) = \int_{D_t} \Big(\frac{1}{2}|u'(t,x)|^2 + F_k(u)\Big)dx, \quad 0 \le t \le T^*, \tag{5.333}$$

$$\mathrm{Flux}(u, M_{t_0}^s) = \int_{M_{t_0}^s} \langle e(u), \nu\rangle d\sigma, \quad 0 \le t_0 < s < T^*, \tag{5.334}$$

where ν denotes the exterior normal of $M_{t_0}^s$, and $d\sigma$ denotes the Lebesgue measure on $M_{t_0}^s$. Integrating the following formula over $\Lambda(\delta, t_0, s)$:

$$0 = \partial_t\Big(\frac{1}{2}|u'(t,x)|^2 + F_k(u)\Big) - \mathrm{div}(\partial_t u \cdot \nabla_x u) \tag{5.335}$$

and using the divergence theorem, we see

$$E(u, D_{t_0}) = E(u, D_s) + \mathrm{Flux}(u, M_{t_0}^s). \tag{5.336}$$

We note that $M^s_{t_0}$ consists of points of the form

$$(\delta + T^* - |x - x_0|, x),\ \delta + T^* - |x - x_0| \in [t_0, s] \tag{5.337}$$

(it is essentially defined by $|x - x_0| = \delta + T^* - t$). At such a point, the exterior normal vector is

$$\nu = \frac{1}{\sqrt{2}}\left(1, \frac{x - x_0}{|x - x_0|}\right). \tag{5.338}$$

Thus,

$$\begin{aligned}\sqrt{2}\langle e(u), \nu\rangle &= \frac{1}{2}|u'|^2 + F_k(u) - \partial_t u \frac{x - x_0}{|x - x_0|}\nabla_{x-x_0} u \\ &= \frac{1}{2}\left|\frac{x - x_0}{|x - x_0|}\partial_t u - \nabla_{x-x_0} u\right|^2 + F_k(u).\end{aligned} \tag{5.339}$$

Noting that

$$\mathrm{Flux}(u, M^s_{t_0}) = \frac{1}{\sqrt{2}}\int_{M^s_{t_0}}\left(\frac{1}{2}\left|\frac{x - x_0}{|x - x_0|}\partial_t u - \nabla_{x-x_0} u\right|^2 + F_k(u)\right)d\sigma \geq 0, \tag{5.340}$$

we conclude from this that local energy function $t \longmapsto E(u, D_t)$ is nonincreasing on $t \in [0, T^*)$, and

$$E(u, D_t) \leq E(u(t), u_t(t)) \leq E(u(0), u_t(0)) < \infty.$$

Therefore, we get

$$\lim_{t \to T^*} \mathrm{Flux}(u, M^{T^*}_t) = 0. \tag{5.341}$$

Proposition 5.4.3. *Let $k = 5$ and suppose that $\varphi(x) \in C^3_c(\mathbb{R}^3)$, $\psi(x) \in C^2_c(\mathbb{R}^3)$ satisfy*

$$\operatorname{supp}\varphi(x),\ \operatorname{supp}\psi(x) \subset \{x \mid |x| \leq R\}.$$

Suppose that $u(t, x) \in C^2([0, T^*) \times \mathbb{R}^3)$ solves (5.288). For a fixed $x_0 \in \mathbb{R}^3$, assume that

$$\int_{|x-x_0| \leq T^* - t_0}\left(\frac{1}{2}|u'(t_0)|^2 + F_k(u(t_0))\right)dx < \varepsilon. \tag{5.342}$$

Then there is an $\varepsilon_0 = \varepsilon_0(R, T^*, E(u(0), u_t(0))) > 0$ such that, if $0 < \varepsilon < \varepsilon_0$ and $0 \leq t_0 < T^*$,

$$u(t) \in L^4_t L^{12}_x(\Lambda(\delta, t_0, T^*)), \tag{5.343}$$

provided that $\delta > 0$, $T^* - t_0 > 0$ are sufficiently small.

Remark 5.4.4. We need only required that t_0 is close to T^* for the proof. Since $u(t) \in C^2([0,T^*) \times \mathbb{R}^3)$ and

$$u(t,x) \equiv 0, \quad |x| > t + R,$$

we know

$$u(t,x) \in L_t^4 L_x^{12}(\Lambda(\delta, 0, T^*)). \tag{5.344}$$

As u is compactly supported, $u(t,x)$ satisfies (5.305).

Proof of Proposition 5.4.3. Under the assumption of (H2), if (5.342) holds, then

$$\sup_{t_0 \le t < T^*} \int_{|x-x_0| \le \delta + T^* - t} |u(t,x)|^6 dx \le 2C_1\varepsilon, \tag{5.345}$$

provided that $\delta > 0$ and $T^* - t > 0$ are sufficiently small. To prove this, we take a sufficiently small $\delta > 0$ and obtain

$$\int_{|x-x_0| \le \delta + T^* - t_0} \left(\frac{1}{2}|u'(t_0)|^2 + F_k(u(t_0))\right) dx < \frac{3}{2}\varepsilon. \tag{5.346}$$

Since $E(u, D_t)$ is a nonincreasing function of t, we have

$$\sup_{t_0 \le t < T^*} \int_{|x-x_0| \le \delta + T^* - t} \left(\frac{1}{2}|u'(t)|^2 + F_k(u(t))\right) dx < \frac{3}{2}\varepsilon. \tag{5.347}$$

Hence, choosing $\delta > 0$ and $T^* - t_0$ are sufficiently small, we get

$$\begin{aligned}
&\int_{|x-x_0| \le \delta + T^* - t} |u(t,x)|^6 dx \\
&\quad \le \frac{4\pi}{3} C_1 (\delta + T_0 - t_0)^3 + C_1 \int_{|x-x_0| \le \delta + T^* - t} F_k(u) dx \\
&\quad \le \frac{4\pi}{3} C_1 (\delta + T_0 - t_0)^3 + \frac{3C_1}{2}\varepsilon \le 2C_1\varepsilon.
\end{aligned} \tag{5.348}$$

We claim that (5.348) implies (5.343) if $\varepsilon > 0$ is small enough. By the Strichartz estimate and Huygen's principle (the estimate of nonlinear function should also be taken in the same domain), we obtain

$$\begin{aligned}
\|u\|_{L_t^4 L_x^{12}(\Lambda(\delta, t_0, s))} &\lesssim \|u'(t_0)\|_{L^2(\mathbb{R}^3)} + \|f_k(u)\|_{L_t^1 L_x^2(\Lambda(\delta, t_0, s))} \\
&\lesssim (2E(\varphi, \psi))^{\frac{1}{2}} + \|f_k(u)\|_{L_t^1 L_x^2(\Lambda(\delta, t_0, s))}.
\end{aligned} \tag{5.349}$$

Noting nonlinear assumption (H1), we have

$$\begin{aligned}\|f_k(u)\|_{L_t^1L_x^2(\Lambda(\delta,t_0,s))} &\lesssim C_1(T^*,R) + \||u|^4u\|_{L_t^1L_x^2(\Lambda(\delta,t_0,s))} \\ &\lesssim C_1(T^*,R) + \|u\|_{L_t^\infty L_x^6(\Lambda(\delta,t_0,s))}\|u\|^4_{L_t^4L_x^{12}(\Lambda(\delta,t_0,s))}.\end{aligned} \tag{5.350}$$

Combining (5.345), (5.349) with (5.350), we obtain

$$\begin{aligned}\|u\|_{L_t^4L_x^{12}(\Lambda(\delta,t_0,s))} &\le C[2E(\varphi,\psi))^{\frac{1}{2}} + C_1(R,T^*)] \\ &\quad + C(2C_0\varepsilon)^{\frac{1}{6}}\|u\|^4_{L_t^4L_x^{12}(\Lambda(\delta,t_0,s))}.\end{aligned} \tag{5.351}$$

Finally, according to Lemma 5.4.1 provided ε satisfies

$$C(2C_0\varepsilon)^{1/2} < 2^{-4}(C(2E(\varphi,\psi))^{1/2} + C_1(R,T^*))^{-3}. \tag{5.352}$$

We can ensure

$$\|u\|_{L_t^4L_x^{12}(\Lambda(\delta,t_0,s))} \le 2(C(2E(\varphi,\psi))^{\frac{1}{2}} + C_1(R,T^*)), \tag{5.353}$$

where ε depends only on T^*, R and $E(\varphi,\psi)$. The proof is completed. □

Remark 5.4.5. Using the last result, we can reduce our task to show that the energy cannot concentrate at any point (T^*,x_0). More precisely, for any x_0, it suffices to show that

$$\lim_{t\nearrow T^*}\int_{|x-x_0|<T^*-t}\left(\frac{1}{2}|u'(x,t)|^2 + F_k(u(t))\right)dx = 0. \tag{5.354}$$

This means that, $\forall\varepsilon > 0$, if $T^* - t_0$ is sufficiently small, we can ensure (5.342) hold in Proposition 5.4.3. Hence, for every fixed $\forall x_0 \in \mathbb{R}^3$, there must be a $\delta > 0$ such that (5.343) holds. Since $u(t,x) \in C^2([0,t_0]\times\mathbb{R}^3)$, $u(t)$ satisfies

$$u(t) \in L_t^4L_x^{12}(\Lambda(\delta,0,T^*)). \tag{5.355}$$

Since

$$\operatorname{supp} u(t) \subset \{(x,t) \mid |x| \le t+R, 0 \le t \le T^*\}, \tag{5.356}$$

we can cover its support by finitely many of these sets $\Lambda_j(\delta,0,T^*)$ such that

$$\bigcup_j \Lambda_j(\delta,0,T^*) \supset \operatorname{supp} u(t,x). \tag{5.357}$$

Thus, we have

$$u(t) \in L_t^4L_x^{12}([0,T^*)\times\mathbb{R}^3), \tag{5.358}$$

which implies the global existence of a smooth solution.

Proposition 5.4.4. *Let $k = 5$. Then the sufficient condition of* (5.354) *is*

$$\lim_{t\nearrow T^*} \int_{|x-x_0|\le T^*-t} F_k(u)dx = 0. \tag{5.359}$$

Proof. This means (5.359) can get rid of the energy concentration. According to (H1) and (H2), (5.359) is equivalent to

$$\lim_{t\nearrow T^*} \int_{|x-x_0|<T^*-t} |u(t,x)|^6 dx = 0. \tag{5.360}$$

The proof of Proposition 5.4.3 shows that (5.360) in turn implies that

$$u(t,x) \in L_t^4 L_x^{12}(\Lambda(0,0,T^*)). \tag{5.361}$$

Therefore, the problem comes down to prove (5.354) holds under the condition (5.360). Applying the Strichartz estimate to the equation,

$$\Box u' = -f_k'(u)u', \tag{5.362}$$

we get (suppose $0 \le t_0 < s < T^*$)

$$\begin{aligned}
\sup_{t_0\le t\le s}\Big(\int_{|x-x_0|<T^*-t} |u'(t,x)|^6 dx\Big)^{1/6} &= \|u'\|_{L_t^\infty L_x^6(\Lambda(0,t_0,s))} \\
&\le C\sum_{|\alpha|=2} \|\partial^\alpha u(t_0)\|_{L^2(\mathbb{R}^3)} + C\|f_k'(u)u'\|_{L_t^1 L_x^2(\Lambda(0,t_0,s))} \\
&\le C\sum_{|\alpha|=2} \|\partial^\alpha u(t_0)\|_{L^2(\mathbb{R}^3)} + C_1(R,T^*) + C\|u^4 u'\|_{L_t^1 L_x^2(\Lambda(0,t_0,s))} \\
&\le C(t_0) + C\|u\|^4_{L_x^4 L_x^{12}(\Lambda(0,t_0,s))}\|u'\|_{L_t^\infty L_x^6(\Lambda(0,t_0,s))}.
\end{aligned} \tag{5.363}$$

Since $u(t,x) \in C^2([0,T^*)\times \mathbb{R}^3)$, $\operatorname{supp} u \subset \{x \,|\, |x-x_0| \le R+T^*\}$, we know that $C(t_0) < \infty$ is finite. In addition, noting that

$$\lim_{t_0\to T^*} \|u\|_{L_t^4 L_x^{12}(\Lambda(0,t_0,T^*))} = 0,$$

we therefore conclude from (5.363) that

$$\sup_{t_0\le t<T^*}\Big(\int_{|x-x_0|<T^*-t} |u'(t,x)|^6 dx\Big)^{\frac{1}{6}} \le 2C(t_0). \tag{5.364}$$

By an application of Hölder's inequality, we obtain

$$\left(\int_{|x-x_0|<T^*-t} |u'(t,x)|^2 dx\right)^{1/2} \le 2C(t_0)\left(\frac{4\pi}{3}(T^*-t)^3\right)^{1/3}, \quad t_0 \le t \le T^*. \tag{5.365}$$

Hence,

$$\lim_{t\nearrow T^*} \int_{|x-x_0|<T^*-t} \frac{1}{2}|u'(t,x)|^2 dx = 0. \tag{5.366}$$

□

Finally, we use the Morawetz–Pohožaev identity to show (5.359) or (5.360). For simplicity, shift (T^*, x_0) in

$$\lim_{t\to T^*} \int_{|x-x_0|<T^*-t} F_k(u(x,t))dx = 0$$

to the origin. More precisely, assume

$$u(t,x) \in C^2([-T^*,0) \times \mathbb{R}^3) \tag{5.367}$$

is the smooth solution of

$$\begin{cases} \Box u + f_k(u) = 0, \\ u(-T^*) = \varphi(x), \quad u_t(-T^*) = \psi(x). \end{cases} \tag{5.368}$$

Under this assumption, we will prove

$$\lim_{t\nearrow 0} \int_{|x|<|t|} F_k(u)dx = 0. \tag{5.369}$$

To derive Morawetz's identity, we shall use Noether's principle. Consider Lagrange's density function as following, which is associated with the wave equation

$$L(q,p) = \frac{1}{2}|p_0|^2 - \frac{1}{2}\sum_{j=1}^{3}|p_j|^2 - F_k(q),$$

where $(q,p) \in \mathbb{R}^{1+3}$. Verifying directly, for $v \in C_c^\infty([-T^*,0) \times \mathbb{R}^3)$, we have

$$\begin{aligned} &\frac{d}{d\varepsilon} \int_{[-T^*,0)\times\mathbb{R}^3} L(u+\varepsilon v, (u+\varepsilon v)')dtdx|_{\varepsilon=0} \\ &= -\int_{[-T^*,0)\times\mathbb{R}^3} (\Box u + f_k(u))v dtdx = 0. \end{aligned} \tag{5.370}$$

Thus, u definitely satisfy the Euler–Lagrange equation corresponding to $L(q,p)$,

$$\frac{\partial L}{\partial q}(u,u') - \sum_{j=0}^{3} \partial_j\left(\frac{\partial L}{\partial p_j}(u,u')\right) = 0, \tag{5.371}$$

where $\partial_0 = \partial_t$.

Suppose u_ε is a one-parameter C^1 deformation of u, then

$$\partial_\varepsilon L(u_\varepsilon, u'_\varepsilon) = \frac{\partial L}{\partial q}(u_\varepsilon, u'_\varepsilon)\partial_\varepsilon u_\varepsilon + \sum_{j=0}^{3} \frac{\partial L}{\partial p_j}(u_\varepsilon, u'_\varepsilon)\partial_j \partial_\varepsilon u_\varepsilon. \tag{5.372}$$

If we assume that $u_{\varepsilon_0} = u$, then we could use the Euler–Lagrange equation (5.371) to obtain

$$\partial_\varepsilon L(u_\varepsilon, u'_\varepsilon)|_{\varepsilon=\varepsilon_0} = \sum_{j=0}^{3} \partial_j\left[\frac{\partial L}{\partial p_j}(u_\varepsilon, u'_\varepsilon)\partial_\varepsilon u_\varepsilon\bigg|_{\varepsilon=\varepsilon_0}\right]. \tag{5.373}$$

In particular, let

$$u_\varepsilon(t,x) = \varepsilon u(\varepsilon t, \varepsilon x), \quad \varepsilon_0 = 1, \tag{5.374}$$

then

$$\partial_\varepsilon u_\varepsilon|_{\varepsilon=\varepsilon_0} = u + \sum_{j=0}^{3} x_j \partial_j u \quad (x_0 = t). \tag{5.375}$$

Consider the dilation of $L(q,p) + F_k(q)$. It is easy to see

$$L(u_\varepsilon, u'_\varepsilon) = \varepsilon^4 L(u,u')(\varepsilon t, \varepsilon x) + \varepsilon^4 F_k(u(\varepsilon t, \varepsilon x)) - F_k(u_\varepsilon(x,t)) \tag{5.376}$$

and

$$\begin{aligned}
\partial_\varepsilon L(u_\varepsilon, u'_\varepsilon)|_{\varepsilon=1} &= \sum_{j=0}^{3} x_j \frac{\partial}{\partial x_j} L(u,u') + 4L(u,u') + 4F_k(u) + \sum_{j=0}^{3} F'_k(u)\frac{\partial u}{\partial x_j}\cdot x_j \\
&\quad - F'_k(u)\cdot u - \sum_{j=0}^{3} F'_k(u)\frac{\partial u}{\partial x_j}\cdot x_j \\
&= \sum_{j=0}^{3} x_j \frac{\partial}{\partial x_j} L(u,u') + 4L(u,u') + 4F_k(u) - uF'_k(u).
\end{aligned} \tag{5.377}$$

Combining this with (5.373), we get

$$\sum_{j=0}^{3} \partial_j\left[\frac{\partial L}{\partial p_j}(u,u')\left(u + \sum_{k=0}^{3} x_k \partial_k u\right) - x_j L(u,u')\right] = 4F_k(u) - uf_k(u). \tag{5.378}$$

Converting the Lagrange expression to divergence expression, (5.378) becomes

$$\operatorname{div}_{t,x}(tQ + \partial_t u \cdot u, -tP) = 4F_k(u) - uf_k(u), \tag{5.379}$$

where

$$Q = \frac{1}{2}|u'|^2 + F_k(u) + t^{-1}\partial_t ux \cdot \nabla_x u, \tag{5.380}$$

$$P = \left(\frac{1}{2}|\partial_t u|^2 - \frac{1}{2}|\nabla u|^2 - F_k(u)\right)\frac{x}{t} + \left(\frac{u}{t} + \partial_t u + \frac{x}{t} \cdot \nabla_x u\right)\nabla_x u. \tag{5.381}$$

Indeed, we can easily verify that

$$\begin{aligned}
&\frac{\partial L}{\partial p_0}(u, u')(u + tu_t + x \cdot \nabla u) - tL(u, u') \\
&\quad = u_t u + tu_t^2 + u_t(x \cdot \nabla u) - t\left[\frac{1}{2}|u_t|^2 - \frac{1}{2}|\nabla u|^2 - F_k(u)\right] \\
&\quad = t\left[\frac{1}{2}|u_t|^2 + \frac{1}{2}|\nabla u|^2 + F_k(u)\right] + u_t(x \cdot \nabla u) + u_t u \\
&\quad = tQ + uu_t,
\end{aligned}$$

and

$$\begin{aligned}
&\frac{\partial L}{\partial p_1}(u, u')\left(u + \sum_{k=0}^{3} x_k \partial_k u\right) - x_1 L(u, u') \\
&\quad = -\frac{\partial u}{\partial x_1}(u + tu_t + x \cdot \nabla u) - x_1\left(\frac{1}{2}|\partial_t u|^2 - \frac{1}{2}|\nabla u|^2 - F_k(u)\right) \\
&\quad = -\frac{\partial u}{\partial x_1}\left(\frac{u}{t} + u_t + \frac{x}{t} \cdot \nabla u\right)t - \left(\frac{1}{2}|\partial_t u|^2 - \frac{1}{2}|\nabla u|^2 - F_k(u)\right)x_1 \\
&\quad = -t\frac{\partial u}{\partial x_1}\left(\frac{u}{t} + u_t + \frac{x}{t} \cdot \nabla u\right) - t\left(\frac{1}{2}|\partial_t u|^2 - \frac{1}{2}|\nabla u|^2 - F_k(u)\right)\frac{x_1}{t},
\end{aligned}$$

which implies that P satisfies (5.381).

Remark 5.4.6.
(a) Note that $t\partial_t + x \cdot \nabla + 1$ is the generator of the transform

$$u \longmapsto u_\varepsilon(t, x) = \varepsilon u(\varepsilon t, \varepsilon x).$$

Hence, we can get (5.379) by multiplying both sides of equation $\Box u + f_k(u) = 0$ by $t\partial_t u + x \cdot \nabla u + u$.

(b) If we replace the transform in (a) with

$$u \longmapsto u_\varepsilon(t, x) = u(t + \varepsilon, x),$$

then by similar derivation as before, we can derive energy identify (i. e., (5.316)) by formula

$$\partial_t\left[\frac{\partial L}{\partial p_0}(u,u')-L(u,u')\right]+\sum_{j=1}^{3}\partial_j\left[\frac{\partial L}{\partial p_j}(u,u')\partial_t u\right]=0. \tag{5.382}$$

Next, we will prove (5.369) by (5.379). If $T^* < T < S \le 0$, denote

$$\begin{aligned} D_T &= \{(T,x),\ |x| \le -T\},\\ \Lambda(T,S) &= \{(t,x):\ T \le t \le S,\ |x| \le -t\},\\ M_T^S &= \{(t,x)\mid T \le t \le S,\ |x| = -t\}. \end{aligned}$$

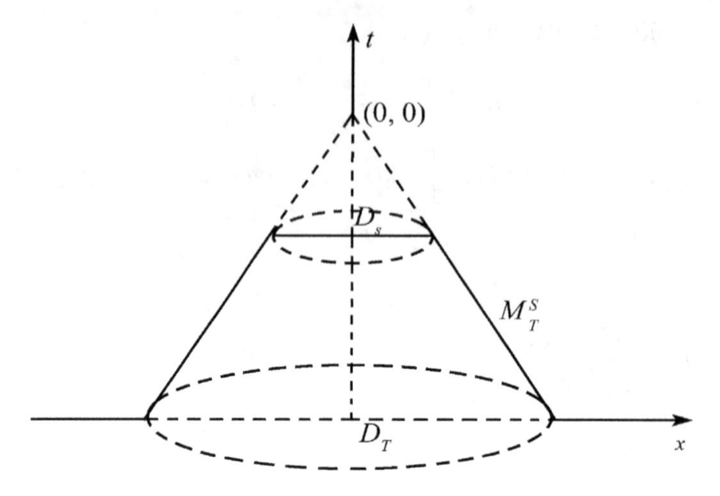

Figure 5.3: The frustum $\Lambda(T,S)$.

Thus, frustum $\Lambda(T,S)$ is the intersection of $[T,S]\times\mathbb{R}^3$ and the backward light cone with vertex $(0,0)$ (see Figure 5.3). The boundary of $\Lambda(T,S)$ can be divided into the following three parts:

$$\partial\Lambda(T,S) = D_T \cup D_S \cup M_T^S. \tag{5.383}$$

Integrating (5.379) and applying the divergence theorem, we obtain

$$\begin{aligned} &\int_{D_S}(SQ+u\partial_t u)dx - \int_{D_T}(TQ+u\partial_t u)dx + \frac{1}{\sqrt{2}}\int_{M_T^S}(tQ+u\partial_t u + x\cdot P)d\sigma\\ &\quad= \iint_{\Lambda(T,S)}(4F_k(u)-uf_k(u))dxdt, \end{aligned} \tag{5.384}$$

by the fact that $|x| = -t$ and $\nu = \frac{1}{\sqrt{2}}(1, \frac{x}{|x|})$. According to the conservation of energy and Hölder's inequality, as $S \nearrow 0$, (5.384) becomes

$$I + II = \iint_{\Lambda(T,0)} (4F_k(u) - uf_k(u))dxdt, \tag{5.385}$$

where

$$I = -\int_{D_T} (TQ + u\partial_t u)dx, \tag{5.386}$$

$$II = \frac{1}{\sqrt{2}} \int_{M_T^0} (tQ + u\partial_t u + x \cdot P)d\sigma. \tag{5.387}$$

By (H3) and the energy inequality, we conclude that

$$I + II \leq CT^4. \tag{5.388}$$

Next, we will estimate precisely I and II (by the expression of Q, we can see that in I there exists the term we desire to control).

Let us start to handle II. Since $|x| = -t$ on M_T^0, we have

$$\begin{aligned} II &= \frac{1}{\sqrt{2}} \int_{M_T^0} \Big[-|x||\partial_t u|^2 + 2(x \cdot \nabla_x u)\partial_t u - \frac{(x \cdot \nabla_x u)^2}{|x|} \\ &\quad - u\frac{x}{|x|} \cdot \nabla_x u + u\partial_t u \Big] d\sigma \\ &= -\frac{1}{\sqrt{2}} \int_{M_t^0} \Big[|x| \Big(\frac{x \cdot \nabla u}{|x|} - \partial_t u \Big)^2 + \Big(\frac{x \cdot \nabla_x u}{|x|} - \partial_t u \Big) u \Big] d\sigma. \end{aligned} \tag{5.389}$$

If we parameterize M_T^0 by

$$y \to (-|y|, y), \quad |y| \leq T,$$

then

$$d\sigma = \sqrt{2}dy.$$

If we set $v(y) = u(-|y|, y)$, then

$$y \cdot \frac{\nabla v}{|y|} = \frac{y \cdot \nabla_y u}{|x|} - \partial_t u.$$

Therefore,

$$II = -\int_{|y|\le|T|}\left(\frac{|y\cdot\nabla v|^2}{|y|} + v\frac{y\cdot\nabla v}{|y|}\right)dy$$
$$= -\int_{|y|\le|T|}\frac{|y\cdot\nabla v + v|^2}{|y|}dy + \int_{|y|\le|T|}\left[\frac{v^2}{|y|} + v\frac{y\cdot\nabla v}{|y|}\right]dy. \tag{5.390}$$

If we use polar coordinates, $y = rw$, then

$$v\frac{y\cdot\nabla v}{|y|} = v\partial_r v = \frac{1}{2}\partial_r v^2.$$

Hence, integration by parts gives

$$\int_{|y|\le T} v\frac{y\cdot\nabla v}{|y|}dy = \frac{1}{2}\int_{\Sigma^3}\int_0^T \partial_r v^2(rw)r^2 dr d\sigma(w)$$
$$= \frac{1}{2}\int_{\Sigma^3} v^2(|T|w)|T|^2 d\sigma(w) - \int_{\Sigma^3}\int_0^T v^2(rw) r dr d\sigma(w)$$
$$= \frac{1}{2}\int_{\partial D_T} u^2 d\sigma - \int_{|y|\le T} v^2\frac{dy}{|y|}. \tag{5.391}$$

Substituting (5.391) into (5.390), and using the surface integral expression, we get

$$II = \frac{1}{\sqrt{2}}\int_{M_T^0} t\left|\frac{x}{|x|}\cdot\nabla_x u - \partial_t u + \frac{u}{t^2}\right|^2 d\sigma + \frac{1}{2}\int_{\partial D_T} u^2 d\sigma. \tag{5.392}$$

To handle I, we first notice that the integrand of I is

$$TQ + u\partial_t u = T\left(\frac{1}{2}|u'|^2 + F_k(u)\right) + \partial_t u(u + x\cdot\nabla_x u). \tag{5.393}$$

Observing that $|x| \le -T$,

$$u + x\cdot\nabla_x u = x\cdot\left(\nabla_x u + \frac{x}{|x|^2}u\right),$$

and

$$|\partial_t u(u + x\cdot\nabla_x u)| \le -T\left[\frac{1}{2}(\partial_t u)^2 + \frac{1}{2}\left|\nabla_x u + \frac{x}{|x|^2}u\right|^2\right], \tag{5.394}$$

we get

$$
\begin{aligned}
I &\geq -T \int_{D_T} F_k(u)dx - T \int_{D_T} \left(\frac{1}{2}|\nabla_x u|^2 - \frac{1}{2}\left|\nabla_x u + \frac{x}{|x|^2}u\right|^2 \right) dx \\
&\geq |T| \int_{D_T} F_k(u)dx + T\left(\int_{D_T} u \cdot \frac{x \cdot \nabla_x u}{|x|^2} dx + \frac{1}{2}\int_{D_T} \frac{|u|^2}{|x|^2} dx \right).
\end{aligned} \tag{5.395}
$$

Integrating by parts as before, we find that

$$
\begin{aligned}
&\int_{D_T} u \cdot \frac{x \cdot \nabla_x u}{|x|^2} dx + \frac{1}{2}\int_{D_T} \frac{|u|^2}{|x|^2} dx \\
&= \frac{1}{2}\int_{\Sigma^3}\int_0^{|T|} r\partial_r u^2(r\omega) dr d\sigma + \frac{1}{2}\int_{D_T} \frac{|u|^2}{|x|^2} dx \\
&= \frac{1}{2}\left(\int_{\Sigma^3} |T| u^2(|T|\omega) d\sigma(\omega) - \int_{\Sigma^3}\int_0^{|T|} u^2(r\omega) dr d\sigma(\omega) \right) + \frac{1}{2}\int_{D_T} \frac{|u|^2}{|x|^2} dx \\
&= \frac{1}{2}\int_{\Sigma^3} |T| u^2 d\sigma(\omega) - \frac{1}{2}\int_{D_T} \frac{|u|^2}{|x|^2} dx + \frac{1}{2}\int_{D_T} \frac{|u|^2}{|x|^2} dx \\
&= \frac{1}{2}\int_{\partial D_T} \frac{u^2}{|T|} d\sigma(\omega).
\end{aligned} \tag{5.396}
$$

which implies that

$$
I \geq |T| \int_{D_T} F_k(u)dx - \frac{1}{2}\int_{\partial D_T} u^2 d\sigma. \tag{5.397}
$$

Combining this with (5.388), (5.392) and (5.397), we see that

$$
\begin{aligned}
|T| \int_{D_T} F_k(u)dx &\lesssim T^4 + \frac{1}{\sqrt{2}}\int_{M_T^0} |t|\left|\partial_t u + \frac{x}{|x|} \cdot \nabla u + \frac{u}{|x|}\right|^2 d\sigma \\
&\lesssim T^4 + |T| \int_{M_T^0} \left|\partial_t u + \frac{x}{|x|} \cdot \nabla_x u\right|^2 d\sigma + \int_{M_T^0} \frac{|u|^2}{|t|} d\sigma \\
&\lesssim T^4 + |T| \,\mathrm{Flux}(u, M_T^0) + \int_{M_T^0} \frac{|u|^2}{|t|} d\sigma \\
&\lesssim T^4 + |T| \,\mathrm{Flux}(u, M_T^0) + \left(\int_{M_T^0} |t|^{-\frac{3}{2}} d\sigma \right)^{\frac{2}{3}} \cdot \left(\int_{M_T^0} |u|^6 d\sigma \right)^{\frac{1}{3}}
\end{aligned}
$$

$$\lesssim T^4 + |T|\operatorname{Flux}(u, M_T^0) + |T|\left(\int_{M_T^0} (1 + F_k(u))d\sigma\right)^{\frac{1}{3}}$$
$$\lesssim T^4 + |T|\operatorname{Flux}(u, M_T^0) + |T|(\operatorname{Flux}(u, M_T^0))^{\frac{1}{3}}$$
$$+ |T|\left(\frac{4\pi|T|^3}{3}\right)^{\frac{1}{3}},$$

and

$$\int_{D_T} F_k(u)dx \le C|T|^3 + \operatorname{Flux}(u, M_T^0) + \operatorname{Flux}(u, M_T^0)^{\frac{1}{3}} + C\left(\frac{4\pi|T|^3}{3}\right)^{\frac{1}{3}}. \tag{5.398}$$

This finally gives us

$$\lim_{T\to 0} \int_{D_T} F_k(u)dx = 0$$

since the Flux $(u, M_T^0) \to 0$ as $T \to 0$.

This completes the proof of Theorem 5.4.1.

Remark 5.4.7.

(i) For the case with $n = 3$, the alternative Theorem 5.4.2 of the existence of smooth solution reduces the problem whether local solution on $I = [0, T^*)$ can be extended to a global smooth solution to whether $u(t, x) \in L^\infty([0, T^*) \times \mathbb{R}^3)$, where the multiple indices α satisfies $|\alpha| \le \frac{n+6}{2}$. The better result is in the following discussion.

(ii) For the case with $n = 3$, we can study directly the well-posedness of solutions in $C^2([0, \infty) \times \mathbb{R}^n)$ using the expression of solution. Actually, if $u(t, x) \in C^2([0, T] \times \mathbb{R}^n)$ is a solution of the nonlinear wave equation (5.285), then by seeking the solution of integral equation

$$u(x, t) = v_1(t, x) + \frac{1}{4\pi}\int_T^t \int_{|x-y|=t-s} \frac{f(u(y, s))}{t - s} dyds, \tag{5.399}$$

we can obtain the extension of the solution in $C^2([0, T] \times \mathbb{R}^n)$, where

$$v_1(t, x) = \partial_t\left(\frac{t}{4\pi}\int_{\Sigma^1} \varphi(x + t\omega)d\omega\right) + \frac{t}{4\pi}\int_{\Sigma^1} \psi(x + t\omega)d\omega$$
$$+ \frac{1}{4\pi}\int_0^T \int_{|x-y|=t-s} \frac{f(u(y, s))}{t - s} dyds \in C^2([0, T] \times \mathbb{R}^n). \tag{5.400}$$

However, this method is not feasible for $n > 3$.

(iii) With the dimensions increasing, it is getting more and more difficult to research on the global smooth solution of wave equation. In the following, we will discuss the mechanism of well-posedness of the global smooth solution of the wave equation. Consider the Cauchy problem of the linear wave equation

$$\begin{cases} u_{tt} - \Delta u + h(x,t) = 0, & (x,t) \in \mathbb{R}^n \times \mathbb{R}^+, \\ u(0) = \varphi(x), \quad u_t(0) = \psi(x), & x \in \mathbb{R}^n. \end{cases} \tag{5.401}$$

Let $R(x,t)$ be the fundamental solution of

$$\begin{cases} u_{tt} - \Delta u = 0, & (x,t) \in \mathbb{R}^n \times \mathbb{R}^+, \\ u(0) = 0, \quad u_t(0) = \delta(x), & x \in \mathbb{R}^n, \end{cases} \tag{5.402}$$

by using characteristic coordinates transformation, spherical means method and method of descent, we can easily get

$$R(x,t) = \frac{1}{2}\chi_{|x|\le t}(x), \quad n = 1, \tag{5.403}$$

and when $n \ge 3$ odd,

$$R(x,t) = A_n\left(\frac{1}{t}\partial_t\right)^{\frac{n-3}{2}} \frac{1}{t}\delta(t-|x|), \quad A_n = \frac{1}{\omega_{n-1}(n-2)\cdots 3\cdot 1}, \tag{5.404}$$

and $n \ge 2$ even,

$$R(x,t) = A_n\left(\frac{1}{t}\partial_t\right)^{\frac{n-2}{2}} \frac{\chi_{B_t(x)}}{\sqrt{t^2-|x|^2}}, \quad A_n = \frac{2}{\omega_n(n-1)\cdots 3\cdot 1}. \tag{5.405}$$

Thus, the solution of (5.401) can be written as

$$v(x,t) = \frac{\partial}{\partial t}(R(x,t) * \varphi) + R(x,t) * \psi + \int_0^t R(x,t-\tau) * h(x,\tau)d\tau. \tag{5.406}$$

Here,

$$R(x,t) * \psi = A_n\left(\frac{1}{t}\partial_t\right)^{\frac{n-3}{2}} \left(t^{n-2}\int_{\Sigma^{n-1}} \psi(x+t\omega)d\omega\right), \quad n \ge 3 \text{ odd}, \tag{5.407}$$

$$R(x,t) * \psi = A_n\left(\frac{1}{t}\partial_t\right)^{\frac{n-2}{2}} \left(t^{n-1}\int_{B_1(0)} \frac{\psi(x+yt)}{\sqrt{1-|y|^2}}dy\right), \quad n \ge 2 \text{ even}. \tag{5.408}$$

Particularly,

$$v(x,t) = \partial_t\left(\frac{t}{4\pi}\int_{\Sigma^1}\varphi(x+t\omega)d\omega\right) + \frac{t}{4\pi}\int_{\Sigma^1}\psi(x+t\omega)d\omega$$

$$+\frac{1}{4\pi}\int_0^t\int_{\partial B_{t-\tau}(x)}\frac{h(y,\tau)}{t-\tau}d\sigma(y)d\tau, \quad n=3,$$

$$v(x,t) = \partial_t\left(\frac{t}{2\pi}\int_{B_1(0)}\frac{\varphi(x+yt)}{\sqrt{1-|y|^2}}dy\right) + \frac{t}{2\pi}\int_{B_1(0)}\frac{\psi(x+yt)}{\sqrt{1-|y|^2}}dy \tag{5.409}$$

$$+\frac{1}{2\pi}\int_0^t\int_{B_1(0)}\frac{(t-\tau)h(x+(t-\tau)y,\tau)}{\sqrt{1-|y|^2}}dyd\tau, \quad n=2. \tag{5.410}$$

From the expression of solution, we can easily find that, in order to obtain the smooth solution $u(x,t) \in C^2([0,\infty)\times\mathbb{R}^n)$, we need

$$\varphi(x) \in C^{[\frac{n}{2}]+2}(\mathbb{R}^n), \quad \psi(x) \in C^{[\frac{n}{2}]+1}(\mathbb{R}^n), \quad h(x,t) \in C^2([0,T];C^{[\frac{n}{2}]+2}(\mathbb{R}^n)), \tag{5.411}$$

which actually means $[\frac{n}{2}]$ loss of derivative. Now, we discuss in $C^2([0,\infty)\times\mathbb{R}^n)$ the smooth solution of the nonlinear wave equation. Let $I=[0,T]$ be the interval obtained the first time we applied the fixed-point method. From the perspective of flow or Picard successive prolongation, we cannot continue, because $u(T)$, $u_t(T)$ do not satisfy (5.411).

(iv) By the Fourier transform, the solution of the linear problem (5.401) can be expressed as

$$u(x,t) = \dot{K}(t)\varphi + K(t)\psi + \int_0^t K(t-\tau)h(x,\tau)d\tau, \quad K(t) = \mathcal{F}^{-1}\frac{\sin|\xi|t}{|\xi|}\mathcal{F}\cdot. \tag{5.412}$$

Multiplying both sides of (5.401) by $u_t(t,x)$, we get

$$\frac{d}{dt}\left[\frac{1}{2}|u_t|^2 + \frac{1}{2}|\nabla u|^2\right] - \operatorname{div}(u_t\nabla u) = h(t,x)u_t. \tag{5.413}$$

Thus, we obtain

$$\|u_t\|_2 + \|\nabla u\|_2 \le \|\varphi\|_2 + \|\psi\|_2 + \sqrt{2}\left(\int_0^T\int_{\mathbb{R}^n}|h|^2dxdt\right)^{\frac{1}{2}}. \tag{5.414}$$

We can see that in H^s, the wave equation does not have any loss derivative. This reminds us that we can research the nonlinear wave equation (5.285) in the form of $C([0,T);H^{s+1}\times H^s)$, and obtain the existence of a classical smooth solution by the Sobolev embedding theorem and the equation itself.

Indeed, by the classical energy method, we have the following.

Proposition 5.4.5. *Assume $s \in \mathbb{N}$, $(\varphi(x), \psi(x)) \in H^{s+1} \times H^s$, $h(t,x) \in L^1([0,T]; H^s)$. Then the solution of* (5.401) $u(t,x) \in C([0,T]; H^{s+1}) \cap C^1([0,T]; H^s)$ *satisfies*

$$\sum_{\alpha|\leq 1} \|\partial^\alpha u(t,\cdot)\|_{H^s} \leq C(s,T) \sum_{|\alpha|\leq 1} \|\partial^\alpha u(0,\cdot)\|_{H^s} + \int_0^T \|f\|_{H^s} ds. \tag{5.415}$$

Proposition 5.4.6. *Assume $s > \frac{n+2}{2}$, $(\varphi(x), \psi(x)) \in H^{s+1} \times H^s, f(u) \in C^\infty, f(0) = 0$. Then there exist $T > 0$, the solution of* (5.285) $u(t,x)$ *satisfying* $(u(t,x), u_t(t,x)) \in C([0,T]; H^{s+1} \times H^s)$ *and*

$$\sum_{|\alpha|\leq s+1} \|\partial^\alpha u(t,x)\|_{L^2(\mathbb{R}^n)} < \infty. \tag{5.416}$$

Denote $T^* = \sup T$, then $T^* = \infty$ or

$$T^* < \infty \quad \text{and} \quad \lim_{t\to T^*} \sum_{|\alpha|\leq 2} \|\partial^\alpha u\|_\infty = \infty. \tag{5.417}$$

The result above provides us an effective way to research the smooth solution of the wave equation in high dimensions.

Remark 5.4.8. For the wave equation in higher dimensions, we can easily obtain Morawetz's dilation identity (see [185, 245]) using the multiplier method or Lagrangian variation principle. Shatah and Struwe [236] proved the global well-posedness of the smooth solution of the critical wave equation when $3 \leq n \leq 7$, with the help of the localized Strichartz's estimate. We can also see Kapitanskii [125]. The basic idea of the proof is:

(i) The interaction part of the energy (caused by nonlinearity) does not "concentrate" at a point.
(ii) Establish the regularity of finite energy solution by using the nonconcentration property of the energy and Strichartz estimate.
(iii) Prove the regularity of solution with smooth initial data by the standard technique.

For details, see [125, 185, 236].

5.5 Low regularity problems for nonlinear Klein–Gordon equations

In this section, we study global well-posedness of the Cauchy problem for the Klein–Gordon equation below the energy norm:

$$\begin{cases} u_{tt} - \Delta u + m^2 u = -|u|^{\rho-1}u, & (t,x) \in \mathbb{R} \times \mathbb{R}^n,\ n \geq 3, m \neq 0, \\ u(x,0) = \phi(x), & x \in \mathbb{R}^n, \\ u_t(x,0) = \psi(x), & x \in \mathbb{R}^n. \end{cases} \tag{5.418}$$

Recently, many mathematicians are devoted to the study of lower regularity problems for evolution equations (systems). For details, see [19, 21, 125, 129, 130, 198, 242, 261], etc. The lower regularity problem is to search for the spaces with regularity as low as possible, such that we can study the well-posedness of corresponding nonlinear evolution equations with the initial data belonging to the spaces. In general, we use the Strichartz estimate and take the fractional derivative to establish the local well-posedness of the equation below energy norm, which is relatively easier. While the global well-posedness is more difficult, since in this situation, we do not have the corresponding energy inequality. Bourgain developed the Fourier truncation method, and established the global well-posedness of the Schrödinger equation in H^s ($s_0 < s < 1$) for the first time. The key of Bourgain's method is dividing the initial data into the high frequency part and the low frequency part. According to the method, we first evolve the high frequency part by free equation and the low frequency part by nonlinear equation, then prove that the difference enjoys some regularities (satisfying the corresponding estimate), and finally show that this progress can be repeated, thus we can establish the global well-posedness of equations with lower regularity. To solve the problem of equation below energy norm, I team's I-energy method is also feasible; see [130] for details.

As for the global well-posedness of the wave equation below energy norm, Kenig, Ponce and Vega gave the results in three dimensions in [137]. They obtained in [197] the results in higher dimensions with the help of Keel–Tao's end-point estimate and the estimate of nonlinear functions in Besov space. Now we take Klein–Gordon equation as an example, to elucidate the skills and ideas of studying the global well-posedness of equations below the energy norm.

First, let us recall the local well-posedness of the equation below the energy norm. Assume $(\phi(x), \psi(x)) \in H^s(\mathbb{R}^n) \otimes H^{s-1}(\mathbb{R}^n)$, where

$$\begin{cases} s \in (\nu(\rho), 1], & n = 3 \text{ and } \rho = 2, \\ s \in [\nu(\rho), 1], & n > 3, \text{ or } n = 3 \text{ and } \rho > 2, \end{cases} \tag{5.419}$$

$$\nu(\rho) = \begin{cases} \frac{n+1}{4} - \frac{1}{\rho-1}, & k_0(n) \leq \rho \leq 1 + \frac{4}{n-1}, \\ \frac{n}{2} - \frac{2}{\rho-1}, & 1 + \frac{4}{n-1} \leq \rho < 1 + \frac{4}{n-2}, \end{cases} \tag{5.420}$$

$$k_0(n) = \frac{(n+1)^2}{(n-1)^2 + 4}, \quad n \geq 3. \tag{5.421}$$

Then the Cauchy problem (5.418) is locally well posed in $[0, T_0)$, where $T_0 = T_0(\|\phi\|_{H^s}, \|\psi\|_{H^{s-1}})$, $s \in (\nu(\rho), 1]$ (see [175, 242]). When $n \geq 4$, we refer to [129] for the case containing the end-point $s = \nu(\rho)$.

Theorem 5.5.1. *Let $s \in (\alpha(\rho), 1)$ satisfy*

$$\alpha(\rho) = \frac{2(\rho-1)^2 + (n+2-\rho(n-2)) \cdot (n\rho - n - \rho - 1)}{2(\rho-1)^2 + 2(\rho-1)(n+2-\rho(n-2))},$$
$$n = 3,\ 2 < \rho < \frac{n+3}{n-1},\ \text{or } n \geq 4,\ k_0(n) \leq \rho < \frac{n-1}{n-3}, \tag{5.422}$$

and

$$\alpha(\rho) = \frac{4(\rho-1) + (n+2-\rho(n-2)) \cdot (n\rho - n - 4)}{2(\rho-1)(n+4-\rho(n-2))},$$
$$n = 3,\ \frac{n+3}{n-1} \leq \rho < \frac{n+2}{n-2}. \tag{5.423}$$

Then for any $T > 0$ and $(\phi(x), \psi(x)) \in \dot{H}^s(\mathbb{R}^n) \cap L^{\rho+1}(\mathbb{R}^n) \otimes \dot{H}^{s-1}(\mathbb{R}^n)$, the Cauchy problem (5.418) *has a unique solution $u(t)$ in $[0, T)$ satisfying*

$$u(t) = \dot{K}(t)\phi + K(t)\psi + z(t) \tag{5.424}$$

and the following estimate:

$$\sup_{[0,T)} \|z(t)\|_{H^1} \leq CT^{\frac{1-s}{1-s-\eta}}, \tag{5.425}$$

where $K(t) = \sin(m^2 - \Delta)^{\frac{1}{2}} t / (m^2 - \Delta)^{\frac{1}{2}}$,

$$\eta = \frac{2(\rho-1)^2(1-s)}{n+2-\rho(n-2)} + \frac{n\rho - (n+\rho-1)}{2} - \rho s,$$
$$n = 3,\ 2 < \rho < \frac{n+3}{n-1},\ \text{or } n \geq 4,\ k_0(n) \leq \rho < \frac{n-1}{n-3}, \tag{5.426}$$
$$\eta = \frac{2(\rho-1)(1-s)}{n+2-\rho(n-2)} + \frac{n\rho - (n+2)}{2} - \rho s,$$
$$n = 3,\ \frac{n+3}{n-1} \leq \rho < \frac{n+2}{n-2}. \tag{5.427}$$

Remark 5.5.1.

(i) Denote $k_1(n) = 1 + 2/(n-2)$. It is easy to verify that when $n = 3, 4$, we have $k_0(n) < k_1(n)$; when $n \geq 5$, we have $k_0(n) > k_1(n)$. We can easily observe that $1 + \frac{2}{n-2} = \frac{n+3}{n-1}$, $n = 3$. Furthermore, we have

$$\begin{cases} \frac{n-1}{n-3} = \frac{n+2}{n-2}, & n = 4, \\ \frac{n-1}{n-3} < \frac{n+2}{n-2}, & n \geq 5. \end{cases}$$

(ii) When $n \geq 5$, because of technical reasons, we only consider the case $k_0(n) \leq \rho < (n-1)/(n-3)$.

(iii) When $n = 3$ and $m = 0$, we can repeat the conclusion in [137] from Theorem 5.5.1.

Before proving Theorem 5.5.1, let us recall the Strichartz estimate. Direct computation yields

$$\begin{aligned} W(x,t) &= \dot{K}(t)\phi(x) + K(t)\psi(x) + \int_0^t K(t-\tau)f(x,\tau)d\tau \\ &= \dot{K}(t)\phi(x) + K(t)\psi(x) + (\mathcal{G}f)(x,t) \end{aligned} \tag{5.428}$$

and solves the following linear Klein–Gordon equation:

$$\begin{cases} W_{tt} - \Delta W + m^2 W = f(x,t), & (t,x) \in \mathbb{R} \times \mathbb{R}^n, \\ W(x,0) = \phi(x), & x \in \mathbb{R}^n, \\ W_t(x,0) = \psi(x), & x \in \mathbb{R}^n. \end{cases} \tag{5.429}$$

From the space-time estimate for the Klein–Gordon equation, we conclude that

$$\|K(t)\psi(\cdot)\|_{B^s_{r,2}} \le C|t|^{-(n-1+\sigma)(\frac{1}{2}-\frac{1}{r})}\|\psi\|_{B^{\tilde{s}}_{r',2}}, \quad t \ne 0 \tag{5.430}$$

and

$$\|K(t)\psi(\cdot)\|_{B^s_{r,2}} \le C|t|^{-(n-1-\sigma)(\frac{1}{2}-\frac{1}{r})}\|\psi\|_{B^{\tilde{s}}_{r',2}}, \quad t \ne 0. \tag{5.431}$$

By the TT^* argument, we give the Strichartz estimate based on (5.430), while the estimate based on (5.431) is presented in the remark. In order to do this, for the given n, $2 \le r \le \infty$, $0 \le \sigma \le 1$, we define

$$\frac{\delta(r,\sigma)}{n+\sigma} = \frac{2\beta(r,\sigma)}{n+1+\sigma} = \frac{\gamma(r,\sigma)}{n-1+\sigma} = \frac{1}{2} - \frac{1}{r}.$$

For $0 < s < 1$, $2 \le r \le \infty$, $0 \le \sigma \le 1$,

$$(n+1+\sigma)\left(\frac{1}{2} - \frac{1}{r}\right) = 2\beta(r,\sigma) = 1 + \tilde{s} - s,$$

and (5.430) becomes

$$\|K(t)\psi(\cdot)\|_{B^s_{r,2}} \le C|t|^{-\gamma(r,\sigma)}\|\psi\|_{B^{\tilde{s}}_{r',2}}, \quad t \ne 0. \tag{5.432}$$

In particular, we have

$$\|K(t)\psi(\cdot)\|_{B^{1-\beta(r,\sigma)}_{r,2}} \le C|t|^{-\gamma(r,\sigma)}\|\psi\|_{B^{\beta(r,\sigma)}_{r',2}}, \quad t \ne 0. \tag{5.433}$$

By the TT^* method, Hardy–Littlewood–Sobelev's inequality and Sobolev's imbedding, we get the following Strichartz estimate.

Proposition 5.5.1. *Let $\rho_1, \rho_2, \mu \in \mathbb{R}$, $2 \le q_1, q_2, r_1$ and $r_2 \le \infty$ satisfy*

$$\begin{aligned}
0 \le \frac{2}{q_j} \le \min(\gamma(r_j,\sigma),1), \quad & n \ge 3,\ j = 1,2, \\
(q_j, r_j, n, \sigma) \ne (2,\infty,3,0), \quad & j = 1,2, \\
\rho_1 + \delta(r_1,\sigma) - \frac{1}{q_1} = \mu, & \\
\rho_2 + \delta(r_2,\sigma) - \frac{1}{q_2} = 1-\mu. &
\end{aligned}$$

Denote $Y^\mu = H^\mu \times H^{\mu-1}$. then we have the following results:
(i) *For any $(\phi,\psi) \in Y^\mu$, $(w, \partial_t w) \in C(I; Y^\mu) \cap L^{q_1}(I; B^{\rho_1}_{r_1,2} \times B^{\rho_1-1}_{r_1,2})$ satisfies*

$$\|w; L^{q_1}(I; B^{\rho_1}_{r_1,2})\| + \|\partial_t w; L^{q_1}(I; B^{\rho_1-1}_{r_1,2})\| \le C\|(\phi,\psi)\|_{Y^\mu}. \tag{5.434}$$

(ii) *For any $f \in L^{q_2'}(I; B^{-\rho_2}_{r_2',2})$, $\mathcal{G}f \in C(I; Y^\mu) \cap L^{q_1}(I; B^{\rho_1}_{r_1,2})$ satisfies*

$$\|\mathcal{G}f; L^{q_1}(I; B^{\rho_1}_{r_1,2})\| \le C\|f; L^{q_2'}(I; B^{-\rho_2}_{r_2',2})\|. \tag{5.435}$$

(iii) *Denote $I = [0,T)$, $0 < T \le \infty$, then*

$$\begin{aligned}
&\|W; L^{q_1}(I; B^{\rho_1}_{r_1,2})\| + \|\partial_t W; L^{q_1}(I; B^{\rho_1-1}_{r_1,2})\| \\
&\quad \le C(\|(\phi,\psi)\|_{Y^\mu} + \|f; L^{q_2'}(I; \dot{B}^{-\rho_2}_{r_2',2})\|).
\end{aligned} \tag{5.436}$$

Definition 5.5.1. We call (q,r) a sharp admissible pair, if $q,\ r \ge 2$, $(q,r,n,\sigma) \ne (2,\infty,3,0)$ and

$$\frac{2}{q} = (n-1+\sigma)\left(\frac{1}{2} - \frac{1}{r}\right) = \gamma(r,\sigma). \tag{5.437}$$

It is obvious to see that if (q,r) is a sharp admissible pair, then q can be determined uniquely by r and σ, denoted $q = q(r,\sigma)$.

Corollary 5.5.1.
(i) *Assume that $0 \le \sigma \le 1$, $(q(r,\sigma), r)$ and $(q_j(r_j,\sigma), r_j)$ $(j = 1,2)$ are sharp admissible. If $(\phi,\psi) \in H^s \times H^{s-1}$, then we have $w = \dot{K}(t)\phi + K(t)\psi \in L^{q(r,\sigma)}(I; B^{s-\beta(r,\sigma)}_{r,2}) \cap C(I; H^s)$ and*

$$\|w\|_{L^{q(r,\sigma)}(I; B^{s-\beta(r,\sigma)}_{r,2})} \le C(\|\phi\|_{H^s} + \|\psi\|_{H^{s-1}}). \tag{5.438}$$

(ii) *Assume that* $f \in L^{q_2'(r_2,\sigma)}(I; B^{s+\beta(r_2,\sigma)-1}_{r_2',2})$, *then we have*

$$\mathcal{G}f \in L^{q_1(r_1,\sigma)}(I; B^{s-\beta(r_1,\sigma)}_{r_1,2}),$$

and

$$\|\mathcal{G}f\|_{L^{q_1(r_1,\sigma)}(I;B^{s-\beta(r_1,\sigma)}_{r_1,2})} \le C\|f\|_{L^{q_2'(r_2,\sigma)}(I;B^{s+\beta(r_2,\sigma)-1}_{r_2',2})}. \tag{5.439}$$

In particular,

$$\|\mathcal{G}f\|_{L^{q(r,\sigma)}(I;B^{s-\beta(r,\sigma)}_{r,2})} \le C\|f\|_{L^{q'(r,\sigma)}(I;B^{s+\beta(r,\sigma))-1}_{r',2})}. \tag{5.440}$$

Corollary 5.5.2. *Suppose* θ, σ *satisfy*

$$0 \le \theta \le \frac{n+1+\sigma}{2(n-1+\sigma)} < 1, \quad n \ge 3,\ 0 \le \sigma \le 1,$$
$$0 \le \theta < 1, \quad n = 3,\ \sigma = 0,$$

then

$$\|w\|_{L^{\frac{n+1+\sigma}{(n-1+\sigma)\theta}}(I;B^{l-\theta}_{\frac{2(n+1+\sigma)}{n+1+\sigma-4\theta},2})} \le C_\theta(\|\phi(x)\|_{H^l} + \|\psi(x)\|_{H^{l-1}}), \tag{5.441}$$

$$\|w\|_{L^{\frac{n+1+\sigma}{(n-1+\sigma)\theta}}(I;L^r)} \le C_\theta(\|\phi\|_{H^l} + \|\psi\|_{H^{l-1}}), \tag{5.442}$$

where $l \ge 1$,

$$\frac{2(n+1+\sigma)}{n+1+\sigma-4\theta} \le r \le \frac{2(n+1+\sigma)n}{(n-2l)(n+1+\sigma)-2(n-1-\sigma)\theta}. \tag{5.443}$$

Corollary 5.5.3. *Let* $w = \dot{K}(t)\phi + K(t)\psi$, *for* $q, r \ge 2$, $0 \le \sigma \le 1$ *satisfying* $(q,r,n,\sigma) \ne (2,\infty,3,0)$, *then*

$$\|w\|_{L^q(I;L^r)} \le C(\|\phi\|_{H^l} + \|\psi\|_{H^{l-1}}), \quad \delta(r,\sigma) - \frac{1}{q} \le l,\ l \ge 1. \tag{5.444}$$

Remark 5.5.2.

(i) When $\sigma \equiv 0$, denote $\delta(r,0) = \delta(r)$, $\beta(r,0) = \beta(r)$, $\gamma(r,0) = \gamma(r)$, then (5.432) becomes

$$\frac{\delta(r)}{n} = \frac{2\beta(r)}{n+1} = \frac{\gamma(r)}{n-1} = \frac{1}{2} - \frac{1}{r}.$$

Here, the notation correspond exactly to those for the space-time estimate of linear wave equations.

(ii) The following is the space-time estimate derived by Marshall:

$$\|w\|_{L^q(\mathbb{R};B^a_{r,2})} \le C(\|\phi\|_{H^1} + \|\psi\|_{L^2}), \tag{5.445}$$

where

$$\begin{cases} \frac{1}{2} - \frac{2}{(n-1)q} < \frac{1}{r} < \frac{1}{2} - \frac{2}{nq}, \\ 0 \le \frac{1}{q} \le \frac{1}{2}, \quad a < \frac{1}{2} + \frac{1}{r} - \frac{1}{q}, \end{cases} \tag{5.446}$$

$$\|w\|_{L^q(\mathbb{R};L^r)} \le C(\|\phi\|_{H^1} + \|\psi\|_{L^2}). \tag{5.447}$$

Here,

$$0 \le \frac{1}{q} \le \frac{1}{2} \quad \frac{n-2}{2n} - \frac{1}{nq} \le \frac{1}{r} < \frac{1}{2} - \frac{1}{nq}. \tag{5.448}$$

It is easy to verify that (5.446) is equivalent to

$$\begin{cases} \frac{2}{q} = (n-1+\sigma)\cdot\left(\frac{1}{2} - \frac{1}{r}\right) \quad 0 < \sigma < 1, \\ 0 \le \frac{1}{q} \le \frac{1}{2}, \quad a < 1 - \frac{n+1+\sigma}{2}\left(\frac{1}{2} - \frac{1}{r}\right). \end{cases} \tag{5.449}$$

Therefore, (5.445) follows from (5.438).

(iii) If we replace (5.430) with (5.431) and define

$$\frac{\tilde\delta(r,\sigma)}{n} = \frac{2\tilde\beta(r,\sigma)}{n+1+\sigma} = \frac{\tilde\gamma(r,\sigma)}{n-1-\sigma} = \frac{1}{2} - \frac{1}{r}, \quad 0 \le \sigma \le 1, \tag{5.450}$$

then we can obtain the result analogous to Proposition 5.5.1, Corollaries 5.5.1–5.5.3. In particular, we have

$$\|w\|_{L^{\tilde q(r,\sigma)}(I;B^{s-\tilde\beta(r,\sigma)}_{r,2})} \le C(\|\phi\|_{H^s} + \|\psi\|_{H^{s-1}}), \tag{5.451}$$

$$\|\mathcal{G}f\|_{L^{\tilde q_1(r_1,\sigma)}(I;B^{s-\tilde\beta(r_1,\sigma)}_{r_1,2})} \le C\|f\|_{L^{\tilde q_2'(r_2,\sigma)}(I;\dot B^{s+\tilde\beta(r_2,\sigma)-1}_{r_2',2})}, \tag{5.452}$$

$$\|\mathcal{G}f\|_{L^{\tilde q(r,\sigma)}(I;B^{s-\tilde\beta(r,\sigma)}_{r,2})} \le C\|f\|_{L^{\tilde q'(r,\sigma)}(I;B^{s+\tilde\beta(r,\sigma))-1}_{r',2})}, \tag{5.453}$$

where $0 < s < 1$, $2 \le r < \infty$, $0 \le \sigma \le 1$ and

$$\frac{2}{\tilde q(r,\sigma)} = (n-1-\sigma)\left(\frac{1}{2} - \frac{1}{r}\right) = \tilde\gamma(r,\sigma). \tag{5.454}$$

According to (5.451)~(5.453) and Sobolev embedding, we can get for $l \ge 1$,

$$\begin{cases} \|w\|_{L^{\frac{n+1+\sigma}{(n-1-\sigma)\theta}}(I;B^{l-\theta}_{\frac{2(n+1+\sigma)}{n+1+\sigma-4\theta},2})} \le C_\theta(\|\phi\|_{H^l} + \|\psi\|_{H^{l-1}}), \\ \|w\|_{L^{\frac{n+1+\sigma}{(n-1-\sigma)\theta}}(I;L^r)} \le C_\theta(\|\phi\|_{H^l} + \|\psi\|_{H^{l-1}}). \end{cases} \tag{5.455}$$

Here,

$$\begin{cases} 0 \le \theta \le \frac{n+1+\sigma}{2(n-1-\sigma)} < 1, & n \ge 4,\ 0 \le \sigma \le 1, \\ 0 \le \theta < 1, & n = 3,\ \text{or } (n,\sigma) = (4,1). \end{cases} \tag{5.456}$$

In order to prove Theorem 5.5.1, we need prepare with a series of nonlinear estimates. Using Bourgain's method, we split the data $(\phi(x), \psi(x))$ into the high frequency part and low frequency part (the regular part). Let $\varphi \in C_c^\infty(\mathbb{R}^n)$ satisfy

$$\varphi(\xi) = \begin{cases} 1, & |\xi| \le 1, \\ 0, & |\xi| \ge 2. \end{cases}$$

Define

$$\phi(x) = \phi_1(x) + \phi_2(x), \quad \psi(x) = \psi_1(x) + \psi_2(x).$$

Here,

$$\begin{cases} (\phi_1(x), \psi_1(x)) = (\mathcal{F}^{-1}(\varphi_N(\xi)\hat{\phi}(\xi)), \mathcal{F}^{-1}(\varphi_N(\xi)\hat{\psi}(\xi))), \\ (\phi_2(x), \psi_2(x)) = (\phi(x) - \phi_1(x), \psi(x) - \psi_1(x)), \end{cases}$$

$\varphi_N(\xi) = \varphi(\xi/N)$, $N > 0$ to be determined. It is obvious that $\phi_1,\ \psi_1 \in \mathcal{S}(\mathbb{R}^n)$ and

$$\begin{aligned} \|(I-\Delta)^{\frac{l}{2}}\phi_1\|_2 &= \Big(\int_{\mathbb{R}^n} |(1+|\xi|^2)^{\frac{l}{2}}\varphi_N(\xi)\hat{\phi}|^2 d\xi\Big)^{\frac{1}{2}} \\ &\le \Big(\int_{\mathbb{R}^n} (1+|\xi|^2)^{\frac{l}{2}-\frac{s}{2}} |(1+|\xi|^2)^{\frac{s}{2}}\varphi_N(\xi)\hat{\phi}|^2 d\xi\Big)^{\frac{1}{2}} \\ &\le C(1+N)^{l-s}\|\phi\|_{H^s} \sim (1+N)^{l-s}, \quad l \ge s, \end{aligned} \tag{5.457}$$

and

$$\begin{aligned} \|(I-\Delta)^{\frac{l}{2}}\phi_2\|_2 &= \Big(\int_{\mathbb{R}^n} |(1+|\xi|^2)^{\frac{l}{2}}(1-\varphi_N(\xi))\hat{\phi}|^2 d\xi\Big)^{\frac{1}{2}} \\ &\le \Big(\int_{\mathbb{R}^n} (1+|\xi|^2)^{\frac{l}{2}-\frac{s}{2}} |(1+|\xi|^2)^{\frac{s}{2}}(1-\varphi_N(\xi))\hat{\phi}|^2 d\xi\Big)^{\frac{1}{2}} \\ &\le C(1+N)^{l-s}\|\phi\|_{H^s} \sim (1+N)^{l-s}, \quad l \in [0,s]. \end{aligned} \tag{5.458}$$

Hence, $f_N \sim g_N$ implies $|f_N| \le C|g_N|$, where C is independent of N. Similarly, we have

$$\|(I-\Delta)^{\frac{l-1}{2}}\psi_1\|_2 \sim (1+N)^{l-s}, \quad l \ge s, \tag{5.459}$$

$$\|(I-\Delta)^{\frac{l-1}{2}}\psi_2\|_2 \sim (1+N)^{l-s}, \quad l \in [0,s]. \tag{5.460}$$

First, we consider the Cauchy problem with the regular data (ϕ_1, ψ_1),

$$\begin{cases} v_{tt} - \Delta v + m^2 v = -|v|^{\rho-1}v, & \rho \in (1, \frac{n+2}{n-2}), \\ v(x,0) = \phi_1(x), \quad v_t(x,0) = \psi_1(x), & x \in \mathbb{R}^n \end{cases} \tag{5.461}$$

and the corresponding integral form

$$v(x,t) = \dot{K}(t)\phi_1(x) + K(t)\psi_1(x) - \int_0^t K(t-\tau)|v|^{\rho-1}v d\tau. \tag{5.462}$$

Obviously, $v(x,t)$ satisfies the following conservation of energy:

$$\begin{aligned} E(v(\cdot,t), \partial_t v(\cdot,t)) &= \left(\int_{\mathbb{R}^n} \left(|v_t|^2 + |\nabla v|^2 + m^2|v|^2 + \frac{2|v|^{\rho+1}}{\rho+1} \right) dx \right)^{\frac{1}{2}} \\ &= E(\phi_1, \psi_1), \quad \forall t > 0. \end{aligned} \tag{5.463}$$

Using (5.457), (5.459) and Sobolev's inequality, we deduce that

$$\|\partial_t v(t)\|_2, \ \|v(t)\|_{H^1}, \ \|v(t)\|_{\rho+1}^{\frac{\rho+1}{2}} \sim (1+N)^{1-s}, \quad \forall t > 0. \tag{5.464}$$

For $l \geq 0, I \subset \mathbb{R}$ satisfy $0 \in \bar{I}$, we define

$$\begin{aligned} |||\cdot|||_l = & \sup_{0\leq\theta\leq\min(l,\frac{n+1}{2(n-1)})} \|\cdot\|_{L^{\frac{n+1}{(n-1)\theta}}(I;L^{\frac{2(n+1)n}{(n+1)(n-2l)-2(n-1)\theta}})} \\ & + \sup_{2\leq r\leq\frac{2(n-1)}{n-3},\, \gamma(r)=2/q} \|\cdot\|_{L^q(I;B^{l-\beta(r)}_{r,2})}, \quad n \geq 4, \end{aligned} \tag{5.465}$$

and

$$\begin{aligned} |||\cdot|||_l = & \sup_{0\leq\theta<\min(l,1)} \|\cdot\|_{L^{\frac{2}{\theta}}(I;L^{\frac{6}{3-2l-\theta}})} \\ & + \sup_{2\leq r<\infty,\, \gamma(r)=2/q} \|\cdot\|_{L^q(I;B^{l-\beta(r)}_{r,2})}, \quad n = 3. \end{aligned} \tag{5.466}$$

Then we have the following result.

Lemma 5.5.1. *Let v solve (5.461) or (5.462). Let $I = (0, \Delta T)$ and $\Delta T \sim (1+N)^{-\frac{2(1-s)(\rho-1)}{n+2-\rho(n-2)}}$. If*

$$\begin{cases} k_0(n) < \rho < (n+2)/(n-2), & n = 3, \\ k_0(n) \leq \rho < (n+2)/(n-2), & 4 \leq n \leq 6, \\ k_0(n) \leq \rho < n/(n-3), n \geq 7, \end{cases}$$

then $|||v|||_1 \sim (1+N)^{1-s}$.

Proof. First, we consider the case when $k_1(n) \le \rho < (n+2)/(n-2)$. Let

$$\theta = \frac{\rho(n-2)-n}{2\rho}\cdot\frac{n+1}{n-1}, \quad \frac{1}{\chi} = \frac{n+2-\rho(n-2)}{2}.$$

It is easy to verify

$$\frac{1}{2} = \rho\frac{(n+1)(n-2)-2(n-1)\theta}{2n(n+1)},$$
$$1 = \rho\cdot\frac{(n-1)\theta}{n+1} + \frac{1}{\chi}.$$

Thus, by (5.457), (5.459), Strichartz estimate and Hölder's inequality, we have

$$\begin{aligned}|||v|||_1 &\le C(\|\phi_1\|_{H^1} + \|\psi\|_{L^2} + \|v^\rho\|_{L^1(I;L^2)}) \\ &\le C[(1+N)^{1-s} + (\Delta T)^{\frac{n+2-\rho(n-2)}{2}} |||v|||_1^\rho].\end{aligned}$$

Noticing that $\Delta T \sim (1+N)^{-\frac{2(\rho-1)(1-s)}{n+2-\rho(n-2)}}$, we have

$$|||v|||_1 \sim (1+N)^{1-s}, \quad \text{for sufficently large } N.$$

According to Remark 5.5.1(i), when $n = 3$, we only need to consider the case with $k_0(n) < \rho \le k_1(n)$; when $n = 4$, it suffices to consider $k_0(n) \le \rho \le k_1(n)$. Taking

$$\lambda = \frac{(\rho+1)(n-\rho(n-2))}{\rho[2n-(\rho+1)(n-2)]},$$

one easily see that $\lambda \in (0,1)$. Noting that $\rho+1 < 2\rho \le 2n/(n-2)$ and by interpolation, we have

$$\begin{aligned}\|f(v)\|_{L^1(I;L^2)} &\le C\Delta T\|v\|^{\lambda\rho}_{L^\infty(I;L^{\rho+1})}\|v\|^{(1-\lambda)\rho}_{L^\infty(I;L^{\frac{2n}{n-2}})} \\ &\le C\Delta T(1+N)^{\frac{2n-2\rho(n-2)}{n+2-\rho(n-2)}(1-s)} N^{\frac{n(\rho-1)}{n+2-\rho(n-2)}(1-s)} \\ &\le C\Delta T(1+N)^{\frac{n-\rho n+4\rho}{n+2-\rho(n-2)}(1-s)} \\ &\sim (1+N)^{1-s}.\end{aligned}$$

Here, we have used (5.464) and the embedding relation $H^1 \hookrightarrow L^{2n/(n-2)}$. From this, we get

$$|||v|||_1 \le C(\|\phi_1\|_{H^1} + \|\psi_1\|_2 + \|v^\rho\|_{L^1(I;L^2)}) \sim (1+N)^{1-s}. \qquad \square$$

Remark 5.5.3.
(i) For any (κ, r, q) satisfying $\kappa + \delta(r) - \frac{1}{q} \leq 1, 2 \leq r \leq \infty$ and

$$\begin{cases} 2 \leq q \leq \infty, & n \geq 4, \\ 2 < q \leq \infty, & n = 3, \end{cases}$$

we deduce from Proposition 5.5.1 that

$$\|v\|_{L^q(I;B^{\kappa}_{r,2})} \sim (1+N)^{1-s}, \quad \text{for sufficiently large } N. \tag{5.467}$$

Here, $I = [0, \Delta T)$ and $\Delta T \sim (1+N)^{-\frac{2(1-s)(\rho-1)}{n+2-\rho(n-2)}}$.

(ii) For the wave equation in three-dimensional space, the similar estimate can be deduced by the following norm (see [137]):

$$|||\cdot|||_l := \|(-\Delta)^{l/2} \cdot\|_{L^\infty(I;L^2)} + \|\cdot\|_{L^{2/l}(I;L^{2/(1-l)})}.$$

The method here seems easier to deal with more general cases.

Next, we consider the Cauchy problem for the difference part $y(t) = u(t) - v(t)$,

$$\begin{cases} y_{tt} - \Delta y + m^2 y = -|y+v|^{\rho-1}(v+y) + |v|^{\rho-1}v, \\ y(x,0) = \phi_2(x), \quad y_t(x,0) = \psi_2(x) \end{cases} \tag{5.468}$$

and the corresponding integral equation

$$\begin{aligned} y(x,t) &= \dot{K}(t)\phi_2(x) + K(t)\psi_2(x) - \int_0^t K(t-\tau)F(\tau)d\tau \\ &= \dot{K}(t)\phi_2(x) + K(t)\psi_2(x) + z(t), \end{aligned} \tag{5.469}$$

where $F(t) = |y+v|^{\rho-1}(v+y) - |v|^{\rho-1}v$. One sees easily

$$|F(t)| \leq C|y|^{\rho} + C|v|^{\rho-1}|y| =: F_1 + F_2. \tag{5.470}$$

Lemma 5.5.2. *Let $0 \leq l \leq s < 1$, ρ satisfy*

$$\begin{cases} k_0(n) < \rho < (n+2)/(n-2), & n = 3, \\ k_0(n) \leq \rho < (n-1)/(n-3), & n \geq 4. \end{cases} \tag{5.471}$$

Let $I = [0, \Delta T]$ satisfy $\Delta T \sim (1+N)^{-\frac{2(1-s)(\rho-1)}{n+2-\rho(n-2)}}$, then $|||y|||_l \sim (1+N)^{l-s}$.

Proof. By Remark 5.5.1(i), if $n \geq 5$, it suffices to consider $\rho \geq k_1(n)(= 1 + 2/(n-2))$. By (5.458), (5.460), (5.470) and Proposition 5.5.1, we have

$$\begin{aligned}|||y|||_l &\le C[\|\phi_2\|_{\dot{H}^l} + \|\psi_2\|_{\dot{H}^{l-1}} + \|F\|_{L^1(I;L^{\frac{2n}{n+2-2l}})}] \\ &\le C[N^{l-s} + \|y^\rho\|_{L^1(I;L^{\frac{2n}{n+2-2l}})} + \|yv^{\rho-1}\|_{L^1(I;L^{\frac{2n}{n+2-2l}})}]. \end{aligned} \tag{5.472}$$

Step 1. The estimate of $\|yv^{\rho-1}\|_{L^1(I;L^{\frac{2n}{n+2-2l}})}$.

We first consider the case of $n = 3$, $k_1(n) \le \rho < (n+2)/(n-2)$ and $n \ge 4$, $k_1(n) \le \rho < (n-1)/(n-3)$. Let $p = 2n/(n+2-2l)$ and $\chi = 2/[n+2-\rho(n-2)]$. Define

$$\theta = \frac{(\rho-1)(n-2)-2}{2(\rho-1)} \cdot \frac{n+1}{n-1},$$
$$\frac{1}{r_1} = \frac{(n+1)(n-2)-2(n-1)\theta}{2n(n+1)},$$
$$\frac{1}{r_2} = \frac{n-2l}{2n}.$$

Then we have $1/p = (\rho-1)/r_1 + 1/r_2$, $1 = (\rho-1)(n-1)\theta/(n+1) + 1/\chi$. According to the generalized Hölder inequality (see Lemma 2.1.3 in Chapter 2), we derive

$$\begin{aligned}\|v^{\rho-1}y\|_{L^1(I;L^{\frac{2n}{n+2-2l}})} &\le C\int_I \|v\|_{r_1}^{\rho-1}\|y\|_{r_2}\,dt \\ &\le C\|v\|^{\rho-1}_{L^{(n+1)/[(n-1)\theta]}(I;L^{r_1})}\Big(\int_I \|y\|_{r_2}^\chi\,dt\Big)^{1/\chi} \\ &\le C|||v|||_1^{\rho-1}\|y\|_{L^\infty(I;L^{r_2})}(\Delta T)^{1/\chi} \\ &\le C(\Delta T)^{\frac{n+2-\rho(n-2)}{2}} N^{(\rho-1)(1-s)}|||y|||_l. \end{aligned} \tag{5.473}$$

Here, we used Lemma 5.5.1.

Second, we consider the case of $n = 3$, $k_0(n) < \rho \le k_1(n)$ and $n = 4$, $k_0(n) \le \rho \le k_1(n)$. Let

$$\kappa = \frac{2(\rho-1)n - 2(\rho+1)}{2n-(\rho+1)(n-2)}.$$

Then

$$\frac{n+2-2l}{2n} = \frac{\rho-1-\kappa}{\rho+1} + \kappa\frac{n-2}{2n} + \frac{n-2l}{2n}.$$

Noting (5.467), Lemma 5.5.1 and the generalized Hölder inequality, we deduce

$$\begin{aligned}\|v^{\rho-1}y\|_{L^1(I;L^{\frac{2n}{n+2-2l}})} &\le \int_I \|v\|_{r_1}^{\rho-1-\kappa}\|v\|_{r_2}^\kappa\|y\|_{r_3}\,dt \\ &\le C(\Delta T)\|v\|^{\rho-1-\kappa}_{L^\infty(I;L^{\rho+1})}|||v|||_1^\kappa|||y|||_l \\ &\le C(\Delta T)N^{\frac{(\rho-1)(2+\kappa)(1-s)}{\rho+1}}|||y|||_l. \end{aligned} \tag{5.474}$$

Step 2. In the case of $n = 3$, $k_0(n) \le \rho \le k_1(n)$ or $n = 4$, $k_0(n) \le \rho \le k_1(n)$, we prove the following estimate:

$$|||y|||_l \sim (1+N)^{l-s}, \quad \Delta T \sim (1+N)^{-\frac{2(1-s)(\rho-1)}{n+2-\rho(n-2)}}. \tag{5.475}$$

Let $l_0 = [n\rho - (n+2)]/[2(\rho-1)]$. Then

$$\frac{n+2-2l}{2n} = (\rho-1)\frac{n-2l_0}{2n} + \frac{n-2l}{2n}.$$

Thus, by the definition of $|||\cdot|||_l$ and generalized Hölder's inequality, we have

$$\|y^\rho\|_{L^1(I;L^{\frac{2n}{n+2-2l}})} \le C(\Delta T)|||y|||_{l_0}^{\rho-1}|||y|||_l. \tag{5.476}$$

Noting that

$$\frac{n+2-2l_0}{2n} = \rho\frac{n-2l_0}{2n},$$

we have

$$\|y^\rho\|_{L^1(I;L^{\frac{2n}{n+2-2l_0}})} \le C(\Delta T)\|y\|^\rho_{L^\infty(I;L^{\frac{2n}{n-2l_0}})} \le C(\Delta T)|||y|||_{l_0}^\rho.$$

Combining this with (5.472) and (5.474), for sufficiently large N, we have the estimate

$$|||y|||_{l_0} \le (1+N)^{l_0-s}, \quad \Delta T \sim (1+N)^{-\frac{2(1-s)(\rho-1)}{n+2-\rho(n-2)}}. \tag{5.477}$$

Combining (5.472), (5.474), (5.476) with (5.477), we get (5.475).

Step 3. When $n = 3$, $k_1(n) \le \rho < (n+2)/(n-2)$ or $n = 4$, $k_1(n) \le \rho < (n-1)/(n-3)$, we prove for any $l_0 \le l \le s < 1$, we get the estimate

$$|||y|||_l \sim (1+N)^{l-s}, \quad \Delta T \sim (1+N)^{-\frac{2(1-s)(\rho-1)}{n+2-\rho(n-2)}}, \tag{5.478}$$

where

$$l_0 = \max\left(\frac{\rho n - n - \rho - 2}{2(\rho-1)}, \frac{n}{2} - \frac{2}{\rho-1}, \frac{(\rho n - n - 2)(n+1)}{2(2\rho n - n - 1)}\right).$$

It is easily seen that for $l_0 \le l \le s < 1$, one has

$$\frac{n+2-2l}{2n} = \rho\frac{(n+1)(n-2l) - 2(n-1)\theta}{2n(n+1)},$$

$$\frac{1}{\rho} - \frac{(n-1)\theta}{n+1} = \frac{(n+4-2l) - \rho(n-2l)}{2\rho} \ge 0.$$

Here,

$$\theta = \frac{\rho(n-2l) - (n+2-2l)}{2\rho} \cdot \frac{n+1}{n-1}.$$

Hence, by the definition of $|||\cdot|||_l$ and the generalized Hölder inequality, we derive that

$$\|y^\rho\|_{L^1(I;L^{\frac{2n}{n+2-2l}})} \le (\Delta T)^{\frac{(n+4-2l)-\rho(n-2l)}{2}} |||y|||_l^\rho. \tag{5.479}$$

Thus, from (5.472)~(5.474) and (5.479), we get (5.478).

Step 4. For $0 \le l \le l_0$, when $n = 3$, $k_1(n) \le \rho < (n+2)/(n-2)$ or $n \ge 4$, $k_1(n) \le \rho < (n-1)/(n-3)$, we prove the estimate

$$|||y|||_l \sim (1+N)^{l-s}, \quad (\Delta T) \sim (1+N)^{\frac{-2(\rho-1)(1-s)}{n+2-\rho(n-2)}}. \tag{5.480}$$

Here, l_0 has the same definition as in Step 3.

Let

$$l_1 = \max\left(\frac{\rho n - n - \rho - 1}{2(\rho-1)}, l_0\right) \ge l_0,$$
$$\theta = \frac{(\rho-1)(n-2l_1) - 2}{2(\rho-1)} \cdot \frac{n+1}{n-1},$$
$$\chi = \frac{2}{n+4-2l_1-\rho(n-2l_1)}.$$

Then

$$\frac{n+2-2l}{2n} = (\rho-1)\frac{(n+1)(n-2l_1) - 2(n-1)\theta}{2n(n+1)} + \frac{n-2l}{2n},$$
$$1 = (\rho-1)\cdot\frac{(n-1)\theta}{n+1} + \frac{1}{\chi}.$$

By a similar proof as for (5.473), Lemma 2.1.4 in Chapter 2, for $0 \le l \le l_0$, we have

$$\|y^\rho\|_{L^1(I;L^{\frac{2n}{n+2-2l}})} \le C(\Delta T)^{\frac{(n+4-2l_1)-\rho(n-2l_1)}{2}} |||y|||_l \cdot |||y|||_{l_1}^{\rho-1},$$

from which using (5.472), (5.473) and step 3, we conclude (5.480). □

Proof of Theorem 5.5.1. Use Bourgain's method (see [137]). The idea is to solve (5.461) in $[0,\Delta T]$, then substitute $v(t)$ in (5.468) and study $y(t) = u(t) - v(t)$ in $[0,\Delta T]$ and get the corresponding nonlinear estimate. Using the inhomogeneous part $z(t)$ of $y(t)$ (see (5.469)) belongs to H^1, we thus can add $z(\Delta T)$ to $v(\Delta T)$ as the initial data of the regularized problem (5.461) in the next step. We repeat the steps in $[\Delta T, 2\Delta T]$, and make sure the process can be repeated by ensuring that the initial data at each time satisfies a uniform estimate and, therefore, complete Theorem 5.5.1. For simplicity, we will prove Theorem 5.5.1 through the following lemmas.

Lemma 5.5.3. *Assume $\Delta T \sim (1+N)^{-\frac{2(1-s)(\rho-1)}{n+2-\rho(n-2)}}$. Then we have the following estimate:*

(i) *When $n = 3$, $(n+3)/(n-1) \le \rho < (n+2)/(n-2)$, we have*

$$\|(\partial_t z(\Delta T), \nabla z(\Delta T))\|_2 \sim (1+N)^{\frac{n\rho-(n+2)}{2}-\rho s}. \tag{5.481}$$

(ii) *When $n = 3$, $k_0(n) < \rho \le (n+3)/(n-1)$ or $n \ge 4$, $k_0(n) \le \rho < (n-1)/(n-3)$, we have the estimate*

$$\|(\partial_t z(\Delta T), \nabla z(\Delta T))\|_2 \sim (1+N)^{-\frac{(3-\rho)(\rho-1)(1-s)}{n+2-\rho(n-2)}} N^{\frac{n\rho-(n+\rho-1)}{2}-\rho s}. \tag{5.482}$$

Proof. Denote $I = [0, \Delta T)$, and we can easily observe

$$\|(\partial_t z(\Delta T), \nabla z(\Delta T))\|_2 \le C(\|y^\rho\|_{L^1(I;L^2)} + \|v^{\rho-1} y\|_{L^1(I;L^2)}). \tag{5.483}$$

(I) The estimate of $\|y^\rho\|_{L^1(I;L^2)}$. Consider first the case of $n = 3$, $(n+3)/(n-1) \le \rho < (n+2)/(n-2)$. Let

$$\theta = \frac{n+1}{(n-1)\rho} = \frac{2}{\rho}, \quad l = \frac{n\rho - (n+2)}{2\rho}.$$

Then $l \ge \theta$ and

$$\frac{1}{2} = \rho\frac{(n+1)(n-2l) - 2(n-1)\theta}{2(n+1)n}, \quad n = 3.$$

Thus, by generalized Hölder inequality and Lemma 5.5.2, we have

$$\|y^\rho\|_{L^1(I;L^2)} \le \|\|y\|\|_l^\rho \le C(1+N)^{\frac{n\rho-(n+2)}{2}-\rho s}. \tag{5.484}$$

Second, consider the case of $n = 3$, $k_0(n) < \rho \le (n+3)/(n-1)$ and $n \ge 4$, $k_0(n) \le \rho < (n-1)/(n-3)$. Define $l = (n\rho - n - \rho + 1)/(2\rho)$, then we have

$$\frac{1}{2} = \frac{n-2l}{2n} + (\rho-1)\frac{n-2l-1}{2n}, \quad 1 = \frac{1}{\infty} + \frac{\rho-1}{2} + \frac{3-\rho}{2}.$$

Thus, by using Lemma 5.5.2, we obtain

$$\begin{aligned} \|y^\rho\|_{L^1(I;L^2)} &\le C(\Delta T)^{\frac{3-\rho}{2}} \|\|y\|\|_l^\rho \le C(\Delta T)^{\frac{3-\rho}{2}} N^{\rho(l-s)} \\ &\le CN^{-\frac{(3-\rho)(\rho-1)(1-s)}{n+2-\rho(n-2)}} N^{\frac{n\rho-(n+\rho-1)}{2}-\rho s}. \end{aligned} \tag{5.485}$$

(II) The estimate of $\|yv^{\rho-1}\|_{L^1(I;L^2)}$. Consider first the case of $n = 3$, $k_0(n) < \rho \le (n+3)/(n-1)$ (i. e., $2 < \rho \le 3$). For any $0 \le \theta < 1$, let $l = \frac{1-\theta}{2}(\rho-1)$. Then

$$\frac{1}{2} = \frac{1}{2} - \frac{l}{3} + (\rho-1)\frac{1-\theta}{6},$$

$$1 = \frac{\theta}{2}(\rho - 1) + \frac{1}{\chi}.$$

Thus, by Lemma 5.5.2, Lemma 5.5.3 and the generalized Hölder inequality, we deduce

$$\begin{aligned}\|v^{\rho-1}y\|_{L^1(I;L^2)} &\le C(\Delta T)^{1-\frac{\theta(\rho-1)}{2}} |||y|||_l \cdot |||v|||_1^{\rho-1} \\ &\le C(\Delta T)^{1-\frac{\theta(\rho-1)}{2}} (1+N)^{l-s} \cdot (1+N)^{(\rho-1)(1-s)} \\ &\le (1+N)^{\frac{\theta(\rho-1)^2(1-s)}{5-\rho} - \frac{2(\rho-1)(1-s)}{5-\rho} + \frac{1-\theta}{2}(\rho-1) - \rho s + \rho - 1},\end{aligned} \tag{5.486}$$

and

$$\lim_{\theta\to 1}\left(\frac{(\theta(\rho-1)-2)(\rho-1)(1-s)}{5-\rho} + \frac{(1-\theta)(\rho-1)}{2} - \rho s + \rho - 1\right) \tag{5.487}$$

$$= \frac{-(3-\rho)(\rho-1)(1-s)}{n+2-\rho(n-2)} + \frac{n\rho-(n+\rho-1)}{2} - \rho s. \tag{5.488}$$

Second, consider the case of $n = 3$, $(n+3)/(n-1) \le \rho < (n+2)/(n-2)$ (i. e., $3 \le \rho < 5$). For any $0 < \theta < 1$, let $l = \frac{\rho-3}{2} + (1-\theta)$, then $0 \le l < 1$ and

$$\begin{aligned}\frac{1}{2} &= \frac{\rho-3}{6} + 2\frac{1-\theta}{6} + \left(\frac{1}{2} - \frac{l}{3}\right), \\ 1 &= 2\cdot\frac{\theta}{2} + \frac{1}{\chi}.\end{aligned}$$

Thus, by Lemma 5.5.1, Lemma 5.5.2 and the generalized Hölder inequality, we deduce

$$\begin{aligned}\|v^{\rho-1}y\|_{L^1(I;L^2)} &\le C(\Delta T)^{1-\theta} |||y|||_l \cdot |||v|||_1^{\rho-1} \\ &\le (1+N)^{-(1-\theta)\frac{2(\rho-1)(1-s)}{5-\rho} + \frac{\rho-3}{2} + (1-\theta) - \rho s + \rho - 1},\end{aligned} \tag{5.489}$$

and

$$\lim_{\theta\to 1}\left(-(1-\theta)\frac{2(\rho-1)(1-s)}{5-\rho} + \frac{\rho-3}{2} + (1-\theta) - \rho s + \rho - 1\right)$$

$$= \frac{n\rho-(n+2)}{2} - \rho s. \tag{5.490}$$

Finally, consider the case of $n \ge 4$, $k_0(n) \le \rho < (n-1)/(n-3)$. Let $l = (\rho-1)(n-3)/2$, then

$$\begin{aligned}\frac{1}{2} &= \frac{n-2l}{2n} + (\rho-1)\frac{n-3}{2n}, \\ 1 &= \frac{\rho-1}{2} + \frac{3-\rho}{2}.\end{aligned}$$

By Lemma 5.5.2, Lemma 5.5.3 and the generalized Hölder inequality, we conclude

$$\begin{aligned}\|v^{\rho-1}y\|_{L^1(I;L^2)} &\le C(\Delta T)^{\frac{3-\rho}{2}}\|\|y\|\|_l\|\|v\|\|_1^{\rho-1}\\ &\le C(\Delta T)^{\frac{3-\rho}{2}}(1+N)^{l-s}\cdot(1+N)^{(\rho-1)(1-s)}\\ &\le C(1+N)^{-\frac{(3-\rho)(\rho-1)(1-s)}{n+2-\rho(n-2)}+\frac{n\rho-(n+\rho-1)}{2}-\rho s}.\end{aligned}\tag{5.491}$$

By combining (5.483)~(5.486), (5.488) and (5.490), we derive (5.481) and (5.482). □

Lemma 5.5.4. *Assume* $\Delta T \sim (1+N)^{-\frac{2(1-s)(\rho-1)}{n+2-\rho(n-2)}}$. *Then we have the estimate:*

(i) *Assume* $n=3$, $(n+3)/(n-1)\le\rho<(n+2)/(n-2)$, *then*

$$\|z(\Delta T)\|_{\rho+1}^{\frac{\rho+1}{2}} \sim (1+N)^{\tau_1-\frac{\rho(\rho+1)s}{2}},\tag{5.492}$$

$$\tau_1=\max\left(\frac{2\rho^2+\rho-8}{6},\frac{(\rho-2)(3\rho+5)}{4},\frac{(\rho-1)(3\rho-4)-6}{4(\rho-1)}\cdot(\rho+1)\right).$$

(ii) *When* $n=3$, $k_0(n)<\rho\le(n+3)/(n-1)$, *then*

$$\|z(\Delta T)\|_{\rho+1}^{\frac{\rho+1}{2}} \sim \max[(1+N)^{\tau_2},(1+N)^{\tau_3}],\tag{5.493}$$

where

$$\begin{aligned}\tau_2 &= -\frac{(3-\rho)(\rho-1)(1-s)}{n+2-\rho(n-2)}\cdot\frac{\rho+1}{2}-\frac{\rho(\rho+1)s}{2}\\ &\quad+\frac{n\rho(\rho+1)-2n-(\rho+1)^2}{4},\\ \tau_3 &= -\left(1-\frac{\rho(\rho+1)-6}{2(\rho+1)}\right)\cdot\frac{(\rho+1)(\rho-1)(1-s)}{n+2-\rho(n-2)}\\ &\quad+\frac{(\rho-1)(\rho+1)}{2}-\frac{\rho(\rho+1)s}{2}.\end{aligned}$$

(iii) *When* $n\ge4$, $k_0(n)\le\rho<(n-1)/(n-3)$, *we have*

$$\|z(\Delta T)\|_{\rho+1}^{\frac{\rho+1}{2}} \sim (1+N)^{\tau_2},\quad \tau \text{ similar to (ii)}.\tag{5.494}$$

Proof. By Minkowski's inequality and Sobolev embedding theorem, we can see

$$\|z(\Delta T)\|_{\rho+1}\le C\left(\|y^\rho\|_{L^1(I;L^{\frac{n(\rho+1)}{\rho+n+1}})}+\|v^{\rho-1}y\|_{L^1(I;L^{\frac{n(\rho+1)}{\rho+n+1}})}\right),\tag{5.495}$$

where $I=[0,\Delta T)$.

(I) The estimate of $\|y^\rho\|_{L^1(I;L^{\frac{\rho(n+1)}{n+\rho+1}})}$.

When $n=3$, $(n+3)/(n-1)\le\rho<(n+2)/(n-2)$ (i. e., $3\le\rho<5$), let

$$\begin{cases} l=1-\frac{2(4+\rho)}{3(\rho+1)\rho}, & \theta=l,\ 3\rho^2-5\rho-14\le0,\\ l=\frac{(3\rho+5)(\rho-2)}{2(\rho+1)\rho}, & \theta=\frac{2}{\rho},\ 3\rho^2-5\rho-14\ge0,\end{cases}$$

then

$$\begin{cases} \frac{\rho+4}{3(\rho+1)} = \rho\frac{1-l}{2}, & \theta = l,\ 3\rho^2 - 5\rho - 14 \le 0, \\ \frac{\rho+4}{3(\rho+1)} = \rho\frac{3-2l-\theta}{6}, & \theta = \frac{2}{\rho},\ 3\rho^2 - 5\rho - 14 \ge 0 \end{cases}$$

and

$$\frac{1}{\chi} := 1 - \frac{l\rho}{2} = 1 - \frac{3\rho^2+\rho-8}{6(\rho+1)}, \quad 3\rho^2 - 5\rho - 14 \le 0.$$

Therefore, by Hölder's inequality and Lemma 5.5.2, we get: when $3\rho^2 - 5\rho - 14 \le 0$,

$$\begin{aligned} \|y^\rho\|_{L^1(I;L^{\frac{\rho(n+1)}{n+\rho+1}})} &\le C(\Delta T)^{\frac{1}{\chi}} \||y\||_l^\rho \\ &\le (1+N)^{-(1-\frac{3\rho^2+\rho-8}{6(\rho+1)})\frac{2(\rho-1)(1-s)}{n+2-\rho(n-2)}+\rho-\frac{2(\rho+4)}{3(\rho+1)}-\rho s}. \end{aligned} \tag{5.496}$$

When $3\rho^2 - 5\rho - 14 \ge 0$, we have

$$\|y^\rho\|_{L^1(I;L^{\frac{\rho(n+1)}{n+\rho+1}})} \le C\||y\||_l^\rho \le (1+N)^{\frac{(3\rho+5)(\rho-2)}{2(\rho+1)}-\rho s}. \tag{5.497}$$

In addition, we have

$$\frac{2\rho^2+\rho-8}{3(\rho+1)} - \rho s \le \frac{n\rho-(n+2)}{2} - \rho s, \tag{5.498}$$

$$\frac{(3\rho+5)(\rho-2)}{2(\rho+1)} - \rho s \le \frac{n\rho-(n+2)}{2} - \rho s. \tag{5.499}$$

When $n = 3$, $k_0(n) < \rho \le (n+3)/(n-1)$ (i. e., $2 < \rho \le 3$), or $n \ge 4$, $\rho \le (n-1)/(n-3)$, take $l = [n\rho(\rho+1) - 2n - (\rho+1)^2]/[2\rho(\rho+1)]$. Then

$$\frac{\rho+n+1}{n(\rho+1)} = \frac{n-2l}{2n} + (\rho-1)\frac{n-2l-1}{2n}, \quad 1 = \frac{\rho-1}{2} + \frac{3-\rho}{2}.$$

By the generalized Hölder inequality and Lemma 5.5.2, we have

$$\begin{aligned} \|y^\rho\|_{L^1(I;L^{\frac{\rho(n+1)}{n+\rho+1}})} &\le C(\Delta T)^{\frac{3-\rho}{2}} \||y\||_l^\rho \le C(\Delta T)^{\frac{3-\rho}{2}} (1+N)^{\rho(l-s)} \\ &\le C(1+N)^{-\frac{(3-\rho)(\rho-1)(1-s)}{n+2-\rho(n-2)}+\frac{n\rho(\rho+1)-2n-(\rho+1)^2}{2(\rho+1)}-\rho s}. \end{aligned} \tag{5.500}$$

(II) The estimate of $\|v^{\rho-1}y\|_{L^1(I;L^{\frac{n(\rho+1)}{n+\rho+1}})}$.

When $n = 3$, $k_0(n) < \rho \le (n+3)/(n-1)$ (i. e., $2 < \rho \le 3$), take $l = \frac{\rho^2+\rho-6}{\rho^2-1}$. Then

$$\frac{1}{3} + \frac{1}{\rho+1} = \frac{1}{2} + (\rho-1)\frac{1-l}{6},$$

$$\frac{1}{\chi} = 1 - (\rho - 1)\frac{l}{2} = 1 - \frac{\rho^2 + \rho - 6}{2(\rho + 1)}.$$

Thus, by Lemma 5.5.1, Lemma 5.5.2 and the generalized Hölder's inequality, we have

$$\begin{aligned} \|v^{\rho-1}y\|_{L^1(I;L^{\frac{\rho(n+1)}{n+\rho+1}})} &\le C(\Delta T)^{1-\frac{\rho(\rho+1)-6}{2(\rho+1)}} |||y|||_0 \cdot |||v|||_1^{\rho-1} \\ &\le C(\Delta T)^{1-\frac{\rho(\rho+1)-6}{2(\rho+1)}} \cdot (1+N)^{(\rho-1)-\rho s} \\ &\le (1+N)^{-\frac{2(\rho-1)(1-s)}{5-\rho} + \frac{\rho(\rho+1)-6}{\rho+1} \cdot \frac{(\rho-1)(1-s)}{5-\rho} + (\rho-1) - \rho s}. \end{aligned} \tag{5.501}$$

In this situation, we can easily see that

$$\begin{aligned} &-\frac{2(\rho-1)(1-s)}{5-\rho} + \frac{\rho(\rho+1)-6}{\rho+1} \cdot \frac{(\rho-1)(1-s)}{5-\rho} + (\rho-1) - \rho s \\ &\le \frac{-(3-\rho)(\rho-1)(1-s)}{n+2-\rho(n-2)} + \frac{n\rho - (n+\rho-1)}{2} - \rho s. \end{aligned} \tag{5.502}$$

For $n = 3$, $(n+3)/(n-1) \le \rho < (n+2)/(n-2)$ (i. e., $3 \le \rho < 5$), let

$$\theta = \frac{2}{\rho - 2}, \quad l = \frac{\rho - 1}{2} - \frac{3}{\rho + 1},$$

then

$$\frac{1}{3} + \frac{1}{\rho + 1} = (\rho + 1)\frac{1-\theta}{2} + \frac{3-2l}{6}.$$

Thus, by Lemma 5.5.1, Lemma 5.5.2 and the generalized Hölder inequality, we have

$$\begin{aligned} \|v^{\rho-1}y\|_{L^1(I;L^{\frac{\rho(n+1)}{n+\rho+1}})} &\le |||y|||_l \cdot |||v|||_1^{\rho-1} \\ &\le (1+N)^{\frac{(\rho-1)(3\rho-4)-6}{2(\rho-1)} - \rho s}. \end{aligned} \tag{5.503}$$

In this situation, we can easily see that

$$\frac{(\rho-1)(3\rho-4)-6}{2(\rho-1)} - \rho s \le \frac{n\rho-(n+2)}{2} - \rho s. \tag{5.504}$$

Finally, for $n \ge 4$, $k_0(n) \le \rho < \frac{n-1}{n-3}$, define

$$l = \frac{n\rho(\rho+1) - 2n - (\rho+1)(3\rho-1)}{2(\rho+1)}.$$

Then we can observe

$$\frac{\rho+n+1}{n(\rho+1)} = \frac{n-2l}{2n} + (\rho-1)\frac{n-3}{2n}, \quad 1 = \frac{\rho-1}{2} + \frac{3-\rho}{2}.$$

Thus, by Lemma 5.5.1, Lemma 5.5.2 and the generalized Hölder inequality, we have

$$\begin{aligned}\|v^{\rho-1}y\|_{L^1(I;L^{\frac{\rho(n+1)}{n+\rho+1}})} &\leq C(\Delta T)^{\frac{3-\rho}{2}}|||y|||_l \cdot |||v|||_1^{\rho-1} \\ &\leq C(\Delta T)^{\frac{3-\rho}{2}}(1+N)^{l-s} \cdot (1+N)^{(\rho-1)(1-s)} \\ &\leq C(1+N)^{-\frac{(3-\rho)(\rho-1)(1-s)}{n+2-\rho(n-2)}+\frac{n\rho(\rho+1)-2n-(\rho+1)^2}{2(\rho+1)}-\rho s}. \end{aligned} \tag{5.505}$$

Thus, by combining (5.495)~(5.497), (5.500)~(5.502), (5.504) and (5.505), we can obtain the desired estimate (5.492)~(5.494). □

Remark 5.5.4. Next, we compare the growth rate of $\|(\partial_t z(\Delta T), \nabla z(\Delta T))\|_2$ and $\|z(\Delta T)\|_{\rho+1}^{\frac{\rho+1}{2}}$.

(i) For the case of $n = 3$, $(n+3)/(n-1) \leq \rho < (n+2)/(n-2)$ (i. e., $3 \leq \rho < 5$), by (5.498), (5.501) and (5.504), we can see

$$\tau_1 \leq \frac{n\rho - (n+2)}{2} - \rho s.$$

(ii) For the case of $n = 3$, $k_0(n) < \rho \leq (n+3)/(n-1)$ (i. e., $2 < \rho \leq 3$), by (5.502), we get

$$\max(\tau_2, \tau_3) \leq -\frac{(3-\rho)(\rho-1)(1-s)}{n+2-\rho(n-2)} + \frac{n\rho - (n+\rho-1)}{2} - \rho s.$$

(iii) For the case of $n \geq 4$, $k_0(n) \leq \rho < (n-1)/(n-3)$, since $s \geq \frac{n}{2} - \frac{2}{\rho-1}$, we can see by direct verification,

$$\tau_2 \leq -\frac{(3-\rho)(\rho-1)(1-s)}{n+2-\rho(n-2)} + \frac{n\rho - (n+\rho-1)}{2} - \rho s.$$

The proof of Theorem 5.5.1. As discussed above, we first solve the regularized problem (5.461) in $[0, \Delta T]$, then substitute the solution into (5.468) and solve (5.468) in $[0, \Delta T]$. Then we prove the estimate in Lemma 5.5.3 and Lemma 5.5.4. Now we study the regularized problem (5.461) in $[\Delta T, 2\Delta T]$, with the initial data replaced by

$$(v(\Delta T) + z(\Delta T), \partial_t v(\Delta T) + \partial_t z(\Delta T)).$$

Then substitute $v(t)$ into (5.468), and study in $[\Delta T, 2\Delta T]$ the Cauchy problem for the difference equation with the initial data in (5.468) replaced by

$$(\dot{K}(\Delta T)\phi_2 + K(\Delta T)\psi_2, -\Delta K(\Delta T)\psi_2 + \dot{K}(\Delta T)\psi_2).$$

By repeating the proof of Lemma 5.5.3 and Lemma 5.5.4, we can obtain the similar estimate.

For any $T > 0$, in order to extend the solution up to T, the times we need repeat the process above is

$$\frac{T}{\Delta T} = T(1+N)^{\frac{2(p-1)(1-s)}{n+2-p(n-2)}}.$$

Therefore, by Lemma 5.5.3 and Lemma 5.5.4, after $T/\Delta T$ times, the increment of energy (see (5.464)) is:

(i) $n = 3$, $k_0(n) < p \leq (n+3)/(n-1)$ (i. e., $2 < p \leq 3$).

$$CT(1+N)^{\frac{2(p-1)(1-s)}{n+2-p(n-2)}}(1+N)^{-\frac{(3-p)(p-1)(1-s)}{n+2-p(n-2)}+\frac{np-(n+p-1)}{2}-ps}. \tag{5.506}$$

(ii) $n = 3$, and $3 \leq p < (n+2)/(n-2)$.

$$CT(1+N)^{\frac{2(p-1)(1-s)}{n+2-p(n-2)}} \cdot (1+N)^{\frac{np-(n+2)}{2}-ps}. \tag{5.507}$$

(iii) $n \geq 4$, $k_0(n) \leq p < (n-1)/(n-3)$.

$$CT(1+N)^{\frac{2(p-1)(1-s)}{n+2-p(n-2)}}(1+N)^{-\frac{(3-p)(p-1)(1-s)}{n+2-p(n-2)}+\frac{np-(n+p-1)}{2}-ps}. \tag{5.508}$$

To achieve our target, we need to ensure that the increment in (5.506)~(5.508) do not exceed $(1+N)^{1-s}$. Therefore, in the case of $n = 3$, $k_0(n) < p \leq (n+3)/(n-1)$ (i. e., $2 < p \leq 3$), s satisfies

$$s > \frac{2(p-1)^2 + (n+2-p(n-2)) \cdot (np-n-p-1)}{2(p-1)^2 + 2(p-1)(n+2-p(n-2))}.$$

In the case of $n = 3$, $3 \leq p < (n+2)/(n-2)$, s satisfies

$$s > \frac{4(p-1) + (n+2-p(n-2)) \cdot (np-n-4)}{2(p-1)(n+4-p(n-2))}.$$

In the case of $n \geq 4$, $k_0(n) \leq p < (n-1)/(n-3)$, s should satisfy

$$s > \frac{2(p-1)^2 + (n+2-p(n-2)) \cdot (np-n-p-1)}{2(p-1)^2 + 2(p-1)(n+2-p(n-2))}.$$

Taking N large enough ($N = T^{\frac{1}{1-s-\eta}}$, η is defined as in Theorem 5.5.1), we finally complete the proof of Theorem 5.5.1. □

5.6 Notes on dynamics of NLW

In this note, we aim to summarize the history of the study to the initial-value problem for nonlinear wave equations (NLW):

$$(\text{NLW}) \quad \begin{cases} \partial_{tt}u - \Delta u = \mu|u|^{p-1}u, & (t,x) \in \mathbb{R} \times \mathbb{R}^n, \\ (u, \partial_t u)(0,x) = (u_0, u_1)(x), \end{cases} \tag{5.509}$$

where $u : \mathbb{R}_t \times \mathbb{R}_x^n \to \mathbb{R}$, $\mu = \pm 1$ with $\mu = -1$ known as the defocusing case and $\mu = 1$ as the focusing case. Equation (5.509) admits a number of symmetries, explicitly:

- *Space-time translation invariance*: if $u(t,x)$ solves (5.509), then so does $u(t+t_0, x+x_0)$, $(t_0, x_0) \in \mathbb{R} \times \mathbb{R}^n$;
- *Lorentz transformations invariance*: if $u(t,x)$ solves (5.509), then so does

$$u\left(\frac{t - \ell \cdot x}{\sqrt{1 - |\ell|^2}}, \left(-\frac{t}{\sqrt{1-|\ell|^2}} + \frac{1}{|\ell|^2}\left(\frac{1}{\sqrt{1-|\ell|^2}} - 1\right)\ell \cdot x\right)\ell + x\right) \tag{5.510}$$

  where $\ell \in \mathbb{R}^n$, $|\ell| < 1$.
- *Scaling invariance*: if $u(t,x)$ solves (5.509), then so does $u_\lambda(t,x)$ defined by

$$u_\lambda(t,x) = \lambda^{\frac{2}{p-1}} u(\lambda t, \lambda x), \quad \lambda > 0. \tag{5.511}$$

  This scaling defines a notion of criticality for (5.509). In particular, one can check that the only homogeneous L_x^2-based Sobolev space that is left invariant under (5.511) is $\dot{H}_x^{s_c}(\mathbb{R}^n)$, where the critical regularity s_c is given by $s_c := \frac{n}{2} - \frac{2}{p-1}$.
- For $s_c = 1$, we call the problem (5.509) *energy critical*;
- For $s_c < 1$, we call the problem (5.509) *energy subcritical*;
- For $s_c > 1$, we call the problem (5.509) *energy supercritical*.

From the Ehrenfest law or direct computation, the above symmetries induce invariance in the energy space, namely

$$\textit{Energy:}\quad E(u)(t) \triangleq \frac{1}{2}\int_{\mathbb{R}^n} \left(|\nabla u|^2 + |\partial_t u|^2\right) dx - \frac{\mu}{p+1}\int_{\mathbb{R}^n} |u(t,x)|^{p+1}\, dx = E(u_0, u_1),$$

$$\textit{Momentum:}\quad P(u)(t) \triangleq \int_{\mathbb{R}^n} \nabla u u_t \, dx = P(u_0, u_1).$$

5.6.1 Energy-critical NLW

We first consider the defocusing energy-critical nonlinear wave equation

$$\begin{cases} u_{tt} - \Delta u + |u|^{\frac{4}{n-2}} u = 0, & (t,x) \in \mathbb{R} \times \mathbb{R}^n,\ n \geq 3, \\ (u, u_t)(0) = (u_0, u_1). \end{cases} \tag{5.512}$$

- For $n = 3$, in 1988, Rauch [224] proved the global existence of the *radial smooth solution* for (5.512). In 1990, Grillakis [111] removed radial assumption and proved the global existence of a *smooth solution* for (5.512) by the classical Morawetz estimate.
- For the higher-dimensional case, such as $3 \leq n \leq 5$, Grillakis [112] proved the global existence of a *smooth solution* for (5.512) by combining the Strichartz estimates with the classical Morawetz estimate.

- Shatah–Struwe [237] proved the global existence of a *smooth solution* for (5.512) for $n \le 7$. Moreover, they proved the global existence of a *finite energy solution* in $H^1 \times L^2$.
- The scattering result is obtained by Bahouri–Gérard [7], and Tao [262]. In particular, Tao in [262] derived a exponential-type space-time bound as follows:

$$\|u\|_{L_t^4 L_x^{12}(\mathbb{R}\times\mathbb{R}^3)} \le C(1+E_0)^{CE_0^{105/2}}, \quad E_0 = E(u_0,u_1).$$

We summarize the above result as follows.

Theorem 5.6.1 (Defocusing energy-critical NLW). *Let $n \ge 3$. Given $(u_0,u_1) \in \dot{H}^1(\mathbb{R}^n) \times L^2(\mathbb{R}^n)$. Then there is a unique global strong solution u to*

$$\begin{cases} u_{tt} - \Delta u + |u|^{\frac{4}{n-2}} u = 0, \\ (u, \partial_t u)(0) = (u_0, u_1). \end{cases} \tag{5.513}$$

Moreover, the solution u obeys the estimate

$$\int_{\mathbb{R}} \int_{\mathbb{R}^n} |u(t,x)|^{\frac{2(n+1)}{n-2}}\, dx\, dt \le C(\|(u_0,u_1)\|_{\dot{H}_x^1 \times L_x^2}). \tag{5.514}$$

So, the solution scatters in the sense, and there exists $(u_0^\pm, u_1^\pm) \in \dot{H}^1(\mathbb{R}^n) \times L^2(\mathbb{R}^n)$ such that

$$\lim_{t\to\pm\infty} \|u(t,x) - S(t)(u_0^\pm, u_1^\pm)\|_{\dot{H}^1(\mathbb{R}^n)\times L^2(\mathbb{R}^n)} = 0, \tag{5.515}$$

where

$$S(t)(f,g) = \cos(t\sqrt{-\Delta}) f + \frac{\sin(t\sqrt{-\Delta})}{\sqrt{-\Delta}} g.$$

Next, we consider the focusing energy-critical nonlinear wave equation

$$\text{(FNLW)} \quad \begin{cases} \partial_{tt} u - \Delta u = |u|^{\frac{4}{n-2}} u, \quad (t,x) \in \mathbb{R} \times \mathbb{R}^n, \\ (u, \partial_t u)(0,x) = (u_0,u_1)(x), \end{cases} \tag{5.516}$$

which is much richer: small data solutions exist globally and scatter; however, large solutions that blow up in finite time may occur [164], where Levine showed that if

$$(u_0,u_1) \in \dot{H}^1 \times L^2, \quad u_0 \in L^2, \quad E(u_0,u_1) < 0,$$

there is always a breakdown in finite time. Recently, Krieger–Schlag–Tataru [155] have constructed explicit radial examples, which breakdown in finite time. On the other hand, it is well known that there exist solutions with compactly supported smooth initial data, which blow up in finite time. This is most easily seen by observing that

$$u(t,x) = \left(\frac{n(n-2)}{4}\right)^{\frac{n-2}{4}} /(1-t)^{\frac{n-2}{2}}$$

is an explicit solution which, by the finite speed of propagation, can be used to construct a blowup solution of the aforementioned type. This kind of breakdown is referred to as a *ODE blowup* and it is conjectured to comprise the "generic" blowup scenario [15].

There exist nonscattering global solutions. Examples of such solutions are the so-called solitary waves and are given by solutions of the elliptic equation:

$$-\Delta W = |W|^{\frac{4}{n-2}} W, \quad W \in \dot{H}^1, \tag{5.517}$$

$$W(x) \triangleq \frac{1}{(1+\frac{|x|^2}{n(n-2)})^{\frac{n-2}{2}}} \in \dot{H}^1, \tag{5.518}$$

but in L^2 only if $n \geq 5$.

Below threshold of ground state

Below, in the threshold of ground state: $E(u_0, u_1) < E(W, 0)$, Kenig–Merle show the scattering/blowup dichotomy.

Theorem 5.6.2 (Scattering/Blowup dichotomy, Kenig–Merle [152]). *Assume* $3 \leq n \leq 5$, $(u_0, u_1) \in \dot{H}^1 \times L^2$ *with*

$$E(u_0, u_1) < E(W, 0).$$

Then:
1. *If* $\|u_0\|_{\dot{H}^1} < \|W\|_{\dot{H}^1}$, *then the solution u of* (5.516) *exists globally and scatters.*
2. *If* $\|u_0\|_{\dot{H}^1} > \|W\|_{\dot{H}^1}$, *then the solutions u of* (5.516) *blows up in finite time in both directions.*

Threshold of solutions

Now, we consider $E(u_0, u_1) = E(W, 0)$. Duyckaerts–Merle [79] and Li–Zhang [167] give the classification of threshold solutions as follows.

Theorem 5.6.3 (Connecting orbits, Duyckaerts–Merle [79] for $3 \leq n \leq 5$, Li–Zhang [167] for $d \geq 6$). *Let $n \geq 3$. There exist radial solutions W^- and W^+ of* (5.516), *with initial conditions* $(W_0^\pm, W_1^\pm) \in \dot{H}^1 \times L^2$ *such that*

$$E(W,0) = E(W_0^+, W_1^+) = E(W_0^-, W_1^-), \tag{5.519}$$

$$T_+(W^-) = T_+(W^+) = +\infty \quad \text{and} \quad \lim_{t\to+\infty} W^\pm(t) = W \quad \text{in} \quad \dot{H}^1, \tag{5.520}$$

$$\|\nabla W^-\|_2 < \|\nabla W\|_2, \quad T_-(W^-) = +\infty, \quad \|W^-\|_{L^{\frac{2(n+1)}{n-2}}((-\infty,0)\times\mathbb{R}^n)} < \infty, \tag{5.521}$$

$$\|\nabla W^+\|_2 > \|\nabla W\|_2, \quad T_-(W^+) < +\infty. \tag{5.522}$$

The next result is that W, W^- and W^+ are, up to the symmetry of the equation, the only examples of new behavior at the critical level.

Theorem 5.6.4 (Dynamical classification at the critical level [79, 167]). *Let $n \geq 3$. Let $(u_0, u_1) \in \dot{H}^1 \times L^2$ such that*

$$E(u_0, u_1) = E(W, 0) = \frac{1}{nC_n^n}. \tag{5.523}$$

Let u be the solution of (5.516) *with initial conditions (u_0, u_1) and I its maximal interval of definition. Then the following holds:*

1. *If $\|\nabla u_0\|_2^2 < \|\nabla W\|_2^2 = 1/C_n^n$, then $I = \mathbb{R}$. Furthermore, either $u = W^-$ up to the symmetries of the equation, or*

$$\|u\|_{L_{t,x}^{\frac{2(n+1)}{n-2}}(\mathbb{R}\times\mathbb{R}^n)} < \infty.$$

2. *If $\|\nabla u_0\|_2^2 = \|\nabla W\|_2^2$, then $u = W$ up to the symmetries of the equation.*
3. *If $\|\nabla u_0\|_2^2 > \|\nabla W\|_2^2$ and $u_0 \in L^2$, then either $u = W^+$ up to the symmetries of the equation, or I is finite.*

In the above theorem, by u equals v up to the symmetry of the equation, we mean that there exists $t_0 \in \mathbb{R}$, $x_0 \in \mathbb{R}^n$, $\lambda_0 > 0$, $\iota_0, \iota_1 \in \{-1, +1\}$ such that

$$u(t, x) = \frac{\iota_0}{\lambda_0^{(n-2)/2}} v\left(\frac{t_0 + \iota_1 t}{\lambda_0}, \frac{x + x_0}{\lambda_0}\right).$$

Beyond threshold of ground state

When $E(u_0, u_1) > E(W, 0)$, there exist two types of blowup solutions. We refer to [72, 77, 121, 154, 155], and [54, 76, 78] for the soliton resolution conjecture.

5.6.2 Energy-subcritical NLW

Consider energy subcritical NLW equation

$$\text{(NLW)} \quad \begin{cases} \partial_{tt} u - \Delta u = \mu |u|^{p-1} u, & (t, x) \in \mathbb{R} \times \mathbb{R}^n, \\ (u, \partial_t u)(0, x) = (u_0, u_1)(x), \end{cases} \tag{5.524}$$

where $u : \mathbb{R}_t \times \mathbb{R}_x^n \to \mathbb{R}$, $s_c = \frac{n}{2} - \frac{2}{p-1} < 1$.

1. For $n = 3$ and $1 < p < p_c = 5$, Jörgen in 1961 proved the global existence of smoothing solution of (5.524) [122].
2. For higher dimensions with $4 \leq n \leq 9$, Brenner, Wahl and Pecher proved the global existence of smoothing solution of (5.524) [26, 215, 282].

3. Ginibre–Velo proved the global well-posedness of the energy solution of (5.524) with initial data $(u_0, u_1) \in H^1 \times L^2$ [102, 103].

Low regularity: $(u_0, u_1) \in H^s \times H^{s-1}$ with $s_c \leq s < 1$
As for Schrödinger equations in Chapter 4, one also has the low regularity problem about wave equations.

Conjecture 5.6.1 (Low regularity conjecture). *Let $s_c \geq 0$, $\mu = -1$ and $(u_0, u_1) \in \dot{H}^{s_c} \times \dot{H}^{s_c-1}$. Then u exists globally and scatters in the sense that there exist unique $(u_0^{\pm}, u_1^{\pm}) \in \dot{H}_x^{s_c}(\mathbb{R}^n) \times \dot{H}^{s_c-1}$ such that*

$$\lim_{t\to\pm\infty} \|(u, \partial_t u)(t) - S(t)(u_0^{\pm}, u_1^{\pm})\|_{\dot{H}_x^{s_c} \times \dot{H}^{s_c-1}} = 0.$$

For the history of the study of this conjecture, we refer to Table 5.1 below.

Table 5.1: The well-known results on the cubic wave equation in $\mathbb{R}^3$.

$s > 3/4$	$s = 3/4$
KPV [137], Gallagher–Planchon [95]	Bahouri–Chemin [6]

Dodson solve Conjecture 5.6.1 for the radial initial data in [65].

Critical norm conjecture: $(u_0, u_1) \in \dot{H}^{s_c} \times \dot{H}^{s_c-1}$
Similar to Schrödinger equations in Chapter 4, one also has the critical norm problem about wave equations.

Conjecture 5.6.2 (Critical norm conjecture). *Let $s_c > 0$ and $\mu = -1$. Suppose $u : I \times \mathbb{R}^n \to \mathbb{R}$ is a maximal-lifespan solution to (5.524) such that*

$$(u, \partial_t u) \in L_t^{\infty}(I; \dot{H}_x^{s_c} \times \dot{H}^{s_c-1}). \tag{5.525}$$

Then u is global and scatters in the sense that there exist unique $(u_0^{\pm}, u_1^{\pm}) \in \dot{H}_x^{s_c}(\mathbb{R}^n) \times \dot{H}_x^{s_c-1}(\mathbb{R}^n)$ such that

$$\lim_{t\to\pm\infty} \|(u, \partial_t u)(t) - S(t)(u_0^{\pm}, u_1^{\pm})\|_{\dot{H}_x^{s_c} \times \dot{H}^{s_c-1}} = 0.$$

We list the history of the study of the above conjecure in the Table 5.2 below.

Table 5.2: Defocusing wave equation.

	$\sqrt{2} < p \leq 2$	$2 < p < 4$
$n = 3$	Dodson–Lawrie [68]	Shen [238]
$n \geq 4$	open	open

For the dynamics of the nonlinear wave equation with inverse square potential

$$\begin{cases} u_{tt} - \Delta u + \frac{a}{|x|^2} u = \mu |u|^{p-1} u, & (t,x) \in \mathbb{R} \times \mathbb{R}^n, \\ (u, u_t)(0,x) = (u_0, u_1)(x), \end{cases} \tag{5.526}$$

we refer to [169, 187, 189, 292].

6 Nonlinear Schrödinger equation with inverse-square potential

In this chapter, we mainly study the initial-value problem for the energy-critical defocusing nonlinear Schrödinger equation with an inverse-square potential in three spatial dimensions

$$\begin{cases} (i\partial_t - \mathcal{L}_a)u = |u|^4 u, & (t,x) \in \mathbb{R} \times \mathbb{R}^3, \\ u(0,x) = u_0(x) \in \dot{H}^1_x(\mathbb{R}^3), \end{cases} \tag{6.1}$$

where $u : \mathbb{R}_t \times \mathbb{R}^3_x \to \mathbb{C}$, and $\mathcal{L}_a = -\Delta + \frac{a}{|x|^2}$ with $a > -\frac{1}{4}$. We refer to Section 1.5 in Chapter 1 for the basic harmonic analysis questions related to the Schrödinger operator $\mathcal{L}_a$.

There are two important conserved quantities:

$$\textit{Mass:}\quad M(u) = \int_{\mathbb{R}^3} |u(t,x)|^2\, dx \equiv M(u_0); \tag{6.2}$$

$$\textit{Energy:}\quad E_a(u) = \int_{\mathbb{R}^3} \left(\frac{1}{2}|\nabla u(t,x)|^2 + \frac{a}{2|x|^2}|u(t,x)|^2 + \frac{1}{6}|u(t,x)|^6 \right) dx \equiv E_a(u_0). \tag{6.3}$$

Equation (6.1) admits a number of symmetries in energy space H^1, explicitly:

- *Time translation invariance:* if $u(t,x)$ solves (6.1), then so does $u(t+t_0,x)$, $t_0 \in \mathbb{R}$;
- *Phase invariance:* if $u(t,x)$ solves (6.1), then so does $e^{i\gamma}u(t,x)$, $\gamma \in \mathbb{R}$;
- *Scaling invariance:* if $u(t,x)$ solves (6.1), then so does $u_\lambda(t,x)$ defined by

$$u_\lambda(t,x) = \lambda^{\frac{1}{2}} u(\lambda^2 t, \lambda x), \quad \lambda > 0. \tag{6.4}$$

This scaling defines a notion of criticality for (6.1). In particular, one can check that the only homogeneous L^2_x-based Sobolev space that is left invariant under (6.4) is $\dot{H}^1_x(\mathbb{R}^3)$. Thus, we call the problem (6.1) energy critical.

Next, we introduce a few definitions.

Definition 6.0.1 (Strong solution). A function $u : I \times \mathbb{R}^3 \to \mathbb{C}$ on a nonempty time interval $I \ni 0$ is a solution to (6.1) if it belongs to $C_t \dot{H}^{s_c}_x(K \times \mathbb{R}^3) \cap L^{10}_{t,x}(K \times \mathbb{R}^3)$ for any compact interval $K \subset I$ and obeys the Duhamel formula

$$u(t) = e^{-it\mathcal{L}_a}u_0 - i\int_0^t e^{-i(t-s)\mathcal{L}_a}|u(s)|^4 u(s)\, ds \tag{6.5}$$

for each $t \in I$. We call I the lifespan of u. We say that u is a maximal-lifespan solution if it cannot be extended to any strictly larger interval. We call u global if $I = \mathbb{R}$.

https://doi.org/10.1515/9783111384726-006

In Section 6.1, we will prove that such solutions exist, at least locally in time, when $a > -\frac{1}{4} + \frac{1}{25}$. The restriction $a > -\frac{1}{4} + \frac{1}{25}$ represents the limit of what can be done within the confines of the usual Strichartz methodology. This breakdown does not originate in the failure of Strichartz estimates for the propagator $e^{-it\mathcal{L}_a}$; indeed, the paper [29] shows that the full range of such estimates holds for $a > -\frac{1}{4}$. It stems from failures in Sobolev embedding; see the details in Remark 6.1.1 below.

Definition 6.0.2 (Scattering size and blowup). For a solution $u : I \times \mathbb{R}^3 \to \mathbb{C}$ to (6.1), we define the scattering size of u on I by

$$S_I(u) := \int_I \int_{\mathbb{R}^3} |u(t,x)|^{10}\, dx\, dt. \tag{6.6}$$

If there exists $t_0 \in I$ such that $S_{[t_0, \sup I)}(u) = \infty$, we say that u blows up forward in time. Similarly, if there exists $t_0 \in I$ such that $S_{(\inf I, t_0]}(u) = \infty$, we say that u blows up backward in time. In particular, a solution may blow up in infinite time.

On the other hand, standard arguments show that if u is a global solution to (6.1) that obeys $S_{\mathbb{R}}(u) < \infty$, then u scatters, that is, there exist unique $u_\pm \in \dot{H}^1_x(\mathbb{R}^3)$ such that

$$\lim_{t\to\pm\infty} \|u(t) - e^{it\mathcal{L}_a} u_\pm\|_{\dot{H}^1_x(\mathbb{R}^3)} = 0. \tag{6.7}$$

The main result of this chapter is to prove global well-posedness of the problem (6.1) and determine the asymptotic behavior of solutions as $t \to \pm\infty$, which was established in Killip–Miao–Visan-Zhang–Zheng [141].

Theorem 6.0.1. *Fix $a > -\frac{1}{4} + \frac{1}{25}$. Given $u_0 \in \dot{H}^1(\mathbb{R}^3)$, there is a unique global solution u to* (6.1) *satisfying*

$$\int_{\mathbb{R}} \int_{\mathbb{R}^3} |u(t,x)|^{10}\, dx\, dt \le C(\|u_0\|_{\dot{H}^1_x}). \tag{6.8}$$

Moreover, the solution u scatters in the sense of (6.7).

The proof of Theorem 6.0.1 employs the induction on energy argument pioneered in [22, 48] and subsequently reimagined in [132]. Some of the ingredients that underlie the Kenig–Merle approach to induction on energy are the ideas of linear and nonlinear profile decompositions from [7] and the notion of a minimal blowup solution from [139]; however, the potential of these ideas for addressing the well-posedness problem for NLS was first realized in [132]. For a pedagogical introduction to the Kenig–Merle argument illustrating how it may be applied to the case of the defocusing energy-critical NLS without potential see, for example, [148].

The argument proceeds by contradiction. Assuming that if Theorem 6.0.1 was to fail, one first demonstrates the existence of a minimal counterexample, that is, a solution to (6.1) that has infinite space-time norm and has minimal energy among all such solutions. The (concentration) compactness argument used to show this existence shows more, namely that such a minimal blowup solution must be almost periodic (cf. Theorem 6.3.1). The second half of the argument is to use monotonicity formulae and/or conservation laws to rule out the existence of such solutions. The almost periodicity is essential here; it provides both a spatial center and a length scale that are intrinsic to the solution at hand. These are essential for using conservation laws and/or monotonicity formulae that are not translation invariant and/or that do not respect the scaling of the equation.

The proof of the existence of a minimal blowup solution hinges on the construction of a nonlinear profile decomposition, which is applied to a sequence of solutions u_n witnessing the supposed failure of Theorem 6.0.1. This decomposition says that (after passing to a subsequence and up to a negligible error) the solutions with initial data $u_n(0)$ can be asymptotically expressed as a linear combination of the nonlinear evolutions of a fixed collection of profiles; the profiles are, however, modified (in an n-dependent manner) by the symmetries of the equation, namely space and time translations, as well as scaling.

Here, we see a new obstacle introduced by the presence of a nonzero potential; it breaks the space translation symmetry. Ultimately, this means that profiles living increasingly far from the origin relative to their intrinsic scale cannot be treated inductively—they cannot be modeled by a single solution of (6.1) up to symmetries of the equation. On the other hand, we may expect that the solution stemming from such data is little affected by the potential and can be approximated by a solution to

$$\begin{cases} (i\partial_t + \Delta)u = |u|^4 u, & (t,x) \in \mathbb{R} \times \mathbb{R}^3, \\ u(0,x) = u_0(x), \end{cases} \tag{6.9}$$

thus inheriting the space-time bounds guaranteed by the result of Colliander–Keel–Staffilani–Takaoka–Tao [48]. This heuristic is realized in Theorem 6.2.1, whose proof fills Section 6.3. Both this result and the proof that such profiles decouple from the other nonlinear profiles rely on various convergence statements for linear operators proved in Section 6.4. The necessity of proving such convergence statements stems from the absence of $L_x^p \to L_x^{p'}$ dispersive estimates for the propagator $e^{-it\mathcal{L}_a}$ when $a < 0$; see [218]. Indeed, one should view (1.123) as a substitute for the dispersive estimate in this setting (cf. [150, Theorem 4.1]); this is essential for the construction of the minimal blowup solution.

6.1 Local well-posedness and stability

In this section, we first show local well-posedness in the inhomogeneous space $H_a^1(\mathbb{R}^3)$. And then, local well-posedness in the larger space $\dot{H}_a^1(\mathbb{R}^3)$ follows as an application of the stability theory Theorem 6.1.1.

Proposition 6.1.1. *Let $a > -\frac{1}{4} + \frac{1}{25}$ and $u_0 \in H_a^1(\mathbb{R}^3)$. For given $A \geq 0$, there exists $\eta = \eta(A)$ so that the following holds: Assume that*

$$\|\sqrt{\mathcal{L}_a}u_0\|_{L^2(\mathbb{R}^3)} \leq A \quad \text{and} \quad \|e^{it\mathcal{L}_a}u_0\|_{L^{10}_{t,x}(I\times\mathbb{R}^3)} \leq \eta \tag{6.10}$$

for some time interval $I \ni 0$. Then there is a unique strong solution u to (6.1) *on the time interval I such that*

$$\|u\|_{L^{10}_{t,x}(I\times\mathbb{R}^3)} \lesssim \eta \quad \text{and} \quad \|\sqrt{\mathcal{L}_a}\, u\|_{C_tL^2_x\cap L^{10}_t L_x^{\frac{30}{13}}(I\times\mathbb{R}^3)} \lesssim A. \tag{6.11}$$

Proof. Throughout the proof, all space-time norms will be on $I \times \mathbb{R}^3$.

We only need to show that the map

$$\Phi : u \mapsto e^{-it\mathcal{L}_a}u_0 - i\int_0^t e^{-i(t-s)\mathcal{L}_a}|u(s)|^4 u(s)\, ds$$

is a contraction on the (complete) space

$$B := \{u \in C_tH^1_a \cap L^{10}_t H_a^{1,\frac{30}{13}}(I\times\mathbb{R}^3) : \|\sqrt{\mathcal{L}_a}\, u\|_{L^{10}_t L_x^{\frac{30}{13}}} \leq CA \text{ and } \|u\|_{L^{10}_{t,x}} \leq 2\eta\}$$

endowed with the metric

$$d(u,v) := \|u-v\|_{L^{10}_t L_x^{\frac{30}{13}}}.$$

The constant C depends only on the dimension and a and it reflects various constants in the Strichartz and Sobolev embedding inequalities.

By Strichartz estimates, (6.10) and fractional Leibnitz rule Lemma 1.5.1, for $u \in B$ we have

$$\|\sqrt{\mathcal{L}_a}\,\Phi(u)\|_{C_tL^2_x\cap L^{10}_tL_x^{\frac{30}{13}}} \lesssim A + \|\sqrt{\mathcal{L}_a}\,(|u|^4u)\|_{L^2_tL_x^{\frac{6}{5}}} \lesssim A + \eta^4\|\sqrt{\mathcal{L}_a}\,u\|_{L^{10}_tL_x^{\frac{30}{13}}}.$$

Thus, choosing η sufficiently small, we get

$$\|\sqrt{\mathcal{L}_a}\,\Phi(u)\|_{C_tL^2_x\cap L^{10}_tL_x^{\frac{30}{13}}} \leq CA. \tag{6.12}$$

Similarly, for $u \in B$,

$$\|\Phi(u)\|_{C_tL^2_x\cap L^{10}_tL_x^{\frac{30}{13}}} \lesssim \|u_0\|_{L^2_x} + \||u|^4u\|_{L^2_tL_x^{\frac{6}{5}}} \lesssim \|u_0\|_{L^2_x} + \eta^4\|u\|_{L^{10}_tL_x^{\frac{30}{13}}} < \infty. \tag{6.13}$$

Proceeding once more in a parallel manner shows

$$\|\Phi(u)\|_{L^{10}_{t,x}} \leq \eta + C'\|u\|^4_{L^{10}_{t,x}} \|\sqrt{\mathcal{L}_a}\, u\|_{L^{10}_t L^{\frac{30}{13}}_x} \leq \eta + C'(2\eta)^4 2A \leq 2\eta,$$

provided η is chosen sufficiently small. This, (6.12) and (6.13) show that Φ does indeed map B into itself.

To see that Φ is a contraction, we argue analogously:

$$\begin{aligned}\|\Phi(u) - \Phi(v)\|_{L^{10}_t L^{\frac{30}{13}}_x} &\lesssim \||u|^4 u - |v|^4 v\|_{L^2_t L^{\frac{6}{5}}_x} \\ &\lesssim \|u - v\|_{L^{10}_t L^{\frac{30}{13}}_x} \left(\|u\|_{L^{10}_{t,x}} + \|v\|_{L^{10}_{t,x}}\right)^4,\end{aligned}$$

which shows that for $u, v \in B$, one has $d(\Phi(u), \Phi(v)) \leq \frac{1}{2} d(u, v)$ provided η is chosen sufficiently small. □

Remark 6.1.1. The condition $a > -\frac{1}{4} + \frac{1}{25}$ expresses the limit of what can be achieved by the method presented above. In order to estimate the nonlinearity in a dual Strichartz space, one must be able to place the solution in a space $L^q_t L^r_x$ with $q \leq 10$. By scaling this then requires $r \geq 10$. On the other hand, if $a \leq -\frac{1}{4} + \frac{1}{25}$ then the Sobolev embedding $\dot{H}^{1,\frac{3r}{r+3}}_a \hookrightarrow L^r_x$ breaks down for all $r \geq 10$.

Next, we turn to the formulation of the stability theory. As noted above, it allows us to remove the finite-mass assumption from Proposition 6.1.1. It also plays an essential role in the implementation of the induction on energy argument employed in this chapter; in particular, it shows that solutions to (6.1) whose initial data are the sum of two parts that live at vastly different length scales can be approximated by the sum of the corresponding solutions to (6.1). Following the general outline in [48, 228, 266], exploiting the same spaces used in the proof of Proposition 6.1.1 and Lemma 1.5.1 provides the key input for differentiating the nonlinearity with respect to the operator $\sqrt{\mathcal{L}_a}$, one has the following.

Theorem 6.1.1 (Stability). *Fix $a > -\frac{1}{4} + \frac{1}{25}$. Let I be a compact time interval and let $\tilde{u}$ be an approximate solution to* (6.1) *on $I \times \mathbb{R}^3$ in the sense that*

$$(i\partial_t - \mathcal{L}_a)\tilde{u} = |\tilde{u}|^4 \tilde{u} + e$$

for some function e. For some positive constants E and L, assume that

$$\|\tilde{u}\|_{L^\infty_t \dot{H}^1_a(I\times\mathbb{R}^3)} \leq E \quad \text{and} \quad \|\tilde{u}\|_{L^{10}_{t,x}(I\times\mathbb{R}^3)} \leq L.$$

Let $t_0 \in I$ and let $u_0 \in \dot{H}^1_a(\mathbb{R}^3)$. Assume the smallness conditions

$$\|u_0 - \tilde{u}(t_0)\|_{\dot{H}^1_a(\mathbb{R}^3)} + \|\sqrt{\mathcal{L}_a}\, e\|_{N^0(I)} \leq \epsilon$$

for some $0 < \epsilon < \epsilon_1 = \epsilon_1(E, L)$. Then there exists a unique strong solution $u : I \times \mathbb{R}^3 \mapsto \mathbb{C}$ to (6.1) *with initial data u_0 at time $t = t_0$ satisfying*

$$\|\sqrt{\mathcal{L}_a}\,(u - \tilde{u})\|_{S^0(I)} \le C(E, L)\,\epsilon,$$
$$\|\sqrt{\mathcal{L}_a}\,u\|_{S^0(I)} \le C(E, L).$$

6.2 Embedding nonlinear profiles

In this section, we prove an analogue of Theorem 6.0.1 for initial data that lives far from the origin relative to its intrinsic length scale. In this setting, one may imagine that the potential plays little role. Indeed, we will use solutions to (6.9), whose existence is guaranteed by the result in [48], to approximate the solutions to (6.1) in this case.

To keep formulas within margins, in this section, we will adopt the following abbreviation:

$$\dot{X}^1(I) := L_t^{10} \dot{H}_x^{1,\frac{30}{13}}(I \times \mathbb{R}^3).$$

Theorem 6.2.1 (Embedding nonlinear profiles). *Let $a > -\frac{1}{4} + \frac{1}{25}$. Let $\{\lambda_n\} \subset 2^{\mathbb{Z}}$ and $\{x_n\} \subset \mathbb{R}^3$ be such that $|x_n|/\lambda_n \to \infty$. Let $\{t_n\} \subset \mathbb{R}$ be such that either $t_n \equiv 0$ or $t_n \to \pm\infty$. Let $\phi \in \dot{H}^1(\mathbb{R}^3)$ and define*

$$\phi_n(x) := \lambda_n^{-\frac{1}{2}} [e^{-it_n \mathcal{L}_a} \phi]\left(\frac{x - x_n}{\lambda_n}\right).$$

Then for n sufficiently large there exists a global solution v_n to (6.1) with initial data $v_n(0) = \phi_n$, which satisfies

$$\|v_n\|_{L_{t,x}^{10}(\mathbb{R}\times\mathbb{R}^3)} \lesssim 1,$$

with the implicit constant depending only on $\|\phi\|_{\dot{H}^1(\mathbb{R}^3)}$. Furthermore, for every $\epsilon > 0$, there exists $N_\epsilon \in \mathbb{N}$ and $\psi_\epsilon \in C_c^\infty(\mathbb{R} \times \mathbb{R}^3)$ such that for all $n \ge N_\epsilon$ we have

$$\|v_n(t - \lambda_n^2 t_n, x + x_n) - \lambda_n^{-\frac{1}{2}} \psi_\epsilon(\lambda_n^{-2} t, \lambda_n^{-1} x)\|_{\dot{X}^1(\mathbb{R})} < \epsilon. \tag{6.14}$$

Proof. The proof of this theorem follows the general outline of the proof of [150, Theorem 6.3]. We will divide to five steps. In the first step, we select appropriate global solutions to the energy-critical NLS without potential. In the second step, we construct a putative approximate solution to (6.1). In the third step, we prove that this asymptotically matches the initial data ϕ_n and in the fourth step we prove that it is indeed an approximate solution to (6.1). In the fifth and last step, we use Theorem 6.1.1 to find v_n and prove the approximation result (6.14).

Step 1: Selection of solutions to (6.9).

Let $\theta := \frac{1}{100}$. The construction of solutions to (6.9) depends on the behavior of t_n. If $t_n \equiv 0$, we let w_n and w_∞ be the solutions to (6.9) with initial data $w_n(0) = \phi_{\le (|x_n|/\lambda_n)^\theta}$ and $w_\infty(0) = \phi$. If $t_n \to \pm\infty$, we let w_n and w_∞ be the solutions to (6.9) satisfying

$$\|w_n(t) - e^{it\Delta}\phi_{\leq(|x_n|/\lambda_n)^\theta}\|_{\dot H^1(\mathbb{R}^3)} \to 0$$

and

$$\|w_\infty(t) - e^{it\Delta}\phi\|_{\dot H^1(\mathbb{R}^3)} \to 0 \quad \text{as} \quad t \to \infty.$$

In all cases, using [48, Theorem 1.1], perturbation theory and persistence of regularity for (6.9), we deduce that w_n and w_∞ are global solutions obeying

$$\begin{cases} \|w_n\|_{\dot S^1(\mathbb{R})} + \|w_\infty\|_{\dot S^1(\mathbb{R})} \lesssim 1, \\ \lim_{n\to\infty} \|w_n - w_\infty\|_{\dot S^1(\mathbb{R})} = 0, \\ \||\nabla|^s w_n\|_{\dot S^1(\mathbb{R})} \lesssim (\frac{|x_n|}{\lambda_n})^{s\theta} \quad \text{for all} \quad s \geq 0. \end{cases} \tag{6.15}$$

Step 2: Constructing the approximate solution to (6.1).
Fix $T > 0$ to be chosen later. We define

$$\tilde v_n(t,x) := \begin{cases} \lambda_n^{-\frac{1}{2}}[\chi_n w_n](\lambda_n^{-2}t, \lambda_n^{-1}(x - x_n)), & |t| \leq \lambda_n^2 T, \\ e^{-i(t-\lambda_n^2 T)\mathcal{L}_a}\tilde v_n(\lambda_n^2 T, x), & t > \lambda_n^2 T, \\ e^{-i(t+\lambda_n^2 T)\mathcal{L}_a}\tilde v_n(-\lambda_n^2 T, x), & t < -\lambda_n^2 T, \end{cases}$$

where χ_n is a family of smooth functions satisfying

$$|\partial^k \chi_n(x)| \lesssim \left(\frac{\lambda_n}{|x_n|}\right)^k \quad \text{and} \quad \chi_n(x) = \begin{cases} 0 & \text{if} \quad |x_n + \lambda_n x| \leq \frac{|x_n|}{4}, \\ 1 & \text{if} \quad |x_n + \lambda_n x| \geq \frac{|x_n|}{2}. \end{cases}$$

By changing variables, Hölder's inequality and (6.15), we get

$$\|\chi_n w_n(\pm T)\|_{\dot H_x^1} \lesssim \|\nabla\chi_n\|_{L_x^3}\|w_n(\pm T)\|_{L_x^6} + \|\chi_n\|_{L_x^\infty}\|\nabla w_n(\pm T)\|_{L_x^2} \lesssim 1.$$

Combining this with Strichartz estimates and Theorem 1.5.3, we obtain

$$\begin{aligned} \|\sqrt{\mathcal{L}_a}\tilde v_n\|_{L_t^{10}L_x^{\frac{30}{13}} \cap L_t^\infty L_x^2} &\lesssim \|\sqrt{\mathcal{L}_a^n}(\chi_n w_n)\|_{L_t^{10}L_x^{\frac{30}{13}} \cap L_t^\infty L_x^2} + \|\chi_n w_n(\pm T)\|_{\dot H_x^1} \\ &\lesssim \|\nabla(\chi_n w_n)\|_{L_t^{10}L_x^{\frac{30}{13}} \cap L_t^\infty L_x^2} + 1 \\ &\lesssim \|w_n\|_{\dot S^1} + 1 \lesssim 1. \end{aligned} \tag{6.16}$$

Step 3: Asymptotic agreement of the initial data:

$$\lim_{T\to\infty} \limsup_{n\to\infty} \|\sqrt{\mathcal{L}_a}[\tilde v_n(\lambda_n^2 t_n) - \phi_n]\|_{L_x^2} = 0. \tag{6.17}$$

We first consider the case when $t_n \equiv 0$. Using the equivalence of Sobolev spaces and performing a change of variables, we obtain

$$\begin{aligned}\|\sqrt{\mathcal{L}_a}[\tilde{v}_n(0)-\phi_n]\|_{L_x^2} &\lesssim \|\nabla[\chi_n\phi_{>(|x_n|/\lambda_n)^\theta}]\|_{L_x^2}+\|\nabla[(1-\chi_n)\phi]\|_{L_x^2}\\ &\lesssim \|\nabla\chi_n\|_{L_x^3}\|\phi_{>(|x_n|/\lambda_n)^\theta}\|_{L_x^6}+\|\nabla\phi_{>(|x_n|/\lambda_n)^\theta}\|_{L_x^2}\\ &\quad+\|\nabla\chi_n\|_{L_x^3}\|\phi\|_{L_x^6(\operatorname{supp}(\nabla\chi_n))}+\|\nabla\phi\|_{L_x^2(\operatorname{supp}(1-\chi_n))},\end{aligned}$$

which converges to zero as $n\to\infty$ since $\chi_n(x)\to 1$ and $\chi_{\operatorname{supp}(\nabla\chi_n)}(x)\to 0$ almost everywhere as $n\to\infty$.

It remains to prove (6.17) when $t_n\to\infty$; the case $t_n\to-\infty$ can be treated analogously. As T is fixed, for sufficiently large n we have $t_n>T$ and so

$$\tilde{v}_n(\lambda_n^2 t_n,x)=e^{-i(t_n-T)\lambda_n^2\mathcal{L}_a}\Big[\lambda_n^{-\frac12}(\chi_n w_n(T))\Big(\frac{x-x_n}{\lambda_n}\Big)\Big].$$

Thus, by a change of variables and the equivalence of Sobolev spaces,

$$\begin{aligned}\|\tilde{v}_n(\lambda_n^2 t_n)-\phi_n\|_{\dot{H}_a^1} &= \|\sqrt{\mathcal{L}_a^n}[e^{iT\mathcal{L}_a^n}(\chi_n w_n(T))-\phi]\|_{L_x^2}\\ &\lesssim \|\nabla[\chi_n(w_n(T)-w_\infty(T))]\|_{L_x^2} &(6.18)\\ &\quad+\|\nabla[\chi_n w_\infty(T)-w_\infty(T)]\|_{L_x^2} &(6.19)\\ &\quad+\|e^{iT\mathcal{L}_a^n}\sqrt{\mathcal{L}_a^n}\,w_\infty(T)-\sqrt{\mathcal{L}_a^n}\,\phi\|_{L_x^2}, &(6.20)\end{aligned}$$

where $\mathcal{L}_a^n$ is as in Definition 1.5.1 and $y_n=x_n/\lambda_n\to\infty$. Note that in this case $\mathcal{L}_a^\infty=-\Delta$.

Using (6.15) and the Sobolev embedding, we see that

$$(6.18)\lesssim\|\nabla\chi_n\|_{L_x^3}\|w_n(T)-w_\infty(T)\|_{L_x^6}+\|\chi_n\|_{L_x^\infty}\|\nabla[w_n(T)-w_\infty(T)]\|_{L_x^2}\to 0$$

as $n\to\infty$.

By Hölder and the dominated convergence theorem,

$$(6.19)\lesssim\|\nabla\chi_n\|_{L_x^3}\|w_\infty(T)\|_{L_x^6(\operatorname{supp}(\nabla\chi_n))}+\|(1-\chi_n)\nabla w_\infty(T)]\|_{L_x^2}\to 0$$

as $n\to\infty$.

Now we look to (6.20). By (1.122), we have

$$\|[\sqrt{\mathcal{L}_a^n}-\sqrt{\mathcal{L}_a^\infty}]w_\infty(T)\|_{L_x^2}+\|[\sqrt{\mathcal{L}_a^n}-\sqrt{\mathcal{L}_a^\infty}]\phi\|_{L_x^2}\to 0\quad\text{as}\quad n\to\infty,$$

while by (1.123) we have

$$\|[e^{iT\mathcal{L}_a^n}-e^{iT\mathcal{L}_a^\infty}]\sqrt{\mathcal{L}_a^\infty}\,w_\infty(T)\|_{L_x^2}\to 0\quad\text{as}\quad n\to\infty.$$

Combining these, we deduce that

$$\limsup_{n\to\infty}(6.20)=\|\sqrt{\mathcal{L}_a^\infty}[e^{iT\mathcal{L}_a^\infty}w_\infty(T)-\phi]\|_{L_x^2},$$

which converges to zero as $T \to \infty$ by the construction of w_∞. This completes the proof of (6.17).

Step 4: We prove that $\tilde{v}_n$ is an approximate solution to (6.1) in the sense required by Theorem 6.1.1:

$$\lim_{T\to\infty} \limsup_{n\to\infty} \| \sqrt{\mathcal{L}_a}\, [(i\partial_t - \mathcal{L}_a)\tilde{v}_n - |\tilde{v}_n|^4 \tilde{v}_n] \|_{N^0(\mathbb{R})} = 0. \tag{6.21}$$

We first verify (6.21) for $|t| > \lambda_n^2 T$. By symmetry, it suffices to consider positive times. For these times, we have

$$e_n := (i\partial_t - \mathcal{L}_a)\tilde{v}_n - |\tilde{v}_n|^4 \tilde{v}_n = -|\tilde{v}_n|^4 \tilde{v}_n.$$

By Lemma 1.5.1 and (6.16),

$$\begin{aligned} \| \sqrt{\mathcal{L}_a} e_n \|_{L_t^2 L_x^{\frac{6}{5}}(\{t>\lambda_n^2 T\}\times\mathbb{R}^3)} &\lesssim \| \sqrt{\mathcal{L}_a} \tilde{v}_n \|_{L_t^{10} L_x^{\frac{30}{13}}(\{t>\lambda_n^2 T\}\times\mathbb{R}^3)} \|\tilde{v}_n\|^4_{L^{10}_{t,x}(\{t>\lambda_n^2 T\}\times\mathbb{R}^3)} \\ &\lesssim \| e^{-it\mathcal{L}_a^n} (\chi_n w_n(T)) \|^4_{L^{10}_{t,x}((0,\infty)\times\mathbb{R}^3)}. \end{aligned}$$

On the other hand, by the analysis of (6.18) and (6.19), we have

$$\| e^{-it\mathcal{L}_a^n} (\chi_n w_n(T)) \|_{L^{10}_{t,x}((0,\infty)\times\mathbb{R}^3)} = \| e^{-it\mathcal{L}_a^n} w_\infty(T) \|_{L^{10}_{t,x}((0,\infty)\times\mathbb{R}^3)} + o(1)$$

as $n \to \infty$.

To continue, let w_+ denote the forward asymptotic state of w_∞; its existence is guaranteed by [48]. By the Strichartz inequality,

$$\begin{aligned} &\| e^{-it\mathcal{L}_a^n} w_\infty(T) \|_{L^{10}_{t,x}((0,\infty)\times\mathbb{R}^3)} \\ &\quad \lesssim \| [e^{-it\mathcal{L}_a^n} - e^{-it\mathcal{L}_a^\infty}] w_\infty(T) \|_{L^{10}_{t,x}((0,\infty)\times\mathbb{R}^3)} + \| w_\infty(T) - e^{iT\Delta} w_+] \|_{\dot{H}^1(\mathbb{R}^3)} \\ &\quad + \| e^{it\Delta} w_+ \|_{L^{10}_{t,x}((T,\infty)\times\mathbb{R}^3)}, \end{aligned}$$

which converges to zero as $n \to \infty$ and then $T \to \infty$ in view of Corollary 1.5.7, the definition of w_+ and the monotone convergence theorem.

Next, we show (6.21) on the middle time interval $|t| \le \lambda_n^2 T$. For these times,

$$\begin{aligned} e_n(t,x) &:= [(i\partial_t - \mathcal{L}_a)\tilde{v}_n - |\tilde{v}_n|^4 \tilde{v}_n](t,x) \\ &= \lambda_n^{-\frac{5}{2}} [(\chi_n - \chi_n^5)|w_n|^4 w_n](\lambda_n^{-2} t, \lambda_n^{-1}(x - x_n)) && (6.22) \\ &\quad + 2\lambda_n^{-\frac{5}{2}} [\nabla\chi_n \cdot \nabla w_n](\lambda_n^{-2} t, \lambda_n^{-1}(x - x_n)) && (6.23) \\ &\quad + \lambda_n^{-\frac{5}{2}} [\Delta\chi_n w_n](\lambda_n^{-2} t, \lambda_n^{-1}(x - x_n)) && (6.24) \\ &\quad - \lambda_n^{-\frac{1}{2}} \frac{a}{|x|^2} [\chi_n w_n](\lambda_n^{-2} t, \lambda_n^{-1}(x - x_n)). && (6.25) \end{aligned}$$

Performing a change of variables and using Lemma 1.5.1 and Hölder, we estimate the contribution of (6.22) by

$$\begin{aligned}&\|(\chi_n - \chi_n^5)|w_n|^4\nabla w_n\|_{L_t^2 L_x^{\frac{6}{5}}} + \|\nabla\chi_n(1-5\chi_n^4)w_n^5\|_{L_t^2 L_x^{\frac{6}{5}}}\\ &\lesssim [\|\nabla w_n\|_{L_t^{10}L_x^{\frac{30}{13}}} + \|w_n\|_{L_{t,x}^{10}}\|\nabla\chi_n\|_{L_x^3}][\|w_n - w_\infty\|_{L_{t,x}^{10}}^4 + \|1_{|x|\sim\frac{|x_n|}{\lambda_n}} w_\infty\|_{L_{t,x}^{10}}^4],\end{aligned}$$

which converges to zero as $n \to \infty$ by (6.15) and the dominated convergence theorem. Similarly, we estimate the contribution of (6.23) and (6.24) by

$$\begin{aligned}&T[\|\nabla\chi_n\|_{L_x^\infty}\|\Delta w_n\|_{L_t^\infty L_x^2} + \|\Delta\chi_n\|_{L_x^\infty}\|\nabla w_n\|_{L_t^\infty L_x^2} + \|\nabla\Delta\chi_n\|_{L_x^3}\|w_n\|_{L_t^\infty L_x^6}]\\ &\lesssim T\Big[\Big(\frac{\lambda_n}{|x_n|}\Big)^{1-\theta} + \Big(\frac{\lambda_n}{|x_n|}\Big)^2 + \Big(\frac{\lambda_n}{|x_n|}\Big)^3\Big] \to 0 \quad \text{as} \quad n \to \infty.\end{aligned}$$

Lastly, we estimate the contribution of (6.25) by

$$\begin{aligned}&T\Big\|\nabla\Big(\frac{\lambda_n^2\chi_n(x)}{|x_n + \lambda_n x|^2}w_n\Big)\Big\|_{L_t^\infty L_x^2}\\ &\lesssim T\Big[\Big\|\nabla\frac{\lambda_n^2\chi_n(x)}{|x_n + \lambda_n x|^2}\Big\|_{L_x^3}\|w_n\|_{L_t^\infty L_x^6} + \Big\|\frac{\lambda_n^2\chi_n(x)}{|x_n + \lambda_n x|^2}\Big\|_{L_x^\infty}\|\nabla w_n\|_{L_t^\infty L_x^2}\Big]\\ &\lesssim T\Big(\frac{\lambda_n}{|x_n|}\Big)^2 \to 0 \quad \text{as} \quad n \to \infty.\end{aligned}$$

This completes the proof of (6.21).

Step 5: Constructing v_n and approximation by C_c^∞ functions.

Using (6.16), (6.17), (6.21) and Theorem 6.1.1, for n sufficiently large we obtain a global solution v_n to (6.1) with initial data $v_n(0) = \phi_n$, which satisfies

$$\|v_n\|_{L_{t,x}^{10}(\mathbb{R}\times\mathbb{R}^3)} \lesssim 1 \quad \text{and} \quad \lim_{T\to\infty}\limsup_{n\to\infty}\|v_n(t - \lambda_n^2 t_n) - \tilde{v}_n(t)\|_{\dot{S}^1(\mathbb{R}\times\mathbb{R}^3)} = 0.$$

It remains to prove the approximation result (6.14).

From the density of $C_c^\infty(\mathbb{R}\times\mathbb{R}^3)$ in $\dot{X}^1(\mathbb{R})$, for any $\epsilon > 0$ there exists $\psi_\epsilon \in C_c^\infty(\mathbb{R}\times\mathbb{R}^3)$ such that

$$\|w_\infty - \psi_\epsilon\|_{\dot{X}^1(\mathbb{R})} < \frac{\epsilon}{3}.$$

Thus, proving (6.14) reduces to showing

$$\|\tilde{v}_n(t,x) - \lambda_n^{-\frac{1}{2}}w_\infty(\lambda_n^{-2}t, \lambda_n^{-1}(x - x_n))\|_{\dot{X}^1(\mathbb{R})} < \frac{\epsilon}{3} \tag{6.26}$$

for n, T sufficiently large. Changing variables, we estimate

$$\text{LHS (6.26)} \le \|\chi_n w_n - w_\infty\|_{\dot X^1([-T,T])} + \|e^{-i(t-T)\mathcal{L}_a^n}[\chi_n w_n(T)] - w_\infty\|_{\dot X^1((T,\infty))}$$
$$+ \|e^{-i(t+T)\mathcal{L}_a^n}[\chi_n w_n(-T)] - w_\infty\|_{\dot X^1((-\infty,-T))}.$$

We consider each of these three terms separately. Using the dominated convergence theorem and (6.15), we see that

$$\|\chi_n w_n - w_\infty\|_{\dot X^1([-T,T])} \lesssim \|(1-\chi_n)w_\infty\|_{\dot X^1(\mathbb{R})} + \|w_n - w_\infty\|_{\dot X^1(\mathbb{R})} \to 0 \tag{6.27}$$

as $n \to \infty$. The second and third term can be treated similarly; we only present the details for the second term. By the Strichartz estimate, we have

$$\|e^{-i(t-T)\mathcal{L}_a^n}[\chi_n w_n(T)] - w_\infty\|_{\dot X^1((T,\infty))}$$
$$\lesssim \|w_\infty\|_{\dot X^1((T,\infty))} + \|\sqrt{\mathcal{L}_a^n}[\chi_n w_n(T) - w_\infty(T)]\|_{L_x^2} + \|[\sqrt{\mathcal{L}_a^n} - \sqrt{\mathcal{L}_a^\infty}]w_\infty(T)\|_{L_x^2}$$
$$+ \|[e^{-i(t-T)\mathcal{L}_a^n} - e^{i(t-T)\Delta}]\sqrt{\mathcal{L}_a^\infty}\, w_\infty(T)\|_{L_t^{10}L_x^{\frac{30}{13}}} + \|e^{it\Delta}w_+\|_{\dot X^1((T,\infty))}$$
$$+ \|e^{-iT\Delta}w_\infty(T) - w_+\|_{\dot H_x^1} \to 0, \quad \text{as} \quad n \to \infty \quad \text{and then} \quad T \to \infty$$

by the monotone convergence theorem, (6.27), Lemma 1.5.4 and the definition of the asymptotic state w_+. This completes the proof of (6.26) and with it, the proof of Theorem 6.2.1. □

6.3 Existence of the minimal blowup solution

In this section, we aim to show that the failure of Theorem 6.0.1 implies the existence of a minimal counterexample that has good compactness properties.

Theorem 6.3.1 (Existence of almost periodic solutions). *Suppose Theorem* 6.0.1 *fails. Then there exist a critical energy* $0 < E_c < \infty$ *and a solution* $u : [0, T^*) \times \mathbb{R}^3 \to \mathbb{C}$ *to* (6.1) *with*

$$E(u) = E_c \quad \text{and} \quad \|u\|_{L_{t,x}^{10}([0,T^*)\times\mathbb{R}^3)} = \infty.$$

Moreover, the orbit of u *is precompact in* $\dot H^1(\mathbb{R}^3)$ *modulo scaling in the sense that there exists* $N(t) : I \to \mathbb{R}^+$ *such that the set* $\{N(t)^{-\frac{1}{2}}u(t, \frac{x}{N(t)}) : t \in [0, T^*)\}$ *is precompact in* $\dot H^1(\mathbb{R}^3)$. *Finally, we may assume that the frequency scale function* $N(t)$ *satisfies* $\inf_{t\in[0,T^*)} N(t) \ge 1$.

Define

$$L(E) := \sup\{\|u\|_{L_{t,x}^{10}(I\times\mathbb{R}^3)} : E(u_0) < E\}.$$

From the small data theory, we know that $L(E) < +\infty$ for all E sufficiently small. Suppose now that Theorem 6.0.1 fails. We see that there must exist a "critical" $E_c \in (0, \infty)$ such that

$$L(E) < +\infty \quad \text{for} \quad E < E_c \quad \text{and} \quad L(E) = \infty \quad \text{for} \quad E > E_c.$$

The key ingredient in the proof of Theorem 6.3.1 is a Palais–Smale condition for minimizing sequences of blowup solutions to (6.1) as follows.

Proposition 6.3.1 (Palais–Smale condition). *Let $u_n : I_n \times \mathbb{R}^3 \to \mathbb{C}$ be a sequence of solutions with $E(u_n) \to E_c$ and let $t_n \in I_n$ so that*

$$\lim_{n\to\infty} \|u_n\|_{L^{10}_{t,x}(\{t\geq t_n\}\times\mathbb{R}^3)} = \lim_{n\to\infty} \|u_n\|_{L^{10}_{t,x}(\{t\leq t_n\}\times\mathbb{R}^3)} = \infty.$$

Then, passing to a subsequence, $\{u(t_n)\}$ converges in $\dot{H}^1(\mathbb{R}^3)$ modulo scaling.

Proof. The proof of Proposition 6.3.1 follows along well-established lines. For a thorough discussion of this result and its relevance in proving Theorem 6.3.1 in the absence of a potential, see [148]. The presence of the potential introduces several new difficulties. These are of a similar nature as those arising for the energy-critical NLS outside a convex obstacle and will be overcome by mimicking the arguments in [150]. In what follows, we will sketch the proof emphasizing the main steps.

By time translation symmetry, we may assume $t_n \equiv 0$; thus,

$$\lim_{n\to\infty} \|u_n\|_{L^{10}_{t,x}(\{t\geq 0\}\times\mathbb{R}^3)} = \lim_{n\to\infty} \|u_n\|_{L^{10}_{t,x}(\{t\leq 0\}\times\mathbb{R}^3)} = \infty. \tag{6.28}$$

Applying Proposition 1.5.7 to the sequence $u_n(0)$ (which is bounded in $\dot{H}^1_a(\mathbb{R}^3)$) and passing to a subsequence if necessary, we have the linear profile decomposition

$$u_n(0) = \sum_{j=1}^{J} \phi_n^j + w_n^J \tag{6.29}$$

satisfying the properties stated in Proposition 1.5.7. In particular, for any finite $0 \leq J \leq J^*$, we have the energy decoupling property

$$\lim_{n\to\infty} \left\{ E(u_n) - \sum_{j=1}^{J} E(\phi_n^j) - E(w_n^J) \right\} = 0. \tag{6.30}$$

To prove Proposition 6.3.1, we need to show that $J^* = 1$, $w_n^1 \to 0$ in $\dot{H}^1(\mathbb{R}^3)$, $t_n^1 \equiv 0$ and $x_n \equiv 0$. To this end, we will show that all other possibilities contradict (6.28). We discuss two cases.

Case I: $\sup_j \limsup_{n\to\infty} E(\phi_n^j) = E_c$.

From the Hardy inequality and the nontriviality of the profiles, we deduce that $\liminf_{n\to\infty} E(\phi_n^j) > 0$ for each finite $1 \leq j \leq J^*$. Thus, (6.30) implies that there is a single profile in the decomposition (6.29) (i. e., $J^* = 1$) and we can write

$$u_n(0) = \phi_n + w_n \quad \text{with} \quad \lim_{n\to\infty} \|w_n\|_{\dot{H}^1(\mathbb{R}^3)} = 0. \tag{6.31}$$

If $\frac{|x_n|}{\lambda_n} \to \infty$, then for n sufficiently large Theorem 6.2.1 guarantees the existence of a global solution v_n to (6.1) with initial data $v_n(0) = \phi_n$ and uniform bounded $L^{10}_{t,x}$-norm. By Theorem 6.1.1, this space-time bound extends to the solution u_n for n sufficiently large, thus contradicting (6.28). Therefore, we must have $x_n \equiv 0$.

To obtain the desired compactness property, it remains to preclude the case $t_n \to \pm\infty$. By symmetry, it suffices to consider $t_n \to \infty$. By the Strichartz inequality, monotone convergence and Corollary 1.5.7, we see that

$$\begin{aligned}
&\|e^{-it\mathcal{L}_a} u_n(0)\|_{L^{10}_{t,x}(\{t>0\}\times\mathbb{R}^3)} \\
&\quad \le \|e^{-it\mathcal{L}_a} w_n\|_{L^{10}_{t,x}(\{t>0\}\times\mathbb{R}^3)} + \|e^{-it\mathcal{L}_a^\infty} \phi\|_{L^{10}_{t,x}(\{t>t_n\}\times\mathbb{R}^3)} \\
&\quad\quad + \|[e^{-it\mathcal{L}_a^n} - e^{-it\mathcal{L}_a^\infty}]\phi\|_{L^{10}_{t,x}(\mathbb{R}\times\mathbb{R}^3)} \to 0 \quad \text{as} \quad n \to \infty.
\end{aligned}$$

By the small data theory, this implies that $\|u_n\|_{L^{10}_{t,x}(\{t>0\}\times\mathbb{R}^3)} \to 0$ as $n \to \infty$, which again contradicts (6.28).

Case II: $\sup_j \limsup_{n\to\infty} E(\phi_n^j) \le E_c - 3\delta$ for some $\delta > 0$.

In this case, for each finite $J \le J^*$, we have $E(\phi_n^j) \le E_c - 2\delta$ for all $1 \le j \le J$ and n sufficiently large.

We will prove that Case II is inconsistent with (6.28). To this end, we first introduce nonlinear profiles associated with each ϕ_n^j.

For those j for which $\frac{|x_n^j|}{\lambda_n^j} \to +\infty$, let v_n^j denote the global solution to (6.1) guaranteed by Theorem 6.2.1.

We next consider those j for which $x_n^j \equiv 0$. If $t_n^j \equiv 0$, let v^j denote the maximal-lifespan solution to (6.1) with initial data $v^j(0) = \phi^j$. If instead $t_n^j \to \pm\infty$, let v^j denote the maximal-lifespan solution to (6.1), which scatters to $e^{-it\mathcal{L}_a}\phi^j$ as $t \to \pm\infty$. In both cases, we define

$$v_n^j(t,x) := (\lambda_n^j)^{-\frac{1}{2}} v^j\left(\frac{t}{(\lambda_n^j)^2} + t_n^j, \frac{x}{\lambda_n^j}\right).$$

Note that v_n^j is also a solution to (6.1) with

$$\lim_{n\to\infty} \|v_n^j(0) - \phi_n^j\|_{\dot{H}^1(\mathbb{R}^3)} = 0.$$

In particular, by the Hardy inequality we have that $E(v_n^j) \le E_c - \delta$ for all $1 \le j \le J$ and n sufficiently large. By the definition of the critical energy E_c, this implies that the solutions v_n^j are global in time and have finite Strichartz norms.

The next step in the argument is to construct an approximate solution to (6.1) with finite $L^{10}_{t,x}$-norm. We define

$$u_n^J := \sum_{j=1}^{J} v_n^j + e^{-it\mathcal{L}_a} w_n^J.$$

By construction, u_n^J is defined globally in time and satisfies

$$\|u_n^J(0) - u_n(0)\|_{\dot{H}^1(\mathbb{R}^3)} \to 0 \quad \text{as} \quad n \to \infty$$

for any J.

The asymptotic decoupling of parameters (1.118) begets (as in [150, Lemma 7.3]) asymptotic decoupling of the nonlinear solutions v_n^j. Using this together with the control, we have over the $L_t^{10} \dot{H}_x^{1,\frac{30}{13}}$-norm of each v_n^j, one can show that

$$\limsup_{n\to\infty} \|u_n^J\|_{L_t^{10} \dot{H}_x^{1,\frac{30}{13}}(\mathbb{R}\times\mathbb{R}^3)} \lesssim_{E_c,\delta} 1,$$

uniformly in J.

Lastly, one can prove that for large enough n and J, u_n^J is an approximate solution to (6.1) in the sense that

$$\lim_{J\to\infty} \limsup_{n\to\infty} \| \sqrt{\mathcal{L}_a}[(i\partial_t - \mathcal{L}_a)u_n^J - |u_n^J|^4 u_n^J]\|_{N^0(\mathbb{R})} = 0.$$

In order to verify this statement, one uses the asymptotic decoupling of the solutions v_n^j together with the fact that, by the linear profile decomposition, $e^{-it\mathcal{L}_a} w_n^J$ converges to zero (as n and J converge to infinity) in energy-critical Strichartz spaces without derivatives. Corollary 1.5.5 is used to control terms involving one derivative of $e^{-it\mathcal{L}_a} w_n^J$; see, for example, the justification of (7.19) in [150].

Putting everything together and invoking Theorem 6.1.1, we deduce that the solutions u_n inherit the space-time bounds of u_n^J for n sufficiently large. This however contradicts (6.28). □

6.4 Precluding the minimal blowup solution

In this section, we aim to preclude the minimal blowup solutions as in Theorem 6.3.1. Assume that Theorem 6.0.1 were to fail, and let $u : [0, T^*) \times \mathbb{R}^3 \to \mathbb{C}$ be a minimal counterexample of the type provided by Theorem 6.3.1. The fact that the orbit of u is precompact in $\dot{H}_x^1$ modulo scaling combined with the lower bound $N(t) \geq 1$ on the frequency scale function yields that for any $\eta > 0$ there exists $C(\eta) > 0$ such that

$$\int_{|x|\geq C(\eta)} |\nabla u(t,x)|^2 + |u(t,x)|^6 \, dx \leq \int_{|x|\geq \frac{C(\eta)}{N(t)}} |\nabla u(t,x)|^2 + |u(t,x)|^6 \, dx \leq \eta, \tag{6.32}$$

uniformly for $t \in [0, T^*)$.

Note also that by the Hardy inequality and the conservation of energy, for all solutions u to (6.1), we have

$$\|\nabla u(t)\|_{L_x^2}^2 \sim E(u) \quad \text{uniformly for} \quad t \in [0, T^*). \tag{6.33}$$

To preclude the existence of the minimal counterexample u guaranteed by Theorem 6.3.1, we will distinguish two cases, namely (1) the solution u is global forward in time (i. e., $T^* = \infty$) or (2) the solution u blows up in finite time (i. e., $T^* < \infty$). We will show that the first scenario is inconsistent with the virial identity; see Theorem 6.4.1. Finally, we will employ a transport of mass argument to preclude the second scenario; see Theorem 6.4.2. This will complete the proof of Theorem 6.0.1.

Theorem 6.4.1. *There are no solutions to* (6.1) *of the form given in Theorem* 6.3.1 *with* $T^* = \infty$.

Proof. Assume toward a contradiction that $u : [0, \infty) \times \mathbb{R}^3 \to \mathbb{C}$ is such a solution. Let ϕ be a smooth radial cutoff function such that

$$\phi(r) = \begin{cases} r & \text{for} \quad r \leq 1 \\ 0 & \text{for} \quad r \geq 2, \end{cases}$$

and define

$$V_R(t) := \int_{\mathbb{R}^3} \psi(x) |u(t,x)|^2 \, dx \quad \text{where} \quad \psi(x) := R^2 \phi\Big(\frac{|x|^2}{R^2} \Big)$$

for some $R > 0$.

Differentiating V_R with respect to the time variable and using Hölder, the Sobolev embedding and (6.33), we find

$$\begin{aligned} |\partial_t V_R(t)| &= \Big| 4 \operatorname{Im} \int_{\mathbb{R}^3} \phi'\Big(\frac{|x|^2}{R^2} \Big) \overline{u(t,x)} \, x \cdot \nabla u(t,x) \, dx \Big| \\ &\lesssim R^2 \|u(t)\|_{L^6} \|\nabla u(t)\|_{L^2} \lesssim_u R^2, \end{aligned} \tag{6.34}$$

uniformly for $t \geq 0$.

Taking another derivative with respect to the time variable and using the Hardy inequality, the Sobolev embedding and (6.32), we obtain

$$\begin{aligned} \partial_{tt} V_R(t) = {} & 4 \operatorname{Re} \int_{\mathbb{R}^3} \psi_{ij}(x) u_i(t,x) \bar{u}_j(t,x) \, dx - \frac{4}{3} \int_{\mathbb{R}^3} (\Delta \psi)(x) |u(t,x)|^6 \, dx \\ & - \int_{\mathbb{R}^3} (\Delta\Delta\psi)(x) |u(t,x)|^2 \, dx - 4a \int_{\mathbb{R}^3} \frac{x}{|x|^4} \nabla \psi(x) |u(t,x)|^2 \, dx \end{aligned}$$

$$= 8 \int_{\mathbb{R}^3} |\nabla u(t,x)|^2 + \frac{a}{|x|^2}|u(t,x)|^2 + |u(t,x)|^6 \, dx$$
$$+ O\Big(\int_{|x|\geq R} |\nabla u(t,x)|^2 + |u(t,x)|^6 \, dx \Big)$$
$$\gtrsim \|\nabla u(t)\|_{L_x^2}^2 - \eta,$$

provided η is small and $R = R(\eta)$ is chosen sufficiently large. Combining this with (6.33) and taking η sufficiently small depending on the energy of u, we get

$$\partial_{tt} V_R(t) \gtrsim_u 1 \quad \text{uniformly for} \quad t \geq 0.$$

Together with (6.34) and the fundamental theorem of calculus on $[0, T]$ for T sufficiently large, this yields the desired contradiction. □

Theorem 6.4.2. *There are no solutions to* (6.1) *of the form given in Theorem* 6.3.1 *with* $T^* < \infty$.

Proof. Assume toward a contradiction that $u : [0, T^*) \times \mathbb{R}^3 \to \mathbb{C}$ is such a solution. By Corollary 5.19 in [148], this implies

$$\liminf_{t \nearrow T^*} N(t) = \infty. \tag{6.35}$$

Consequently, the mass of u leaves any fixed ball as $t \to T^*$; specifically,

$$\limsup_{t \nearrow T^*} \int_{|x|\leq R} |u(t,x)|^2 \, dx = 0 \quad \text{for any} \quad R > 0. \tag{6.36}$$

Indeed, for $0 < \eta < 1$ and $t \in [0, T^*)$, using Hölder, we estimate

$$\int_{|x|\leq R} |u(t,x)|^2 \, dx \leq \int_{|x|\leq \eta R} |u(t,x)|^2 \, dx + \int_{\eta R \leq |x| \leq R} |u(t,x)|^2 \, dx$$
$$\lesssim \eta^2 R^2 \|u(t)\|_{L_x^6}^2 + R^2 \Big(\int_{|x|\geq \eta R} |u(t,x)|^6 \, dx \Big)^{1/3}.$$

By (6.33) and the Sobolev embedding, we can make the first term arbitrarily small by letting $\eta \to 0$. Moreover, by (6.32) and (6.35), we see that the second term converges to zero as $t \to T^*$.

To continue, for $t \in [0, T^*)$ we define

$$M_R(t) := \int_{\mathbb{R}^3} \phi\Big(\frac{|x|}{R}\Big) |u(t,x)|^2 \, dx, \tag{6.37}$$

where ϕ is a smooth radial function such that $\phi(r) = 1$ for $r \leq 1$ and $\phi(r) = 0$ for $r \geq 2$. By (6.36),

$$\limsup_{t \nearrow T^*} M_R(t) = 0 \quad \text{for all} \quad R > 0. \tag{6.38}$$

On the other hand, by the Hardy inequality and (6.33),

$$|\partial_t M_R(t)| \lesssim \|\nabla u(t)\|_{L_x^2} \left\| \frac{u(t)}{|x|} \right\|_{L_x^2} \lesssim_u 1.$$

Thus, by the fundamental theorem of calculus,

$$M_R(t_1) = M_R(t_2) + \int_{t_2}^{t_1} \partial_t M_R(\tau)\, d\tau \lesssim_u M_R(t_2) + |t_1 - t_2|$$

for all $t_1, t_2 \in [0, T^*)$ and $R > 0$. Letting $t_2 \nearrow T^*$, we deduce that

$$M_R(t_1) \lesssim_u |T^* - t_1|.$$

Now letting $R \to \infty$ and invoking the conservation of mass, we derive

$$\|u(0)\|_{L_x^2}^2 \lesssim_u |T^* - t_1|.$$

Finally, letting $t_1 \nearrow T^*$ we obtain $u(0) \equiv 0$, which contradicts the fact that the $L_{t,x}^{10}$-norm of u is infinite. □

6.5 Focusing case

In this section, we consider the analogue of (6.1) with focusing nonlinearity, posed in $\mathbb{R}^n$ with $n \geq 3$:

$$\begin{cases} (i\partial_t - \mathcal{L}_a)u = -|u|^{\frac{4}{n-2}}u, & (t,x) \in \mathbb{R} \times \mathbb{R}^n, \\ u(t=0) \in \dot{H}_a^1(\mathbb{R}^n), \end{cases} \tag{6.39}$$

which conserves the energy

$$E_a(f) := \int_{\mathbb{R}^n} \left\{ \frac{1}{2}|\nabla f(x)|^2 + \frac{a}{2|x|^2}|f(x)|^2 - \frac{n-2}{2n}|f(x)|^{\frac{2n}{n-2}} \right\} dx. \tag{6.40}$$

Definition 6.5.1. Given $a > -(\frac{n-2}{2})^2$, we define $\beta > 0$ via $a = (\frac{n-2}{2})^2[\beta^2 - 1]$, or equivalently, $\sigma = \frac{n-2}{2}(1 - \beta)$. We then define the ground state soliton by

$$W_a(x) := \left[n(n-2)\beta^2\right]^{\frac{n-2}{4}} \left[\frac{|x|^{\beta-1}}{1+|x|^{2\beta}}\right]^{\frac{n-2}{2}}. \tag{6.41}$$

One can verify that

$$\mathcal{L}_a W_a = |W_a|^{\frac{4}{n-2}} W_a$$

and, using a standard variant of Euler's Beta integral (cf. (1.1.20) in [4]), that

$$\|W_a\|^2_{\dot{H}^1_a(\mathbb{R}^n)} = \int_{\mathbb{R}^n} |W_a(x)|^{\frac{2n}{n-2}}\, dx = \frac{\pi n(n-2)}{4}\left[\frac{2\sqrt{\pi}\beta^{n-1}}{\Gamma(\frac{n+1}{2})}\right]^{\frac{2}{n}}. \tag{6.42}$$

Thus, W_a is a ground state soliton in the sense of being a radial nonnegative static solution to (6.39). As we will see in Proposition 6.5.1 below, these solitons occur as optimizers in Sobolev embedding inequalities when $a \le 0$, but not when $a > 0$. The proof of that proposition will also explain how we derived formula (6.41).

Proposition 6.5.1 (Sharp Sobolev embedding, [141]). *Fix $n \ge 3$ and $a > -(\frac{n-2}{2})^2$.*
(i) *If $-(\frac{n-2}{2})^2 < a < 0$, then*

$$\|f\|_{L_x^{\frac{2n}{n-2}}(\mathbb{R}^n)} \le \|W_a\|_{L_x^{\frac{2n}{n-2}}(\mathbb{R}^n)} \|W_a\|^{-1}_{\dot{H}^1_a(\mathbb{R}^n)} \|f\|_{\dot{H}^1_a(\mathbb{R}^n)}. \tag{6.43}$$

Moreover, the equality holds in (6.43) if and only if $f(x) = \alpha W_a(\lambda x)$ for some $\alpha \in \mathbb{C}$ and some $\lambda > 0$.
(ii) *The inequality (6.43) is valid also when $a = 0$; however, the equality now holds if and only if $f(x) = \alpha W_0(\lambda x + y)$ for some $\alpha \in \mathbb{C}$, some $y \in \mathbb{R}^n$ and some $\lambda > 0$.*
(iii) *If $a > 0$, then*

$$\|f\|_{L_x^{\frac{2n}{n-2}}(\mathbb{R}^n)} \le \|W_0\|_{L_x^{\frac{2n}{n-2}}(\mathbb{R}^n)} \|W_0\|^{-1}_{\dot{H}^1(\mathbb{R}^n)} \|f\|_{\dot{H}^1_a(\mathbb{R}^n)}. \tag{6.44}$$

In this case, the equality never holds (for $f \not\equiv 0$); however, the constant in (6.44) cannot be improved.

Corollary 6.5.1 (Sharp Sobolev embedding). *Fix $n \ge 3$ and $a > -(\frac{n-2}{2})^2$. Then*

$$\frac{2E_a(f)}{\|W_{a\wedge 0}\|^2_{\dot{H}^1_{a\wedge 0}(\mathbb{R}^n)}} \ge \frac{\|f\|^2_{\dot{H}^1_a(\mathbb{R}^n)}}{\|W_{a\wedge 0}\|^2_{\dot{H}^1_{a\wedge 0}(\mathbb{R}^n)}} - \frac{n-2}{n}\left[\frac{\|f\|^2_{\dot{H}^1_a(\mathbb{R}^n)}}{\|W_{a\wedge 0}\|^2_{\dot{H}^1_{a\wedge 0}(\mathbb{R}^n)}}\right]^{\frac{n}{n-2}}.$$

This formulation of sharp Sobolev embedding is well suited to proving energy trapping and for demonstrating coercivity of the virial.

Corollary 6.5.2 (Coercivity). *Fix $n \ge 3$ and $a > -(\frac{n-2}{2})^2$. Let $u : I \times \mathbb{R}^n \to \mathbb{C}$ be a solution to (6.39) with initial data $u(t_0) = u_0 \in \dot{H}^1_a(\mathbb{R}^n)$ for some $t_0 \in I$. Assume $E_a(u_0) \le (1 -$*

$\delta_0)E_{a\wedge0}(W_{a\wedge0})$ for some $\delta_0 > 0$. Then there exist positive constants δ_1 and c (depending on d, a and δ_0) such that:

(a) *If $\|u_0\|_{\dot H^1_a(\mathbb{R}^n)} \le \|W_{a\wedge0}\|_{\dot H^1_{a\wedge0}(\mathbb{R}^n)}$, then for all $t \in I$,*

$$\text{(1)}\quad \|u(t)\|_{\dot H^1_a(\mathbb{R}^n)} \le (1-\delta_1)\|W_{a\wedge0}\|_{\dot H^1_{a\wedge0}(\mathbb{R}^n)};$$

$$\text{(2)}\quad \int_{\mathbb{R}^n} |\nabla u(t,x)|^2 + \frac{a}{|x|^2}|u(t,x)|^2 - |u(t,x)|^{\frac{2n}{n-2}}\,dx \ge c\|u(t)\|^2_{\dot H^1_a};$$

$$\text{(3)}\quad c\|u(t)\|^2_{\dot H^1_a} \le 2E_a(u) \le \|u(t)\|^2_{\dot H^1_a}.$$

(b) *If $\|u_0\|_{\dot H^1_a(\mathbb{R}^n)} \ge \|W_{a\wedge0}\|_{\dot H^1_{a\wedge0}(\mathbb{R}^n)}$, then for all $t \in I$,*

$$\text{(1)}\quad \|u(t)\|_{\dot H^1_a(\mathbb{R}^n)} \ge (1+\delta_1)\|W_{a\wedge0}\|_{\dot H^1_{a\wedge0}(\mathbb{R}^n)};$$

$$\text{(2)}\quad \int_{\mathbb{R}^n} |\nabla u(t,x)|^2 + \frac{a}{|x|^2}|u(t,x)|^2 - |u(t,x)|^{\frac{2n}{n-2}}\,dx \le -c < 0.$$

Proof. Combining Corollary 6.5.1 with the assumption $E_a(u_0) \le (1-\delta_0)E_{a\wedge0}(W_{a\wedge0})$ and (6.42), we obtain

$$\frac{2}{n}(1-\delta_0) \ge \frac{\|u(t)\|^2_{\dot H^1_a}}{\|W_{a\wedge0}\|^2_{\dot H^1_{a\wedge0}}} - \frac{n-2}{n}\left[\frac{\|u(t)\|^2_{\dot H^1_a}}{\|W_{a\wedge0}\|^2_{\dot H^1_{a\wedge0}}}\right]^{\frac{n}{n-2}}.$$

Claims (a)(1) and (b)(1) now follow by using a simple continuity argument together with the conservation of energy and the elementary inequality,

$$\frac{2}{n}(1-\delta_0) \ge y - \frac{n-2}{n}y^{\frac{n}{n-2}} \quad\Longrightarrow\quad |y-1| \ge \delta_1$$

for some $\delta_1 = \delta_1(n,\delta_0)$.

To verify items (a)(2) and (b)(2), we first write

$$\int_{\mathbb{R}^n}\left(|\nabla u(t,x)|^2 + \frac{a}{|x|^2}|u(t,x)|^2 - |u(t,x)|^{\frac{2n}{n-2}}\right)dx = \frac{2n}{n-2}E_a(u) - \frac{2}{n-2}\|u\|^2_{\dot H^1_a}. \tag{6.45}$$

We will also need that by (6.42),

$$\frac{2n}{n-2}E_{a\wedge0}(W_{a\wedge0}) = \frac{2}{n-2}\|W_{a\wedge0}\|^2_{\dot H^1_{a\wedge0}}.$$

In the setting of (b)(2), these two ingredients yield

$$\begin{aligned}\text{LHS (6.45)} &\le \frac{2n}{n-2}E_{a\wedge0}(W_{a\wedge0}) - \frac{2}{n-2}(1+\delta_1)^2\|W_{a\wedge0}\|^2_{\dot H^1_{a\wedge0}}\\ &\le -\frac{4\delta_1}{n-2}\|W_{a\wedge0}\|^2_{\dot H^1_{a\wedge0}} < 0,\end{aligned}$$

which resolves this case.

In the setting of (a)(2), rearranging Corollary 6.5.1 yields

$$\|u(t)\|^2_{\dot H^1_a(\mathbb{R}^n)} - 2E_a(u) \le \frac{n-2}{n}(1-\delta_1)^{\frac{4}{n-2}}\|u(t)\|^2_{\dot H^1_a(\mathbb{R}^n)}.$$

In this way, we deduce that

$$\text{LHS of (6.45)} = \|u(t)\|^2_{\dot H^1_a} - \frac{n}{n-2}\big[\|u(t)\|^2_{\dot H^1_a} - 2E_a(u)\big] \ge \big[1-(1-\delta_1)^{\frac{4}{n-2}}\big]\|u(t)\|^2_{\dot H^1_a}.$$

This leaves us with the task of verifying claim (a)(3), which is straightforward. As the nonlinearity is focusing, $2E_a(u) \le \|u\|^2_{\dot H^1_a}$. The other inequality follows from (a)(2) since LHS (6.45) $\le 2E_a(u)$. This completes the proof of the lemma. □

Using the usual virial argument and Corollary 6.5.2(b), one can prove that finite-time blowup occurs for (6.39) for $a > -(\frac{n-2}{2})^2 + (\frac{n-2}{d+2})^2$; see Proposition 6.5.2 below. The restriction on a stems from the local well-posedness theory; it is needed to ensure that solutions can be constructed, at least locally in time. The proof of Proposition 6.5.2 follows from a straightforward adaptation of the arguments in, say, [145, Section 9], relying on Corollary 6.5.2(b) for the requisite concavity.

Proposition 6.5.2 (Blowup). *Fix $n \ge 3$ and $a > -(\frac{n-2}{2})^2 + (\frac{n-2}{n+2})^2$. Let $u_0 \in \dot H^1_x(\mathbb{R}^n)$ be such that $E_a(u_0) < E_{a\wedge 0}(W_{a\wedge 0})$ and $\|u_0\|_{\dot H^1_a(\mathbb{R}^n)} \ge \|W_{a\wedge 0}\|_{\dot H^1_{a\wedge 0}(\mathbb{R}^n)}$. Assume also that either $xu_0 \in L^2_x(\mathbb{R}^n)$ or $u_0 \in H^1_x(\mathbb{R}^n)$ is radial. Then the corresponding solution u to* (6.39) *blows up in finite time.*

For the Cauchy problem (6.39) for the focusing energy-critical nonlinear Schrödinger equation with an inverse-square potential, we refer to [291].

6.6 Notes on dynamics of NLS with inverse square potential

In this section, we summarize the history of the study for the 3d focusing cubic NLS with inverse square potential of the form

$$\begin{cases} (i\partial_t - \mathcal{L}_a)u = -|u|^2u, & (t,x) \in \mathbb{R}\times\mathbb{R}^3, \\ u(0,x) = u_0(x) \in H^1(\mathbb{R}^3), \end{cases} \tag{6.46}$$

with $a > -\frac{1}{4}$. When $a = 0$, (6.46) reduces to the standard focusing cubic NLS:

$$(i\partial_t + \Delta)u = -|u|^2u. \tag{6.47}$$

As in (6.47), the equation (6.46) enjoys several symmetries and conservation laws. First, the class of solutions is invariant under rescaling

$$u(t,x) \mapsto u^\lambda(t,x) := \lambda u(\lambda^2 t, \lambda x), \tag{6.48}$$

which identifies $\dot{H}_x^{\frac{1}{2}}(\mathbb{R}^3)$ as the scaling-critical space of initial data. Second, solutions to (6.46) conserve their mass and energy, defined respectively by

$$M(u(t)) := \int_{\mathbb{R}^3} |u(t,x)|^2\,dx,$$

$$E_a(u(t)) := \int_{\mathbb{R}^3} \left\{\frac{1}{2}|\nabla u(t,x)|^2 + \frac{a}{2|x|^2}|u(t,x)|^2 - \frac{1}{4}|u(t,x)|^4\right\} dx.$$

Initial data belonging to $H_x^1(\mathbb{R}^3)$ have finite mass and energy, as is evident from

$$\|\sqrt{\mathcal{L}_a} f\|^2_{L_x^2(\mathbb{R}^3)} \sim \|\nabla f\|^2_{L_x^2(\mathbb{R}^3)} \quad \text{for} \quad a > -\frac{1}{4}, \tag{6.49}$$

and the following variant of the Gagliardo–Nirenberg inequality:

$$\|f\|^4_{L_x^4(\mathbb{R}^3)} \le C_a \|f\|_{L_x^2(\mathbb{R}^3)} \|f\|^3_{\dot{H}_a^1(\mathbb{R}^3)}, \tag{6.50}$$

where C_a denotes the sharp constant in the inequality above, that $0 < C_a < \infty$ follows from the standard Gagliardo–Nirenberg inequality and (6.49). By using the variational analysis for the sharp Gagliardo–Nirenberg inequality (6.50), Killip–Murphy–Visan–Zheng [143] showed the following theorem, which leads naturally to the thresholds appearing in Theorem 6.6.3.

Theorem 6.6.1 (Sharp Gagliardo–Nirenberg inequality). *Fix $a > -\frac{1}{4}$ and define*

$$C_a := \sup\{\|f\|^4_{L_x^4} \div [\|f\|_{L_x^2} \|f\|^3_{\dot{H}_a^1}] : f \in H_a^1 \backslash \{0\}\}.$$

Then $C_a \in (0, \infty)$ and the following hold:

(i) *If $a \le 0$, then equality in the Gagliardo–Nirenberg inequality* (6.50) *is attained by a function $Q_a \in H_a^1$, which is a nonzero, nonnegative, radial solution to the elliptic problem*

$$-\mathcal{L}_a Q_a - Q_a + Q_a^3 = 0. \tag{6.51}$$

(ii) *If $a > 0$, then $C_a = C_0$, but equality in* (6.50) *is never attained.*

Global existence, scattering and blow up for (6.47) were studied in [75, 118]. The authors identified a sharp threshold between scattering and blowup, described in terms of the ground state Q_0, which is the unique, positive, radial, decaying solution to the elliptic problem

$$\Delta Q_0 - Q_0 + Q_0^3 = 0.$$

Theorem 6.6.2 (Scattering/blowup dichotomy [75, 118]). *Let $u_0 \in H_x^1(\mathbb{R}^3)$ satisfy $M(u_0) \times E_0(u_0) < M(Q_0)E(Q_0)$.*

(i) *If $\|u_0\|_{L_x^2}\|u_0\|_{\dot{H}_x^1} < \|Q_0\|_{L_x^2}\|Q_0\|_{\dot{H}_x^1}$, then the solution to (6.47) with initial data u_0 is global and scatters.*

(ii) *If $\|u_0\|_{L_x^2}\|u_0\|_{\dot{H}_x^1} > \|Q_0\|_{L_x^2}\|Q_0\|_{\dot{H}_x^1}$ and u_0 is radial or $xu_0 \in L_x^2(\mathbb{R}^3)$, then the solution to (6.47) with initial data u_0 blows up in finite time in both time directions.*

Killip–Murphy–Visan–Zheng [143] extends the above result to the equation (6.46). Define the following thresholds:

$$\mathcal{E}_a := M(Q_{a\wedge 0})E_{a\wedge 0}(Q_{a\wedge 0}) \quad \text{and} \quad \mathcal{K}_a := \|Q_{a\wedge 0}\|_{L_x^2}\|Q_{a\wedge 0}\|_{\dot{H}^1_{a\wedge 0}}. \tag{6.52}$$

Theorem 6.6.3 (Scattering/blowup dichotomy [143]). *Fix $a > -\frac{1}{4}$. Let $u_0 \in H_x^1(\mathbb{R}^3)$ satisfy $M(u_0)E_a(u_0) < \mathcal{E}_a$.*

(i) *If $\|u_0\|_{L_x^2}\|u_0\|_{\dot{H}_a^1} < \mathcal{K}_a$, then the solution to (6.46) with initial data u_0 exists globally and scatters.*

(ii) *If $\|u_0\|_{L_x^2}\|u_0\|_{\dot{H}_a^1} > \mathcal{K}_a$ and u_0 is radial or $xu_0 \in L_x^2(\mathbb{R}^3)$, then the solution to (6.46) with initial data u_0 blows up in finite time in both time directions.*

For $a > 0$, Miao–Murphy–Zheng [188] showed the threshold scattering result.

Theorem 6.6.4 (Threshold scattering [188]). *Let $a > 0$. Suppose $u_0 \in H^1(\mathbb{R}^3)$ satisfies*

$$M(u_0)E_a(u_0) = M(Q_0)E_0(Q_0) \quad \textit{and} \quad \|u_0\|_{L^2}\|u_0\|_{\dot{H}_a^1} < \|Q_0\|_{L^2}\|Q_0\|_{\dot{H}^1}. \tag{6.53}$$

Then the corresponding solution u to (6.46) exists globally with $u \in L_{t,x}^5(\mathbb{R} \times \mathbb{R}^3)$. In particular, u scatters in H^1.

For the orbital stability of standing waves, we refer to [272].

Bibliography

[1] S. Agmon, Spectral properties of Schrödinger operators and scattering theory, Ann. Sc. Norm. Super. Pisa, Cl. Sci. (4), **2** (1975), 151–218.

[2] D. Albritton, E. Brué and M. Colombo. Non-uniqueness of Leray solutions of the forced Navier–Stokes equations, Ann. Math., **196** (2022), no. 1, 415–455.

[3] G. Alexopoulos, Spectral multipliers for Markov chains, J. Math. Soc. Jpn., **56** (2004), 833–852.

[4] G. Andrews, R. Askey and R. Roy, Special Functions. Encyclopedia of Mathematics and its Applications, Vol. 71. Cambridge University Press, Cambridge, 1999.

[5] T. Aubin, Problémes isopérimétriques et espaces de Sobolev, J. Differ. Geom., **11** (1976), 573–598.

[6] H. Bahouri and J. Chemin, On global well-posedness for defocusing cubic wave equation, Int. Math. Res. Not., **2006** (2006), 54873.

[7] H. Bahouri and P. Gérard, High frequency approximation of solutions to critical nonlinear wave equations, Am. J. Math., **121** (1999), no. 1, 131–175.

[8] V. Banica, Remarks on the blow-up for the Schrödinger equation with critical mass on a plane domain, Ann. Sc. Norm. Super. Pisa, Cl. Sci. (5), **3** (2004), no. 1, 139–170.

[9] J. E. Barab, Nonexistence of asymptotic free solutions for a nonlinear Schrödinger equation, J. Math. Phys., **25** (1984), 3270–3273.

[10] T. Barker, Localized quantitative estimates and potential blow-up rates for the Navier–Stokes equations, SIAM J. Math. Anal., **55** (2023), no. 5, 5221–5259.

[11] T. Barker and C. Prange, Quantitative regularity for the Navier–Stokes equations via spatial concentration, Commun. Math. Phys., **385** (2021), no. 2, 717–792.

[12] M. Beceanu and W. Schlag, Structures formulas for wave operators, Am. J. Math., **142** (2020), no. 3, 751–807.

[13] H. Berestycki, P.-L. Lions and L. A. Peletier, An ODE approach to the existence of positive solutions for semilinear problems in $\mathbb{R}^N$, Indiana Univ. Math. J., **30** (1981), no. 1, 141–157.

[14] J. Bergh and J. Löfström, Interpolation Spaces, Springer, New York, 1976.

[15] P. Bizoń, T. Chmaj and Z. Tabor, On blowup for semilinear wave equations with a focusing nonlinearity, Nonlinearity, **17** (2004), no. 6, 2187–2201.

[16] J. M. Bouclet and H. Mizutani, Uniform resolvent and Strichartz estimates for Schrödinger equations with scaling critical potentials, Trans. Am. Math. Soc., **370** (2018), 7293–7333.

[17] J. Bourgain, A remark on Schrödinger operators, Isr. J. Math., **77** (1992), 1–16.

[18] J. Bourgain, Some new estimates on oscillatory integrals. In Essays on Fourier Analysis in Honor of Elias M. Stein, Princeton, NJ 1991. Princeton Math. Ser., Vol. 42, Princeton University Press, New Jersey, (1995), pp. 83–112.

[19] J. Bourgain, Scatering in energy space and below for 3D NLS equations, J. Anal. Math., **75** (1998), 267–297.

[20] J. Bourgain, Refinements of Strichartz' inequality and applications to 2D-NLS with critical nonlinearity, Int. Math. Res. Not., **5** (1998), 253–283.

[21] J. Bourgain, The Global Solution of Nonlinear Schrödinger Equations, AMS, Providence, 1999.

[22] J. Bourgain, Global well-posedness of defocusing 3D critical NLS in the radial case, J. Am. Math. Soc., **12** (1999), 145–171.

[23] J. Bourgain, On the Schrödinger maximal function in higher dimensions, Proc. Steklov Inst. Math., **280** (2013), no. 1, 46–60.

[24] J. Bourgain, A note on the Schrödinger maximal function, J. Anal. Math., **130** (2016), 393–396.

[25] P. Brenner, On scattering and everywhere defined scattering operator for nonlinear Klein–Gordon equations, J. Differ. Equ., **56** (1985), 310–344.

[26] P. Brenner and W. von Wahl, Global classical solution of non-linear wave equations, Math. Z., **176** (1981), 87–121.

https://doi.org/10.1515/9783111384726-007

[27] T. Buckmaster, M. Colombo and V. Vicol. Wild solutions of the Navier–Stokes equations whose singular sets in time have Hausdorff dimension strictly less than 1, J. Eur. Math. Soc., **24** (2021), no. 9, 3333–3378.
[28] T. Buckmaster and V. Vicol, Nonuniqueness of weak solutions to the Navier–Stokes equation, Ann. Math., **189** (2019), no. 2, 101–144.
[29] N. Burq, F. Planchon, J. Stalker and A. S. Tahvildar-Zadeh, Strichartz estimates for the wave and Schrödinger equations with the inverse-square potential, J. Funct. Anal., **203** (2003), 519–549.
[30] L. Caffarelli, R. Kohn and L. Nirenberg, Partial regularity of suitable weak solutions of Navier–Stokes equations, Commun. Pure Appl. Math., **35** (1982), 771–831.
[31] C. P. Calderón, Existence of weak solutions for the Navier–Stokes equations with initial data in L^p, Trans. Am. Math. Soc., **318** (1990), 179–200.
[32] M. Cannone, A generalization of a theorem by Kato on Navier–Stokes equations, Rev. Mat. Iberoam., **13** (1997), 515–541.
[33] M. Cannone, Harmonic analysis tools for solving the incompressible Navier–Stokes equations. Handbook of Mathematical Fluid Dynamics III, Edited by S. J. Friedlander and D. Serre, Elsevier, 2004.
[34] M. Cannone, Q. Chen and C. Miao, A losing estimate for the ideal MHD equations with application to Blow-up criterion, SIAM J. Math. Anal., **38** (2007), 1847–1859.
[35] M. Cannone and Y. Meyer, Littlewood–Paley decomposition and Navier–Stokes equations, Methods Appl. Anal., **2** (1995), 307–319.
[36] L. Carleson, Some analytic problems related to statistical mechanics. In Euclidean Harmonic Analysis. Lecture Notes in Math., Vol. 779, Springer, Berlin, 1979, pp. 5–45.
[37] T. Cazenave and F. B. Weissler, The Cauchy problem for critical nonlinear Schrödinger equation in H^s, Nonlinear Anal., Theory Methods Appl., **14** (1990), 807–836.
[38] Q. Chen, C. Miao and Z. Zhang, A new Bernstein's inequality and the 2D dissipative quasi-geostrophic equation, Commun. Math. Phys., **271** (2007), 821–838.
[39] Q. Chen, C. Miao and Z. Zhang, The Beale–Kato–Majda criterion to the 3D magneto-hydrodynamics equations, Commun. Math. Phys., **275** (2007), 861–872.
[40] Q. Chen, C. Miao and Z. Zhang, Global well-posedness for the compressible Navier–Stokes equations with the highly oscillating initial velocity, Commun. Pure Appl. Math., **LXIII** (2010), 1173–1224.
[41] Q. Chen, C. Miao and Z. Zhang, Well-posedness in critical spaces for the compressible Navier–Stokes equations with density dependent viscosities, Rev. Mat. Iberoam., **26** (2010), 915–946.
[42] Q. Chen, C. Miao and Z. Zhang, On the well-posedness of the ideal MHD equations in the Triebel–Lizorkin spaces, Arch. Ration. Mech. Anal., **195** (2010), 561–578.
[43] Q. Chen, C. Miao and Z. Zhang, On the ill-posedness of the compressible Navier–Stokes equations in the critical Besov spaces, Rev. Mat. Iberoam., **31** (2015), 1375–1402.
[44] A. Cheskidov and X. Luo. Sharp nonuniqueness for the Navier–Stokes equations, Invent. Math., **229** (2022), 987–1054.
[45] M. Christ and A. Kiselev, Maximal function associated to filtrations, J. Funct. Anal., **179** (2001), 409–425.
[46] J. Colliander, M. Grillakis and N. Tzirakis, Improved interaction Morawetz inequalities for the cubic nonlinear Schrödinger equation on $\mathbb{R}^2$, Int. Math. Res. Not., **23** (2007), 90–119.
[47] J. Colliander, M. Keel, G. Staffilani, H. Takaoka and T. Tao, Global well-posedness for Schrödinger equations with derivatives, SIAM J. Math. Anal., **33** (2001), 649–669.
[48] J. Colliander, M. Keel, G. Staffilani, H. Takaoka and T. Tao, Almost conservation laws and global rough solutions to a nonlinear Schrödinger equation, Math. Res. Lett., **9** (2002), 659–682.
[49] J. Colliander, M. Keel, G. Staffilani, H. Takaoka and T. Tao, Polynomial upper bounds for the orbit instability of the 1D cubic NLS below the energy norm, Discrete Contin. Dyn. Syst., **9** (2003), 31–54.
[50] J. Colliander, M. Keel, G. Staffilani, H. Takaoka and T. Tao, Polynomial upper bounds for the instability of the nonlinear Schrödinger equation below the energy norm, Commun. Pure Appl. Anal., **2** (2003), 33–50.

[51] J. Colliander, M. Keel, G. Staffilani, H. Takaoka and T. Tao, Global existence and scattering for rough solutions to a nonlinear Schrödinger equation on $\mathbb{R}^3$, Commun. Pure Appl. Math., **57** (2004), 987–1014.
[52] J. Colliander, M. Keel, G. Staffilani, H. Takaoka and T. Tao, Resonant decompositions and the I-method for cubic nonlinear Schrödinger equation on $\mathbb{R}^2$, Discrete Contin. Dyn. Syst., Ser. A, **21** (2007), 665–696.
[53] J. Colliander, M. Keel, G. Staffilani, H. Takaoka and T. Tao, Global well-posedness and scattering for the energy-critical nonlinear Schrödinger equation in $\mathbb{R}^3$, Ann. Math., **167** (2008), 767–865.
[54] C. Collot, T. Duyckaerts, C. Kenig and F. Merle, Soliton resolution for the radial quadratic wave equation in space dimension 6, Vietnam J. Math., **52** (2024), 735–773.
[55] B. Dahlberg and C. E. Kenig, A note on the almost everywhere behavior of solutions to the Schrödinger equations. In Harmonic Analysis. Lecture Notes in Math., Vol. 908, Springer, Berlin, 1982, pp. 205–209.
[56] G. David and J. L. Journé, A boundedness criterion for generalized Calderón–Zygmund operators, Ann. Math., **120** (1984), 371–397.
[57] Y. Ding and X. Sun, Strichartz estimates for parabolic equations with higher order differential operators, Sci. China Math., **58** (2015), 1047–1062.
[58] D. B. Dix, Nonuniqueness and uniqueness in the initialvalue problem for the Burger's equation, SIAM J. Math. Anal., **27** (1996), 208–224.
[59] B. Dodson, Global well-posedness and scattering for the defocusing, L^2-critical, nonlinear Schrödinger equation when $d \geq 3$, J. Am. Math. Soc., **25** (2012), 429–463.
[60] B. Dodson, Global well-posedness and scattering for the defocusing, cubic nonlinear Schrödinger equation when $n = 3$ via a linear-nonlinear decomposition, Discrete Contin. Dyn. Syst., Ser. A, **33** (2013), 1905–1926.
[61] B. Dodson, Global well-posedness and scattering for the mass critical nonlinear Schrödinger equation with mass below the mass of the ground state, Adv. Math., **285** (2015), 1589–1618.
[62] B. Dodson, Global well-posedness and scattering for the defocusing, L^2-critical, nonlinear Schrödinger equation when $d = 2$, Duke Math. J., **165** (2016), 3435–3516.
[63] B. Dodson, Global well-posedness and scattering for the defocusing, L^2-critical, nonlinear Schrödinger equation when $d = 1$, Am. J. Math., **138** (2016), no. 2, 531–569.
[64] B. Dodson, Global well-posedness and scattering for the focusing, energy-critical nonlinear Schrödinger problem in dimension $d = 4$ for initial data below a ground state threshold, Ann. Sci. Éc. Norm. Supér., **52** (2019), 139–180.
[65] B. Dodson, Global well-posedness and scattering for the radial, defocusing, cubic nonlinear wave equation, Duke Math. J., **170** (2021), no. 15, 3267–3321.
[66] B. Dodson, A determination of the blowup solutions to the focusing, quintic NLS with mass equal to the mass of the soliton, Ann. PDE, 9 (2023), no. 1, paper no. 3, 86 pp.
[67] B. Dodson, A determination of the blowup solutions to the focusing NLS with mass equal to the mass of the soliton, Anal. PDE, **17** (2024), 1693–1760.
[68] B. Dodson and A. Lawrie, Scattering for the radial 3D cubic wave equation, Anal. PDE, **8** (2015), 467–497.
[69] B. Dodson, C. Miao, J. Murphy and J. Zheng, The defocusing quintic NLS in four space dimensions, Ann. Inst. Henri Poincaré, Anal. Non Linéaire, **37** (2020), 417–456.
[70] B. Dodson and J. Murphy, A new proof of scattering below the ground state for the 3d radial focusing cubic NLS, Proc. Am. Math. Soc., **145** (2017), no. 11, 4859–4867.
[71] B. Dodson and J. Murphy, A new proof of scattering below the ground state for the non-radial focusing NLS, Math. Res. Lett., **25** (2018), no. 6, 1805–1825.
[72] R. Donninger and B. Schörkhuber, Stable blowup for wave equations in odd space dimensions, Ann. Inst. Henri Poincaré, Anal. Non Linéaire, **34** (2017), 1181–1213.

[73] X. Du, L. Guth and X. Li, A sharp Schrödinger maximal estimate in $\mathbb{R}^2$, Ann. Math., **188** (2017), 607–640.
[74] X. Du and R. Zhang, Sharp L^2 estimate of Schrödinger maximal function in higher dimensions, Ann. Math., **189** (2019), 837–861.
[75] T. Duyckaerts, J. Holmer and S. Roudenko, Scattering for the non-radial 3D cubic nonlinear Schrödinger equation, Math. Res. Lett., **15** (2008), no. 6, 1233–1250.
[76] T. Duyckaerts, H. Jia, C. Kenig and F. Merle, Soliton resolution along a sequence of times for the focusing energy critical wave equation, Geom. Funct. Anal., **27** (2017), 798–862.
[77] T. Duyckaerts, C. Kenig and F. Merle, Universality of blow-up profile for small radial type II blow-up solutions of the energy-critical equation, J. Eur. Math. Soc., **13** (2011), 533–599.
[78] T. Duyckaerts, C. Kenig and F. Merle, Soliton resolution for the radial critical wave equation in all odd space dimensions, Acta Math., **230** (2023), 1–92.
[79] T. Duyckaerts and F. Merle, Dynamics of threshold solutions for energy-critical wave equation, Int. Math. Res. Pap., 2008 (2008), art. ID rpn002, 67 pp.
[80] T. Duyckaerts and F. Merle, Dynamic of threshold solutions for energy-critical NLS, Geom. Funct. Anal., **18** (2009), 1787–1840.
[81] T. Duyckaerts and F. Merle, Scattering norm estimate near the threshold for energy-critical focusing semilinear wave equation, Indiana Univ. Math. J., **58** (2009), no. 4, 1971–2001.
[82] T. Duyckaerts and S. Roudenko, Threshold solutions for the focusing 3D cubic Schrödinger equation, Rev. Mat. Iberoam., **26** (2010), 1–56.
[83] T. Duyckaerts and S. Roudenko, Going beyond the threshold: scattering and blow-up in the focusing NLS equation, Commun. Math. Phys., **334** (2015), 1573–1615.
[84] L. Escauriaza, G. A. Seregin and V. Sverák, $L^{3,\infty}$-solutions to the Navier–Stokes equations and backward uniqueness, Russ. Math. Surv., **58** (2003), 211–250.
[85] M. Escobedo and E. Zuazua, Large time behavior for convection-diffusion equation in $\mathbb{R}^n$, J. Differ. Equ., **100** (1991), 119–161.
[86] G. Evéquoz, Existence and asymptotic behavior of standing waves of the nonlinear Helmholtz equation in the plane, Analysis (Berlin), **37** (2017), 55–68.
[87] E. B. Fabes, B. F. Jones and N. M. Riviere, Initial value problem for Navier–Stokes equations with data in L^p, Arch. Ration. Mech. Anal., **45** (1972), 222–240.
[88] L. Fanelli, V. Felli, M. A. Fontelos and A. Primo, Time decay of scaling critical electromagnetic Schrödinger flows, Commun. Math. Phys., **324** (2013), 1033–1067.
[89] L. Fanelli, V. Felli, M. A. Fontelos and A. Primo, Time decay of scaling invariant electromagnetic Schrödinger equations on the plane, Commun. Math. Phys., **337** (2015), 1515–1533.
[90] C. Fefferman and E. M. Stein, H^p spaces of several variables, Acta Math., **129** (1972), 137–193.
[91] G. B. Folland, Introduction to Partial Differential Equations, 2nd edition, Princeton University Press, 1993.
[92] H. Fujita, On the blowing up solutions of the Cauchy problem for $u_t = \Delta u + u^{\alpha+1}$, J. Fac. Sci., Univ. Tokyo, Sect. 1A, Math., **13** (1966), 109–124.
[93] D. Fujiwara and H. Morimoto, An L_r-theorem of the Helmholtz decomposition of vector fields, J. Fac. Sci., Univ. Tokyo, Sect. 1A, Math., **24** (1977), 685–700.
[94] I. Gallagher, G. Koch and F. Planchon, A profile decomposition approach to the $L_t^\infty(L_x^3)$ Navier–Stokes regularity criterion, Math. Ann., **355** (2013), no. 4, 1527–1559.
[95] I. Gallagher and F. Planchon, On global solutions to a dofocusing semi-linear wave equation, Rev. Mat. Iberoam., **19** (2003), 161–177.
[96] B. Gidas, W. M. Ni and L. Nirenberg, Symmetry and related properties via the maximum principle, Commun. Math. Phys., **68** (1979), 209–243.
[97] Y. Giga, Analyticity of the semigroup generated by the Stokes operator in L^r spaces, Math. Z., **178** (1981), 297–329.

[98] Y. Giga, Solutions for Semilinear parabolic equations in L^p and regularity of weak solutions of the Navier–Stokes system, J. Differ. Equ., **61** (1986), 186–212.
[99] Y. Giga and T.Miyakawa, Solutions in L^r of the Navier–Stokes initial value problem, Arch. Ration. Mech. Anal., **89** (1985), 267–281.
[100] J. Ginibre and T. Ozawa, Long range scattering for nonlinear Schrödinger equations and Hartree equations in the space dimension $n \geq 2$, Commun. Math. Phys., **151** (1993), 615–645.
[101] J. Ginibre, T. Ozawa and G. Velo, On the existence of the wave operators for a class of nonlinear Schrödinger equations, Ann. IHP, Phys. Théor., **60** (1994), 211–239.
[102] J. Ginibre and G. Velo, On a class of nonlinear Schrödinger equations, J. Funct. Anal., **32** (1979), 1–71.
[103] J. Ginibre and G. Velo, Scattering theory in energy space for a class nonlinear Schrödinger equations, J. Math. Pures Appl., **64** (1985), 363–401.
[104] J. Ginibre and G. Velo, The global Cauchy problem for some non linear Schrödinger equation revisited, Ann. Inst. Henri Poincaré, Anal. Non Linéaire, **2** (1985), 309–327.
[105] J. Ginibre and G. Velo, Time decay of finite energy solutions of nonlinear Klein–Gordon equation and nonlinear Schrödinger equations, Ann. Inst. Henri Poincaré, Anal. Non Linéaire, **43** (1985), 399–422.
[106] J. Ginibre and G. Velo, The global Cauchy problem for nonlinear Klein–Gordon equations, Math. Z., **189** (1985), 487–505.
[107] J. Ginibre and G. Velo, Reguality of solution of critical and sub-critical nonlinear wave equations, Nonlinear Anal., Theory Methods Appl., **22** (1994), 1–19.
[108] J. Ginibre and G. Velo, The Cauchy problem in local spaces for complex Ginzburg–Landau equation, Commun. Math. Phys., **187** (1997), 45–79.
[109] R. Glassey, On the Blowing-up of solution to the Cauchy problem for the nonlinear Schrödinger equations, J. Math. Phys., **18** (1977), 1794–1797.
[110] R. Glassey and M. Tsutsumi, On uniqueness of weak solutions to semilinear wave equations, Commun. Partial Differ. Equ., **7** (1981), 153–195.
[111] M. Grillakis, Regularity and asymptotic behaviour of nonlinear wave equations with critical nonlinearity, Ann. Math., **132** (1990), 485–505.
[112] M. Grillakis, Regularity for nonlinear wave equation with critical nonlinearity, Commun. Pure Appl. Math., **45**, 1992, 749–774.
[113] L. Guth and N. Katz, On the Erdös distinct distance problem in the plane, Ann. Math., **181** (2015), 155–190.
[114] S. Gutiérrez, Non trivial L^q solutions to the Ginzburg–Landau equation, Math. Ann., **328** (2004), 1–25.
[115] N. Hayashi and Y. Tsutsumi, Scattering theory for the Hartree type equations, Ann. IHP, Phys. Théor., **61** (1987), 187–213.
[116] W. Hebisch, A multiplier theorem for Schrödinger operators, Colloq. Math., **60/61** (1990), no. 2, 659–664.
[117] H. Hirata and C. Miao, Space-time estimates of linear flow and application to some nonlinear integral-differential equations corresponding to fractional order time derivative, Adv. Differ. Equ., **7** (2001), 217–236.
[118] J. Holmer and S. Roudenko, A sharp condition for scattering of the radial 3D cubic nonlinear Schrödinger equation, Commun. Math. Phys., **282** (2008), no. 2, 435–467.
[119] E. Hopf, Über die anfang swetaufgabe für die hydrodynamischer grundgleichungan, Math. Nachr., **4** (1951), 213–231.
[120] L. Hörmander, The Analysis of Linear Partial Differential Operator (I–IV), Springer-Verlag, 1983.
[121] J. Jendrej, Construction of type II blow-up solutions for the energy-critical wave equation in dimension 5, J. Funct. Anal., **272** (2017), 866–917.
[122] K. Jörgen, Das Anfangswert problem im Grossen für eine nichlineare, Wellengleichungen, Math. Z., **77** (1961), 295–308.
[123] K. Jörgens, Das Angfangswertproblem im grossen für eine klasse nichtlinearer wellengleichungen, Math. Z., **77** (1961), 295–307.

[124] H. Kalf, U. W. Schmincke, J. Walter and R. Wüst, On the spectral theory of Schrödinger and Dirac operators with strongly singular potentials. In Spectral Theory and Differential Equations. Lect. Notes in Math., Vol. 448, Springer, Berlin, 1975, pp. 182–226.
[125] L. Kapitanskii, Weak and yet weak solutions of semilinear wave equations, Commun. Partial Differ. Equ., **19** (1994), 1629–1676.
[126] T. Kato, Strong L^p-solutions of the Navier–Stokes equation in $\mathbb{R}^m$, with applications to weak solutions, Math. Z., **187** (1984), 471–480.
[127] T. Kato and S. T. Kuroda, The abstract theory of scattering, Rocky Mt. J. Math., **1**, (1971), no. 1, 127–171.
[128] T. Kato and G. Ponce, The Navier–Stokes equation with weak initial data, Int. Math. Res. Not., **10** (1994), 435–444.
[129] M. Keel and T. Tao, The endpoint Strichartz estimates, Am. J. Math., **120** (1998), 955–980.
[130] M. Keel and T. Tao, Local and global well-posedness of wave maps in $\mathbb{R}^{1+1}$ for rough data, Int. Math. Res. Not., **21** (1998), 1117–1156.
[131] C. Kenig and G. Koch, An alternative approach to regularity for the Navier–Stokes equations in critical spaces, Ann. Inst. Henri Poincaré, Anal. Non Linéaire, **28** (2011), no. 2, 159–187.
[132] C. Kenig and F. Merle, Global well-posedness, scattering, and blow-up for the energy-critical focusing nonlinear Schrödinger equation in the radial case, Invent. Math., **166** (2006), 645–675.
[133] C. Kenig and F. Merle, Scattering for $\dot{H}^{\frac{1}{2}}$-bounded solutions to the cubic, defocusing NLS in 3 dimensions, Trans. Am. Math. Soc., **362** (2010), 1937–1962.
[134] C. E. Kenig, Harmonic Analysis Techniques for Second Order Elliptic Boundary Value Problems, CBMS of Amer. Math. Soc., Vol. 83, 1994.
[135] C. E. Kenig, G. Ponce and L. Vega, Oscillatory integral and regularity of dispersive equations, Indiana Univ. Math. J., **40** (1991), 33–69.
[136] C. E. Kenig, G. Ponce and L. Vega, Small solutions to nonlinear Schrödinger equations, Ann. Inst. Henri Poincaré, Anal. Non Linéaire, **10** (1993), 255–288.
[137] C. E. Kenig, G. Ponce and L. Vega, Global well-posedness for nonlinear wave equations, Commun. Partial Differ. Equ., **25** (2000), 683–705.
[138] C. E. Kenig, A. Ruiz and C. D. Sogge, Uniform Sobolev inequalities and unique continuation for second order constant coefficient differential operators, Duke Math. J., **55** (1987), 329–347.
[139] S. Keraani, On the blow up phenomenon of the critical nonlinear Schrödinger equation, J. Funct. Anal., **235** (2006), no. 1, 171–192.
[140] R. Killip, D. Li, M. Visan and X. Zhang, The characterization of minimal-mass blowup solutions to the focusing mass-critical NLS, SIAM J. Math. Anal., **41** (2009), 219–236.
[141] R. Killip, C. Miao, M. Visan, J. Zhang and J. Zheng, The eneregy critical NLS with inverse square potential, Discrete Contin. Dyn. Syst., Ser. A, **37** (2017), 3831–3866.
[142] R. Killip, C. Miao, M. Visan, J. Zhang and J. Zheng, Sobolev space adapted to the Schrödinger operator with inverse-square potential, Math. Z., **288** (2018), 1273–1298.
[143] R. Killip, J. Murphy, M. Visan and J. Zheng, The focusing cubic NLS with inverse-square potential in three space dimensions, Differ. Integral Equ., **30** (2017), 161–206.
[144] R. Killip, T. Tao and M. Visan, The cubic nonlinear Schrödinger equation in two dimensions with radial data, J. Eur. Math. Soc., **11** (2009), 1203–1258.
[145] R. Killip and M. Visan, The focusing energy-critical nonlinear Schrödinger equation in dimensions five and higher, Am. J. Math., **132** (2010), no. 2, 361–424.
[146] R. Killip and M. Visan, Energy-supercritical NLS: critical $\dot{H}^s$-bounds imply scattering, Commun. Partial Differ. Equ., **35** (2010), 945–987.
[147] R. Killip and M. Visan, Global well-posedness and scattering for the defocusing quintic NLS in three dimensions, Anal. PDE, **5** (2012), 855–885.
[148] R. Killip and M. Visan, Nonlinear Schrödinger equations at critical regularity. In Evolution Equations. Clay Math. Proc., Vol. 17, Amer. Math. Soc., Providence, RI, 2013, pp. 325–437.

[149] R. Killip, M. Visan and X. Zhang, The mass-critical nonlinear Schrödinger equation with radial data in dimensions three and higher, Anal. PDE, **1** (2008), no. 2, 229–266.

[150] R. Killip, M. Visan and X. Zhang, Quintic NLS in the exterior of a strictly convex obstacle, Am. J. Math., **138** (2016), 1193–1346.

[151] S. Klainerman, The null condition and global existence to nonlinear wave equations, Lect. Appl. Math., **23** (1986), 293–326.

[152] S. Klainerman and M. Machedon, Space–time estimates for null forms and the local existence theorem, Commun. Pure Appl. Math., **46** (1993), 1221–1268.

[153] H. Koch and D. Tataru, Well-posedness for the Navier–Stokes equations, Adv. Math., **157** (2001), 27–35.

[154] J. Krieger and W. Schlag, Full range of blow up exponents for the quintic wave equation in three dimensions, J. Math. Pures Appl., **101** (2014), no. 6, 873–900.

[155] J. Krieger, W. Schlag and D. Tataru, Slow blow-up solutions for the $H^1(\mathbb{R}^3)$ critical focusing semilinear wave equation, Duke Math. J., **147**, 2009, no. 1, 1–53.

[156] M. K. Kwong, Uniqueness of positive solutions of $\Delta u - u + u^p = 0$ in $\mathbb{R}^n$, Arch. Ration. Mech. Anal., **105** (1989), no. 3, 243–266.

[157] O. Ladyzhenskaya and G. Seregin, On partial regularity of suitable weak solutions to the three-dimensional Navier–Stokes equations, J. Math. Fluid Mech., **1** (1999), 356–387.

[158] O. A. Ladyzhenskaya, Unique global solvability of the three-dimensional Cauchy problem for the Navier–Stokes equations in the presence of axial symmetry, Zap. Naucn. Sem. Leningrad. Otdel. Mat. Inst. Steklov, **7** (1968), 155–177.

[159] O. A. Ladyzhenskaya, On the Theory of Mathematical in the Incompressible Fluid, 2nd edition, Gordon and Breach, New York, 1969.

[160] S. Lee, On pointwise convergence of the solutions to Schrödinger equations in $\mathbb{R}^2$, Int. Math. Res. Not., **2006** (2006), 32597.

[161] P. G. Lemarié-Rieusset, Recent Developments in the Navier–Stokes Problem, Chapman and Hall/CRC Press, Boca Raton, FL, 2002.

[162] P. G. Lemarié-Rieusset, The Navier–Stokes Problem in the 21st Century, CRC Press, Boca Raton, FL, 2016.

[163] J. Leray, Sur le mouvement d'um liquide visqieux emlissant l'space, Acta Math., **64** (1934), 193–284.

[164] H. A. Levine, Instability and nonexistence of global solutions to nonlinear wave equations of the form $Pu_{tt} = -Au + F(u)''$, Trans. Am. Math. Soc., **192**, (1974), 10–21.

[165] D. Li, C. Miao and X. Zhang, The focusing energy-critical Hartree equation, J. Differ. Equ., **246** (2009), 1139–1163.

[166] D. Li and X. Zhang, Dynamics for the energy critical nonlinear Schrödinger equation in high dimensions, J. Funct. Anal., **256** (2009), 1928–1961.

[167] D. Li and X. Zhang, Regularity of almost periodic modulo scaling solutions for mass-critical NLS and application, Anal. PDE, **3** (2010), 175–195.

[168] D. Li and X. Zhang, On the rigidity of solitary waves for the focusing mass-critical NLS in dimensions $d \geq 2$, Sci. China Math., **55** (2012), 385–434.

[169] X. Li, C. Miao and L. Zhao, Soliton resolution for the energy critical wave equation with inverse-square potential in the radial case, Sci. China Math., DOI: https://doi.org/10.1007/s11425-023-2320-y.

[170] Y. Li, Z. Zeng and D. Zhang, Non-uniqueness of weak solutions to 3D generalized magnetohydrodynamic equations, J. Math. Pures Appl., **165** (2022), 232–285.

[171] E. Lieb and M. Loss, Analysis, 2nd edition. Graduate Studies in Mathematics, Vol. 14, American Mathematical Society, Providence, RI, 2001.

[172] F.-H. Lin, A new proof of the Caffarelli–Kohn–Nirenberg theorem, Commun. Pure Appl. Math., **51** (1998), 241–257.

[173] J. L. Lin and W. Strauss, Decay and scattering of nonlinear Schrödinger equation, J. Funct. Anal., **30** (1978), 245–263.

[174] H. Lindblad and C. D. Sogge, On existence and scattering with minimal regularity for semilinear wave equations, J. Funct. Anal., **130** (1995), 357–426.

[175] H. Lindblad and C. D. Sogge, Restriction theorems and semi-linear Klein–Gordon equations in $(1+3)$ dimensions, Duke Math. J., **85** (1996), 227–252.

[176] J. L. Lions, Quelques méthodes de résolution des problémes aux limites non-linéaires, Dunod, Paris, 1969.

[177] V. Liskevich and Z. Sobol, Estimates of integral kernels for semigroups associated with second order elliptic operators with singular coefficients, Potential Anal., **18** (2003), 359–390.

[178] C. Lu and J. Zheng, The radial defocusing energy-supercritical NLS in dimension four, J. Differ. Equ., **262** (2017), 4390–4414.

[179] J. Lu, C. Miao and J. Murphy, Scattering in H^1 for the intercritical NLS with an inverse-square potential, J. Differ. Equ., **264** (2018), no. 5, 3174–3211.

[180] R. Luca and M. Rogers, An improved necessary condition for Schrödinger maximal estimate, arXiv: 1506.05325.

[181] F. Merle, Construction of solutions with exact k blow-up points for the Schrödinger equation with critical power nonlinearity, Commun. Math. Phys., **149** (1992), 205–214.

[182] F. Merle, Determination of blow-up solutions with minimal mass for nonlinear Schrödinger equation with critical power, Duke Math. J., **69** (1993), 427–453.

[183] Y. Meyer, Remarques sur un thèorém de J.B.Bony, Rend. Circ. Mat. Palermo Suppl., **1** (1981), 1–20.

[184] C. Miao, Harmonic Analysis and Applications to Partial Differential Equations (in Chinese), Science Press, Beijing, 1999.

[185] C. Miao, Time–space estimates of solutions to general semi-linear parabolic equations, Tokyo J. Math., **24** (2001), 247–278.

[186] C. Miao, J. Murphy and J. Zheng, The defocusing energy-supercritical NLS in four space dimensions, J. Funct. Anal., **267** (2014), 1662–1724.

[187] C. Miao, J. Murphy and J. Zheng, The energy-critical nonlinear wave equation with an inverse-square potential, Ann. Inst. Henri Poincaré, Anal. Non Linéaire, **37** (2020), 417–456.

[188] C. Miao, J. Murphy and J. Zheng, Threshold scattering for the focusing NLS with a repulsive potential, Indiana Univ. Math. J., **72** (2023), 409–453.

[189] C. Miao, R. Shen and T. Zhao, Scattering theory for the subcritical wave equation with inverse square potential, Sel. Math. (N. S.), **29** (2023), no. 3, paper no. 44, 30pp.

[190] C. Miao, X. Su and J. Zheng, The $W^{s,p}$-boundedness of stationary wave operators for the Schrödinger operator with inverse-square potential, Trans. Am. Math. Soc., **376** (2023), 1739–1797.

[191] C. Miao, G. Xu and J. Yang, Global well-posedness for the defocusing Hartree equation with radial data in $\mathbb{R}^4$, Commun. Contemp. Math., **22** (2020), 1950004, 35 pp.

[192] C. Miao, G. Xu and L. Zhao, Global well-posedness and scattering for the energy-critical, defocusing Hartree equation for radial data, J. Funct. Anal., **253** (2007), 605–627.

[193] C. Miao, G. Xu and L. Zhao, Global well-posedness and uniform bound for the defocusing $H^{1/2}$-subcritical Hartree equation in $\mathbb{R}^d$, Ann. Inst. Henri Poincaré, Anal. Non Linéaire, **26** (2009), 1831–1852.

[194] C. Miao, G. Xu and L. Zhao, Global well-posedness and scattering for the mass-critical Hartree equation with radial data, J. Math. Pures Appl., **91** (2009), 49–79.

[195] C. Miao, G. Xu and L. Zhao, Global well-posedness and scattering for the energy-critical, defocusing Hartree equation in $\mathbb{R}^{1+n}$, Commun. Partial Differ. Equ., **36** (2011), 729–776.

[196] C. Miao and W. Ye, On the weak solutions for the MHD systems with controllable total energy and cross helicity, J. Math. Pures Appl., **181** (2024), 190–227.

[197] C. Miao and B. Zhang, H^s-global well-posedness for semilinear wave equations, J. Math. Anal. Appl., **283** (2003), 645–666.

[198] C. Miao and B. Zhang, The Cauchy problem for the semilinear parabolic equations in Besov spaces, Houst. J. Math., **30** (2004), 829–878.

[199] C. Miao, J. Zhang and J. Zheng, Maximal estimates for Schrödinger equation with inverse square potential, Pac. J. Math., **273** (2015), 1–20.

[200] P. D. Milman and Yu. A. Semenov, Global heat kernel bounds via desingularizing weights, J. Funct. Anal., **212** (2004), 373–398.

[201] H. Mizutani, Uniform Sobolev estimates for Schrödinger operators with scaling-critical potentials and applications, Anal. PDE, **13** (2020), 1333–1369.

[202] H. Mizutani, J. Zhang and J. Zheng, Uniform resolvent estimates for Schrödinger operator with an inverse-square potential, J. Funct. Anal., **278** (2020), 108350.

[203] S. J. Montgomery-Smith, Time decay for the bounded mean oscillation of solutions of the Schrödinger and wave equations, Duke Math. J., **25** (1998), 1–31.

[204] C. Morawetz and W.Srauss, Decay and scattering of solutions of a nonlinear relativistic wave equations, Commun. Pure Appl. Math., **25** (1972), 1–31.

[205] A. Moyua, A. Vargas and L. Vega, Schrödinger maximal function and restriction properties of the Fourier transform, Int. Math. Res. Not., **1996** (1996), 793–815.

[206] J. Murphy, Intercritical NLS: critical H^s-bounds imply scattering, SIAM J. Math. Anal., **46** (2014), 939–997.

[207] J. Murphy, The defocusing $\dot{H}^{1/2}$-critical NLS in high dimensions, Discrete Contin. Dyn. Syst., Ser. A, **34** (2014), 733–748.

[208] J. Murphy, The radial defocusing nonlinear Schrödinger equation in three space dimensions, Commun. Partial Differ. Equ., **40** (2015), 265–308.

[209] M. Nakamura and T. Ozawa, Low energy scattering for nonlinear Schrödinger equations, Rev. Math. Phys., **9** (1997), 397–410.

[210] M. Nakamura and T. Ozawa, Nonlinear Schrödinger equations in the Sobolev space of critical order, J. Funct. Anal., **155** (1998), 365–380.

[211] K. Nakanishi and W. Schlag, Global dynamics above the ground state energy for the cubic NLS equation in 3D, Calc. Var. Partial Differ. Equ., **44** (2012), no. 1–2, 1–45.

[212] T. Ozawa, Long range scattering theory for nonlinear Schrödinger equations in one space dimension, Commun. Math. Phys., **139** (1991), 479–493.

[213] S. Palasek, Improved quantitative regularity for the Navier–Stokes equations in a scale of critical spaces, Arch. Ration. Mech. Anal., **242** (2021), 1479–1531.

[214] A.Pazy, Semigroups of Linear Operator and Applications to Partial Differential Equations, Springer-Verlag, 1983.

[215] H. Pecher, L^p Abschätzungen und klassiche Lösungen hürnichr lineare wellengleichungen I, Math. Z., **150** (1976), 159–183.

[216] H. Pecher, Nonlinear small data scattering for the wave and Klein–Gordon equations, Math. Z., **185** (1984), 261–270.

[217] F. Planchon, Existence Globale et scattering pour les solutions de masse fine de l'équation de Schrödinger cubique en dimension deux [d'après Benjamin Dodson, Rowan Killip, Terence Tao, Monica Visan et Xiaoyi Zhang], séminaire Bourbaki, vol 2010/2011, n 1042.

[218] F. Planchon, J. Stalker and A. S. Tahvildar-Zadeh, Dispersive estimates for wave equation with the inverse-square potential, Discrete Contin. Dyn. Syst., **9** (2003), 1387–1400.

[219] F. Planchon and L. Vega, Bilinear virial identities and applications, Ann. Sci. Éc. Norm. Supér. (4), **42** (2009), 261–290.

[220] G. Ponce, On the well posedness of some nonlinear evolution equations. In Nonlinear Wave, Proceedings of the Fourth MSJ International Research Institute, Edited by R. Agemi, Y. Giga and T. Ozawa, 1996, pp. 379–409.

[221] G. Ponce and T. Sideris, Local regularity of nonlinear wave equations in three space dimensions, Commun. Partial Differ. Equ., **18** (1993), 169–177.

[222] G. Prodi, Un teorema di unicità per le equazioni di Navier–Stokes, Ann. Mat. Pura Appl. (4), **48** (1959), 173–182.

[223] P. Raphael, Concenration compacité à la Kenig–Merle. In Séminaire Bourbaki, vol 2011/2012, Exposes, 1043–1058. Asterisque, Vol. 352, 2013.

[224] J. Rauch, The u^5-Klein–Gordon equation. In Non-Linear PDE and Their Applications. Pitman Res. Notes in Math., Vol. 53, Longman, Harlow 1980, pp. 335–364.

[225] M. Reed, Abstract Nonlinear Wave Equation. Lecture Notes in Math., Vol. 507, Springer-Verlag, 1976.

[226] M. Reed and B. Simon, Methods of Modern Mathematical Physics, I–IV, Academic Press Inc., 1975, 1979.

[227] F. Ribaud, Cauchy problem for semilinear parabolic equation with data in $H_p^s(\mathbf{R}^n)$ spaces, Rev. Mat. Iberoam., **14** (1998), 1–45.

[228] E. Ryckman and M. Visan, Global well-posedness and scattering for the defocusing energy-critical nonlinear Schrödinger equation in $\mathbb{R}^{1+4}$, Am. J. Math., **129** (2007), 1–60.

[229] V. Scheffer, Turbulence and Hausdorff dimension. In Turbulence and the Navier–Stokes Equations. Lecture Notes in Math., Vol. 565, Springer-Verlag, 1976, pp. 94–112.

[230] V. Scheffer, The Navier–Stokes equations in space dimension four, Commun. Math. Phys., **61** (1978), 41–68.

[231] I. E. Segal, The global Cauchy problem for relativistic scalar field with power interaction, Bull. Soc. Math. Fr., **91** (1963), 129–135.

[232] I. E. Segal, Non-linear semigroups, Ann. Math., **78** (1963), 339–364.

[233] I. E. Segal, The space-time decay for solution of wave equations, Adv. Math., **22** (1976), 302–311.

[234] J. Serrin, The initial value problem for Navier–Stokes equations. In Non-linear Problems, Edited by R. E. Langer, Wisconsin Press, 1963, pp. 69–98.

[235] S. Shao, On localization of the Schrödinger maximal operator, arXiv:1006.2787v1.

[236] J. Shatah and M. Struwe, Regularity result for nonlinear wave equation with critical nonlinearity, Ann. Math., **138** (1993), 503–515.

[237] J. Shatah and M. Struwe, Well-posedness in the energy space for semilinear wave equation with critical growth, Int. Math. Res. Not., **7** (1994), 303–309.

[238] R. Shen, On the energy subcritical nonlinear wave equation in $\mathbb{R}^3$ with radial data, Anal. PDE, **6** (2013), 1929–1987.

[239] P. Sjölin, Regularity of solutions to nonlinear Schrödinger equation, Duke Math. J., **55** (1987), 699–715.

[240] H. Smith and C. D. Sogge, Global Strichartz estimates for nontrapping perturbations of Laplacian, Commun. Partial Differ. Equ., **25** (2001), 2171–2183.

[241] C. D. Sogge, Fourier Integral in Classical Analysis, Cambridge University Press, 1993.

[242] C. D. Sogge, Lecture on Nonlinear Wave Equations, Int. Press, Boston, 1995.

[243] E. M. Stein, Harmonic Analysis: Real-Variable Methods, Orthogonality, and Oscillatory Integrals, Princeton Univ. Press, 1993.

[244] E. M. Stein and G. Weiss, Introduction to Fourier Analysis in Euclidean Spaces, Princeton Univ. Press, 1970.

[245] W. Strauss, On the weak solutions of semilinear hyperbolic equations, An. Acad. Bras. Ciênc., **42** (1970), 645–651.

[246] W. Strauss, Nonlinear scattering at low energy, J. Funct. Anal., **41** (1981), 110–133; Sequel **43** (1981), 281–293.

[247] W. Strauss, Nonlinear Wave Equations. Monograph in the CBM-AMS, Vol. 73, 1989.

[248] R. Strichartz, Multipliers in fractional Sobolev spaces, J. Math. Mech., **16** (1967), 461–471.

[249] R. Strichartz, A priori estimates for the wave equations and some applications, J. Funct. Anal., **5** (1970), 218–235.

[250] R. Strichartz, Restrictions of Fourier transforms to quadratic surfaces and decay of solution of wave equations, Duke Math. J., **44** (1977), 705–714.

[251] M. Struwe, Globally regular solutions to the u^5 Klein–Grodon equation, Ann. Sc. Norm. Super. Pisa, **15** (1988), 495–513.

[252] M. Struwe, Semilinear wave equations, Bull. Am. Math. Soc., **26** (1992), 53–85.

[253] Q. Su, Global well-posedness and scattering for the defocusing cubic NLS in $\mathbb{R}^3$, Math. Res. Lett., **19** (2012), 431–451.

[254] Q. Su and Z. Zhao, Dynamics of subcritical threshold solutions for energy-critical NLS, Dyn. Partial Differ. Equ., **20** (2023), 37–72.

[255] K. Taira, Diffusion Process and Partial Differential Equations, Academic Press, San Diego, New York, London, Tokyo, 1985.

[256] G. Talenti, Best constant in Sobolev inequality, Ann. Mat. Pura Appl., **110** (1976), 353–372.

[257] T. Tao, From rotation needle to stability of waves emerging connections between combinations, analysis and PDEs, Not. Am. Math. Soc., **48** (2001), 294–303.

[258] T. Tao, Local well-posedness of the Yang–Mills equations in the temporal gauge below the enery norm, J. Differ. Equ., **189** (2003), 366–382.

[259] T. Tao, A sharp bilinear restriction estimate for paraboloids, Geom. Funct. Anal., **13** (2003), no. 6, 1359–1384.

[260] T. Tao, Global well-posedness and scattering for higher-dimensional energy-critical non-linear Schrödinger equation for radial data, N. Y. J. Math., **11** (2005), 57–80.

[261] T. Tao, A counterexample to an endpoint bilinear Strichartz inequality, Electron. J. Differ. Equ., 2006 (2006), paper no. 151, 6 pp.

[262] T. Tao, Spacetime bounds for the energy-critical nonlinear wave equation in three spatial dimensions, Dyn. Partial Differ. Equ., **3** (2006), 93–110.

[263] T. Tao, Nonlinear Dispersive Equations, Local and Global Analysis. CBMS Regional Conference Series in Mathematics, Vol. 106, 2006.

[264] T. Tao, Quantitative bounds for critically bounded solutions to the Navier–Stokes equations. In Nine Mathematical Challenges: An Elucidation. Proc. Sympos. Pure Math., Vol. 104, Amer. Math. Soc., Providence, RI, 2021, pp. 149–193.

[265] T. Tao and A. Vargas, A bilinear approach to cone multipliers. II. Applications, Geom. Funct. Anal., **10** (2003), no. 1, 216–258.

[266] T. Tao and M. Visan, Stability of energy-critical nonlinear Schrödinger equations in high dimensions, Electron. J. Differ. Equ., **2005** (2005), paper no. 118, 28 pp.

[267] T. Tao, M. Visan and X. Zhang, Global well-posedness and scattering for the mass-critical nonlinear Schrödinger equation for radial data in high dimensions, Duke Math. J., **140** (2007), 165–202.

[268] T. Tao, M. Visan and X. Zhang, The nonlinear Schrödinger equation with combined power-type nonlinearities, Commun. Partial Differ. Equ., **32** (2007), 1281–1343.

[269] T. Tao, M. Visan and X. Zhang, Minimal-mass blowup solutions of the mass-critical NLS, Forum Math., **20** (2008), 881–919.

[270] M. E. Taylor, Tools for PDE. Mathematical Surveys and Monographs, Vol. 81. American Mathematical Society, Providence, RI, 2000.

[271] P. Tomas, Restriction theorems for Fourier transform, Proc. Symp. Pure Math., **35** (1979), 111–114.

[272] G. P. Trachanas and N. B. Zographopoulos, Orbital stability for the Schrödinger operator involving inverse square potential, J. Differ. Equ., **259** (2015), 4989–5016.

[273] H. Triebel, Theory of Function Spaces, Springer, New York, 1983.

[274] H. Triebel, Theory of Function Spaces (II), Birkhäuser Verlag, Basel–Boston–Berlin, 1992.

[275] Y. Tsutsumi, L^2-solutions for nonlinear Schrödinger equations and nonlinear groups, Ann. IHP, Phys. Théor., **43** (1985), 321–347.

[276] Y. Tsutsumi, K. Yajima, The asymptotic behavior of nonlinear Schrödinger equations, Bull. Am. Math. Soc., **11** (1984), 186–188.

[277] A. Vasseur, A new proof of partial regularity of solutions to Navier–Stokes equations, NoDEA Nonlinear Differ. Equ. Appl., **14** (2007), 753–785.

[278] M. Visan, The defocusing energy-critical nonlinear Schrödinger equation in dimensions five and higher. Ph.D. Thesis, UCLA, 2006.

[279] M. Visan, The defocusing energy-critical nonlinear Schrödinger equation in higher dimensions, Duke Math. J., **138** (2007), 281–374. MR2318286.

[280] M. Visan, Global well-posedness and scattering for the defocusing cubic NLS in four dimensions, Int. Math. Res. Not., **2012** (2012), 1037–1067.

[281] M. Visan and X. Zhang, Global well-posedness and scattering for a class of nonlinear Schrödinger equations below the energy space, Differ. Integral Equ., **22** (2009), 99–124.

[282] W. Wahl, L^p-decay rates for homogeneous wave equation, Math. Z., **120** (1971), 93–106.

[283] W. Wahl, The Equation of the Navier–Stokes and Abstract Parabolic Equations, Vieweg Verlag, Braunschwei–Wiesbaden, 1985.

[284] W. Wahl, Regularity of weak solution of the Navier–Stokes equations. In Nonlinear Functional Analysis and Its Applications, Part 2. Proc. of Symp. in Pure Math., Vol. 45, Amer. Math. Soc., Providence, RI, 1986, pp. 497–503.

[285] H. Wang and S. Wu, Restriction estimates using decoupling theorems and two-ends Furstenberg inequalities, arXiv:2411.08871.

[286] K. Yajima, The $W^{k,p}$-continuity of wave operators for Schrödinger operators, Proc. Jpn. Acad., Ser. A, Math. Sci., **69**, (1993), no. 4, 94–98.

[287] K. Yajima, The $W^{k,p}$-continuity of wave operators for Schrödinger operators. II. Positive potentials in even dimensions $m \geq 4$. In Spectral and Scattering Theory (Sanda, 1992). Lecture Notes in Pure and Appl. Math., Vol. 161, Dekker, New York, 1994, pp. 287–300.

[288] K. Yajima, The $W^{k,p}$-continuity of wave operators for Schrödinger operators, J. Math. Soc. Jpn., **47** (1995), no. 3, 551–581.

[289] K. Yajima, The $W^{k,p}$-continuity of wave operators for Schrödinger operators, III. Even-dimensional cases $m \geq 4$, J. Math. Sci. Univ. Tokyo, **2** (1995), no. 2, 311–346.

[290] K. Yajima, L^p-boundedness of wave operators for two-dimensional Schrödinger operators, Commun. Math. Phys., **208** (1999), no. 1, 125–152.

[291] K. Yang, C. Zeng and X. Zhang, Dynamics of threshold solutions for energy critical NLS with inverse square potential, SIAM J. Math. Anal., **54** (2022), no. 1, 173–219.

[292] K. Yang and X. Zhang, Dynamics of threshold solutions for energy critical NLW with inverse square potential, Math. Z., **302** (2022), no. 1, 353–389.

[293] K. Yosida, Functional Analysis, Springer-Verlag, Berlin, 1980.

[294] X. Yu, Global well-posedness and scattering for the defocusing $\dot{H}^{\frac{1}{2}}$-critical nonlinear Schrödinger equation in $\mathbb{R}^2$, Anal. PDE, **14** (2021), 2225–2268.

[295] J. Zhang and J. Zheng, Scattering theory for nonlinear Schrödinger equations with inverse-square potential, J. Funct. Anal., **267** (2014), no. 8, 2907–2932.

Index

https://doi.org/10.1515/9783111384726-008

De Gruyter Studies in Mathematics

Volume 101
Marko Kostić
Almost Periodic Type Solutions. to Integro-Differential-Difference Equations, 2025
ISBN 978-3-11-168728-5, e-ISBN 978-3-11-168974-6, e-ISBN (ePUB) 978-3-11-169021-6

Volume 64
Dorina Mitrea, Irina Mitrea, Marius Mitrea, Michael Taylor
The Hodge–Laplacian. Boundary Value Problems on Riemannian Manifolds, 2nd Edition, 2025
ISBN 978-3-11-148098-5, e-ISBN 978-3-11-148140-1, e-ISBN (ePUB) 978-3-11-148389-4

Volume 100
Changxing Miao, Ruipeng Shen
Regularity and Scattering of Dispersive Wave Equation. Multiplier Method and Morawetz Estimate, 2025
ISBN 978-3-11-148754-0, e-ISBN 978-3-11-148835-6, e-ISBN (ePUB) 978-3-11-148940-7

Volume 99
Marcus Laurel, Marius Mitrea
Weighted Morrey Spaces. Calderón-Zygmund Theory and Boundary Problems, 2024
ISBN 978-3-11-145816-8, e-ISBN 978-3-11-145827-4, e-ISBN (ePUB) 978-3-11-146145-8

Volume 98
Peter J. Brockwell, Alexander M. Lindner
Continuous-Parameter Time Series, 2024
ISBN 978-3-11-132499-9, e-ISBN 978-3-11-132503-3, e-ISBN (ePUB) 978-3-11-132520-0

Volume 97
Ştefan Ovidiu I. Tohăneanu
Commutative Algebra Methods for Coding Theory, 2024
ISBN 978-3-11-121292-0, e-ISBN 978-3-11-121479-5, e-ISBN (ePUB) 978-3-11-121538-9

Volume 25
Karl H. Hofmann, Sidney A. Morris
The Structure of Compact Groups. A Primer for the Student – A Handbook for the Expert, 5th Edition, 2023
ISBN 978-3-11-117163-0, e-ISBN 978-3-11-117260-6, e-ISBN (ePUB) 978-3-11-117405-1

De Gruyter Studies in Mathematics

www.ingramcontent.com/pod-product-compliance
Lightning Source LLC
LaVergne TN
LVHW082039240725
817018LV00006B/164

* 9 7 8 3 1 1 1 3 8 4 5 1 1 *